Black Holes, Gravitational Waves and Cosmology: An Introduction to Current Research

Topics in Astrophysics and Space Physics

Edited by A. G. W. Cameron, *Yeshiva University*, and
George B. Field, *University of California at Berkeley*

Volume 10

M. Rees, R. Ruffini, and J. A. Wheeler *Black Holes, Gravitational Waves and Cosmology: An Introduction to Current Research*

Additional volumes in the series (published or in preparation):

1 V. L. Ginzburg *Elementary Processes for Cosmic Ray Astrophysics*
2 V. L. Ginzburg *The Origin of Cosmic Rays*
3 C. R. Cowley *The Theory of Stellar Spectra*
4 S. Glasstone *The Book of Mars*
5 T. Arny *Star Formation and Interstellar Clouds*
6 D. B. Melrose *Plasma Astrophysics*
7 H. Reeves *Nuclear Reactions in Stellar Surfaces and their Relations with Stellar Evolution*
8 K. Greisen *The Physics of Cosmic x-ray, γ-ray and Particle Sources*
9 A. Lenchek *The Physics of Pulsars*
11 K. Apparao *Composition of Cosmic Radiation*
12 P. W. Hodge *Interplanetary Dust*

Black Holes, Gravitational Waves and Cosmology: An Introduction to Current Research

MARTIN REES
Institute of Astronomy
Cambridge, England

REMO RUFFINI

and

JOHN ARCHIBALD WHEELER
Joseph Henry Laboratories
Princeton University, U.S.A.

GORDON AND BREACH SCIENCE PUBLISHERS
NEW YORK LONDON PARIS

Gordon and Breach, Science Publishers, Inc.
One Park Avenue
New York, N.Y. 10016

Editorial office for the United Kingdom
Gordon and Breach, Science Publishers Ltd.
42 William IV Street
London W.C. 2.

Editorial office for France
Gordon & Breach
7–9 rue Emile Dubois
Paris 14e

Library of Congress catalog card number 72–78921. ISBN 0 677 04580 8.
Printed in the German Democratic Republic by Offizin Andersen Nexö, Leipzig

Preface

THIS BOOK is intended as an introduction to the rapidly developing field of relativistic astrophysics and cosmology. Our aim is to introduce the basic concepts on a level comprehensible to beginning graduate students and to summarize relevant observations. We do not claim to survey the field in the same detail and depth as, for example, Zel'dovich and Novikov, *Relativistic Astrophysics* (University of Chicago Press, Vol. 1, 1971; Vol. 2 in press), Weinberg, *Gravitation and Cosmology* (Wiley, New York, 1972), Hawking and Ellis, *The Large-scale Structure of Spacetime* (Cambridge University Press, 1973), or Misner, Thorne and Wheeler, *Gravitation* (Freeman, San Francisco, 1973). The reader is referred to these treatises for more extensive discussion of topics which are only touched on here.

Since the early chapters were written†, there have been important new developments in the theory of black holes and gravitational radiation as well as on the observational front, especially in X-ray astronomy. We have therefore added an appendix which reprints some important recent papers.

† Chapters 1–10 are based on a report ,"Relativistic cosmology and space platforms" by R. Ruffini and J. A. Wheeler, a chapter in A. F. Moore and V. Hardy, eds., *The Significance of Space Research for Fundamental Physics*, European Space Research Organization (ESRO) book No. SP-52, Paris, 1971, as updated for the present book.

Acknowledgements

WE WISH to express appreciation to many colleagues for discussion and communications, among them John Bahcall, James Bardeen, Brandon Carter, Robert Dicke, Rolf Hagedorn, James Hartle, Stephen Hawking, James LeBlanc, Malcolm Longair, Charles Misner, Jan Oort, and Bruce Partridge as well as James Peebles, Roger Penrose, David Pines, Malvin Ruderman, Allan Sandage, Dennis Sciama, Kip Thorne, Joseph Weber, David Wilkinson, and James Wilson.

We thank the European Space Research Organization and J. R. U. Page, A. F. Moore, and V. Hardy of that center for editorial collaboration on, and for permission to reproduce, selected parts of SP-52. We also thank Peter Walsh and David Wright for help with the proofs and index.

We are grateful to the authors of the reprints included in this book for the permission they have given for reproduction here, and to many other authors for allowing us to use figures from their publications.

M. R., R. R., and J. A. W.

Contents

List of Figures

List of Tables

CHAPTER I

Introduction

1.1 THE DOUBLE CHALLENGE: — MORE PRECISION IN KNOWN EFFECTS: THE SEARCH FOR NEW EFFECTS

"...THE OPINION seems to have got abroad, that in a few years all the great physical constants will have been approximately estimated, and that the only occupation which will then be left to men of science will be to carry on these measurements to another place of decimals". These words were pronounced by James Clerk Maxwell at the inauguration of the Devonshire (Cavendish) Physical Laboratory at Cambridge[1]. Decades later Albert A. Michelson made the idea of "the search for the next decimal place" even more famous to a still wider audience. This great tradition of precision physics flourishes still more actively in our own time, thanks not least to the development of the atomic clock, radar, the Mössbauer effect and the laser. Hopes are high that within a few years we shall have tests, with greatly improved accuracy, of the three famous predictions of general relativity, the precession of the orbit of Mercury, the bending of light by the Sun and the red shift of light from the Sun; at present none of these effects is reliably established with a precision better than 10 per cent. Ideas for these and other precision tests of Einstein's theory today receive more attention from more skilled experimenters than ever before.

Measure known effects with improved precision; or explore for new effects. Those who prefer the one activity find relativity as challenging as those who prefer the other. The reason is clear. Einstein's standard 1915 geometrodynamics makes new predictions not only about the numbers but also about the nature of physics. The geometry of space is dynamic. The Universe is closed[2]. The Universe starts its life unbelievably small, reaches a maximum dimension and recontracts. It undergoes complete gravitational collapse. In many ways a similar gravitational collapse overtakes certain stars with big dense cores. This collapse terminates in some cases ("supernova events") with the formation of a neutron star. In other cases, where the star core is more massive, the collapse goes to completion. A black hole is formed. Given such a black hole, one has no way, even in principle, of measuring or even defining—one believes—how many baryons and leptons this object contains. In this sense one thinks of the law of conservation of baryons (and leptons) as being transcended in the phenomenon of complete collapse. Black holes formed in the gravitational collapse of stars with big dense cores are dotted here and there about this and other galaxies, if present expectations are correct. A star in the final stages of collapse to a neutron star or a black hole is a powerful source of gravitational radiation: waves in the geometry of space. Gravitational radiation also emerges with great strength from two compact masses in close gravitational interaction, whether the two objects are neutron stars or black holes, or one of each.

1.2 SPACE FROM RIEMANN TO EINSTEIN

Space itself, until Bernhard Riemann's Göttingen inaugural lecture[3] of 10June 1854, could be viewed as an ideal Euclidean perfection standing unmoved high above the battles of matter and energy. Einstein translated Riemann's vision of a dynamic geometry into clear-cut mathematical terms. Einstein's geometrodynamics has withstood test and attack for over half a century. In the conception of Riemann and Einstein space tells matter how to move. But matter in turn tells space how to curve. Only so can the principle of action and reaction be upheld! In a revolution, chained-up space broke loose and caesed to be the passive arena for physics. It became an active participant. The agent of this revolution, the Einstein who endowed geometry with a life of its own, also established rules for the governance of this new dynamic entity. Those rules tell us what we have now known for a decade, the structure of an edifice more magnificent than space: superspace. Superspace is the rigid and perfect arena, infinite in number of dimensions, in which the geometry of the Universe executes its dynamic changes: expansion, vibration, undulation, attainment of maximum dimensions, recontraction and collapse. Important as it is to mention superspace for a fuller perspective of what Einstein's geometrodynamics is today, it is also true that one can dispense with mention of superspace so long as one looks apart, as we shall here, from quantum fluctuations in the geometry of space, and from the quantum effects that dominate the final stages of gravitational collapse. Classical geometrodynamics is rich enough as a field of investigation!

1.3 THE LOCAL CHARACTER OF GRAVITATION

Matter gets its moving orders from spacetime; and spacetime is curved by matter: these are the two central principles of classical general relativity. Each admits of ready illustration. Observe the pea "dropped" at lunchtime in the spaceship. It stays in the center of the craft. Seen by X-ray vision from outside, it is travelling around the Earth in the same Kepler orbit as the ship itself. How in this miracle possible? Shut away inside the vehicle, it can see neither Earth, Sun nor any star. From what source then does it derive its movement? From the geometry of spacetime right where it is, says Einstein; from far away, says Newton.

Newtonian theory says the satellite moves relative to an ideal God-given never-changing Euclidean reference frame that pervades all space and endures for all time. That theory goes on to say that the satellite and the pea would have moved along an ideal straight line in this global frame had not "forces of gravitation" deflected them. It adds the postulate that the "force of gravitation" acting on each object or, more directly, the "gravitational mass" of each object, is proportional to the inertial mass of that object. By this combination of postulates it ensures that both objects have the same acceleration relative to the ideal straight line. But nothing explains why gravitational mass should be equal to inertial mass. And nothing, not even light, ever moves along the ideal straight line—it is a purely theoretical line.

Einstein's theory says there is no ideal Euclidean reference that extends all over space, despite anything to the contrary that Euclid wrote 2270 years ago. And why say there is when there is nothing that directly evidences that hypothesis? To try to describe motion relative to far-away objects is the wrong way to do physics! Physics is simple only when analysed locally. And locally the world line that a satellite follows is already as straight as any world line can be. The pea feels no "force of gravitation". The traveller feels no "force of gravitation". The space ship feels no "force of gravitation". So let us forget about "force of gravitation". Recognise that every one of these objects has simple moving orders: "Follow a straight line in the local inertial reference frame". Each has only to sense the local structure of spacetime, right where it is, in order to follow the correct track. No more talk of "inertial mass" and "gravitational mass"; and no more talk of gravitation, so long as one follows the motion of a single test object.

1.4 TIDE-PRODUCING ACCELERATION AND SPACETIME CURVATURE

One has to observe the relative acceleration of two nearby test objects to have any proper local measure of a gravitational effect. Denoting their separation by $\eta^\alpha(\alpha = 0, 1, 2, 3)$, the 4-velocity of the fiducial particle relative to the local Lorentz reference system by u^α, and absolute differentiation with respect to proper time by $D/D\tau$, one has the effect of geometry on *relative* motion expressed in the socalled law of geodesic deviation,

$$D^2\eta^\alpha/D\tau^2 + R^\alpha_{\beta\gamma\delta}u^\beta\eta^\gamma u^\delta = 0 \tag{1}$$

For objects at rest or moving very slowly compared to the speed of light in the local Lorentz frame, one has

$$D^2\eta^m/D\tau^2 + R^m_{0n0}\eta^n = 0 \tag{2}$$

Here the quantities R^m_{0n0} ($m, n = 1, 2, 3$) are the components of the tide-producing acceleration or, otherwise stated, components of the Riemannian curvature of spacetime, at the location in question.

The contrast with electromagnetism is striking. There one speaks of the deflection of a single test particle,

$$D^2x^\alpha/D\tau^2 = (e/m)\, F^\alpha_\beta u^\beta \tag{3}$$

or, for a slowly moving test particle,

$$D^2x^n/D\tau^2 = (e/m)\, F^n_0, \tag{4}$$

where the F^0_n ($n = 1, 2, 3$) are the components (E^x, E^y, E^z) of the electric field. To speak of the force of gravity on a single test mass has no well-defined meaning in Einstein's geometrical theory of gravitation; one has in it the identity

$$D^2x^\alpha/D\tau^2 \equiv Du^\alpha/D\tau \equiv 0 \tag{5}$$

It takes two nearby test particles to measure even a single component of the meaningful local feature of gravitation, the tide-producing acceleration.

1.5 SPACETIME *VS* SPACE

Why curvature of spacetime? Why not curvature of space? If space curvature had sufficed to tell the story of gravitation, Riemann would have discovered that story long ago. Great geometer that he was, he was working on a theory for gravitation when he died at 39. It took 1905 and special relativity to open the door to spacetime; and it took spacetime to open the door to a geometrical explanation of gravitation. No one could account for gravity who considered only curvature in space. The ball that crosses the 10 m-wide room in 1 sec has to arch up the center of its track by 1.2 m. The photon that crosses the 10 m-wide room has to arch up the center of its track by the tiniest amount. The curvatures of the two tracks evidently differ fantastically in order of magnitude. No one curvature of space is going to account for these completely different deflections. However, turn from space to spacetime (Figure 1). The baseline of the photon arch remains of the order of 10 m. The baseline of the ball's arch is stretched out in the time dimension (geometrical units!) to

$$(1 \text{ sec}) \times (3 \times 10^8 \text{ m/sec}) = 3 \times 10^8 \text{ m}$$

Even with its higher arch this track now has a curvature of the same order as that of the photon track. Curvature of spacetime is the geometrical machinery behind gravitation!

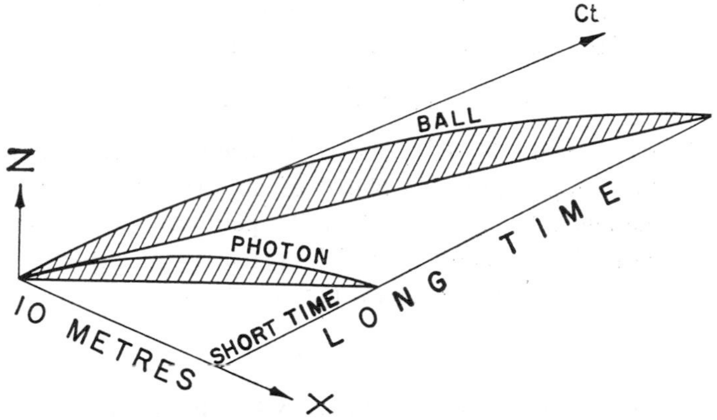

Figure 1 Tracks of ball and photon through space have very different curvatures; through spacetime, comparable curvature.

1.6 CURVATURE AND DENSITY

How much does matter curve spacetime? Newtonian theory gives a quick route to the answer. Put a test particle in a close-in orbit around a planet of radius a and of uniform density

$$\varrho\ (\text{cm}^{-2}) = (G/c^2)\ \varrho_{\text{conv}}\ (\text{g/cm}^3) = (0.742 \times 10^{-28}\ \text{cm/g})\ \varrho_{\text{conv}} \tag{6}$$

Equate the kinematic acceleration of the test particle, going around with angular velocity $\omega(\text{cm}^{-1}) = \omega$ (radian/cm of light travel time) $= (1/c)$ $\omega_{\text{conv}}(\text{sec}^{-1})$, to Newton's inverse square law acceleration of gravity; thus,

$$\omega^2 a = (4\pi a^3 \varrho/3)/a^2 \tag{7}$$

Find

$$\omega^2 = 4\pi\varrho/3 \tag{8}$$

a circular frequency independent of the radius of the planet—whether its radius be 10,000 km or 10 m! Bore holes through the center of the planet along the x-axis and along the z-axis. Test masses dropped in these holes execute simple harmonic motion with circular frequency ω. The circular motion of the orbiting test particle can even be regarded as the superposition, with 90-degree phase differences, of two such vibrations along the x- and z-axes. Once one notes the simple harmonic character of the motion of one test particle, one also recognises the simple harmonic character of the relative motion of two nearby particles oscillating along the x-axis. Their separation η satisfies the equation,

$$D^2\eta^1/D\tau^2 + \omega^2\eta^1 = 0 \tag{9}$$

Comparing with the equation of geodesic deviation, one finds the (x, x) components of the tide-producing acceleration:

$$R^1_{010} = \omega^2 = 4\pi\varrho/3. \tag{10}$$

Identical values obtain for R^2_{020} and R^3_{030} throughout the interior of the planet. In this simple case of a planet of uniform density and spherical symmetry one has only to observe the full repetition time T between crossing of the world lines of two test particles (T(cm) $= cT_{\text{conv}}$(sec)) in order to have at once ($T \to \omega \to \varrho$) a value for the density.

1.7 EINSTEIN'S EQUATION CONNECTING CURVATURE WITH DENSITY

In the case of an attracting object of no special symmetry and of non-uniform density, ordinarily no single one of the indicated components of the tide-producing acceleration has the value $4\pi\varrho/3$. Only a certain combination of the components of the curvature tensor is fixed by the density. Compare Newtonian theory, where ϱ determines neither $(\partial^2/\partial x^2) \times$ (gravitational potential) $= \partial^2\varphi/\partial x^2$ nor $\partial^2\varphi/\partial y^2$ individually, but only the combination

$$\partial^2\varphi/\partial x^2 + \partial^2\varphi/\partial y^2 + \partial^2\varphi/\partial z^2 = 4\pi\varrho \tag{11}$$

The principle of correspondence with Newtonian theory, plus other compelling considerations of principle, gave Einstein a unique equation. It ties the density to none of the cited components of the curvature tensor directly. Rather, it gives as basic formula the following:

$$\hat{R}^{12}_{12} + \hat{R}^{23}_{23} + \hat{R}^{31}_{31} = 8\pi\varrho \tag{12}$$

Here the caret superscripts indicate use of a local Lorentz reference system ($-\hat{g}_{00} = \hat{g}_{11} = \hat{g}_{22} = \hat{g}_{33} = 1$; $\hat{g}_{mn} = 0$ for $m \neq n$) rather than a completely general coordinate system. (In a general coordinate system Einstein's equation reads

$$R_{\alpha\beta} - \tfrac{1}{2}g_{\alpha\beta}R = 8\pi T_{\alpha\beta} \tag{13}$$

where $T_{\alpha\beta}$ is the tensor measuring the density of mass-energy, momentum, and stress). When spacetime is static or, even if not static, when spacetime admits a moment of time symmetry (as for example a model universe at the phase of maximum expansion) then the key equation takes the simple form

$$^{(3)}R = 16\pi\varrho \tag{14}$$

Here $^{(3)}R$ is the scalar curvature invariant of the 3-dimensional space-like hyper-surface that slices through spacetime at the moment of time symmetry. When, in addition, this 3-geometry possesses spherical symmetry, one can write the element of distance in the form

$$ds^2 = \frac{dr^2}{1 - 2m(r)/r} + r^2(d\theta^2 + \sin^2\theta\, d\varphi^2) \tag{15}$$

Then the equation relating curvature and density takes a form easy to remember,

$$dm(r)/dr = 4\pi r^2 \varrho(r) \tag{16}$$

So much for recalling what Einstein's geometrodynamics has to say about the effect of curvature on matter and the effect of matter on curvature.

CHAPTER 2

The Physics of a Superdense Star

Mass-energy curves space, and enough mass-energy curves space up into closure. Space, curved up into a 3-sphere of uniform curvature and radius a, has a scalar curvature invariant of magnitude

$$^{(3)}R = 6/a^2 = 16\pi\varrho \tag{17}$$

From this formula one finds how large a system can be before it can't be (Table I). An object of solar mass will not curve space up into closure even when compacted to the density of nuclear matter. In such a system general relativity effects have not yet reached the point where they are all-powerful; but they are nevertheless significant. These effects force one to give up the Newtonian equation of hydrostatic equilibrium for the variation of pressure with depth,

$$-dp(r)/dr = \varrho(r)\, m(r)/r^2 \tag{18}$$

and replace it by the corresponding Tolman-Oppenheimer-Volkoff (T-O-V) equation[5],

$$-dp(r)/dr = \frac{(\varrho + p)\,(m + 4\pi r^3 p)}{r(r - 2m)} \tag{19}$$

Table I Radius of curvature of space, static or at phase of maximum expansion, for selected values of the density of mass-energy (the last column refers to a closed three-sphere universe of the given density)

Illustrative system	An average or representative density	Same, in geometrical units	Critical radius	Critical Mass, $2\pi^2a^3\varrho$ / Mass of Sun
Model universe at phase of maximum expansion	10^{-30} g/cm^3	0.742×10^{-58} cm^{-2}	4×10^{28} cm	6×10^{23}
Sun (80 g/cm^3 at center; 1 g/cm^3 average)	1 g/cm^3	0.742×10^{-28} cm^{-2}	4×10^{13} cm	6×10^{8}
White dwarf star	10^7 g/cm^3	0.742×10^{-21} cm^{-2}	10^{10} cm	2×10^{5}
Neutron star	10^{14} g/cm^3	0.742×10^{-14} cm^{-2}	4×10^{6} cm	60

To integrate the system of Eqs. (16) and (18) or (16) and (19) we need to have a relation between pressure and density, $p = p(\varrho)$. Once this additional equation is given we choose a value for the central density $\varrho(0) = \varrho_0$ with corresponding pressure $p(\varrho_0) = p_0$. We assume the boundary condition $m(0) = 0$ and integrate outward from $r = 0$. For every value of r we find the value of the pressure $p(r)$, of the density $\varrho(r)$ and of the mass $m(r)$ contained inside a sphere of radius r. The integration is continued to the surface of the star, defined as the place where the pressure drops to zero.

We restrict attention for simplicity to "cold catalyzed matter"; that is, matter which has reached the endpoint of thermonuclear burning. We consider densities so great that the mass-energy of compression is appreciable in comparison to the rest mass of the individual baryons (special relativity effects in both pressure and density!). Thus ϱ includes not "matter density" alone but density of mass-energy from all local sources: rest mass, kinetic energy, short range particle-particle interactions.

Question: For a star model of a given central density, how sensitive is the mass to (1) the difference between Einstein's geometrostatic and Newton's gravitational theory; (2) the difference between Einstein's theory and the scalar-tensor theory of gravitation of Brans and Dicke[6]; and (3) uncertainties in the equation of state? To clarify issues (1) and (2) first, we pick one equation of state and compare the predictions of the three theories. Then we pick one theory and compare the predictions of different equations of state. In the first part of our investigation we pick for the sake of definiteness the Harrison-Wheeler equation of state (Table II) the derivation and details of which are given in the literature[5].

2.1 RESULTS OF NEWTONIAN TREATMENT

The Newtonian equilibrium configuration for any selected value of the central density is immediately obtained by the integration of Eqs. (16) and (18). In strict Newtonian theory the outcome of the integration out to the surface of the star (place where p goes to zero) is threefold: (1) a value for the radius, R, of the star; (2) a value for the amount of "matter" in the star; and (3) enough supplementary information to calculate a value for the energy of the star. However, according to special relativity energy has mass. This blurs the distinction between (3) and (2). The distinction is re-established already in special relativity when we give up the idea of measuring amount of "matter" by amount of mass and use instead for the measure of that quantity the number of baryons (law of conservation of baryons!). Here we have not taken the trouble to calculate the baryon number for Newton-

Table II Equation of state of cold catalyzed matter as given by Harrison and Wheeler (47 entries) as extended by B. K. Harrison to 72 entries for the purpose of numerical calculations (first quoted by Hartle and Thorne[7])

P	*E*	ε	*P*	*E*	ε
8.31 E-41	5.82 E-28	1.00 E-45	2.73 E-21	2.34 E-18	9.21 E-21
4.17 E-40	5.84 E-28	7.11 E-43	6.49 E-21	4.68 E-18	2.25 E-20
8.31 E-40	5.86 E-28	2.77 E-42	1.10 E-20	7.41 E-18	4.05 E-20
4.17 E-39	5.97 E-28	4.16 E-41	1.88 E-20	1.17 E-17	7.21 E-20
8.31 E-39	6.04 E-28	1.12 E-40	3.05 E-20	1.86 E-17	1.28 E-19
2.79 E-38	6.32 E-28	8.64 E-40	4.58 E-20	2.95 E-17	2.25 E-19
2.38 E-37	8.52 E-28	3.36 E-38	6.59 E-20	4.68 E-17	3.88 E-19
1.37 E-36	1.23 E-27	3.21 E-37	9.55 E-20	7.41 E-17	6.60 E-19
7.00 E-36	2.34 E-27	3.47 E-36	1.50 E-19	1.17 E-16	1.11 E-18
6.96 E-35	6.54 E-27	5.01 E-35	2.54 E-19	1.86 E-16	1.88 E-18
4.79 E-34	1.56 E-26	3.76 E-34	4.49 E-19	2.95 E-16	3.17 E-18
1.74 E-33	2.95 E-26	1.52 E-33	9.14 E-19	4.68 E-16	5.39 E-18
5.95 E-33	5.22 E-26	5.18 E-33	1.88 E-18	7.41 E-16	9.28 E-18
1.56 E-32	8.52 E-26	1.45 E-32	6.09 E-18	1.48 E-15	2.18 E-17
4.62 E-32	1.53 E-25	4.73 E-32	2.63 E-17	3.71 E-15	7.27 E-17
2.67 E-31	3.71 E-25	2.72 E-31	8.23 E-17	7.41 E-15	1.90 E-16
9.63 E-31	7.41 E-25	1.05 E-30	2.60 E-16	1.48 E-14	5.16 E-16
4.83 E-30	1.86 E-24	5.82 E-30	1.09 E-15	3.71 E-14	2.02 E-15
2.32 E-29	4.68 E-24	3.03 E-29	3.25 E-15	7.41 E-14	5.70 E-15
5.19 E-29	7.41 E-24	6.82 E-29	9.71 E-15	1.48 E-13	1.61 E-14
1.65 E-28	1.48 E-23	2.27 E-28	3.93 E-14	3.71 E-13	6.30 E-14
8.23 E-28	3.71 E-23	1.11 E-27	9.71 E-14	7.41 E-13	1.69 E-13
2.37 E-27	7.41 E-23	3.59 E-27	2.42 E-13	1.48 E-12	4.34 E-13
7.19 E-27	1.48 E-22	1.12 E-26	7.34 E-13	3.71 E-12	1.43 E-12
2.69 E-26	3.71 E-22	4.84 E-26	1.65 E-12	7.41 E-12	3.37 E-12
7.86 E-26	7.41 E-22	1.42 E-25	3.60 E-12	1.48 E-11	7.71 E-12
1.93 E-25	1.48 E-21	4.03 E-25	1.01 E-11	3.71 E-11	2.23 E-11
6.60 E-25	3.71 E-21	1.53 E-24	2.08 E-11	7.41 E-11	4.87 E-11
1.65 E-24	7.41 E-21	4.07 E-24	4.28 E-11	1.48 E-10	1.04 E-10
4.18 E-24	1.48 E-20	1.07 E-23	1.10 E-10	3.71 E-10	2.82 E-10
1.35 E-23	3.71 E-20	3.77 E-23	2.26 E-10	7.41 E-10	5.89 E-10
3.29 E-23	7.41 E-20	9.57 E-23	4.64 E-10	1.48 E-09	1.22 E-09
8.07 E-23	1.48 E-19	2.41 E-22	1.19 E-09	3.71 E-09	3.19 E-09
2.67 E-22	3.71 E-19	8.17 E-22	2.43 E-09	7.41 E-09	6.52 E-09
6.53 E-22	7.41 E-19	2.04 E-21	4.91 E-09	1.48 E-08	1.33 E-08
1.21 E-21	1.17 E-18	3.72 E-21	1.23 E-08	3.71 E-08	3.41 E-08

ian configurations. We do calculate the mass, with all its contributions: (1) rest mass of the particles; (2) mass-energy of compression (positive); and (3) gravitational binding (negative). In actuality (1) and (2) are automatically combined in our definition of the density ϱ of mass-energy (rest mass plus kinetic energy plus energy of local interactions) in the equation of state. Therefore the "mass" that comes out of our Newtonian analysis, $M = \int_0^R 4\pi\varrho r^2\, dr$, is neither the strict Newtonian mass (including as it does mass-energy of compression) nor the strict special relativity value for the mass. It fails to correct for the mass-energy equivalent of gravitational binding,

$$E_{gr} = -\int^R (m(r)/r)\, dm(r) \tag{20}$$

It is simple to make the necessary correction; thus,

$$M_{tot} = M + E_{gr} \tag{21}$$

2.2 EQUILIBRIUM CONFIGURATIONS FOR GENERAL RELATIVITY

The general relativity equilibrium configuration for any ϱ_0 is obtained by integrating Eqs. (16) and (19). The results are given in Figure 2 and Table III. Comparison of the Newtonian and general relativity results in Table III shows that in the white dwarf region the corrections introduced by general relativity are extremely small (0.01 per cent at the first maximum). General relativity has a much greater effect at the second maximum, moving it from $\sim 10^{16}$ g/cm³ to $\sim 6 \times 10^{15}$ g/cm³. Moreover, the location of the maximum indicates the point of change of stability, according to the general relativity theory of equlibrium configurations. Thus general relativity lowers the central density required to reach instability by a factor of the order of two. In addition it reduces the critical mass for the largest stable neutron star from $1.2M_\odot$ to $0.7M_\odot$. Thus, in the physics of a neutron star general relativity does not introduce those small effects so well known in the three traditional tests of general relativity. Instead it makes a difference of the order of a factor of two both in density and in critical mass.

To Table II

Both the pressure and the density are given in geometrical units (cm^{-2}). Multiply p(cm^{-2}) by $c^4/G = 1.21 \times 10^{49}$ g cm/sec² to obtain pressure in conventional units (g/cm sec²) and ϱ (cm^{-2}) by $c^2/G = 1.35 \times 10^{28}$ g/cm to obtain pressure in g/cm³ (see text for discussion of range of validity). For densities over $\varrho = 3.7 \times 10^{-8}$ cm^{-2} ($\varrho_{conv} = 5 \times 10^{20}$ g/cm³) in default of other information the simplest procedure is to assume $p = \varrho/3$.

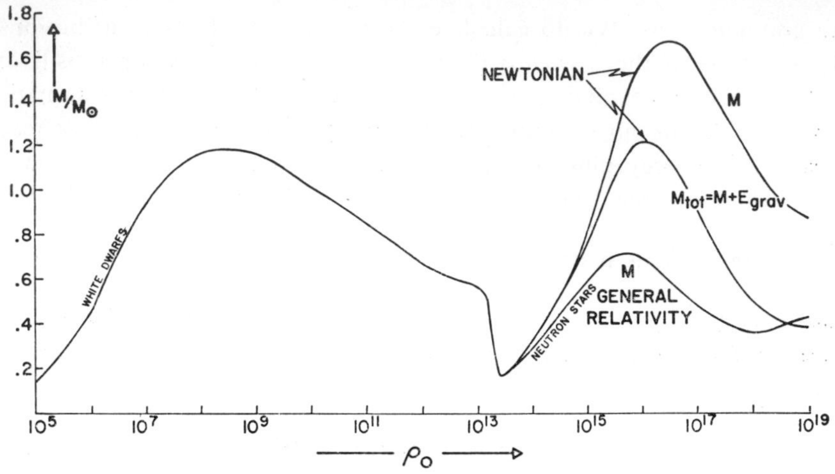

Figure 2 Mass of a cold star calculated by numerical integration from center to surface for selected values of the central density. Upper curve, Newtonian equation of hydrostatic equilibrium with mass $M = \int_0^R 4\pi\varrho r^2\, dr$ including rest mass and mass-energy of compression but omitting mass-energy of gravitational binding. Middle curve, same corrected for mass-energy of gravitational binding ($M_{\text{tot}} = M + E_{\text{gr}}$). Lower curve, result of integrating T-O-V general relativity equation of hydrostatic equilibrium, including effect of space curvature. The mass is again given by the formula $M = \int_0^R 4\pi\varrho r^2\, dr$ but this expression now automatically includes the contributions of rest mass, energy of compression, and energy of gravitational binding. To make the comparison of general relativity and Newtonian theory meaningful, the same equation of state was used in both integrations (Harrison-Wheeler equation).

Table III General relativity versus Newtonian gravitation theory for configurations of hydrostatic equilibrium, in both cases for the same equation of state (Harrison-Wheeler, Table II)

ϱ (g/cm^3)	General relativity		Newtonian theory	
	$M/M_\odot$	R (km)	$M/M_\odot$	R (km)
7.781×10^{13}	0.260	41.9	0.279	42.1
2.729×10^{14}	0.423	21.4	0.495	22.3
6.127×10^{14}	0.528	16.7	0.671	17.8
1.082×10^{15}	0.593	14.1	0.816	15.4
1.660×10^{15}	0.642	12.3	0.961	13.8

Table III (*cont.*)

ϱ (g/cm^3)	General relativity		Newtonian theory	
	$M/M_{\odot}$	R (km)	$M/M_{\odot}$	R (km)
2.361×10^{15}	0.674	11.0	1.090	12.6
3.214×10^{15}	0.700	10.0	1.232	11.7
4.400×10^{15}	0.715	9.2	1.355	11.0
5.852×10^{15}	0.715	8.6	1.445	10.4
7.653×10^{15}	0.708	8.0	1.516	9.9
9.993×10^{15}	0.694	7.6	1.564	9.4
1.238×10^{16}	0.676	7.0	1.598	8.9
1.543×10^{16}	0.657	6.7	1.627	8.5
1.927×10^{16}	0.637	6.4	1.648	8.2
2.362×10^{16}	0.616	6.2	1.660	7.9
2.895×10^{16}	0.595	5.9	1.666	7.6
3.558×10^{16}	0.574	5.7	1.666	7.3
4.379×10^{16}	0.554	5.6	1.658	7.1
5.358×10^{16}	0.534	5.4	1.644	6.9
6.463×10^{16}	0.515	5.3	1.624	6.6
7.834×10^{16}	0.497	5.2	1.601	6.4
9.518×10^{16}	0.479	5.1	1.574	6.2
1.148×10^{17}	0.463	5.1	1.514	6.1
1.386×10^{17}	0.448	5.0	1.512	5.9

Column 1, central density in g/cm^3; columns 2 and 4, mass-energy as sensed at infinity, in units of the solar mass $M_{\odot} = 1.987\times 10^{33}$ g; columns 3 and 5, radius in km (integrations by R.R.).

2.3 SCALAR TENSOR THEORY AND EQUILIBRIUM CONFIGURATIONS

The scalar-tensor theory of Jordan and Brans and Dicke[6] leads to a system of equations for hydrostatic equilibrium more complicated than either the Newtonian or the simple standard Einstein theory. The necessary equations, previously given in isotropic coordinates by Salmona[8], take the following simplified form in Schwarzschild coordinates:

$$\frac{dm(r)}{dr} = 4\pi r^2\left[\frac{\varrho}{\varphi} + \frac{\omega\varphi'^2}{16\pi\varphi^2}\,\frac{r-2m}{r} + \frac{1}{\varphi}\,\frac{(3p(r)-\varrho(r))}{3+2\omega}\right]r^2 \tag{22}$$

[the J.B.D. version of Eq. (16)]

$$-\frac{dp(r)}{dr} = \frac{p+\varrho(p)}{r(r-2m)}\left[m + 4\pi r^3\left(\frac{p}{\varphi} - \frac{\omega}{4\varphi^2}\left(1-\frac{2m(r)}{r}\right)\varphi'^2\right)\right] \tag{23}$$

[the J.B.D. version of Eq. (19)] and finally we have the scalar wave equation with source term

$$\varphi'' + \left[\frac{3}{r} - \frac{r - 2m}{r^2} + \frac{4\pi}{\varphi}(p - \varrho) - \frac{\omega\varphi'^2}{2\varphi^2}(r - 2m) - \frac{8\pi}{\varphi}\frac{r(3p - \varrho)}{(3 + 2\omega)}\right]\varphi'$$

$$= \frac{8\pi}{(3 + 2\omega)}(3p - \varrho)\frac{r}{(r - 2m)} \tag{24}$$

Here we indicate the scalar field with φ and with the prime its derivative with respect to r. The quantity ω is a dimensionless constant for which Dicke favours a value in the range $4 \leqq \omega \leqq 6$. To recover the general relativistic equations it is sufficient to make $\omega \rightarrow \infty$.

The integration of this system of equations with $\omega = 4$ gives results which are qualitatively identical to the general relativistic ones. The only quantitative difference is an increase of approximately two per cent ($M_{\mathrm{crit}} = 0.730 M_{\odot}$) in the value of the critical mass for a neutron star and an increase of a similar order of magnitude in the radius of the neutron star.

2.4 EQUATION OF STATE

If general relativity makes a difference by a factor of two in the critical mass for a neutron star, compared to Newtonian theory, what are the prospects for measuring this effect as a new test of Einstein's theory? Poor today; perhaps better tomorrow. The principal difficulty is an uncertainty by a factor of the order of two in the critical mass arising from ignorance about the equation of state at supranuclear densities (from 10^{14} g/cm^3 (nuclear matter) to 5×10^{15} g/cm^3 (central density for initiation of instability)). No such uncertainties apply to the equation of state at nuclear and sub-nuclear densities where interpolations and extrapolations of existing experimental and theoretical evidence can be applied with some confidence. A look at this region of lower densities will illustrate the kind of information that one would also like to have for the high-density region. The entire discussion will refer to the idealised and well-defined case of matter at the end point of thermonuclear evolution. How close to this end point does matter end up after the violent and high temperature implosion

white dwarf $\rightarrow$ neutron star?

Remnants of past history may persist in the topmost fraction of a kilometre of a neutron star of nearcritical mass. However, below that depth the density exceeds 10^{13} g/cm^3. The pressure is enormous. All traces of the past history of thermonuclear reactions in the material are wiped out. The idealisation of "cold catalyzed matter" would therefore appear to be legitimate throughout the interior.

For a first study of this situation it is helpful to define a "local gamma law" in the equation of state by the equation

$$\gamma = \frac{p + \varrho}{p} \frac{dp}{d\varrho}$$

and this quantity is illustrated as a function of density in Figure 3. Values of $\gamma > 4/3$ are regions of stability. On the right-hand side we show the regions of stability and instability as obtained by integration of the equation of hydrostatic equilibrium correlated with the regions where γ is less than 4/3 and greater than 4/3. Now to more detailed considerations!

The following regimes are encountered as one goes down in depth from the surface: First,

$$7.8 \text{ g/cm}^3 \leqq \varrho \leqq 15 \text{ g/cm}^3,$$

densities where one knows the equation of state of iron from laboratory measurements. Second,

$$15 \text{ g/cm}^3 \leqq \varrho \leqq 10^4 \text{ g/cm}^3,$$

the region where considerations based on the Fermi-Thomas statistical atom

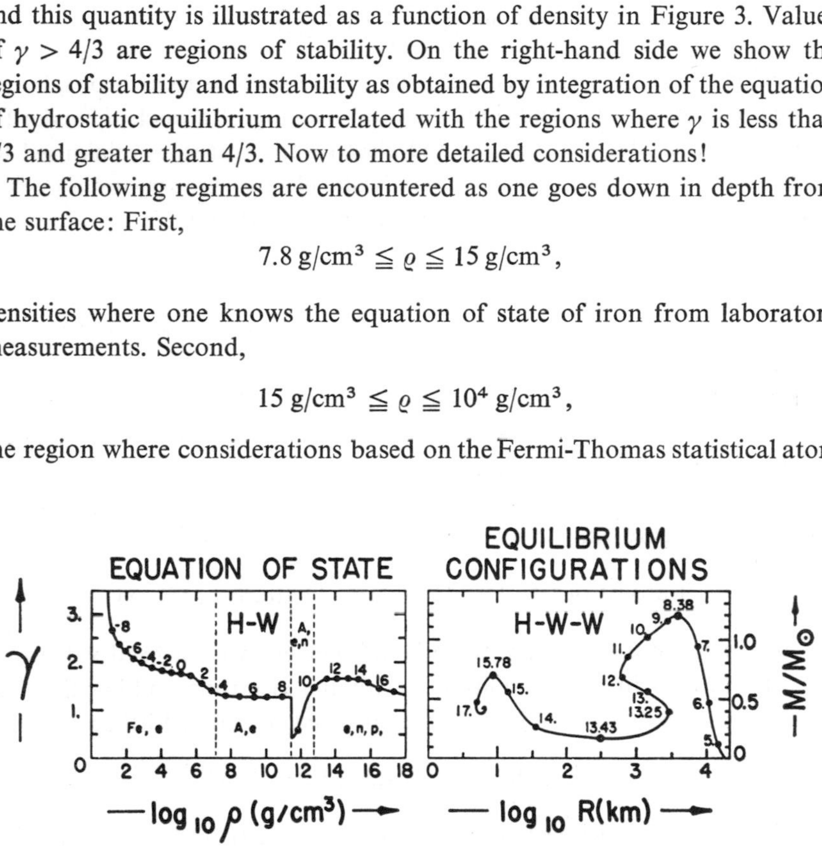

Figure 3 Equilibrium configurations of cold catalyzed matter (diagram adapted from K. S. Thorne, Varenna Lectures 1966). On the right-hand side the total mass (mass at infinity in units of solar mass) is given as a function of the radius (Schwarzschild coordinate of the surface). On the curve values for $\log_{10} \varrho_0$ (central density) are shown. The black dots appearing under 15.78, 13.43 and 8.38 mark a place of change in the stability of the equilibrium configurations. On the left-hand side the factor $\gamma = (p + \varrho)(dp/d\varrho)/p$ is plotted as a function of the density. The dots on the curve represent $\log_{10} p/c^2$ measured in g/cm³.

model suffice to give the equation of state (Feynman, Metropolis and Teller) Third,

$$10^4 \text{ g/cm}^3 \leqq \varrho \leqq 10^7 \text{ g/cm}^3,$$

the region where the electron Fermi energy exceeds the atomic binding, and we have a gas of iron nuclei and electrons with pressure arising almost exclusively from the electrons. These electrons become relativistic at the upper end of this density regime. Fourth,

$$10^7 \text{ g/cm}^3 \leqq \varrho \leqq 10^{11} \text{ g/cm}^3,$$

the region where the electrons transform bound protons to neutrons, bound or free. Under normal conditions the total packing of a nucleus, under the two conflicting effects of nuclear and electrostatic forces, is minimised for a value of $Z = 28$ and $A = 56$. A relativistic electron transmutes a nucleus of charge Z and atomic number A by inverse beta decay

$$e + (Z, A) \rightarrow (Z - 1, A) + \nu$$

The nuclei become neutron-rich compared to nuclei unpressured by electrons. For such neutron-rich nuclei the mass number $A = 56$ no longer represents the point of maximum stability. Stability shifts to higher A-values. The detailed mechanisms of shift are complicated and numerous and have not yet been analysed in detail. However, they are irrelevant for the determination of the final equilibrium itself. Symbolically—and only symbolically—we can represent the relevant transformation in

$$A \text{ atoms of } (Z, A - 1) \rightarrow (A - 1) \text{ atoms of } (Z, A)$$

At any one electron pressure there is one nucleus with a fixed value of Z and A which is in beta equilibrium with the electrons and has the most favourable packing fraction under the specified pressure. Fifth, the regime

$$10^{11} \text{ g/cm}^3 \leqq \varrho \leqq 4.5 \times 10^{12} \text{ g/cm}^3$$

At 10^{11} g/cm^3 nuclei become so heavy ($A \sim 122$) and so neutron-rich ($N/Z \sim 83/39$) that they "drip" neutrons and form an atmosphere of unbound neutrons. With a furtheɹ increase in the electron pressure the nuclei become still richer in neutrons and the numder of unbound neutrons increases. The new system now has three components: (1) degenerate relativistis electron gas; (2) nuclei; and (3) degenerate neutron gas. In the formula for the pressure the nuclei alwais give a very small contribution. However, the contribution of the neutrons does not remain negligible as the density rises. Already at a density of $\varrho \sim 4.5 \times 10^{12}$g/cm^3 the pressure due to the neutrons, already large in comparison to that due to the nuclei, is also larger than the pressure due to the electrons. Sixth, the regime

$$4.5 \times 10^{12}\ \text{g/cm}^3 \leqq \varrho \leqq 10^{14}\ \text{g/cm}^3,$$

where individual nuclei disappear and the center of the star becomes itself in effect one giant nucleus, composed of a mixture of electrons, neutrons and protons. The condition of electrical neutrality imposes n_e (density of electrons) $= n_p$ (density of protons). This condition immediately implies identity in the Fermi momentum of electrons and protons. Enough is known empirically and theoretically about nuclear matter ($\varrho \sim 10^{14}$ g/cm^3) to enable one to estimate with some confidence the pressure-density for this kind of matter up to 10^{14} g/cm^3 and a little beyond. In the simplest approximation one neglects interactions between the particles and considers a mixture of three ideal non-interacting Fermi gases. Although the result of this calculation is not badly out of line with experience even for $\varrho \sim 10^{14}$ g/cm^3, one can do better by allowing for the influence of nuclear forces. Seventh, the regime

$$10^{14}\ \text{g/cm}^3 \leqq \varrho$$

where the Fermi momenta of even the baryons move up towards relativistic values. Different conclusions emerge about the equation of state in this domain according as the emphasis is put on one or another of three important physical effects: relativistic baryons; nuclear interactions; and creation of new particles. One does not know how to treat either of the latter two effects from first principles. Therefore, for the sake of simplicity and definiteness Harrison and Wheeler neglected altogether both of these complications. They considered a mixture of three ideal non-interacting Fermi gases (e^-, p, n) even up to the highest densities. At extreme relativistic conditions the condition of beta equilibrium,

$$E_e + E_p = E_n \text{ (total mass-energy)}$$

implies that the neutron Fermi momentum is equal to twice the Fermi momentum of either charged particle. Consequently the calculated particle abundances stand to each other in the ratio

$$n_e/n_p/n_n = 1/1/8$$

For such a relativistic gas the equation of state could not be more simple,

$$p = \varrho/3$$

(asymptotic limit of H-W equation of state).

The effects of particle-particle interactions on the equation of state ("hard core" and other assumptions about the nucleon-nucleon coupling) have been treated by a number of workers, with varied results! As an example, Cameron, Cohen, Langer and Rosen[9] consider a two-particle interaction between nucleons of the type given by Brueckner[10]. A certain difficulty arises with their calculated pressure, in that it gives a speed of sound greater than the speed of light ("violation of causality"!). However, this happens for densities $\sim 10^{16}$ g/cm^3, beyond the domain of interest for stable neutron star configurations. For the peak mass of a neutron star they obtain the figure $M = 2.4M_\odot$, and for the minimum mass $M = 0.065M_\odot$ (Figure 4). As another example, Wang, Rose and Schlenker[11] making use of the phase shifts observed in high energy nucleon-nucleon scattering experiments, and making appropriate corrections to the effective baryon mass by reason of the finite density of the nuclear medium, obtain a "softer" equation of state. From it they calculate a minimum mass for a neutron star of $M = 0.13M_\odot$ or $M = 0.18M_\odot$ (depending upon details). Their analysis does not go to high enough densities to allow them to evaluate the maximum mass with any reliability. However, at the highest density at which they can rely upon their analysis, $\varrho = 10^{15}$ g/cm^3, the calculated mass is $0.16M_\odot$ or $0.26M_\odot$, again depending upon the details of the analysis. They give reasons based upon the careful treatment of nuclear forces by R. Reid[12], to doubt previous analyses of the effect of nuclear forces upon the equation of state.

The third effect in the high-density regime, particle creation, was first considered in any detail by Ambartsumyan and Saakyan[13]. They also neglected potential energy of interaction between particle and particle, about which one knows so little, and treated the medium as a mixture of ideal non-interacting gases, many more than three in number. The new feature of their work was the allowance for such particle production reactions as

$$e^- + n \rightarrow \Sigma^- + \nu$$

$$e^- + n \rightarrow \mu^- + n$$

$$e^- + n \rightarrow \pi^- + \nu + n$$

and their inverses. The existence of these reactions opens up more cells in phase space and lowers Fermi energies and pressures. They calculate equilibrium configurations with this equation of state and find limiting masses for a neutron star of

$$M_{\min} = 0.136M_\odot \quad \text{and} \quad M_{\max} = 0.634M_\odot$$

At higher and higher densities new particles of higher and higher rest masses can be formed. Ambartsumyan and Saakyan cut off the analysis with the known particles. Hagedorn, in a recent CERN preprint[14], assumes

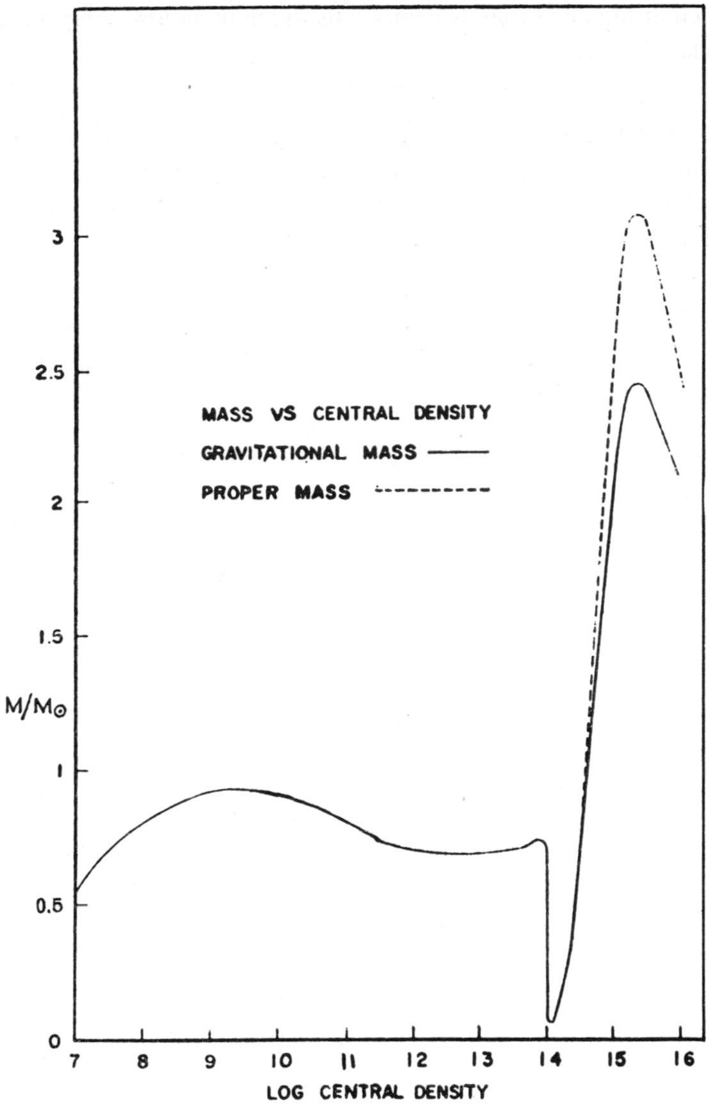

Figure 4 Equilibrium configurations for the Cameron-Cohen-Langer-Rosen equation of state, as reproduced from their paper[9]. "Gravitational mass" is the mass as seen from infinity; "proper mass" the product of the number of baryons by the mass of the proton plus the mass of the electron.

an infinite spectrum of particle masses, governed by a simple statistical law. In this way he arrives at an asymptotic equation of state of the form

$$p = \varrho_0 \ln (\varrho/\varrho_0) + p_0$$

one which at high densities evidently falls indefinitely low compared to the usual relativistic law

$$p = \varrho/3$$

Rhoades and Ruffini[15] have integrated the general equation of hydrostatic equilibrium for this equation of state ($\varrho > 10^{15}$ g/cm^3; pure neutron gas without interaction, for simplicity (Harrison-Wheeler), for $\varrho < 10^{15}$ g/cm^3). They calculated a critical mass of $M = 0.67M$ (Figure 5).

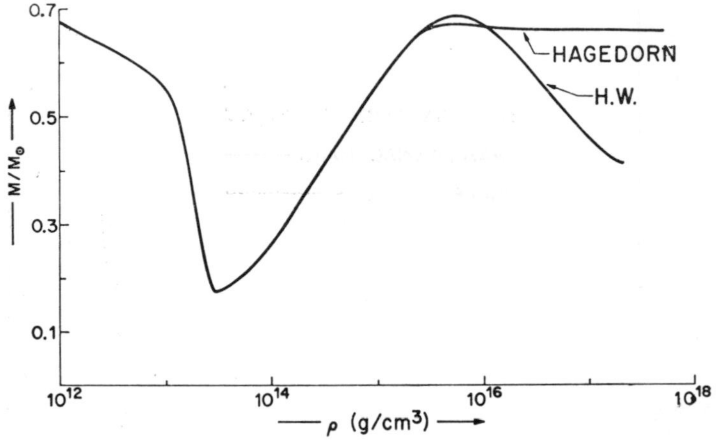

Figure 5 The mass at infinity (measured in units of solar mass) for the equilibrium configurations of the Hagedorn equation of state are given as a function of the central density. For a direct comparison the results of the treatment based on the Harrison-Wheeler equation of state for a degenerate non-interacting $e^- - n - p$ gas are given.

A few details of the Hagedorn picture of high-density matter may be in order:

1) Mixture of pions, kaons, nucleons, hyperons and all their possible resonances.

2) These hadrons—non-elementary hadrons; each is viewed as consisting of the others.

3) A universal effective highest "temperature" $T_0 = 160$ MeV.

4) Spectrum of hadronic masses m growing exponentially with a law

$$dN = \frac{a}{(m_0^2 + m^2)^{5/4}} e^{m/T_0}\, dm$$

where the constants a and m_0 are determined by fitting the 1432 particles and resonances known at January 1967: $a = 2.63 \times 10^4$ $(\text{MeV})^{2/3}$ and $m_0 = 500$ MeV.

Under these assumptions the constants in the Hagedorn equation of state have the values

$$p_0 = 1.344 \times 10^8 \text{ MeV}^4 \qquad (p_{0\,\text{conv}} = 0.314 \times 10^{14} \text{ g/cm}^3)$$

$$\varrho_0 = 5.415 \times 10^8 \text{ MeV}^4 \qquad (\varrho_{0\,\text{conv}} = 1.253 \times 10^{14} \text{ g/cm}^3)$$

Out of this review of equations of state in the regime of supranuclear densities we conclude (1) the existence of a stable family of neutron stars is invariant with respect to any change in the equation of state and (2) the minimum mass of this family lies very close to $M_{\text{min}} = 0.16 M_\odot$ but (3) the maximum mass is uncertain by a factor of the order of four. This uncertainty makes it impossible to look forward to an immediate test of general relativity *vs* newtonian or other theories of gravity, despite the factor ~ 2 in critical mass between the different gravitational theories. However, if we *assume* the correctness of Einstein theory and can work out a way to measure the masses and radii of a considerable number of neutron stars, then Gerlach[16] supplies us with a prescription to work backward from this information to deduce the equation of state. Thus, over the longer time-scale, observations of neutron stars offer a prospect of a deeper understanding of the behaviour of large collections of matter at nuclear density.

Let us make a final consideration on the use of an equation of state. In the preceding paragraphs we have seen how in the treatment of superdense stars two different physics have to be considered:

a) A "local physics" which is determined by the constitution of matter, fully described by giving the equation of state. All the details of solid state physics, nuclear, electromagnetic interactions, etc. are here contained.

b) A "non-local physics" which is determined by the generation of the gravitational field from this matter and is described by solving the Einstein equations or an equivalent set of equations (depending on the particular gravitational theory adopted).

Is this separation legitimate or do we have to consider a direct influence of the gravitational field on the local properties of matter? This problem has recently been treated in detail by Ruffini and Bonazzola[17]. They describe a neutron star by a self-consistent field approach without any use of an eqnation of state. Their main assumptions are:

1) Consider the relativistic Dirac equation for a free neutron field in a curved background metric.

2) Evaluate the expectation value of the energy momentum tensor operator of the spinor field relative to a state vector of a degenerate system of neutrons.

3) Make the limit for high quantum numbers, using a JWKB approximation.

4) Use the expectation value of the momentum energy tensor obtained by this limiting process as a source of Einstein equations.

They conclude that the use of an equation of state is perfectly legitimate up to densities of the order of 10^{32} g/cm^3!

CHAPTER 3

Pulsars

THERE are no richer signals from neutron stars nor signals more amenable to study from a space platform than those that come from pulsars. The pulsating sources of radiation, found by ground-based radio observations, may also give off X-rays, a valuable clue to the physics of neutron stars and only to be picked up by detectors based above the atmosphere.

The identification of pulsars was a lively issue for months following their discovery. Toward the end of 1967 Anthony Hewish and his co-workers at Cambridge University[18] discovered extraordinary astronomical radio sources characterised by the emission of sharp pulses of radio energy at exactly spaced intervals of time. To these objects they gave the name of pulsars. In Table IV, reproduced from ref. 19, we list the 61 sources known as of December 1971. For each pulsar we give the position, the period and its rate of increase, the approximate epoch at which the observations were made, the so-called "dispersion measure" (electrons per square centimeter in the intervening space along the line of sight), the approximate pulse width in milliseconds, and the strength of the observed pulses at 400 MHz. In the following a few experimental facts are summarised.

One of the most surprising features of these objects is the extreme sharpness (one part in 10^6, or better, over a one-day interval for all known pulsars) with which the repetition period of the signal is defined. The value of the periods (Table IV) ranges from a minimum of 33 ms in the case of PSR 0531 + 21 to a maximum value of 3.7 sec in the case of PSR 0525 + 21. For twenty-five of the sixty-one pulsars known today we have the value of the time derivative of the period. In every case the period is increasing. The most rapid increase occurs for the pulsars with the shortest periods. The shape of the pulse detected optically and averaged over 10^3–10^4 pulses is different from pulsar to pulsar but is sharply constant in time for every single pulsar. Associated with the main pulse there are in general one or two sub-pulses with different spectral and polarization properties. The width of the main pulse is usually very short in comparison with the repetition period and is larger for longer period pulsars. The radio pulses are often nearly 100 per cent linearly polarized and the power emitted at radio wave-lengths by pulsars of period $\sim$ 1 sec is of the order of 10^{30} erg/sec[19]. In the case of PSR 0833—43, the pulsar associated with the Vela supernova remnant, the pulse has

Table IV Parameters of 61 pulsars (from Manchester and Taylor[19])

(1) PSR	(2) Right ascension (1950.0)	(3) Declination (1950.0)	(4) l	(5) b	(6) Period (A.1)	(7) dP/dt (10^{-15} s/s)	(8) Epoch (JD-2400000)	(9) Dispersion (cm^{-3} pc)	(10) W_e (ms)	(11) E_{400} (10^{-29} J m^{-2} Hz^{-1})
0031−07	00^h 31^m 36.4^s ± 1.0^s	−07° 38′ 26″ ± 32″	110.4°	−69.8°	0.942950785^s ± 2	0.40 ± 0.11	40690.17	10.89 ± 0.01	42	25
0254−54	02 54 24 ± 1	−54 ± 2	271	−55	0.448 ± 3			10 ± 5	(10)	(50)
0301+19	03 01 45 ± 6	+19 42 ± 60	161	−33	1.39 ± 1				(20)	(20)
0329+54	03 29 11.00 ± 0.01	+54 24 36.7 ± 0.3	145.0	−1.2	0.71451866388 ± 3	2.0510 ± 0.0014	40621.50	26.776 ± 0.005	8.7	1200
0450−18	04 50 22 ± 2	−18 04 14 ± 20	217.1	−34.1	0.54893507 ± 3		40930	25 ± 10	27	70
0525+21	05 25 52.08 ± 0.07	+21 56 32 ± 16	183.9	−6.9	3.745491520 ± 15	39.95 ± 0.06	40400.50	50.955 ± 0.003	75	350
0531+21	05 31 31.428 ± 0.005	+21 58 54.40 ± 0.06	184.6	−5.8	0.03309756505419*	422.6889764*	40352.31*	56.791 ± 0.001	1.9	16
0628−28	06 28 50.8 ± 0.3	−28 34 08.1 ± 1.5	237.0	−16.8	1.2444148960 ± 5	2.51 ± 0.03	40243.44	34.36 ± 0.08	57	(500)
0736−40	07 36 51 ± 1	−40 34.5 ± 2.0	254.2	−9.2	0.37491832 ± 3		40220	100 ± 10	18	(50)
0740−28	07 40 47.5 ± 1.5	−28 15 16 ± 20	243.8	−2.4	0.166750167 ± 5		41020	80 ± 15	8	40
0809+74	08 09 03.0 ± 0.3	+74 38 12.2 ± 1.2	140.0	31.6	1.2922413240 ± 8	0.16 ± 0.04	40689.47	5.757 ± 0.002	45	105
0818−13	08 18 06 ± 2	−13 40 57 ± 20	235.9	12.6	1.23812811 ± 2		41089	40.99 ± 0.03	20	70
0823+26	08 23 50.52 ± 0.02*	+26 47 18.1 ± 0.8*	197.0	31.7	0.53065959904 ± 5	1.664 ± 0.004	40264.93	19.463 ± 0.001	6	22
0833−45	08 33 39.01 ± 0.03*	−45 00 06 ± 2*	263.6	−2.8	0.08920930095 ± 2	125.264 ± 0.008	40304.75	69.2 ± 0.1	1.71*	400
0834+06	08 34 26.15 ± 0.03	+06 20 43.0 ± 0.6	219.7	26.3	1.27376353580 ± 8	6.798 ± 0.004	40625.69	12.8550 ± 0.0005	17	100
0835−40	08 35 34 ± 1	−40 ± 2	260	0	0.765 ± 5			120 ± 12	(20)	(100)
0904+77	09 04 ± 10	+77 40 ± 50	135	34	1.57905 ± 6				(80)	
0940−56	09 40 40 ± 2	−56 ± 2	279	−3	0.662 ± 3			145 ± 15	(30)	(100)
0943+10	09 43 37 ± 6	+10 05 ± 2	225.4	43.2	1.097707 ± 3		40520.01	15.35 ± 0.01	30	
0950+08	09 50 30.60 ± 0.05	+08 09 44.5 ± 1.5	228.9	43.7	0.25306503679 ± 2	0.2321 ± 0.0005	39924.70	2.969 ± 0.002	9.5	65
0959−54	09 59 52 ± 2	−54 37 ± 15	280.1	0.3	1.436551 ± 2			90 ± 10	(50)	110
1112+50	11 12 49 ± 10	+50 18 ± 60	155	61	1.66 ± 2					(10)
1133+16	11 33 27.45 ± 0.01	+16 07 35.4 ± 0.2	241.9	69.2	1.18791119969 ± 3	3.734 ± 0.001	40621.82	4.8479 ± 0.0005	18	120
1154−62	11 54 45 ± 5	−62 ± 2	297	0	0.400 ± 5			270 ± 35		(30)
1237+25	12 37 11.99 ± 0.02	+25 10 16.6 ± 0.6	252.5	86.5	1.38244861342 ± 12	0.956 ± 0.006	40625.87	9.296 ± 0.005	25	55
1240−63	12 40 21 ± 2	−63 36 ± 12	302.0	−1.0	0.388 ± 4			220 ± 20	(60)	(50)
1359−50	13 59 43 ± 5	−50 ± 2	314	11	0.690 ± 5			20 ± 10	(20)	(50)
1426−66	14 26 35 ± 9	−66 ± 2	312	−6	0.788 ± 2			60 ± 7	(10)	(100)
1449−65	14 49 22 ± 5	−65 ± 1	315	−5	0.180 ± 2			90 ± 10	(5)	(50)
1451−68	14 51 29 ± 1	−68 32 ± 1	313.9	−8.6	0.26337676 ± 2		40550	8.60 ± 0.04	13*	(200)

1508+55	15 08 03.74±0.18	+55 42 55.8±0.2	91.3	52.3	0.73967789848±3	5.0389±0.0014	40625.96	19.599±0.005	13	35
1530−53	15 30 23±1	−53 30±10	325.7	1.9	1.368852±14			20±5	(25)	(100)
1541+09	15 41 10±15	+09 38 29±20	17.8	45.8	0.748442±5			34.99±0.02	44	40
1604−00	16 04 39±3	−00 25 08±25	10.7	35.5	0.42181607±3		41006	10.72±0.05	8	10
1642−03	16 42 24.65±0.09	−03 12 31.1±0.3	14.1	26.1	0.38768879129±2	1.7803±0.0006	40622.04	35.71±0.01	4.0	135
1706−16	17 06 33.23±0.05	−16 37 12±3	5.8	13.7	0.65305047396±10	6.369±0.005	40622.07	24.88±0.02	11	45
1727−47	17 27 47±1	−47 40±10	342.6	−7.6	0.829683±4			121±4	(30)	180
1747−46	17 47 57±1	−46 56±10	345.0	−10.2	0.742349±3			40±10	(20)	90
1749−28	17 49 49.27±0.07	−28 06 01±4	1.5	−1.0	0.56255316830±4	8.154±0.002	40128.19	50.88±0.14	6	(500)
1818−04	18 8 13.61±0.02	−04 29 03.3±0.3	25.5	4.7	0.59807263977±3	6.3210±0.0015	40622.10	84.38±0.02	11	55
1845−01	18 45±4	−01 27±1	34	2	0.659475			90	(80)	(30)
1845−04	18 45 10±45	−04 05 32±25	28.9	−1.0	0.59773452±6		41000	141.9±0.6	20	40
1857−26	18 57 45±2	−26 04 49±20	10.3	−13.5	0.6122083±5			35±10	15	80
1858+03	18 58 40±45	+03 27 02±25	37.2	−0.6	0.655444±1		40754	402±2	35	20
1911−04	19 11 15.14±0.02	−04 45 59.4±0.6	31.3	−7.1	0.82593368981±6	4.062±0.003	40624.13	89.43±0.02	7.5	50
1914+13	19 14±4	+13 50±1	48	1	0.194635			90	(17)	(10)
1919+21	19 19 36.20±0.03	+21 47 15.2±0.8	55.8	3.5	1.33730110168±7	1.345±0.003	39912.08	12.4309±0.0005	25	190
1929+10	19 29 51.90±0.03	+10 53 03.5±0.6	47.4	−3.9	0.22651704512±4	1.1580±0.0017	40625.15	3.176±0.003	5.5	15
1933+16	19 33 31.86±0.10	+16 09 58.3±0.2	52.4	−2.1	0.35873543128±2	6.0041±0.0008	40689.97	158.53±0.05	6.5	35
1944+17	19 44 39±3	+17 58 44±20	55.3	−3.5	0.4406179±7			35±10	23	30
1946+35	19 46 34.5±2.0	+35 25±15	70.6	5.0	0.717306±1		40560	129.1±0.1	21	(50)
1953+29	19 53±1	+29 15 03±20	66	1	0.426676±1		40754	20	13	8
2003+31	20 03±1	+31 30±12	69	0	2.111206±1		40756	225±10	25	42
2016+28	20 16 00.18±0.06	+28 30 30.2±0.3	68.1	−4.0	0.55795340727±4	0.1495±0.0018	40689.00	14.176±0.008	14	105
2020+28	20 20 32±2	+28 44 30±25	68.9	−4.7	0.34340072±1		40965	44±10	15	70
2021+51	20 21 25.30±0.03	+51 45 07.9±0.3	87.9	8.4	0.52919532808±4	3.0448±0.0019	40625.17	22.580±0.004	11.5	70
2045−16	20 45 46.85±0.17	−16 27 48±8	30.5	−33.1	1.96156687960±14	10.965±0.009	40695.01	11.51±0.01	42	120
2111+46	21 11 40.5±2.5	+46 27±10	89.0	−1.3	1.01468455±3		41092	141.50±0.04	29	50
2217+47	22 17 45.90±0.09	+47 39 48.1±0.2	98.4	−7.6	0.53846739459±3	2.7635±0.0012	40624.26	43.54±0.01	7.3	40
2303+30	23 03 30±10	+30 45±4	97.7	−26.6	1.57588443±2		41006	49.9±0.2	17	40
2319+60	23 19 42±5	+60 00±6	112.0	−0.7	2.256483±1		40734	96±3	140	40

been found to be completely polarized (linear polarization > 95%) with the direction of the plane of polarization varying systematically with time through the main peak of the pulse[20].

The distance of a pulsar is established by measuring the dispersion of the pulse. More, precisely it has been noticed that the time of arrival of a given pulse at the Earth's surface is a function of the frequency and is delayed by the amount

$$\tau = \frac{\alpha}{\nu^2} \int_0^L n_e \, dl \quad \text{where} \quad \alpha = \frac{e^2}{2\pi mc}$$

From the value of the integral $\int n_e dl$ (for example, 20 pc cm³ or 6×10^{19} electrons/cm²), by assuming a value for the density of the electrons (for example $n_e = 0.1$ electrons/cm³), we can obtain an estimate of the distance of the source (here 200 pc). Knowing the value of the integral $\int n_e dl$ we can also evaluate the component of the mangetic field in the direction of the line of sight of the pulsar. In fact, linearly polarized radiation passing through an ionized medium changes its angle of polarization by an angle

$$\Delta\theta \propto \lambda^2 \int n_e B_{\parallel} \, dl$$

where λ is the wavelength of the radiation and $B_{\parallel}$ is the component of the magnetic field parallel to the line of sight.

The majority of the pulsars are concentrated near the galactic plane. Some pulsars are associated with supernovae remnants and some do not have any presently visible trace of debris in their neighbourhood. Some supernova remnants show no trace of a pulsar; but the two shortest period pulsars PSR 0531+21 and PSR 0833—45 are both associated with supernova remnants: with the Crab Nebula and with Vela X respectively.

The Crab Nebula pulsar is the first and only one to be detected both in the optical region (16.6 magnitude star suggested as central to the Crab Nebula by Minkowski in 1934[21]) and in the X-ray region. Optical reception not only explores an extremely important region of the spectrum but also improves the timing of the period and its first and second time derivative for PSR 0531. The second derivative of the period $\ddot{P}$, obtained from observations of optical data over six weeks in the months of June and July 1969, has given[22] a value of

$$\ddot{P} = 0.110 \pm 0.02 \text{ ps/day}^2$$

Measurements over a longer period of time (November 1968 to July 1969) using radio techniques have given[23] a value of

$$\ddot{P} = 0.024 \pm 0.006 \text{ ps/day}^2$$

Unfortunately, no accurate measurement of the second derivative of any other pulsar is available at the moment. This measurement is extremely important for theoretical models; it gives the so-called "slowing exponent" n, such that the power of the frequency to which the rate of loss of rotational energy is proportional is $(n + 1)$:

$$n = \frac{\omega\ddot{\omega}}{\dot{\omega}^2} = 2 - P\ddot{P}/\dot{P}^2$$

Discontinuous changes in the period and in the slowing down rate have been observed both for PSR 0531 and for PSR 0833. The observed speeding up has been $\Delta\omega/\omega \sim 2 \times 10^{-6}$ in the case of the Vela pulsar[24,25] and $\Delta\omega/\omega \sim 10^{-8}$ for the Crab pulsar. The possibility of a "wobbling" in the period of PSR 0531 with a period of three months has also been reported[26] but not yet confirmed by a more accurate data analysis. Subsequent work suggests that the index n may not be a well-defined quantity, perhaps because of frequent small changes in $\dot{P}$. Its value generally lies in the range 2–3.

On 13 March 1969 an Aerobee rocket flight provided forty seconds of observations of the Crab pulsar in the X-ray region at wavelengths in the range 1.2 to 13 keV and also at 0.65 keV. The X-ray flux emitted by the pulsar amounts to approximately five per cent of the integrated X-ray flux of the entire nebula[27]. Moreover, the power emitted by the pulsar in the X-ray region is about 200 times the optical power and about 2×10^4 times the radio power.

3.1 PULSARS AS NEUTRON STARS

Nothing did more in the end to drive one to neutron stars as being the explanation for pulsars than the order of magnitude of the repetition period and the slow increase of this period with time. However, many other possible interpretations were proposed and require mention.

That a pulsar cannot be a body of planetary mass follows from its energy output. Thus the "dispersion measure" gives the order of magnitude of the distance (a few pc to several thousand pc). This distance allows the flux of radio energy observed at the Earth to be translated into an absolute output of radio energy of the order of 10^{30} erg/sec, as compared to the output of the Sun, integrated over all frequencies, of 2×10^{33} erg/sec.

That a pulsar cannot be a body as large as the Sun follows from the existence of periods of the order of a second. This fixes an upper limit to the dimensions of the order of one light second or $\sim 10^5$ km. A much more stringent limit comes from the discovery of the Crab pulsar (PSR 0531), with a period of 33 ms, or corresponding light travel distance of 10^4 km.

No one has ever been able to propose any object other than a white dwarf or a neutron star compatible with these energy and size requirements; and no mechanism for its precise time-keeping other than pulsation or rotation has received general attention.

The white dwarf model, attractive at first, was soon abandoned because of difficulties about the period and its time rate of change (quite apart from the failure to see any white dwarf in the telescope at any pulsar location!). If the white dwarf turned once a second, its equatorial velocity would exceed the critical value for loss of mass. Therefore a rotating white dwarf could be excluded at once. A vibrating white dwarf has a fundamental pulsational period larger than two seconds, and therefore too large to explain the pulsar periods[28]. However, there is abundant evidence for the excitation of overtone modes in certain stars. Such a mode could have a period of the right order of magnitude. However, in the usual cases where such an overtone is observed in stars, it is mixed with other vibrational modes. Such a mixing would be incompatible with the sharply defined repetition period of the pulses. Moreover, a vibratory motion as it loses energy almost always increases in frequency (molecular vibrations; vibrations of springs that are not perfectly elastic; vibrations of gas spheres and globes of fluid) contrary to the observed increase in period. Therefore the interpretation of a pulsar as a vibrating white dwarf was given up.

Still another way to drive a white dwarf is to put it into orbit around another white dwarf. However, the period calculated for this motion ($P \sim 4$ sec or more) also makes this model inadequate to explain the pulsars.

White dwarfs being excluded, only neutron stars remain. A pulsational model was not possible. The pulsational period of a neutron star is in the millisecond region[28] far away from the observed period of pulsars. A model based on orbital motions was also easily excluded. A double neutron-star system could have a period of revolution in the region of milliseconds to seconds. However, emission of gravitational radiation would be extremely large, the lifetime of the system quite short[29] and most important the period should *decrease* in time in contradiction with the experimental evidence.

The only model left is therefore a rotating neutron star. With this model the observed periods are easily compatible in four respects.

1) The surface velocity even for the most rapidly turning neutron star (~ 20 km $2\pi/0.033$ sec ~ 4000 km/sec) is negligible in comparison with the speed required for mass loss

$$(v \sim (2GM/R)^{1/2} \sim 10^5 \text{ km/sec})$$

2) The angular momentum calculated on the neutron star model

$$(I\omega \sim 4 \times 10^{44} \text{ g cm}^2 \times 200 \text{ rad/sec} \sim 10^{47} \text{ g cm}^2/\text{sec})$$

is smaller than the angular momentum of such a typical star as the Sun (1.7×10^{48} g cm^2/sec, or more, according to Dicke) and therefore compatible with the idea of formation of a neutron star by stellar collapse.

3) The gradual slowing of the rate of pulsation in every instance is exactly what one would expect from the loss of rotational energy.

4) A survey of all mechanisms for powering the Crab Nebula, published shortly before the discovery of pulsars[30], led to the conclusion that, "Energy of rotation of the [neutron] star itself or energy of bulk motion of the ion clouds ejected at the time of its formation seems to be a stockpile of energy more rewarding [than vibration or thermal energy] for further investigation... Presumably this mechanism [energy of rotation] can only be effective, if then, when the magnetic field of the residual neutron star is well coupled to the surrounding ion clouds".

Accepting the interpretation of pulsars as neutron stars, we have to ask three central questions: (1) what is the dominant mechanism for the loss of rotational energy; (2) what is the mechanism for the emission of the pulses; and (3) why did the period suddenly drop in two instances?

Two mechanisms for the loss of rotational energy immediately come to attention: magnetic dipole radiation and gravitational quadrupole radiation. The one puts energy into the surrounding plasma; the other radiates straight out into space. That some energy must go into the plasma is suggested by the great output of the Crab Nebula. Radiating at the present rate, the high energy electrons should be emptied of energy in a time of the order of a year, according to simple estimates; but the nebula has been radiating for 900 years. Is not therefore the rotating neutron star the source of much of this energy? (Table V). Moreover, such a star, formed by collapse of a star with white dwarf core, will be expected to conserve the original magnetic flux, as its precursor will be expected to conserve in turn the flux of the original main sequence star. Thus 100 gauss for a star with core 7×10^{10} cm in diameter will become 100 gauss $(7 \times 10^{10}\text{ cm}/2 \times 10^{6}\text{ cm})^2$ or 1.2×10^{11} gauss. Magnetohydrodynamical stability considerations also raise the question whether an alignment of the magnetic moment parallel to the spin axis is stable. Thus a magnetic moment may be anticipated of the very rough order of $m_\perp \sim 10^{25}$ gauss cm^2 perpendicular to the rotational axis, and turning with the neutron star. Such a rotator will act as a magnetic dipole radiator, losing energy at the rate

$$-(dE/dt)_{\text{mag}} = -(d/dt)\left(\tfrac{1}{2}I\omega^2\right)_{\text{mag}} = (2m_\perp^2/3c^3)\,\omega^4 = \frac{1}{6c^3}\,a^6 B^2 \omega^4 \sin^2 \alpha$$

Here B is the polar magnetic field, α the angle between the rotation axis and the axis of the magnetic dipole, and a the radius of the neutron star[31].

Table V Energy balance for Crab pulsar

Effect	Power assuming distance 5×10^{21} cm	Power assuming distance 3×10^{21} cm
Observed radiative output from the Crab pulsar itself (primarily in X-ray region; less by factor 10^2 in visible; less by factor 10^4 in radio)[27]	0.48×10^{36} erg/sec	0.17×10^{36} erg/sec
Observed radiation output from Crab Nebula ($\sim$3 lyr radius; believed to be driven by Crab pulsar)[30]	6×10^{37} erg/sec	2×10^{37} erg/sec
Calculated loss of energy from a neutron star of mass $0.405 M_\odot$, radius 20.8 km, central density 3×10^{14} g/cm^3, moment of inertia 4.4×10^{44} g/cm^2, turning with a rotation period of $P = 33$ m sec, and rotation energy of 7.9×10^{48} erg slowing at a rate $\dot{P} = 13.5$ μsec/yr[30]	2.0×10^{38} erg/sec	
Component of magnetic dipole moment perpendicular to the rotation axis required to give the observed rate of slowing down if 30-cycle magnetic radiation alone is to account for the observed rate of slowing down[32]	2.5×10^{30} gauss cm^3	
Equivalent "effective magnetic field"[31] ($\alpha = \pi/4$)	1.2×10^{11} gauss	
Eccentricity $\varepsilon = (a - b)/\sqrt{ab}$ of neutron star in equatorial plane required to provide the observed rate of slowing down assuming that gravitational radiation is the predominant mode of loss of energy[32] (simplified model of uniform density!)	$\varepsilon = 1.1. \times 10^{-3}$ implying $(a - b) \sim 23$ m	

The value of $m_\perp$ required to account for the observed slowing of the Crab pulsar is of a reasonable order of magnitude (Table V); so also in the case of other pulsars[32]. The "slowing exponent" for this mechanism of retardation is 3 (power radiated $\propto \omega^4$).

The intensity of any electromagnetic wave can be characterised by a parameter $f = eE/m_e c\omega$, where E is the electric field of the wave. In most astronomical or laboratory situations, $f \ll 1$; but in the case of 30 Hz waves emerging from the Crab pulsar this "strength parameter" varies from $f \simeq 10^{11}$ at the light cylinder (radius $c/\omega \simeq 1.5 \times 10^8$ cm) to $f \simeq 10$ at the boundary of the nebula (radius $\sim 3 \times 10^{18}$ cm). In this situation, a test electron exposed to the waves attains a relativistic energy with Lorentz factor $\gamma \simeq f$. 30 Hz waves can propagate through a plasma of density n_e provided that $n_e \lesssim 10^{-5} f$ cm^{-3}. Otherwise their pressure will expel the plasma, perhaps at a speed $\sim c$.

Thus the radiated energy will accelerate particles; but where, to what energies, and with what consequences, are complicated questions of plasma physics still unanswered. Most puzzling of all is how the short, sharp pulses are produced with their characteristic shapes: are the pulse shapes determined by magnetic field geometry close to the stellar surface; or does every pulsar produce the same kind of pulse wave front which, however, has different wave forms for different angles between the rotation axis and the line of sight? Almost all theories for the radiation mechanism require the presence of some plasma in a "magnetosphere" around the pulsar. The formulae for emission by a spinning dipole "in vacuo" therefore cannot be exactly valid. Theoretical studies suggest that they nevertheless remain true in order of magnitude: the expected braking index n is, however, *less* than 3 if the plasma pressure is sufficient to distort the field towards a radial configuration.

Additional to the loss of energy by magnetic coupling to the outside is gravitational radiation. It takes place only when the neutron star departs from rotational symmetry about the axis of rotation ("time varying quadrupole moment"). Denote by

$$\varepsilon = (a - b)/(ab)^{1/2}$$

the eccentricity in the equatorial plane and by I the moment of inertia. Then the rate of emission on the simplified model of a system of uniform density is

$$-(dE/dt) = (288G/45c^5)\,(I\varepsilon\omega^3)^2$$

and the slowing exponent is 5.

To measure the slowing exponent is a simple way to get a first indication whether magnetic dipole or gravitational quadrupole radiation dominates the slowing process. Unfortunately the value of the second derivative of the period is know today only for one pulsar, that in the Crab. For this object we have two different values of the second derivative. One, by radio observations on a time-scale of $\sim$ 8 months gives[23]

$$n = 2.6 \pm 0.6$$

compatible with dominance of magnetic but not of gravitational radiation. The second, by optical reception on a time-scale of $\sim$ 2 years gives[22]

$$n = 2.42 \pm 0.22$$

Too much significance should not be attached to either number because both measurements were made before one had recognised—thanks most of all to the later work of the observing teams—the possibility that a star quake, even a very small star quake, could affect the apparent values of P and n.

A slowing exponent of 5 is one test for gravitational radiation. Reasonableness of the value of the eccentricity required to account for the observed energy loss is another test. What does this test give? An equatorial eccentricity of ~ 10 m (Table V) is required for the Crab pulsar if gravitational radiation dominates in its slowing. A magnetic field of 10^{11} gauss acting on a star with the density of nuclear matter is estimated to produce an "upwelling" of material in the strong field region less than a micron in height! Nobody yet knows how to calculate from first principles whether 10 m is an eccentricity naturally to be expected in some other way. We can certainly say that if the neutron star were completely liquid and turning as "slowly" as it is today, it would have no equatorial eccentricity. However, it turned much more rapidly in the past and it is not liquid. Moreover, a globe of ideal incompressible fluid in sufficiently fast rotation has a natural equilibrium confiugration which is a prolate spheroid with its principal axis perpendicular to the axis of rotation[32] (ε positive!). The crust of the neutron star (density from 10^9 to 5×10^{13} g/cm^3) has a calculated depth from a few hundred metres to a few kilometres, depending on the mass of the neutron star. The calculated melting temperature for a Coulomb lattice, of the rough order of 10^{9}°K[33], exceeds the temperature estimated for a short time after formation. Thus an initial deformation could well be frozen in. But could a 10 m high bulge remain frozen in?

One way to find out is to look directly for gravitational radiation from the Crab pulsar. An emission of 10^{38} erg/sec at a distance of 5×10^{21} cm implies a flux at the Earth of 3×10^{-7} erg/cm^2 sec gravitational radiation of very sharply defined frequency—a radiation quite susceptible to being detected.

Another way eventually to judge the reasonableness of a 10 m bulge is to investigate star quakes due to possible discontinuous changes in the shape of a neutron star as it slows down and transforms from an oblate spheroid towards a sphere. The discontinuities in period observed in the Vela pulsar (PSR 8033—45) and in the Crab pulsar (PSR 0531+21) have been interpreted as "quakes" caused by sudden release of stress in the crust. From the observed change in period one deduces the change in moment of inertia and from this the change in dimensions: of the order of 1 cm in one case and tenths of a mm in the other. If a star cannot support more than this much departure from equilibrium one can ask, how can it possibly support a 23 m bulge? Though the question is simple the answer may not be as simple as it seems!

3.2 THE CRUST AND INTERIOR OF A NEUTRON STAR

So much for gravitational radiation; now for the crust and interior of a neutron star as matters of interest in their own right. Table VI shows the calculated variation of pressure, density and composition with depth going down step by step into the interior of a neutron star under two assumptions: (1) the idealised case of complete combustion to the endpoint of thermonuclear evolution and (2) some incompleteness of combustion (as, for example, H burned to He but not burned to iron) and some formation of compounds (as, for example, FeHe in upper layers). The FeHe compound listed in the last column is purely illustrative; nobody has yet calculated the course of thermonuclear combustion in the collapse of a white dwarf in sufficient detail to state precisely what nuclear species will be produced and in what amounts in the upper layers of the resulting neutron star. The important point is the incompleteness of the combustion, implying the presence in any given layer of more than one nuclear species. Moreover, the pressures are so high and densities so great (10^8 g/cm^3 $< \varrho <$ 10^{12} g/cm^3) that these nuclei find themselves in a practically ideally degenerate electron gas. Under such conditions they can and will crystallize[34]. Multiple-component lattices probably complicate typical *n*-star "geo" logy. However, the simpler case of a binary lattice lends itself to detailed analysis. Dyson[35] finds that a particularly stable configuration is given by a lattice with NaCl structure and with a ratio of 0.07 between electric charges of the two kinds of constituent nuclei; hence the use as a purely illustrative example of FeHe in Table VI.

The memory of the past history of combustion, carried by the fossil composition of the crust at various levels, is wiped out below a certain depth. Nuclear reactions intermediated by neutron loss and neutron pick-up become important below a certain depth ($\varrho \sim 3 \times 10^{11}$ g/cm^3) and bring the material quickly to a standard "end point condition", uniquely determined by the density (Table VI).

At densities of the order of 5×10^{13} and more nuclei as such disappear and neutrons dominate. Protons (and electrons) are less numerous by about one order of magnitude. Ginzburg and Kirzhnitz[36] pointed out that the neutrons should form a Bardeen-Cooper-Schrieffer superfluid. The calculated temperature of transition from superfluid to normal fluid, governed by the neutron-neutron pairing force, $T \sim 3 \times 10^9$ °K or more, is smaller than the temperature to be expected at the center of the neutron star after formation ($\sim 10^{12}$ °K) but larger than the temperature estimated for one year after formation ($\sim 10^7$ °K[37]).

The neutron superfluid carries a substantial fraction of the angular momentum of the *n*-star. However, this part of the angular momentum is not distributed smoothly over the fluid. It is concentrated in quantised vortex lines, with separation from line to line of the order of 10^{-2} cm. Clear evidence for similar vortex lines has been found in superfluid helium and there is every reason to believe they should exist in a neutron superfluid. They make possible the propagation of Tkachenko-Dyson[38] waves through the neutron superfluid at a velocity of the order of 1 cm/sec (stacking density wave, similar to the effect of a puff of wind passing over a wheat field).

Table VI Properties of a superdense star of central density 3.5×10^{13} g/cm³. (Calculations based on HTWW equation of state.[5] The calculated mass is $0.18 M_\odot$)

Schwarzschild radial coordinate (km) (approx.)	Density (g/cm³)	Pressure (g/cm sec²)	Dominant nucleus in an idealised neutron star at the absolute endpoint of thermonuclear evolution	Sample of conceivable constitution for an actual neutron star (incomplete thermonuclear combustion)
210 (top ~1 cm)	gaseous	gaseous	26 Fe 56	gaseous Fe
210 (next few cm)	7.85	"0"	26 Fe 56	Fe—He compound
170	8.00 E 6	8.96 E 23	26 Fe 56	crust { Fe—He compound
50	1.67 E 10	1.56 E 28	31 Ga 78	Fe—He compound
30	3.18 E 11	5.83 E 29	39 Y 122	Fe—He compound
20	4.5 E 12	6.62 E 30	Fermi gas { neutrons protons electrons	core { Superfluid neutrons plus superconducting protons plus degenerate electrons
2	3.49 E 13	1.85 E 32		

The proton gas is similarly coupled by proton-proton pairing forces and should therefore be a Bardeen-Cooper-Schrieffer superconductor, according to Baym, Pethick and Pines[39]. The calculated temperature for the transition to superconductivity is in excess of 10^9 °K. Baym and his collaborators reason that this superconducting proton medium does not expel the magnetic field of the neutron star (estimated to be $\sim 10^{11}$ or 10^{12} gauss) but "channels" it within the star (type II superconductor; estimated lifetime for expulsion or decay of the field of the order of 10^{13} yr).

Electron gas, proton superconductor, neutron superfluid and the crystalline crust are coupled together by the magnetic field. This coupling is calculated to have an important effect on the rate of rotation of the crust ("starquake"[40]). The crust should not immediately re-adjust its rate of rotation to the new and smaller moment of inertia. Instead, there should be relaxation phenomena. The diagnosis of the several relaxation times should do much to test and extend our preliminary information about the internal constitution of the neutron star. It should also yield information on the mass and radius of the *n*-star. Central to such diagnosis are high precision measurements of the timing of the pulses by optical and radio observations.

On-the-ground measurements thus hold forth prospects of learning about the deep interior of a neutron star. By contrast, measurements from a space platform in the infrared (dip in Crab pulsar spectrum[41]), the X-ray region and the gamma-ray region, safe from atmospheric absorption (Figure 6) promise insight into what goes on outside the neutron star. From the spectral distribution and polarization of this radiation one hopes to learn something about the mechanism of emission of the observed pulses themselves, the

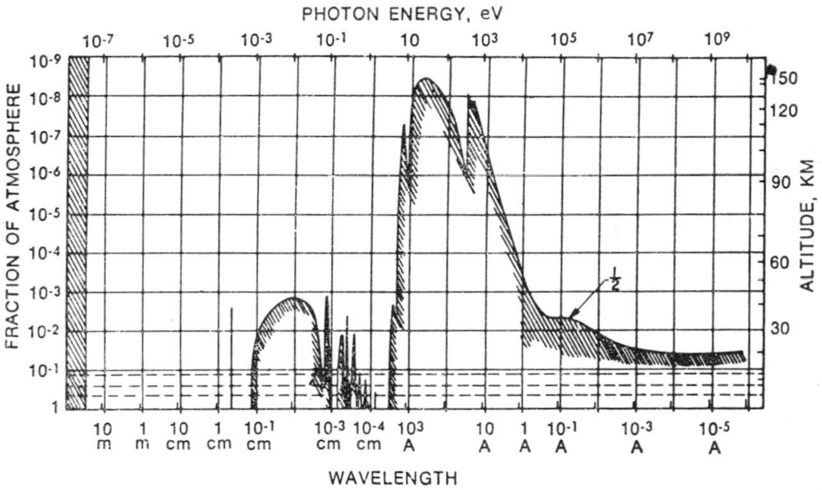

Figure 6 Height in the Earth's atmosphere to which detecting apparatus must be carried in order to detect more than 50% of the incident cosmic electromagnetic radiations. (We are indebted to B. Rossi for this figure.)

intensity and shape of the magnetic field involved, and the density of the electrons. The Crab pulsar puts out more energy in X-rays than in any other form of radiation of which we have evidence. If the same is true of other pulsars, then an X-ray telescope in space via this one application alone may come to dominate a whole rich new field of physics. Already X-ray telescopes in brief flight above the atmosphere have discovered a whole family of new objects. Some of these so-called X-ray stars may be related to neutron stars or other collapsed objects.

CHAPTER 4

Supernovae

THE PHYSICS of the formation of a neutron star is more complicated than the physics of the neutron star itself. It is believed that in this process a star (late giant or other) having a dense core with a radius of a few thousands of kilometres collapses to a compact object with a radius of a few tens of kilometres. The dense core has slowly evolved over thousands of years to a degree of compaction where it is unstable against gravitational collapse. This does not necessarily mean that its mass lies precisely at $M \sim 1.3 M_{\odot}$, the first peak in Figure 2. It may be two, or five, or ten times more massive and still not collapse when sufficiently inflated by sufficiently high temperature. But cooling of such a system will automatically bring it to the point of collapse. Colgate, May and White at Livermore[42,43] have made computer investigations of what then happens, under the simplifying assumption of spherical symmetry. The material of the star starts moving inwards slowly at first, then more and more rapidly, with a characteristic time of speed-up of less than a tenth of a second. Soon the substantial inner portion of this mass, the "core" (*of* the dense core!), becomes sufficiently compacted to greatly increase the strength of the gravitational fields acting to draw this core together. In consequence the core accelerates more rapidly than the surrounding envelope. Two very different results ensue according as the core mass and its kinetic energy of implosion do, or do not, suffice to drive the system on beyond nuclear densities to the point of complete gravitational collapse. Complete collapse produces a black hole (Chapter 5). On the other hand, when the mass is too small or the velocity of implosion is too low, or both, the collapse is halted at nuclear or near-nuclear density. The stopping of so large a mass implies the sudden conversion of an enormous kinetic energy into an enormous heating ($\sim 10^{12}$ °K), as if a charge of dynamite had been set off at the center of the system. The high temperature develops high pressure. The envelope, falling in on a slow time-scale, suddenly feels this pressure. The implosion is reversed. The envelope is propelled outward to give cosmic rays (the outermost 10^{-4} or 10^{-5} of the mass) and an expanding ion cloud (as, for example, the Crab Nebula itself; estimated mass some substantial fraction of a solar mass). Some details of this process as calculated by Colgate and White appear in Figure 7. These and other investigators are working towards calculations that will include still more

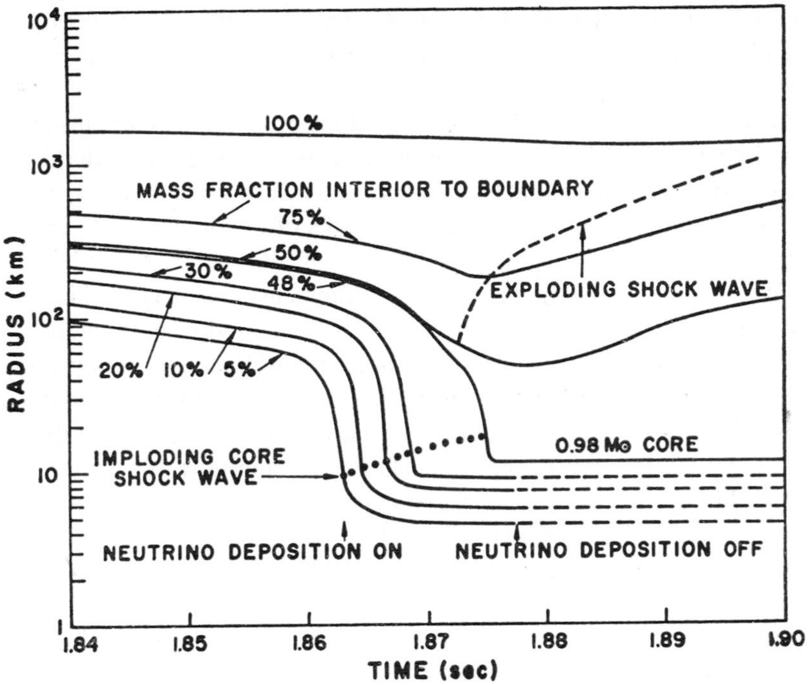

Figure 7 Formation of a neutron star by the implosion of the inner 49% of a dense star of mass $2M_{\odot}$ (core of a late giant or other star), according to Colgate and White[42]. The inner region falls inward more and more rapidly. The energy set free goes out as neutrinos. At first this radiation easily penetrates the outer layers of the dense star: soon those outer layers fall inward and become sufficiently compact to absorb neutrinos emitted from the core. A shock is generated in the material just outside the core, the explosion releasing enough energy to form a supernova.

detail. Of special interest are the relative importance and the consequences of two conflicting factors: (1) nuclear reactions in the core and the envelope which could go far enough under certain circumstances to heat up the system prematurely, thereby reinflating the star and spoiling the implosion, and (2) angular momentum.

Recent work of LeBlanc and Wilson[44], also at Livermore, shows in a simple situation how important rotation, and magnetic fields coupled to rotation, can be for changing the character of the implosion. As the center shrinks, it turns faster and faster by reason of the law of conservation of angular momentum, winding up the magnetic lines of force like string on a spool. The Faraday-Maxwell repulsion between the lines of force causes the spool to elongate. The lines of force carry matter with them. Thus, jets shoot out from the two poles. It would be interesting to see how these effects are modified when the calculations include all the physical details that were taken into account in the Colgate-May-White analysis and nuclear reactions as well!

The more supernova implosion calculations progress the more insight one gains into the rich phenomenology of these events, both those that lead to neutron stars and those that lead to black holes. For surveys of relevant physics and literature one may refer to the books of Zel'dovich and Novikov[45] and of Shklovskii[46] or the review article of Wheeler[30]. Excluding from attention the debris of the explosion and the neutron star left behind, what can one learn from the explosion itself and the period of some months following? The light curve in the visible has been well studied not only for the numerous supernovae located during this century in other galaxies but also a few historic supernovae in our own galaxy via old records of comparisons from night to night with other stars of known magnitude. If a nuclear bomb exploded in our atmosphere reaches a maximum brightness only ten seconds later, because of the need for the fireball to open out, it is not surprising that a supernova reaches maximum brightness in the visible only after a few days. A more detailed study of supernova hydrodynamics indicates, according to Colgate[47], that layers at successive depths in the envelope should become the chief emitters of light at successive times, with successive temperatures characterising each one in turn. Thus the time-temperature curve should provide a means of diagnosing the hydrodynamics of the envelope. The envelope first commences opening out in a substantial way when the shock wave from the core reaches the outer surface of the envelope and is returned towards the interior as a rarefaction. The outermost layer goes into particles of relativistic velocities, but each layer deeper down is impeded in its outward propulsion by more and more matter in its way and reaches lower and lower velocities.

It is natural to consider as pre-supernova star a late giant star with dense core plus very thin but enormous envelope. Simplifying, Colgate[47] has analysed the dynamics of a bare core with mass $M \sim 5M_{\odot}$. The surface layer, by reason of shock wave temperature plus superposed Doppler effect is calculated to emit a $\simeq 10^{-5}$ sec flash of photons at temperature of ~ 2 GeV. Following this peak the calculated radiation temperature drops off from moment to moment. The integrated flux of energy expected at a receptor above the Earth's atmosphere from a supernova of this kind at a distance typical of our galaxy, 10^4 pc, is ~ 50 erg/cm^2. Colgate estimates that a properly designed detector of gamma rays located on a space platform should detect such flashes from supernovae in nearby galaxies, with an expected rate of the order of ten events a year.

In addition to gamma rays and X-rays produced in the flash and immediately thereafter by direct thermal radiation, other gamma rays of highly characteristic energies will be given off in the radioactive decay of products of thermonuclear combustion. These gamma rays are attractive as tools to diagnose the nucleosynthesis. Clayton, Colgate and Fishman[48] take as a starting point a star already very rich in silicon (as a consequence of earlier thermonuclear evolution). This star is considered as undergoing collapse to a neutron star, throwing off in the process a shell of mass $\sim 0.5M_\odot$. During the shock developed in throwing off this mass, a substantial portion of the shell-to-be is calculated to be shocked to a temperature $\sim 5 \times 10^9$ °K, a density $\sim 1.3 \times 10^7$ g/cm^3 and to expand in a time ~ 0.1 sec. These conditions are estimated to be favourable to a process of burning of ^{28}Si to Ni^{56}, followed by the beta decays,

$$\begin{array}{c} ^{56}\text{Ni (6.1 days)} \rightarrow {}^{56}\text{Co (77 days)} \rightarrow {}^{56}\text{Fe} \\ 0.163 - 1.56\ \text{MeV}, \qquad 0.511 - 3.47\ \text{MeV} \\ \gamma\text{-rays} \end{array}$$

The estimated production of ^{56}Ni is $\sim 0.16M_\odot$. The estimated output of gamma rays is $\sim 10^{49}$ ergs, released over $\sim 10^6$ sec. From the observations made at the time of the Tycho supernova one independently deduces energy in the visible of this order of magnitude in the same time interval.

One kind of detector for such characteristic gamma rays would have a "window" extending from 0.5 MeV to 3 MeV, and be located on a satellite in order to escape the absorption of the Earth's atmosphere (Figure 6). It is estimated by Colgate *et al.*[48] to be capable of detecting 4×10^{-5} photons/cm^2 sec. Such a detector should pick up $\sim 10^{-3}$ gamma quanta/cm^2 sec of characteristic energy from a supernova at a distance of 10^6 pc, and pick up a detectable signal from a supernova source as far away as 10^7 pc. Recalling that there are of the order of 3000 galaxies out to this distance, and taking the rate of supernova events in one galaxy to be of the order of one in thirty years[49], one estimates of the order of tens of useful events a year, or perhaps even a hundred. This is an example of the kind of observation from a space platform that would contribute to knowledge of nucleosynthesis and hydrodynamics in supernova events.

CHAPTER 5

Black Holes

WHEN THE core of the collapsing star is too massive, or imploding with too much kinetic energy, or both, the implosion may still slow down as nuclear densities are encountered; but nuclear forces will not stop the implosion. Gravitational forces become more and more overwhelming, the system zooms through the neutron star stage, and complete collapse ensues. The resulting system has been variously termed "continuing collapse", a "frozen star" and a "black hole". Each name emphasises a different aspect of the collapsed system. The collapse is "continuing" because even in an infinite time, as seen by a far-away observer, the collapse is still not complete. However, the departure from a static configuration of Schwarzschild radius $r = 2$ m, still as seen by a far-away observer, diminishes exponentially in time with a characteristic time of the order of 2 m (3 km light travel time, or 10 μsec, for an object of one solar mass). In this sense the system is a "frozen star". In another sense—as followed by someone moving in with the collapsing matter—the system is not frozen at all. On the contrary, the dimensions shrink in a finite and very short proper time to indefinitely small values. Moreover, a spherical system appears black from outside—no light can escape. Light shot at it falls in. A particle shot at it falls in. A "metre stick" let down upon it attempts in vain to measure the dimensions of the object. The stick is pulled to pieces by tidal forces and the broken-off pieces fall in without a trace. In all these senses the system is a black hole.

At least three processes offer themselves for the production of a black hole: (1) direct catastrophical collapse of a star with a dense core, a collapse that goes through neutron star densities without a stop; (2) a two-step process, with first the collapse of a star with a dense core to a hot neutron star, then cooling, and finally collapse to a black hole; and (3) a multi-step process, with first the formation of a stable neutron star and then the accretion bit by bit of enough matter to raise the mass above the critical value for collapse.

What happens in the collapse has been well analysed in the case of a system of spherical symmetry (see following three Sections: Collapse of a Cloud of Dust; the Kruskal Diagram; "No Bounce"). Also, small departures from spherical symmetry lend themselves to analysis by perturbation methods (subsequent Section). However, in the general and very important case of large departures from spherical symmetry only a few highly idealised and simplified situations have so far been treated. This fascinating field is largely unexplored. The central question is easily stated: does every system after complete gravitational collapse go to a "standard final state", uniquely fixed by its mass, angular momentum and charge, and by no other, adjustable parameter whatsoever (see Section 5.7)? The detection from a space platform of collapsed objects by X-ray, γ-ray and infra-red emission from accreting matter seems attractive and feasible.

5.1 COLLAPSE OF A SPHERICALLY SYMMETRICAL CLOUD OF DUST

Start with a cloud of dust of density 10^{-16} g/cm^3 and radius 1.7×10^{19} cm (mass $= 2 \times 10^{42}$ g $= 10^9 M_\odot = 1.5 \times 10^{14}$ cm). Let the cloud be imagined as drawing itself together by its own gravitational attraction until its radius falls to 10^{-5} of its original value, or 1.7×10^{14} cm. The density rises by a factor of 10^{15} to 10^{-1} g/cm^3. The dust is still dust. No pressure will arise to prevent the continuing collapse. However, despite the everyday nature of the local dynamics, the global dynamics has clearly reached extreme relativistic conditions. How then does one properly describe what is going on?

Various treatments of this problem have been given, from the original analysis of Oppenheimer and Snyder[50], to the treatments of O. Klein and others[51,52]. The simplest case for our purpose is that of Beckedorff and Misner[53] in which the geometry interior to the cloud of dust is identical with that of a Friedmann universe: a 3-sphere of uniform curvature. The radius of curvature, a, is connected with proper time τ (in a frame attached to any test particle) by the parametric relation

$$a = (a_0/2)(1 + \cos\eta)$$
$$\tau = (a_0/2)(\eta + \cos\eta)$$

Here the increment $d\eta$ of the "time parameter" η measures the "arc distance" covered on the 3-sphere within the corresponding time interval:

$$d\,(\text{arc distance in radians}) = \frac{d\,(\text{distance travelled by photon})}{\text{radius}} = d\tau/a = d\eta$$

The geometry within the 3-sphere is

$$ds^2 = a^2(\eta)\,[-d\eta^2 + d\chi^2 + \sin^2\chi(d\theta^2 + \sin^2\theta\, d\varphi^2)]$$

Here the hyperspherical angle would go from $\chi = 0$ to $\chi = \pi$ if the sphere were complete. It is not. It extends from $\chi = 0$ (center of cloud) out to

$\chi = \chi_0$ (surface of cloud), where

$$r_0 = a_0 \sin \chi_0$$

is the radius, in Schwarzschild coordinates, of the cloud of dust at the instant when it is "released from rest" and starts to fall ($\tau = 0$).

The density of the cloud at the starting instant is given by the standard formula for the Friedmann universe,

$$\varrho_0 = 3/8\pi a_0^2$$

The mass of the cloud is

$$m = 4\pi \int_0^{r_0} \varrho_0 r^2\, dr = (3/2)\, a_0 \int_0^{\chi_0} \sin^2 \chi\, d\chi$$
$$= (3a_0/8)\,(2\chi_0 - \sin 2\chi_0) \simeq a_0\chi_0^3/2$$

As the collapse proceeds, more and more of the gravitational potential energy of the cloud of dust is converted into kinetic energy. However, the total mass-energy remains constant. The quantity m does not change (measurable, for example, by time of revolution of a planet in a Keplerian orbit). Outside the cloud of dust the geometry remains the static geometry of Schwarzschild (Birkhoff theorem):

$$ds^2 = -(1 - 2m/r)\, dt^2 + (1 - 2m/r)^{-1}\, dr^2 + r^2(d\theta^2 + \sin^2\theta\, d\varphi^2)$$

The Friedmann geometry and the Schwarzschild geometry match at the boundary of the cloud of dust. A particle located at this boundary is predicted to fall according to one law,

$$r = r(\tau)$$

as calculated from the Friedmann solution; and according to another law,

$$r = r(\tau)$$

as calculated from the Schwarzschild geometry. But the two must agree—and they do! This checks the concordance of the two geometries. Specifically, the law of fall of a test particle in the Schwarzschild geometry is given by the same "cycloidal law" which applies to the Friedmann universe and even to the fall of a test particle in elementary Newtonian physics:

$$r = (r_0/2)\,(1 + \cos\eta)$$
$$\tau = (r_0/2m)^{1/2}\,(r_0/2)\,(\eta + \sin\eta) \qquad \text{Schwarzschild}$$

$$r = (a_0/2)\sin\chi_0(1 + \cos\eta)$$
$$\tau = (a_0/2)\,(\eta + \sin\eta) \qquad \text{Friedmann}$$

The identity of the two expressions gives

$$r_0 = a_0 \sin \chi_0$$
$$m = r_0^3/2a_0^2 = 4\pi\varrho_0 r_0^3/3.$$

Although proper time is the same in the two geometries, for a test particle that remains at the interface, coordinate time is very different (Figure 8). Thus in the Friedmann geometry, coordinate time t agrees with proper time τ; but in the Schwarzschild geometry, where the test particle is moving, we have

$$d\tau^2 = (1 - 2m/r)\, dt^2 - dr^2/(1 - 2m/r)$$

and

$$t = 2m\left\{\ln\frac{(r_0/2m - 1)^{1/2} + \operatorname{tg}(\eta/2)}{(r_0/2m - 1)^{1/2} - \operatorname{tg}(\eta/2)} + \left(\frac{r_0}{2m} - 1\right)^{1/2} \times\left[\eta + \frac{r_0}{4m}(\eta + \sin\eta)\right]\right\}$$

For the analysis of the final stages of collapse it is often sufficient to abstract away from the precise value $r = r_0$ at the start of the collapse. Thus we go to the limit $r_0 \to \infty$. In this case it is also convenient to displace the zero of proper time to the instant of final collapse. In this limit we have

$$r = r$$
$$\tau/2m = -(2/3)\,(r/2m)^{3/2}$$
$$t/2m = -(2/3)\,(r/2m)^{3/2} - 2(r/2m)^{1/2} + \ln\,[(r/2m)^{1/2} + 1]/[(r/2m)^{1/2} - 1]$$

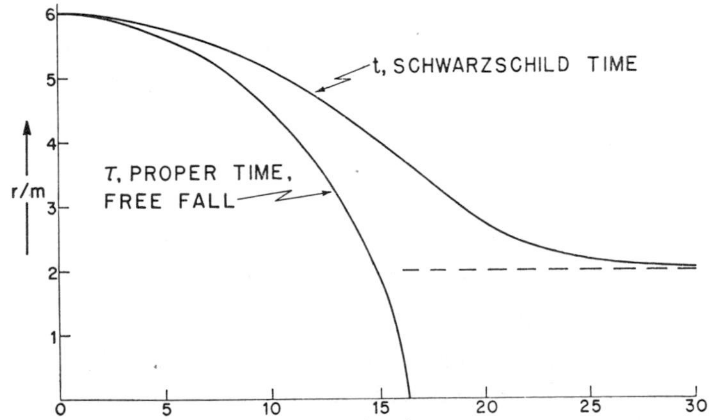

Figure 8 Fall towards a Schwarzschild black hole as seen by co-moving (proper time τ) and far-away observers (Schwarzschild time t).

Thus, for large negative time (particle far away and approaching only very slowly) we have

$$r = (9m\tau^2/2)^{1/3} \simeq (9mt^2/2)^{1/3}$$

whether we refer to coordinate time or to proper time. However, there is a great difference between the complete infall that is seen to occur in test particle proper time

$$r = (9m\tau^2/2)^{1/3} \quad \text{at} \quad \tau \to 0^-$$

and the slower and slower approach to $r = 2m$ that shows up in the Schwarzschild time coordinate t (time as appropriate for a far-away observer),

$$(r/2m) = 1 + 4e^{-8/3}e^{-t/2m}$$

The metric coefficients, following the part of the motion than can be seen from far away, approach

$$e^{\lambda} \simeq (1/4)\, e^{8/3}e^{t/2m} \to \infty$$

and

$$e^{\nu} \simeq 4e^{-8/3}e^{-t/2m} \to 0$$

(mere coordinate singularities). Despite these singularities in the metric no physical singularity occurs at $r = 2m$. Thus the bilinear scalar of curvature approaches the finite value

$$R_{\alpha\beta\gamma\delta}\, R^{\alpha\beta\gamma\delta} = (3/4m^4)$$

and the density of the cloud of matter approaches

$$\varrho = (2 \times 10^{16}\ \text{g/cm}^3)\, (M_{\odot}/M)^2$$

Light given off from a particle at the periphery of the cloud of dust, before arrival at the Schwarzschild radius, will always escape if emitted radially outward. However, if it makes an angle to the radial direction in its own local Lorentz frame, it will make a still larger angle to the radial direction in a local Lorentz frame that happens to have zero radial velocity (and, of course, zero tangential velocity) at the moment in question. The photon will be trapped unless emitted in the allowed cone shown in Figure 9. The allowed cone shrinks to extinction when the cloud of dust contracts within the Schwarzschild radius. Light that emerges radially "outward" from a particle after the cloud has contracted within the Schwarzschild radius never escapes to a far-away observer. It is caught, not in the matter, but in the collapse of the geometry that surrounds the matter.

A photon emitted radially outward from a peripheral atom in the cloud of dust will experience a red shift first, because the component g_{00} of the

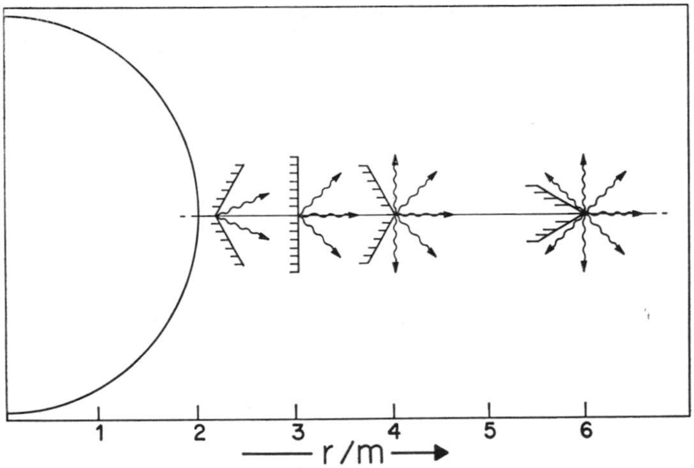

Figure 9 Allowed cone as seen in a local Lorentz frame having zero radial velocity. A photon emitted in a direction not included in the allowed cone falls into the Schwarzschild black hole (adapted from Zel'dovich and Novikov[45]).

metric goes to zero; and second, because the source is falling inward. The emergent frequency is given by the formula

$$\nu/\nu_0 = \omega/\omega_0 = \lambda_0/\lambda = [(1 - \beta)/(1 + \beta)]^{1/2} e^{\nu/2}$$

with

$$\beta = (2m/r)^{1/2}$$

and approaches

$$\nu = \nu_0 2e^{-8/3} e^{-t/2m}$$

The luminosity of the source (erg/sec) goes to zero because of the contraction of the light cone and because of the red shift. Podurets[54] gives a formula for the luminosity as a function of time that takes into proper account (*a*) the contraction of the light cone; (*b*) the red shift; and (*c*) the fact that a photon emitted in a non-radial direction requires longer to reach the far-away observer than one emitted radially. He finds for late times a formula of the type

$$L = L_0 e^{-\frac{4}{3\sqrt{3}}\frac{t}{2m}}$$

The system goes exponentially quickly to darkness. It no longer emits but it still absorbs. In this sense it is really a black hole. Table VII lists characteristic *e*-folding times.

Table VII Times for relevant effects to change by a factor $1/e = 1/2.718$ in final stages of approach to Schwarzschild radius as seen by a far-away observer

Mass of cloud of dust	Distance from $r = 2m$	Red shift from falling object	Red shift from object at fixed r	Luminosity (ergs/sec)
General case	$t_{1/e} = 2m$	$t_{1/e} = 2m$	$t_{1/e} = 4m$	$t_{1/e} = 3\sqrt{3}m/2$
$m = 1M_{\odot}$	9.8 μs	9.8 μs	19.7 μs	12.8 μs
$m = 10^3 M_{\odot}$	9.8 ms	9.8 ms	19.7 ms	12.8 ms

5.2 THE KRUSKAL DIAGRAM

Fall ends at $r = 2m$ as seen by a far-away observer. Fall ends at $r = 0$ according to someone falling with the test mass itself. How can two such different versions of the truth be compatible? For answer it is enough to focus attention on the Schwarzschild geometry itself, and a test particle falling in this geometry. It makes no difference for this question whether the Schwarzschild geometry is "freshly created" at the r-values in question by a cloud of dust falling in just ahead of the test particle, or whether the pure Schwarzschild geometry existed as such for all time. The motion of the test particle is the same in either case. However, the discussion will be simplified if in the beginning we think of the Schwarzschild geometry as having been present for all time.

The central point is simple. The range of coordinates $2m \lesssim r < \infty$, $-\infty < t < +\infty$ fails to cover all the Schwarzschild spacetime. Time "goes on beyond $t = \infty$" just as Achilles "goes on beyond the tortoise" in the famous paradox of Zeno. In no way can one see the incompleteness of the traditional coordinate range more clearly than by reference to the Kruskal coordinates[55]:

$$u = \text{space-like coordinate}$$

$$v = \text{time-like coordinate}$$

In the Kruskal coordinates the Schwarzschild metric takes the form

$$ds^2 = f^2(-dv^2 + du^2) + r^2(d\theta^2 + \sin^2\theta \, d\varphi^2)$$

Here

$$f^2 = (32m^3/r)\, e^{-r/2m}$$

and

$$\left.\begin{aligned} u &= (r/2m - 1)^{1/2} e^{r/4m} \cosh(t/4m) \\ v &= (r/2m - 1)^{1/2} e^{r/4m} \sinh(t/4m) \end{aligned}\right\} \quad \text{for } r > 2m$$

and

$$u = (1 - r/2m)^{1/2} e^{r/4m} \sinh(t/4m)$$
$$v = (1 - r/2m)^{1/2} e^{r/4m} \cosh(t/4m) \qquad \text{for } r < 2m$$

The inverse transformation is given by the formulas

$$(r/2m - 1)e^{r/2m} = u^2 - v^2 \quad \text{for all } r$$

$$v/u = \begin{cases} \coth(t/4m) & \text{for } r < 2m \\ \tanh(t/4m) & \text{for } r > 2m \\ 1 & \text{for } r = 2m \end{cases}$$

Thus, in the (u, v) diagram of Kruskal (Figure 10) points of the same t-value fall on the straight line $v/u = \text{const}$. Points of the same r-value fall on the hyperbola $u^2 - v^2 = \text{const}$, with asymptote $u = \pm v$. A light ray travelling radially outward is always represented by a straight line of slope $dv/du = +1$; one travelling radially inward, by a line of slope $dv/du = -1$.

It can be seen that r is a reasonable "position coordinate" for values of r greater than $2m$; but for values of r less than $2m$ this coordinate changes character; it becomes a time coordinate rather than a space coordinate. The reverse happens to t; it changes from a time coordinate to a position coordinate. A fixed value of r, with $r > 2m$, can be maintained by means of a rocket lift or otherwise. However, a fixed value of $r < 2m$ cannot be maintained any more than time can be made to stand still. One is forced by the evolution of time from $r = 1.9m$ to $r = 1.8m$ and so on all the way to $r = 0$. Hemmed in by the light cone no escape is possible.

The freely falling test particle (heavy world line in Figure 10) falls in to $r = 0$, but no information after stages A, B, C of the fall ever gets to a far-away observer; hence the different conclusions of such an observer, and of the particle itself, about what is going on. To go from the Schwarzschild coordinates ($2m < r < \infty$; $-\infty < t < +\infty$) to the Kurskal coordinates is to extend a geodesically incomplete geometry, free of singularity at its boundary $r = 2m$, to a geometry that does have a singularity, at $r = 0$, but still no singularity at $r = 2m$.

Now turn from the pure Schwarzschild geometry to that associated with the falling cloud of dust, Schwarzschild outside, Friedmann inside. We cut out from the Kruskal diagram in Figure 10 everything to the left of the world line $A, B, \ldots F$ of the test particle. What remains on the right is the external solution—still a static Schwarzschild geometry. On the left is the Friedmann geometry, depicted in Figure 11.

Slices through this Friedmann-Schwarzschild geometry are shown in

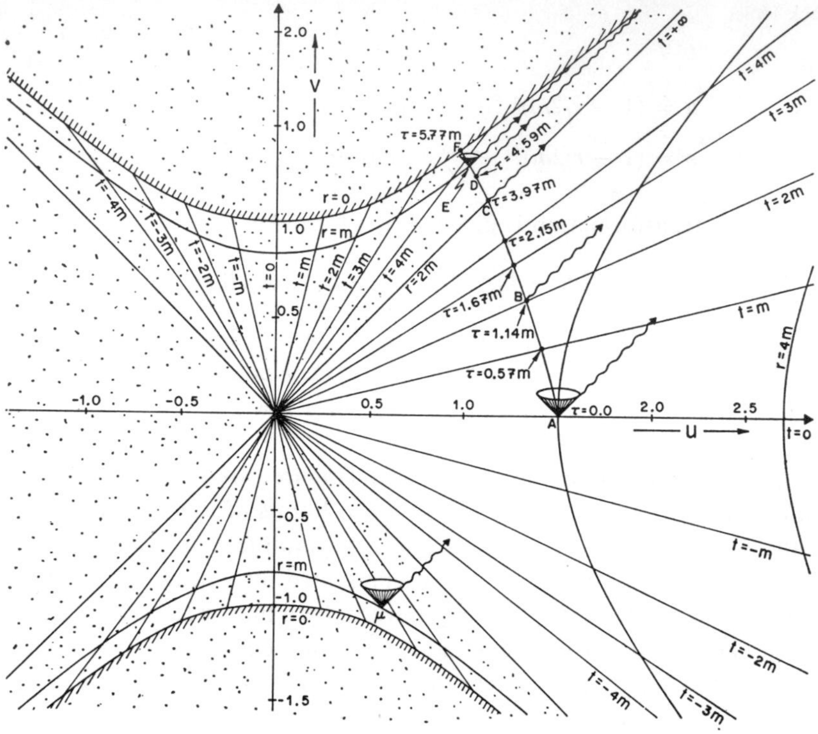

Figure 10 Section of Schwarzschild spacetime $\theta = \text{const}$, $\varphi = \text{const}$, as depicted in terms of the Kruskal coordinates u (space-like) and v (time-like). Radial light rays are straight lines with slope $dv/du = \pm 1$. The relation to the usual Schwarzschild coordinates (r, t) is shown. Only the region free of dots is covered by the usual range of Schwarzschild coordinates, $2m < r < \infty$, $-\infty < t < +\infty$. The heavy line is the world line of a particle that starts at rest at A and falls straight in. A far-away observer receives the signals given out at A and B. The ray C is the "last ray" that ever escapes to infinity—and it only gets to a far-away observer after an infinite Schwarzschild coordinate time t. Rays D and E get caught in the collapse of the geometry; they never reach a far-away observer. The curvature and geometry of space are perfectly normal on the cross-over from B through C to D. However, curvatures and tide-producing forces rise to infinite values on approach to $r = 0$ (point F). This point is reached in a perfectly finite proper time:

	A	B	C	D	E	F
t (in units m)	0	2	∞	—	—	—
τ (in units m)	0	1.14	3.97	4.59	5.24	5.77

A naked Schwarzschild black hole could only be self-luminous if a photon were coming out of it (μ). There is as little reason to expect such radiation to travel outward across the inner boundary of everyday space, $r = 2m$, as there is to expectradiation ("advanced waves") to travel inward across the outer boundary of everyday space, $r = \infty$.

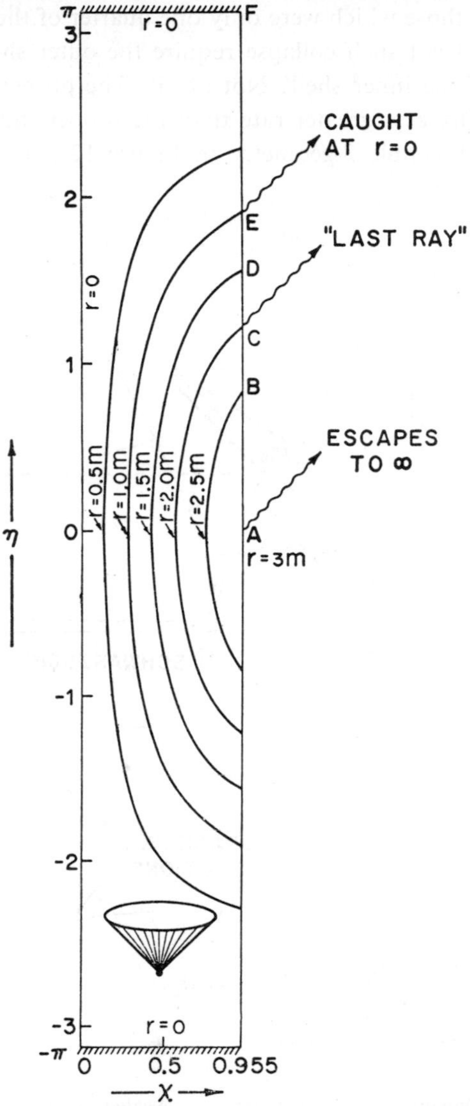

Figure 11 Friedmann geometry representing the interior of a cloud of dust. It joins at right to the Schwarzschild geometry. In this join the points A, B, C, D, E, F are to be identified with the points with the corresponding names in Figure 10. Light rays are still lines inclined at $\pm 45°$.

Figure 12. One sees that collapse takes place. However, *where* collapse "first" occurs is not a well defined notion; it depends upon the choice of the space-like slice. It may at first sight seem preposterous to conceive of particles originally three-quarters of the way from center to surface as collapsing before those which were only one quarter of the way from center to surface. Would not such collapse require the outer shell of particles to "move through" the inner shell? Not at all! The proper *circumference* of the shell can shrink at a greater rate than the proper radius, as illustrated by the uppermost sample 3-geometry in Figure 12 ("tying the neck of a bag").

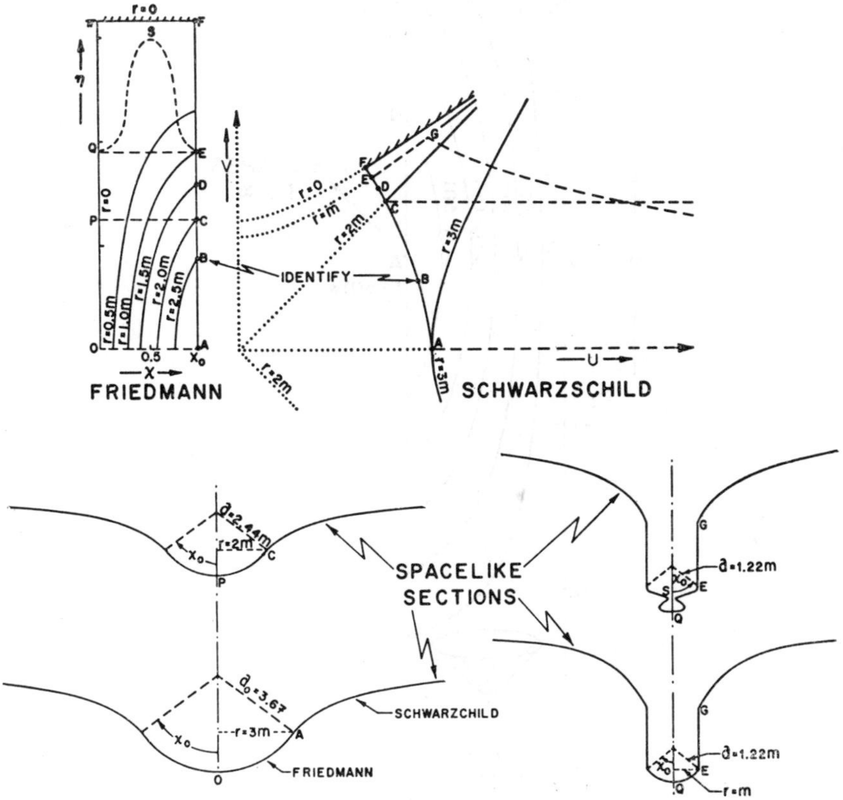

Figure 12 Dynamics of 3-geometry as revealed by making space-like slices through Friedmann-Schwarzschild spacetime (problem of collapsing cloud of dust). "Time" in general relativity has a "many-fingered quality". The slice can be pushed forward in time at one rate at one place, at another rate at another place. There is a multiple infinity of ways of taking space-like slices. Only a few are illustrated here to stress four points (1) join between Friedmann and Schwarzschild geometries (2) collapse of geometry as time advances (3) tube-like quality of Schwarzschild geometry on a space-like surface of constant r when $r < 2m$ (slice EG in diagram) (4) option of he who explores the geometry as to where the space-like hypersurface will first experience collapse (slice QE *vs* slice QSE).

5.3 "NO BOUNCE"

When we turn from a cloud of dust to matter endowed with pressure, but not enough pressure to support it, collapse again ensues. Again the region that "first" experiences total collapse is not a well-defined concept. One has the choice of how fast to arrange to push ahead in time the space-like hypersurface upon which the 3-geometry is to be registered. According as it is pushed ahead faster here or faster there, collapse will seem to occur first here or first there. This circumstance helps one to interpret the May and White diagrams of geometry during collapse of a neutron star[43]. One realises he should not ask the question "why did such and such a region of the star turn out in the calculations to experience total collapse first"!

For matter to be endowed with pressure complicates the calculation of collapse hardly at all when the system under study is a completely closed Friedmann universe. The entire computation is focused on the single equation

$$\begin{pmatrix}\text{extrinsic}\\ \text{curvature}\end{pmatrix} + \begin{pmatrix}\text{intrinsic}\\ \text{curvature}\end{pmatrix} = \begin{pmatrix}\text{energy}\\ \text{density}\end{pmatrix}$$

or

$$(6/a^2)\,(da/dt)^2 + (6/a^2) = 16\pi\varrho(a)$$

Here ϱ is the density of mass-energy as a function of the volume compression factor, a_0^3/a^3. The time from maximum expansion to complete collapse depends in only a minor way on the exact form of the equation of state (Table VIII).

When we turn back from the case of matter enough to curve space up into closure to the more limited amount of matter in a star at the start of collapse, we have to allow for the pressure differential between interior and surface. In the center that differential may not greatly alter the already calculated time of collapse but at the surface it causes a new effect: ejection of a shell of matter.

A closer look (Colgate, May and White[42,43] Zel'dovich and Novikov[56]) reveals some of the same factors at work that control the disassembly of a fission bomb. The center, already denser at the start than the outer parts, by gravitational collapse is still further compacted than its surroundings. In consequence a shock wave runs from the center outward. On arrival at the surface the wave is returned to the interior as a rarefaction. If this rarefaction arrives at the center before the latter has collapsed, the center may never undergo complete collapse. If the rarefaction arrives after the center has undergone complete collapse it can at most prevent some of the outer parts of the core from being caught in the collapse. Thus the critical factor is the time from center to outside and back compared to the time for collapse.

Table VIII Time to go from condition of uniform density, at rest, to complete collapse, as affected by equation of state, when the amount of material is enough to curve up the homogeneous and isotropic space into closure

Description	Equation of state	Radius	Radius at moment of maximum expansion	$8\pi\varrho/3$	Time	Time to collapse from the expansion to the singularity in units of a_0
Dust (Friedmann)	$p = 0$	$a = a_0 \sin^2(\eta/2)$ $0 \leqq \eta \leqq 2\pi$	$a_0(\eta = \pi)$	a_0/a^3	$\tau = \frac{a_0}{2}(\eta - \sin\eta)$	1.5708
Hagedorn	$p = p_0 + \varrho_0 \log \varrho/\varrho_0$	$a = a_0 F(\eta)$[b]	$a_0(\eta = 0)$	[b]	$\tau = a_0 \int_0^\eta F(\chi)\, d\chi$	[b]
Radiation (Tolman)	$p = \varrho/3$	$a = a_0 \sin\eta$ $0 \leqq \eta \leqq \pi$	$a_0(\eta = \pi/2)$	a_0^2/a^4	$\tau = a_0(1 - \cos\eta)$	1.0000
Zel'dovich	$p = \varrho$	$a = a_0\sqrt{\sin 2\eta}$ $0 \leqq \eta \leqq \pi/2$	$a_0(\eta = \pi/4)$	a_0^2/a^6	$\tau = a_0 \int_0^\eta \sqrt{\sin 2\chi}\, d\chi$	0.598[(a)]

[a] $\dfrac{\pi\Gamma(5/2)}{2^{1/2} \times 3 \times \Gamma^2(5/4)}$

[b] cf. material below:

The metric is

$$-ds^2 = c^2\,d\tau^2 - a^2(\tau)\,(d\chi^2 + \sin^2\chi(d\theta^2 + \sin^2\theta\,d\varphi^2)) \tag{1}$$

and assuming

$$c^2\,d\tau^2 = a^2(\eta)\,d\eta^2 \tag{2}$$

we have

$$-ds^2 = a^2(\eta)\,(d\eta^2 - d\chi^2 - \sin^2\chi(d\theta^2 + \sin^2\theta\,d\varphi^2)). \tag{3}$$

The Einstein field equations reduce to

$$\frac{3}{a^4}(\dot{a}^2 + a^2) = \frac{8\pi G}{c^4}\varrho; \qquad 3\frac{da}{a} = -\frac{d\varrho}{\varrho + p} \tag{4}$$

Here the dot indicates derivations with respect to the parameter η. In the case of the Hagedorn equation of state we have introduced two new dimensionless variables $F(\eta) = a(\eta)/a_0$, $Y(\eta) = \varrho(\eta)/\varrho_0$ and we have assumed that $\eta = 0$ corresponds to the moment of maximum expansion ($F(0) = 1$, $Y(0) = 1$). The Eqs. (4) reduce to

$$\frac{dF}{d\eta} = -F\sqrt{F^2 Y - 1}; \qquad \frac{dY}{d\eta} = -\frac{3(Y + A + \log Y)}{F}\frac{dF}{d\eta} \tag{4'}$$

and the time τ from the moment of maximum expansion to the point of complete gravitational collapse is given (in units of a_0/c) by

$$\tau = \int_0^{\eta_c} F(\eta)\,d\eta \quad \text{where} \quad F(\eta_c) = 0.$$

In choosing the value of a_0 or, equivalently the value of ϱ_0, (in fact $\varrho_0 = 3/(8\pi a_0^2)$), we have to take into account that both the Zel'dovich and the Hagedorn equations of state are supposed to be valid asymptotically for large densities; therefore an appropriately small value for a_0 has to be considered. In the following we tabulate the values of the function $F(\eta)$ for selected values of η. (We have chosen for $A = p_0/\varrho_0$ the value 1/4).

η	$F(\eta)$	η	$F(\eta)$	η	$F(\eta)$	η	$F(\eta)$	η	$F(\eta)$
0.00	1.000	0.10	0.984	0.20	0.932	0.30	0.826	0.40	0.585
0.02	0.999	0.12	0.977	0.22	0.916	0.32	0.794	0.42	0.485
0.04	0.997	0.14	0.968	0.24	0.897	0.34	0.757	0.44	0.306
0.06	0.994	0.16	0.958	0.25	0.877	0.36	0.712	0.46	0.123
0.08	0.990	0.18	0.946	0.28	0.853	0.38	0.656		

The collapsing star differs in one essential way from the bomb. Once the core has collapsed inside its Schwarzschild radius, no pressure whatsoever that may subsequently develop can possibly reverse the collapse. This conclusion can be seen in at least three ways:

i) No equation of state compatible with the principle of causality (speed of sound less than speed of light) can give more pressure than $p = \varrho$; and even this critical pressure (cf. last entry in Table VIII) does not help in avoiding the collapse to a singularity.

ii) Every particle (cf. Kruskal diagram) has to move along a time-like world line and *every* time-like world line is forced in the region $r < 2m$ to converge to $r = 0$.

iii) Under well-specified and reasonable conditions, matter located inside a "trapping surface" necessarily undergoes complete collapse to a singularity.

A trapping surface is a closed, space-like two surface with the property that the two systems of null geodesics which meet the surface orthogonally converge locally in future directions. Given Cauchy data for matter and geometry on an initial space-like hypersurface C, and given the 4-manifold M which is the causally determined future time development of this initial value data, Penrose[57] has proved that it is impossible for M to fulfil simultaneously all five of the following conditions: (1) a trapping surface can be found in the empty region surrounding the matter; (2) the manifold M is a non-singular Riemann manifold with the null half cones forming two separate eystems: past and future; (3) for every point of the manifold M and for svery time-like vector t^{μ} we have

$$(R_{\mu\nu} - \tfrac{1}{2}g_{\mu\nu}R)\, t^{\mu}t^{\nu} \geqq 0$$

(non-negative density of mass-energy); (4) every null geodesic in M can be extended into the future to an infinite value of the affine parameter; and (5) every time-like or null geodesic can be extended into the past until it meets C. This theorem has been proved under general conditions and does not require the distribution of matter to be spherically symmetric. However, it requires the existence of a trapping surface. So, and only so, does a singularity seem to be unavoidable, provided that general relativity is correct, that the manifold is maximally extendable, and that the density of mass-energy is positive definite.

Matter on its way in from a finite spread to complete collapse at $r = 0$ appears to a far-away observer as "arrested in flight" ("frozen star"; Zel'dovich and Novikov) as it approaches the Schwarzschild radius $r = 2m$.

What density the matter happens to have at this phase of its motion is very different according as the original system was a dilute cloud (10^{-16} g/cm^3) of mass $10^9 M_\odot$ (density of rough order of 0.02 g/cm^3 on reaching Schwarzschild radius) or a white dwarf or neutron star of mass $\sim M_\odot$ (average density of rough order of 10^{16} g/cm^3 at phase of passage within the Schwarzschild radius; central density higher by one or two factors of 10). In the one case one can expect to learn nothing new about the properties of matter at high density by looking in from outside. In the other case one can, but only by looking at radiation with sufficient penetrating power to go through the supervening cloud of ejected matter: gravitational radiation (Chapter 7) and neutrinos (output in case of formation of a black hole of solar mass comparable to output in case of formation of a neutron star; cf. Colgate, May and White for details[42,43]).

To speak about what can be seen from outside is not to exhaust the physics of the situation. Also of interest is what is to be seen by an observer moving with the matter. He will follow on a "laboratory scale" a gravitational collapse little different from what the Universe itself is expected to undergo in the final stages of its calculated recontraction. Densities will go to unlimited values. For densities of the order of 10^{49} g/cm^3 the calculated scale of the curvature of spacetime becomes comparable to the Compton wavelength of an elementary particle. Under such conditions a direct influence of general relativity on the structure and internal constitution of elementary particles is to be anticipated.

Another critical point is reached on the way to complete collapse when the Planck density is attained,

$$(\text{Planck density}) \sim \frac{(\text{Planck mass})}{(\text{Planck length})^3}$$

$$\sim \frac{(\hbar c/G)^{1/2}}{(\hbar G/c^3)^{3/2}} \sim \frac{2.2 \times 10^{-5}\ \text{g}}{(1.6 \times 10^{-33}\ \text{cm})^3} \sim 10^{94}\ \text{g/cm}^3$$

Here the dimensions of the system as calculated classically become of the same order as the characteristic scale of quantum fluctuations in the geometry. Quantum effects dominate. Any classical analysis of this and subsequent features of the collapse would seem to be completely out of place. One is forced at this stage, if not earlier, to go to the language of quantum geometrodynamics. There one no longer speaks of geometry as undergoing a deterministic evolution with "time". Instead, one has to do with a probability amplitude for this, that and the other 3-geometry[58]. Thus gravitational collapse takes us to one of the great unexplored frontiers of physics.

5.4 SMALL DEPARTURES FROM SPHERICAL SYMMETRY

A spherical cloud of dust falls into a Schwarzschild black hole. What happens if the cloud departs in a minor way from sphericity? It still collapses to a Schwarzschild black hole provided only that it is not endowed with angular momentum; if it has less than a critical angular momentum it still ends up in a uniquely defined but deformed standard black hole configuration (Kerr geometry; see next Section): this is the conclusion to which one has been led by the analysis of Israel[59] and of Doroshkevich, Zel'dovich and Novikov[60]. The latter ask what would happen if the system ended up in a final state which its not the standard configuration; a configuration which though static and axially symmetric is endowed with a quadrupole moment, Q (positive Q, cigar-shaped; negative Q, pancake-shaped). They solve the equations for a sequence of quasi-static small perturbations in the Schwarzschild geometry[61], and find

$$\begin{aligned} g_{00} - (g_{00})_{\text{Sch}} = h_{00} &\sim Q(1 - 2m/r)\ln(1 - 2m/r) \\ h_{11} &\sim Q(1 - 2m/r)\ln(1 - 2m/r) \\ h_{22} \sim h_{33} &\sim Q\ln(1 - 2m/r) \end{aligned}$$

They assume that near the surface of the collapsing star, that is, in the near zone, the metric has approximately the same form as for this sequence of quasi-static configurations. The appearance of this perturbation to the infalling matter is to be found by a transformation to the co-moving coordinate system (fall in the radial or 1-direction). This transformation leaves unchanged the "transverse" components h_{22} and h_{33} of the perturbation. If Q remains constant, the calculated perturbation goes to infinity as the bit of matter under study approaches the Schwarzschild radius. But from the physical point of view the geometry during this phase of the collapse is always perfectly regular. The co-moving observer sees the matter in his neighbourhood as having a perfectly finite density. Perturbations in uniformity that were small a little before this instant remain small; they have no time to grow. Therefore h_{22} and h_{33} have to be free of singularity. This is only possible if the quadrupole moment Q goes to zero at least as fast as $\ln(1 - 2m/r)$ goes to infinity. In terms of the Schwarzschild coordinate time t, the measure of time appropriate for a far-away observer, one has that Q falls off as fast as, or faster than,

$$1/\ln(1 - 2m/r) \propto 1/t.$$

A full dynamical analysis by Price[62] reveals that Q actually falls off as t^{-6}. In contrast to the quadrupole moment, which fades away as $1/t$, and to higher moments, which also go to zero as time runs on, the angular momentum is conserved, and its effect on the metric does not decay with time. Neither does it lead to any singularity near the Schwarzschild surface:

$$h_{03} = -2\sin^2\theta\,(\text{angular momentum})/r$$

(compare with $g_{33} = r^2 \sin^2 \theta$ and $g_{03} = 0$).

All details of the gravitational field get washed out except mass and angular momentum, provided that the original perturbation is not too large. How this comes about can be seen in a little more detail by considering an already existent Schwarzschild black hole, towards the North Pole of which we then drop a small mass δm. The departure δr from the Schwarzschild radius decreases with time as measured by a far-away observer according to the formula

$$\delta r \sim \text{const}\; e^{-t/2m}$$

This perturbation in the distribution of mass evidently dies away very rapidly; and with it die away the quadrupole moment and all the higher mass moments that it generates.

If no trace remains on the outside of where the new mass δm went in, one might try to keep track of its location by attaching to it a small charge q. The electric lines of force lend themselves to simple observation from a distance. The analysis of these lines of force is simplified by considering the charge to be lowered so slowly that each successive configuration can be treated as static. At each level the determination of the lines of force is a simple problem of electrostatics in a static curved space, with the potential satisfying the equation[63]

$$\frac{1}{\sin\theta}\frac{\partial}{\partial\theta}\sin\theta\frac{\partial\psi}{\partial\theta} + \left(1 - \frac{2m}{r}\right)\frac{\partial}{\partial r}\left(r^2\frac{\partial\psi}{\partial r}\right) = 0 \text{ (except at source)}$$

and the field components given by the appropriately expressed gradient of ψ. The lines of force run qualitatively as shown in Figure 13. When the charge is close to the Schwarzschild radius the extreme gravitational field deforms the lines of force very far from the normal pattern. As examined far away they appear to diverge from a point far closer to the center of the sphere than is the charge itself. The dipole moment goes to zero as the charge approaches $2m$. Nothing in the final pattern of the emergent lines of force reveals the location of the charge. We see simply black hole—mass plus charge—and no other details.

It will be more natural to drop the charged object than to lower it slowly. Then electromagnetic radiation will come out[64]. However, the charge as viewed from far away eventually slows down, the emission of radiation

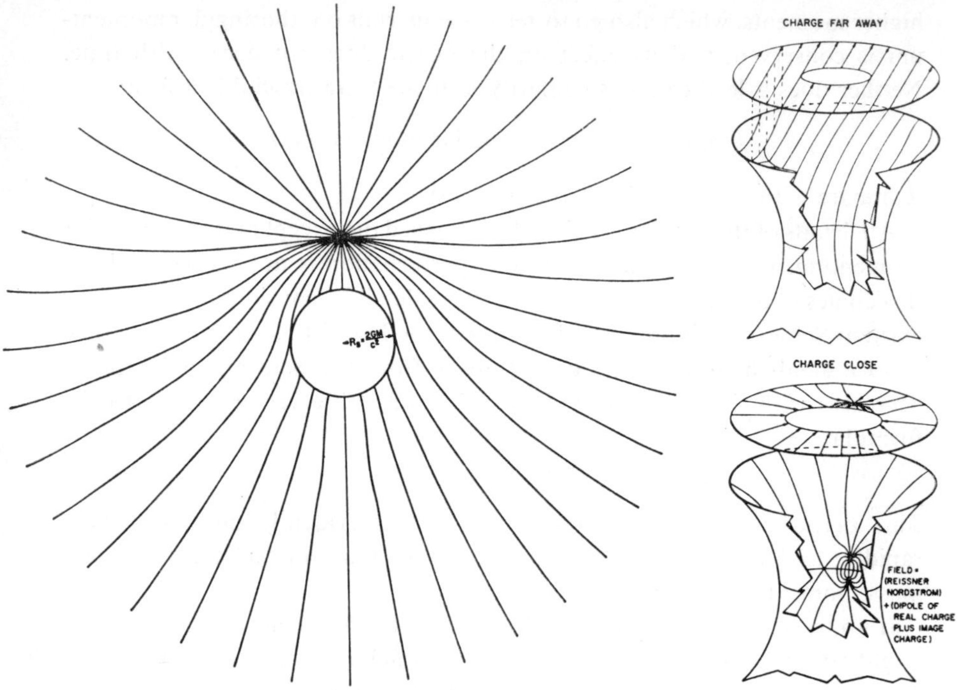

Figure 13 Point charge supported at rest at fixed distance from Schwarzschild black hole, and resulting pattern of lines of force, in the three cases of charge very far away (upper right) and very close (lower right) and at the intermediate point $r = 4m$ (left). At the upper right the top diagram (flat) treats the Schwarzschild coordinates r, θ as if they were standard polar coordinates in a Euclidean plane, whereas the diagram beneath deals with proper distances on the Schwarzschild throat (viewed as a curved geometry imbedded in an enveloping Euclidean space).

ceases, and the fade-away of the dipole moment goes as if the charge were being slowly lowered to the Schwarzschild radius. The dipole moment as one would evaluate it from flat space arguments is

$$p_{\text{class}} \sim q(2m + 8me^{-8/3}e^{-t/2m})$$

—never disappearing. It actually disappears according to a law of the form

$$p_{\text{gen rel}} \sim q \text{ const}/t^4 \quad (\text{Price}^{62})$$

This disappearance of the electric dipole moment takes place for the same reasons as the fade-out of the mass quadrupole moment and the higher moments of the mass asymmetry.

Electromagnetism provides a simple way to calculate the far-away field produced by a charge, either at rest near, or falling into, a black hole. No such option is available when one wants to calculate the gravitational perturbation produced by the small mass. The mass cannot be conceived to be held stationary without doing violence to the field equations[56,65]. The radiation of gravitational wave energy in the act of infall has been treated by Zel'dovich and Novikov[56] and, in more detail, by Zerilli[66] (see A.8). However, no complete calculation is yet available for the coefficients of $1/t^n$ in the expression for all the mass moments in the final stages of the fall of the supplementary mass δm.

The collapse leads to a black hole endowed with mass and charge and angular momentum but, so far as we can now judge, no other adjustable parameters ("a black hole has no hair"; see A.13). Make one black hole out of matter; another, of the same mass, angular momentum and charge, out of anti-matter. No one has ever been able to propose a workable way of telling which is which. Nor is any way known of distinguishing either from a third black hole, formed by collapse of a much smaller amount of matter, and then built up to the specified mass and angular momentum by firing in enough photons, or neutrinos, or gravitons. And on an equal foot-

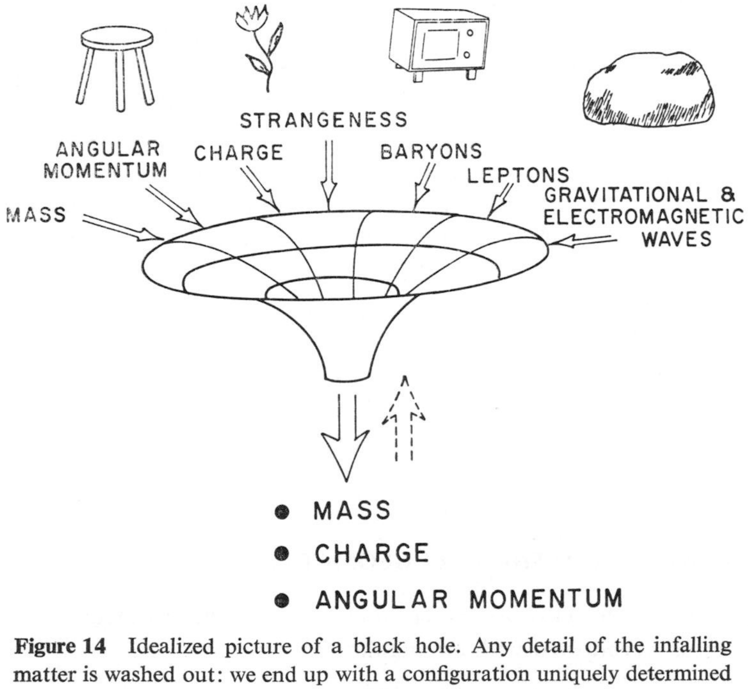

Figure 14 Idealized picture of a black hole. Any detail of the infalling matter is washed out: we end up with a configuration uniquely determined by mass, charge and angular momentum (see text).

ing is a fourth black hole, developed by collapse of a cloud of radiation altogether free from any "matter"[67,68]. (See Figure 14.)

Electric charge is a distinguishable quantity because it carries a long-range force (conservation of flux; the law of Gauss). Baryon number and strangeness carry no such long-range force. They have no Gauss law. It is true that no attempt to observe a change in baryon number has ever succeeded[96,70]. Nor has anyone ever been able to give a convincing reason for expecting a direct and spontaneous violation of the principle of conservation of baryon number. In gravitational collapse, however, that principle is not directly violated; it is transcended. It is transcended because in collapse one loses the possibility of measuring baryon number and loses therefore the possibility of giving this quantity a well-defined meaning for a collapsed object. Similarly for strangeness.

For the determination of lepton number there is a force that might at first sight be thought to provide a measurement tool. A collection of L_1 leptons at rest interact with a collection of L_2 leptons at rest, at a distance r, by exchange of neutrino/anti-neutrino pairs, with an energy of interaction of[71]

$$V(r) = (G^2/4\pi^3)\, L_1 L_2 / r^5$$

where $G = 10^{-5} M_p^{-2}$ is the Fermi constant, and M_p is the proton mass. Could one not measure the force exerted by a black hole on a test body containing L_2 leptons, then the force on a test body of the same mass containing—L_2 leptons and take the difference, thereby finding the "unknown lepton number" L_1 of the black hole? James Hartle, who suggested this procedure to define and determine the leptonic number of a black hole, also pointed out that it cannot work as stated. The collapse brings about a greater and greater red shift of all effects coming from the lepton source L_1. Moreover, there is no Gauss law for the $1/r^5$ interaction. Therefore, in advance of a detailed calculation, one has every reason to expect the lepton-lepton interaction to fade away as gravitational collapse proceeds. Thus one thinks of the law of conservation of lepton number as being transcended as completely as the law of baryon conservation. The black hole ends up by being characterised, one expects, by mass, charge and angular momentum, and by nothing more.

5.5 ROTATION AND THE KERR GEOMETRY

R. P. Kerr in 1963 gave an exact solution[72] for the geometry associated with a mass m endowed with an angular momentum

$$(\text{angular momentum}) = J = -ma$$

and Newman and collaborators in 1965 generalised this solution[73] to encompass charge. A general account of the properties of this "generalised Kerr geometry" has been given by Carter[74]. It is enough for our purpose to write the metric for a black hole in the case when there is no charge,

$$ds^2 = -dt^2 + (r^2 + a^2)\sin^2\theta\, d\varphi^2 + \frac{2mr\,(dt + a\sin^2\theta\, d\varphi)^2}{r^2 + a^2\cos^2\theta}$$

$$+ (r^2 + a^2\cos^2\theta)\left(d\theta^2 + \frac{dr^2}{r^2 - 2mr + a^2}\right)$$

Figure 15 shows the two most interesting surfaces associated with this geometry[75,76]:

1) the "surface of infinite redshift or stationary", where $g_{00} = 0$,

$$r_+ = m + (m^2 - a^2\cos^2\theta)^{1/2}$$

2) the "event horizon" or "one-way membrane",

$$r_0 = m + (m^2 - a^2)^{1/2}$$

(angular momentum parameter a less than, or equal to, the "maximum allowable value", $a = m$). An object that has arrived at, or within, this horizon can send no photons whatsoever to a far-away observer, no matter what the direction of motion of this object, and no matter in what direction it emits the photons. The light cone points inward. Collapse is inescapable for a particle that has fallen into this surface.

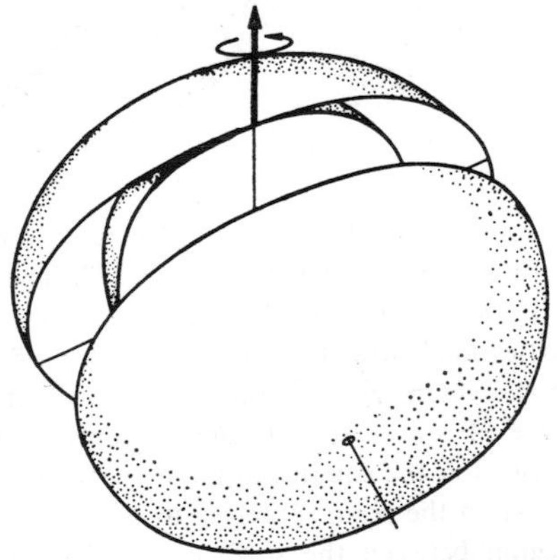

Figure 15 "Ergosphere" of Kerr geometry, the region between the surface of stationarity (flattened figure of revolution, $r = m + (m^2 - a^2\cos^2\theta)^{1/2}$) and the "one-way membrane" or horizon (inner sphere $r = m + (m^2 - a^2)^{1/2}$). A particle entering the ergosphere and emitting a disintegration product can emerge with more energy (rest-plus-kinetic) than it had when it entered.

In contrast, a particle that has come into the region between surfaces (1) and (2) ("ergosphere") can still, if it is properly powered, escape again to infinity. However, its life in the region between the two surfaces has this unusual feature: there is no way for it to sit at rest, rocket-powered or not! The line r = const, θ = const, φ = const, only t increasing (vertical arrow in Figure 16) lies outside the light cone. It is space-like. It is not a possible world line. A real particle in the ergosphere must always change position, regardless of whether it eventually escapes to infinity or crosses the one-way membrane to collapse. From the ergosphere the particle can always send a signal out to infinity. How is this possible? Is it not unreasonable to expect a photon to cross a surface of stationary? The answer is simple. The

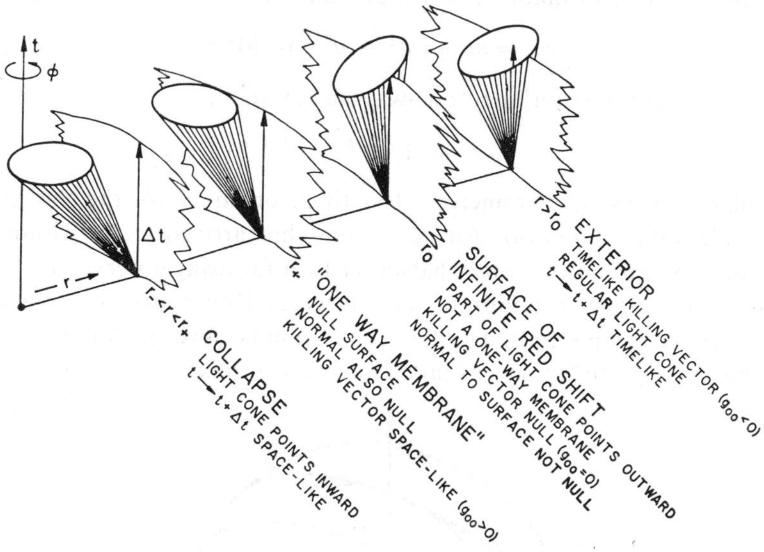

Figure 16 Light cone and time-like Killing vector for Kerr geometry at selected locations in the equatorial plane, $\theta = \pi/2$. The Killing vector is most readily visualised as the transformation $\Delta r = 0$, $\Delta\theta = 0$, $\Delta\varphi = 0$, $t \to t + \Delta t$. It carries one's attention from a given point in spacetime to a point with identical geometry. It is time-like in the exterior. It becomes null on the surface of stationarity (outer flattened figure of revolution in Figure 15; described in the equatorial plane by the circle $r = r_0 = 2m$). It becomes space-like in the region interior to this surface. A particle can come into the region between the surface of infinite red shift and the horizon and get away again to the exterior, and this in a time which is not only finite on its own proper timescale but also finite as seen by a far-away observer. However, to reach the horizon itself takes an infinite time as seen by the far-away observer. For the particle itself it takes only a finite time to cross the horizon. Once across, it can never escape. The light cone points inward ("trapping surface").

light cone reaches outside this surface (Figure 16). Properly directed photons can escape. Is there not a difficulty in this answer? If one source located on this surface can emit a photon to infinity by properly directing this photon, cannot *any* source located on this surface emit a photon to infinity by properly directing it (principle of Lorentz transformation from one local frame to another)? If so, how can we speak of an infinite red shift of the photons emerging from a "source at rest"? This is because on this surface (of stationarity) asource "at rest" in the technical sense of the word

$$(r = \text{const}, \theta = \text{const}, \varphi = \text{const}, t \text{ increasing})$$

is actually not at rest at all—it is moving with the speed of light (arrow located on the light cone in Figure 16).

Only in the case of the Schwarzschild geometry ($a = 0$, charge $= e = 0$) does the surface of stationarity coincide with the "horizon" or "one-way membrane". In the general Kerr geometry the two surfaces are separated everywhere except at the poles. Incidentally, two new surfaces interior to the horizon show up in the Kerr geometry, which coalesce to the singularity, $r = 0$, of the Schwarzschild geometry as $a \to 0$ and $e \to 0$. One is the "interior null surface", $r = m - (m^2 - a^2)^{1/2}$; the other is the "interior $g_{00} = 0$ surface", $r = m - (m^2 - a^2 \cos^2 \theta)^{1/2}$. Neither of these surfaces can make itself felt to a far-away observer.

The dynamics of the geometry interior to the surface of stationarity cannot be probed from outside in the Schwarzschild case but can in the Kerr case. Moreover, one can profit from this accessible zone of dynamic geometry and extract energy out of it, according to Penrose[77]. For this purpose he has suggested (1) shooting a small object in from outside with rest plus kinetic energy E_1; (2) letting it explode (or turn on its rocket engine) in the dynamic zone in such a way that the disintegration product (or, equivalently, the rocket ejecta) crosses the one-way membrane and gets taken up in the black hole; (3) letting the residual mass re-emerge from the surface of infinite red shift with total energy E_2; (4) then so arranging the process that E_2 exceeds E_1 (for details of this process, see Figures 17, 18, 19 and 20). The energy $E_2 - E_1$ can be said to have been extracted from the rotational energy of the black hole in this sense, that the angular momentum of the black hole always decreases in such a process.

If an object by going in and out can pick up energy and angular momentum from a black hole, it is also true that a particle that goes in and gets caught gives energy and angular momentum to the black hole. The process of capture of a particle or photon coming in from infinity has been studied by Doroshkevich[78] and Godfrey[79]. A Schwarzschild black hole captures a photon which approaches with impact parameter b less than $3^{3/2}m$, and

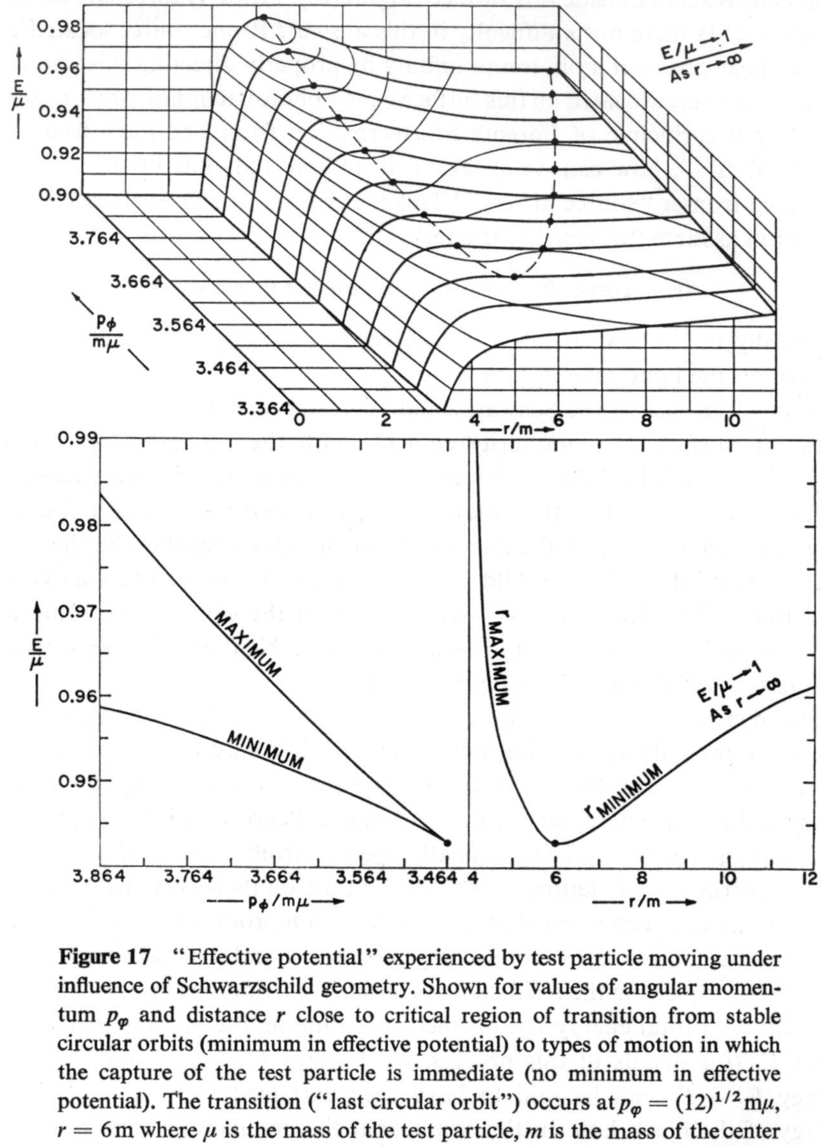

Figure 17 "Effective potential" experienced by test particle moving under influence of Schwarzschild geometry. Shown for values of angular momentum p_φ and distance r close to critical region of transition from stable circular orbits (minimum in effective potential) to types of motion in which the capture of the test particle is immediate (no minimum in effective potential). The transition ("last circular orbit") occurs at $p_\varphi = (12)^{1/2}m\mu$, $r = 6\,\mathrm{m}$ where μ is the mass of the test particle, m is the mass of the center of attraction in geometrical units ($m = 1.47$ km for one solar mass; more generally, $m = (0.742\times 10^{-28}\ \mathrm{cm/g})\, m_{\mathrm{conv}}$) and r is the "Schwarzschild distance" (proper circumference/2π). The energy at this point is $E = (8/9)^{1/2}\mu = 0.943\mu$, corresponding to a binding of 5.7 per cent of the rest mass. The effective potential is defined here as that value of E which annuls the expression

$$E^2 - (\mu^2 + p_\varphi^2/r^2)(1 - 2m/r).$$

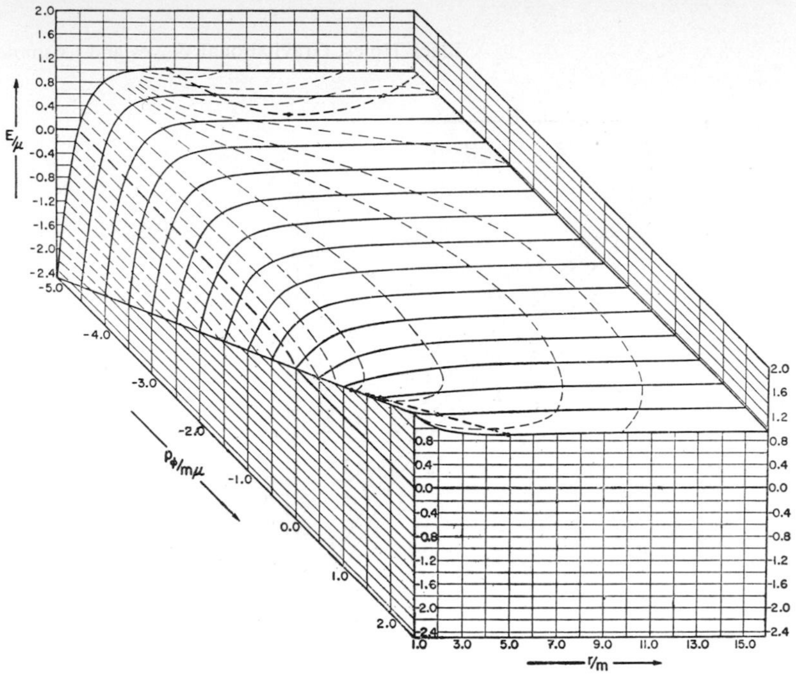

Figure 18 "Effective potential" experienced by a test particle moving in the equatorial plane of an extreme Kerr black hole $a/m = 1$. Here m is the mass of the black hole expressed in geometrical units $m = (G/c^2)\, m_{\text{conv}}$ and a is the angular momentum per unit of mass. We have considered values of angular momentum of the test particles ($p_\varphi/m\mu$) and distance r close to the critical region of transition from stable circular orbit (minimum in effective potential) to types of motion in which the capture of the test particle is immediate (no minimum in effective potential). In contrast to the case of the Schwarzschild black hole (cf. Figure 17) the behaviour of the effective potential is now considerably different depending on whether the angular momentum p_φ has positive or negative values (cf. Figure 19 where the Newtonian, Schwarzschild and Kerr cases are compared and contrasted). For negative angular momentum the last circular orbit occurs at $p_\varphi/m\mu = -22/(3\sqrt{3})$. Here μ is the mass of the test particle and r is the "Schwarzschild distance" (proper circumference/2π). The binding at this point is 3.78 per cent. For positive angular momentum we have circular orbits all the way down to $\hat{r} = r/m = 1$. The last circular orbit occurs at $\hat{p}_\varphi = p_\varphi/m\mu = 2\sqrt{3}$, $\hat{r} = 1$ with a value of the total energy $E = \mu/\sqrt{3}$. The binding energy is at this point $\sim 42\%$! (cf. Figure 19). The effective potential is defined as the value of the energy which annuls the expression

$$\hat{E}^2(\hat{r}^3 + \hat{r} + 2) - 4\hat{E}\hat{p}_\varphi - \hat{r}(\hat{r} - 1)^2 + (2 - \hat{r})\hat{p}_\varphi^2 = 0$$

Of the two solutions of this equation we have here considered the one with the positive sign in front of the radical. The other solution (not shown) is easily obtained by noticing that the expression of the effective potential is unchanged with respect to a simultaneous change of E into $-E$ and p_φ into $-p_\varphi$. The potential energy surfaces for the two solutions ("positive" and "negative" energies) meet at the "knife edge" $\hat{r} = 1$, $\hat{E} = E/\mu = \hat{p}_\varphi/2$.

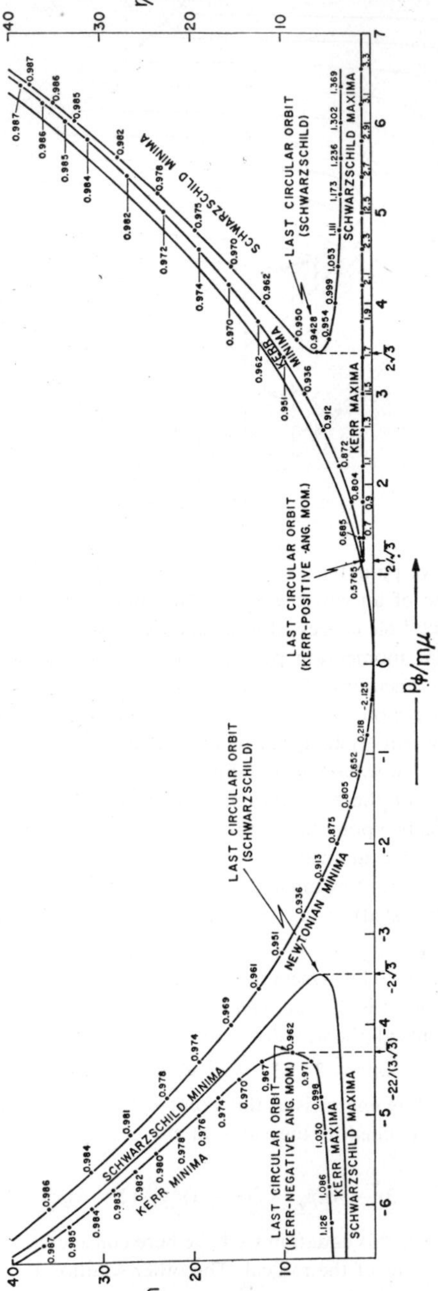

r/m
$p_\phi/m\mu$
SCHWARZSCHILD MINIMA
KERR MINIMA
KERR MAXIMA
SCHWARZSCHILD MAXIMA
NEWTONIAN MINIMA
LAST CIRCULAR ORBIT (SCHWARZSCHILD)
LAST CIRCULAR ORBIT (KERR-POSITIVE ANG. MOM.)
LAST CIRCULAR ORBIT (KERR-NEGATIVE ANG. MOM.)

Figure 19

Figure 19 The minima (stable circular orbit) and maxima (unstable circular orbit) of the "effective potential" experienced by a test particle moving in the field of a Newtonian, a Schwarzschild and an extreme Kerr black hole are here compared and contrasted. The distance r is given in units of the mass of the black hole in geometrical units ($m = (G/c^2)\, m_{\text{conv}}$) and the angular momentum of the incoming particle p_φ in units of the product of the mass m of the black hole and of the mass μ of the incoming particle, both expressed in geometrical units. Numerical values of the effective potential E in units μ are displayed on the representative curves of the maxima and minima for selected values of the distance and of the angular momentum. For any given distance and angular momentum the effective potential is in the general Kerr case defined by the value of the energy which satisfies

$$E^2(r^3 + a^2(r + 2m)) - 4mEap_\varphi + (2m - r)p_\varphi^2 - \mu^2 r^2(r - 2m) - a^2\mu^2 r = 0$$

where a is the angular momentum per unit mass in geometrical units. In the particular case considered in this figure, as well as in Figure 18, we are dealing with an "extreme" Kerr black hole, namely $a/m = 1$. Setting $a = 0$ we obtain the expression for the Schwarzschild black hole

$$E^2 - (1 - 2m/r)((p_\varphi/r)^2 + \mu^2) = 0$$

and finally going to the limit $m/r \ll 1$ and $p_\varphi/r \ll 1$ we obtain the Newtonian case

$$E - \mu + \frac{m\mu}{r} - \frac{p_\varphi^2}{\mu}\frac{1}{2r^2} = 0.$$

The last circular orbit occurs in the three cases at the following values of the parameters (cf. also Figures 17 and 18).

	Newtonian	Schwarzschild	Kerr (Extreme case $a = m$)	
r/m	0.0	6.0	1.0	9.0
E/μ	$-\infty$	$(2\sqrt{2})/3$	$1/\sqrt{3}$	$5/(3\sqrt{3})$
$(\mu - E)/\mu$	$+\infty$	5.72%	42.35%	3.77%
$p_\varphi/m\mu$	0.0	$2\sqrt{3}$	$2/\sqrt{3}$	$-22/(3\sqrt{3})$

In the Newtonian case for any selected value of the angular momentum p_φ there exists a circular stable orbit (minimum in the effective potential) with a value of $r/m = (p_\varphi/m\mu)^2$ and a value of the total energy $E_{\text{cir}} = \mu - m^2\mu^3/2p_\varphi^2$. The maximum of the effective potential, for any value of the angular momentum different from zero, is at $r = 0$, where E diverges to $+\infty$. The value of the effective potential is independent from the sign of p_φ. For the discussion of the case of Schwarzschild and Kerr black holes cf. Figure 17 and Figure 18 and text.

merely deflects a photon that approaches with larger impact parameter. Capture augments the angular momentum by the amount

$$\Delta J = b_{\perp} p$$

where p is the linear momentum of the photon and $b_{\perp}$ is the component of the impact parameter perpendicular to the axis of rotation. A Kerr black hole that captures a photon experiences a change in angular momentum given by the same formula. Now, however, the maximum reach of values, $b_{\perp}$, is greatest (most negative; more negative than $-3^{3/2}m$) for photons that will cut the angular momentum of the system, and least (less than $+3^{3/2}m$; details given by Godfrey[79]) for photons that will augment the angular momentum. Thus capture from a random environment leads to a gradual decrease of the angular momentum of the system.

The opposite of random accretion is bombardment with photons having the maximum positive impact parameter compatible with capture. This process augments the angular momentum along with the mass; but the ratio of angular momentum to the square of the mass approaches only the limiting value

$$(J/m^2)_{\lim} = (a/m)_{\lim} = 3^{3/2}/2^{5/2} = 0.918$$

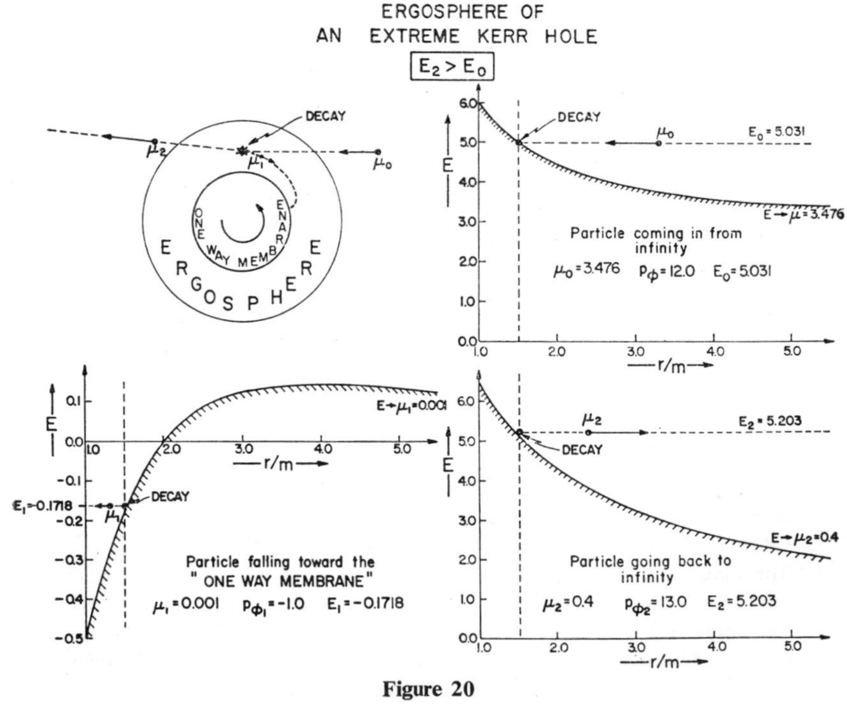

Figure 20

Figure 20 A particle of mass μ_0 coming from infinity with total energy E_0 and a positive value of the angular momentum $p_{\varphi 0}$ can penetrate into the ergosphere of an extreme Kerr hole and here decay into two particles. One particle of mass μ_1, negative value of the angular momentum $p_{\varphi 1}$ and a negative value of the total energy E_1, falls towards and penetrates the one-way membrane. The second particle of mass μ_2, positive value of the angular momentum $p_{\varphi 2}$ and a positive value of the total energy E_2, goes back to infinity. The remarkable new feature in this process is that the energy E_2 of the particle coming back to infinity is larger than the energy E_0 of the particle coming in (R. Penrose). For simplicity we consider a decay process in the equatorial plane of a Kerr hole. The required conservation of the 4 momentum gives

$$\text{(a)}\quad E_0 = E_1 + E_2$$

$$\text{(b)}\quad p_{\varphi 0} = p_{\varphi 1} + p_{\varphi 2}$$

$$\text{(c)}\quad \begin{cases} [(E_0^2 - \mu_0^2)\, r^3 + 2\mu_0^2 m r^2 + (a^2 E_0^2 - p_{\varphi 0}^2 - a^2\mu_0^2)\, r \\ \qquad + 2m(E_0 a - p_{\varphi 0})^2]^{1/2} \\ = -[(E_1^2 - \mu_1^2)\, r^3 + 2\mu_1^2 m r^2 + (a^2 E_1^2 - p_{\varphi 1}^2 - a^2\mu_1^2)\, r \\ \qquad + 2m(E_1 a - p_{\varphi 1})^2]^{1/2} \\ + [(E_2^2 - \mu_2^2)\, r^3 + 2\mu_2^2 m r^2 + (a^2 E_2^2 - p_{\varphi 2}^2 - a^2\mu_2^2)\, r \\ \qquad + 2m(E_2 a - p_{\varphi 2})^2]^{1/2} \end{cases}$$

Here a is the angular momentum J per unit of mass of the Kerr hole $a = J/m$. We have further simplified the problem by considering an extreme Kerr hole $a/m = 1$ and a decay process in which the incoming particle is at a turning point of its trajectory. Equations (a) and (b) are unaltered and equation (c) (conservation of radial momentum) simplifies to

$$\text{(c}'\text{)}\quad [(E_1^2 - \mu_1^2)\, \hat r^3 + 2\mu_1^2 \hat r^2 + (E_1^2 - \hat p_{\varphi 1}^2 - \mu_1^2)\, \hat r + 2(E_1 - \hat p_{\varphi 1})^2]^{1/2} =$$
$$[(E_2^2 - \mu_2^2)\, \hat r^3 + 2\mu_2^2 \hat r^2 + (E_2^2 - \hat p_{\varphi 2}^2 - \mu_2^2)\, \hat r + 2(E_2 - \hat p_{\varphi 2})^2]^{1/2}$$

where $r = r/m$ and $\hat p_\varphi = p_\varphi/m$. The system of equations (a) and (b), (c′) has been solved assuming $r = 1.5$, $\mu_1 = 0.01$, $p_{\varphi 1} = -1.0$, $\mu_2 = 0.40$, $p_{\varphi 2} = 12.0$. An additional condition is the requirement that the total energies of the two particles E_1 and E_2 be larger than or equal to the respective turning point energies. One of the numerical solutions is given in this figure. On the upper left side a qualitative diagram shows the main feature of the decay process in the equatorial plane of the ergosphere of a Kerr hole. In the upper right side is the effective potential (energy required to reach r as a turning point) for the incoming particle. The effective potential is plotted at the lower left and at the lower right side for the particle falling toward the one-way membrane and for the particle going back to infinity.

This "energy gain process" critically depends on the existence and on the size of the ergosphere which in turn depends upon the value a/m of the hole. In the equatorial plane we have for the ergosphere $1 + (1 - a^2/m^2)^{1/2} \leqq \hat r \leqq 2$; when $a/m \to 0$ (Schwarzschild hole) the one-way membrane expands and coalesces with the infinite red shift surface, wiping out the ergosphere. The particle falling towards the one-way membrane will in general alter and reduce the ratio a/m of the black hole.

significantly short of the critical value $a = m$. Very fast particles do no better than photons in raising the angular momentum towards the critical value. Slower particles do better. The cross-section for capture goes up, and the contribution to the angular momentum goes up also. Thus, for photons and relativistic particles the capture cross-section (in the simplest case of small rotation) is

$$\sigma_{\mathrm{capt}} = 27\pi m^2$$

whereas the Newtonian formula for the cross-section for capture of a particle of velocity β (measured at infinity) by a center of radius R is

$$\sigma_{\mathrm{capt}} = \pi R^2(1 + 2m/\beta^2 R)$$

and in general relativity the cross-section of a Schwarzschild black hole for capture of a slow particle ($\beta \ll 1$) is[80]

$$\sigma_{\mathrm{capt}} = 16\pi m^2/\beta^2$$

The maximum impact parameter for capture in this limit is,

$$(b_\perp)_{\mathrm{max}} = 4m/\beta$$

Thus a particle of mass δm, captured with maximum impact parameter, brings in an angular momentum

$$\delta J = (b_\perp)_{\mathrm{max}}\, \delta m\, \beta = 4m\, \delta m$$

At the same time it brings in an increment in mass energy of δm. The ratio of angular momentum to square of the mass rises by the amount

$$\delta(J/m^2) = (2\delta m/m)\,(2 - J/m^2)$$

Taken literally, this formula would imply that selective accretion could eventually raise the J/m^2 ratio up to 2, twice the critical value for the Kerr geometry. However, the increasing angular momentum of the black hole has far-reaching effects on the impact parameter for capture. It is altered[79] from $4m/\beta$ to

$$(b_\perp)_{\mathrm{max}} = (2m/\beta)\,[1 + (1 - J/m^2)^{1/2}]$$

Thus, a particle approaching with 300 km/sec ($\beta = 10^{-3}$) a black hole of solar mass ($m = 1.47$ km) will be caught if it has an impact parameter less than 6000 km; but if the black hole has the critical angular momentum $J = m^2$, the distance for capture decreases to 3000 km. This result applies to capture that augments the angular momentum; for capture that decreases the angular momentum of the black hole, the impact parameter is increased to

$$b_{\perp} = (2m/\beta)\,[1 + (1 + J/m^2)^{1/2}] = 7200 \text{ km}$$

The corrected formula for the increase of the angular momentum in the selective accretion of slow and small particles is

$$\delta(J/m^2) = (2\delta m/m)\,[1 + (1 - J/m^2)^{1/2} - J/m^2]$$

Thus the ratio J/m^2 can be brought by selective accretion in this way asymptotically close to unity, but never above unity[81,82].

5.6 ENERGY SENT OUT WHEN MASS FALLS IN

Neglecting the small amount of energy given off in the dynamic zone, a particle spiralling into a $J = m^2$ black hole radiates a fraction $(1 - 3^{-1/2})$ or 42 per cent of the rest mass of the particle. In contrast, a particle spiralling in towards a Schwarzschild black hole, and arriving at length at the last stable circular orbit, at $r = 6m$, will depart exponentially fast from circular motion and fall in in one leap, again with only a brief snatch of radiation given out in this leap. The important part of this radiation, emitted on passage from $r = \infty$ to $r = 6m$, takes away only the fraction $1 - \sqrt{8}/3$ or 5.7 per cent of the rest mass.

The discovery in quasars of objects with enormous energy release led many workers to investigate gravitational collapse as a mechanism superior to fission or fusion ($\delta m/m \lesssim 0.01$) for converting mass to energy. A closer look caused discouragement. Six per cent of the rest mass, emitted in a form so poorly coupled to the surroundings as gravitational radiation, is no improvement over nuclear energy as a source of power! An alternative was to consider the effect of the tidal forces of a big black hole on a smaller object falling into it. These forces squeeze the object transverse to the direction of fall and stretch it out in the radial direction (effect of squeezing a tube of toothpaste!). However, only something of the order of one per cent of the rest mass of the infalling object re-appears as kinetic energy of the ejected "toothpaste", according to one unpublished estimate[83]. Thus gravitational collapse appeared to be no better than nuclear reaction as a source of energy, so long as one restricted attention to a Schwarzschild black hole. However, a completely different order of magnitude of energy release is evidently possible with a Kerr black hole. This circumstance gives incentive for re-examining the "tube of toothpaste" and other mechanisms of energy release in the context of the Kerr geometry—a task for the future! It also raises this question with special force: is it natural for a black hole to arise which is endowed with critical, or near-critical angular momentum?

5.7 FINAL OUTCOME OF COLLAPSE OF A ROTATING BODY

Selective accretion is not a reasonable process to consider for a black hole immersed in a random cloud of interstellar dust and atoms. Selective accretion is a reasonable process to consider for a black hole immersed in the debris from the collapse of a rotating star. Rotation flattens the original star. In consequence the polar regions are drawn in by stronger gravitational forces than the equatorial regions. Moreover, they have a smaller distance to go before they meet. It is therefore reasonable to expect that the system first contracts to a pancake configuration with a characteristic time of the order of the time of free fall,

$$\tau_1 \sim (\text{effective radius of original system})^{3/2}/(\text{mass})^{1/2}$$

(geometrical units, such as km of light travel time!). This expectation has been confirmed by direct calculation[84] in the idealised case of a cloud of dust which, though not rotating, has an initial flattening by virtue of the arbitrarily specifiable initial conditions themselves. In the case of the real star core, the diameter of the pancake will be expected to exceed the thickness by a substantial factor if the angular momentum of the original system is significant in comparison with the quantity

$$(\text{mass})^{3/2}\ (\text{effective radius of original system})^{1/2}$$

What now happens will be very different according to the mass of the "pancake" and its angular momentum. Consider the case where the "pancake" has a mass of order of $5M_\odot$. To attain nuclear density ($\varrho_{\text{conv}} = 1.5 \times 10^{14}$ g/cm³ or $\varrho = 1.1 \times 10^{-14}$ cm^{-2}) in the median plane will require a pressure $p_{\text{conv}} = 2 \times 10^{33}$ dy/cm² (or $p = 1.6 \times 10^{-16}$ cm^{-2}), implying a surface density

$$\sigma = (2p/\pi)^{1/2} = 1.0 \times 10^{-8}\ \text{cm}^{-1}$$

$$\sigma_{\text{conv}} = 1.4 \times 10^{20}\ \text{g/cm}^2$$

The calculated thickness is of the order of 15 km and the diameter of the order of 100 km. If the angular momentum is not too great, and the collapse can proceed in the equatorial plane, the time for this second phase of the dynamics will be of the order

$$\tau_2 \sim (\text{radius of freshly formed "pancake"})^{3/2}/(\text{residual mass})^{1/2}$$

(residual mass = mass left after expulsion of shell by shock[42]). The system will then end up as a Kerr black hole.

If the angular momentum of the pancake is larger than the critical amount $J = m^2$, considerations are not so simple. At least four possibilities suggest themselves: (1) fragmentation into several neutron stars plus other orbiting material; (2) fragmentation into one or more black holes plus one or more subsidiary neutron stars plus other orbiting material; (3) collapse into a toroidal geometry; and (4) collapse into a geometry without any symmetry at all.

Cases (1) *and* (2), the system is evidently endowed with a large mass quadrupole moment which moreover will be changing rapidly with time as the larger fragments revolve about their common center of gravity. Gravitational radiation will take place at such a rate

$$-dE/dt = (32/5)\, m_1^2 m_2^2 (m_1 + m_2)/r^5$$

(cf. Chapter 7 on gravitational radiation) that these larger fragments will quickly (less than a minute) coalesce to one large black hole. Thus, for example, the angular momentum of two like masses m,

$$J = m^{3/2} r^{1/2} / 2^{1/2}$$

will be radiated away until J falls to some value comparable to m^2 and the system finally collapses to a black hole.

The residual orbiting material, in the form of smaller fragments and dust, will radiate less rapidly and fall in later. Idealise a portion of this residual material as a test particle. What happens if (*a*) the black hole already has the critical value of the angular momentum, $J = m^2$ and (*b*) the additionnal small mass is revolving in the equatorial plane in the sense suited to increase thc angular momentum of the black hole? Bardeen answers[81] that the particle will slowly spiral in, losing energy by gravitational radiation, right up to moment when it arrives at the surface of infinite red shift, $r = 2m$ (Figures 15 and 16). Then it will quickly pass through the dynamic zone, emitting a brief snatch of radiation, and then cross the horizon or one-way membrane, thereby becoming completely assimilated into the black hole. While the particle spirals in, its angular momentum decreases. When it is finally assimilated into the black hole, it raises the angular momentum and the mass of the system to values which still satisfy the limiting condition,

$$J + \delta J = (m + \delta m)^2$$

Case (3), collapse to a toroidal geometry, is most appropriately discussed in connection with the collapse of an infinitely long cylinder of mass ("torus of infinite major radius"). Thorne[85] has shown in simple examples that the collapse can go to completion by: (a) formation of a line singularity; (*b*) radiation of practically all the mass-energy (so-called *C*-energy, definable for cylindrical geometry; relation to the Schwarzschild energy of a torus of finite major radius, as defined by the period of a far-away test object, still uncertain and the subject of investigation) of the system in the form of gravitational waves; and (*c*) no horizon formed (all the process observable by a far-away observer in a perfectly finite time). One can expect that an ideal torus of finite major radius will undergo a qualitatively similar collapse: (i) collapse of the minor radius; (ii) formation of a ring-shaped singularity, or near-singularity; (iii) if any line-density of mass-energy still remains at the end of phase (ii), then collapse of the major radius on a longer time-scale; (iv) possible decrease of the large quadrupole moment and higher moments of the mass distribution, similar to the $\sim 1/(\text{time})^n$ decrease of the moments in the case of small moments (cf. discussion above), in the event that this extrapolation is justified (alternative: a very asymmetric type of collapse—cf. Case 4); (v) approach to a standard Kerr black hole, with (vi) formation of an event horizon and (vii) trapping of a finite portion of the original mass energy of the system in this black hole. How the torus is to be formed in the first place in the initial collapse of the original star core is not in our province to study here. The hydrodynamics of a drop of milk falling into a pan of milk, as revealed by high-speed photography[86], is so fantastic in its complexity and beauty that one could hardly have imagined it, still less supplied a detailed analysis of how it came about. However, small perturbations from ideality will be expected in any such torus; and some of the modes of deformation will grow exponentially with time. This kind of behaviour is well known in the case of a jet of water[87,88]. An analogous break-up into a collection of separate masses can be envisaged here. However, we are then back to Case (1) or (2). Therefore the new element in the picture is obtained, not by looking further into this break-up into separate masses, but by taking the torus to be ideal, as envisaged throughout in Case (3).

Case (4), a highly unsymmetrical distribution of mass at the end of the first stage of collapse, can come about in many ways. It is hardly profitable to look into the details of that part of the hydrodynamics. What happens next is powerful gravitational radiation (time-changing quadrupole moments). The ensuing stages can be envisaged to follow the scenario of Cases (1) and (2). An alternative possibility claims attention. The asymmetry could be of a very special character, as treated originally by Kasner[89], taken up by Khalatnikov and Lifschitz[90], and especially analysed by Misner[91] in connection with his so-called "mixmaster model universe". The special feature of the dynamics here is the anisotropy oscillation of the Universe as a whole. The three principal radii of curvature oscillate in synchronism and with identical amplitudes and phase at every point of the Universe ("excitation by longest standing wave which will fit into closed Universe"; "homogeneous anisotropy"). The gravitational wave in this case is "trapped in a wave guide"—the Universe itself—and neither carries in nor carries away energy except as the "size of the wave-guide" itself changes in time. The same is not true when a system of stellar dimensions undergoes anisotropy oscillations. It becomes a powerful emitter of gravitational radiation and loses mass-energy rapidly. At least two very different outcomes suggest themselves as conceivable: (*a*) the damping is so powerful that the vibration stops before all the mass-energy is lost in gravitational radiation—the system settles down to a spherical configuration (neutron star or black hole); (*b*) the damping is not enough to kill the vibration. Energy continues to pour out. The mass of the system continues to go down. The frequency of the anisotropy vibration rises. The energy of the anisotropy vibration, in the absence of gravitational damping, increases as the frequency (principle of adiabatic invariance of ratio E_{vib}/ν). The amplitude goes up even faster (more energy in a smaller system). The rate of radiation of mass goes up without limit as the mass goes down. Or it reaches the kind of limit that could be imagined to come out from dimensional arguments; for example, to give one simple illustration (simple because independent of m!)

$$-dE/dt \sim \text{(cm of mass-energy)/(cm of light travel time)} \sim \text{(numerical factor, } k\text{)}$$

$$-dE_{\text{conv}}/dt_{\text{conv}} \sim kc^5/G \sim 10^{60} \text{ erg/sec}$$

In a finite time the system exhausts all the energy available. The system ends up with zero mass. The mass it once had has all been transformed into radiation, electromagnetic as well as gravitational, travelling off to great distances in a sudden burst of luminosity. There is not yet any theoretical analysis whatsoever that would say whether such a "complete gravitational fade-out" can really take place. It is important for physics and for astrophysics to know whether a mechanism exists for the complete conversion of "stable matter" to energy!

5.8 SEARCH FOR BLACK HOLES AND THEIR EFFECTS

Black holes have been predicted to exist for over thirty years. No one accepting general relativity has seen any way of denying their existence. Moreover, a black hole is a characteristic geometrodynamical entity. In Newtonian theory a neutron star could still exist, though with altered properties; not so a black hole.

From the prediction of a neutron star (Zwicky, 1934[92]) to its discovery (as a pulsar, Hewish *et al.*, 1968[18]) was 34 years. If from the prediction of a black hole (Oppenheimer and Snyder, 1939[50]) to its discovery is also 34 years, then what will be the tool by which it is detected in 1973?

Of all objects one can conceive to be travelling through empty space, few offer poorer prospects of detection than a solitary black hole of solar mass. No light comes directly from it. Can it be seen by its lens action or other effect on a more distant star? Not by any simple means! It is difficult enough to see a Venus 12,000 km in diameter swimming across the disk of the Sun without looking for a $2(27)^{1/2}\, m$ or 15 km object moving across a stellar light source almost infinitely far away. Turn therefore to a black hole that is not isolated: (*a*) a black hole that affects a companion normal star only through its gravitational pull (Zel'dovich and Guseynov[93]); (*b*) one close enough to draw in matter from the normal star (Shklovskii[94]); (*c*) one actually imbedded in a star; or (*d*) a black hole moving through a cloud of diffuse matter.

Trimble and Thorne[95] review the available evidence on double star systems, one component of which is invisible (alternating Doppler effect seen only in the other, visible, component). They conclude that "only few, if any, of the systems contain collapsed or neutron star secondaries". Moreover, no pulsar is known to be a member of a binary system (sine wave phase modulation of signal from pulsar, with period of one year or less). With a neutron star an infrequent component of a binary system, one can believe that a black hole also appears at most occasionally linked to another star. Nevertheless, just this occasional case is the case we are looking for.

The possibility to capitalise on such a double-star system is more favourable (case *b*) when the black hole is so near to a normal star that it draws in matter from its companion. Such a flow from one star to another is well known in close binary systems[96,97] but no unusual radiation emerges. However, when one of the components is a neutron star or a black hole, a strong emission in the X-ray region is expected[98,99].

Gas being funnelled down into a black hole undergoes heating by compression. Zel'dovich and Novikov[56] devote a whole chapter to the physics of pick up of ("accretion") and radiation by, this gas as they are affected by (*a*) the density of the gas far away from the black hole (*b*) the temperature there (*c*) the effective adiabatic exponent γ of the gas during compression and (*d*) the mass of the black hole. Regarding one other physical factor, the velocity of the star relative to the far-away gas, they recall a result of Salpeter[100]: "A body moving at supersonic speed relative to the gas decelerates in a time so short that accretion cannot appreciably change its mass during the phase of supersonic motion". This circumstance gives reason for focusing attention on the case where the black hole (or neutron star) moves at subsonic velocity relative to the gas.

The density of the gas around a slow-moving black hole is little affected by the pull of the star except at distances of the order of the "critical distance"

r_{crit} and less, where

$$r_{\text{crit}} \sim m/\beta^2_{\text{sound at } \infty}$$

or

$$r_{\text{crit}} \sim Gm_{\text{conv}}/(\text{velocity of sound far away})^2$$

(Example: $m = M_\odot$; ionized hydrogen gas at 10,000°K; $\gamma = 1.5$; $r_{\text{crit}} = 10^{14}$ cm; ionization brought about not least by radiation pouring out as the gas pours in). For distances less than r_{crit} the density goes up approximately as $r^{-3/2}$ for a wide range of conditions of physical interest; and provided angular momentum is unimportant the radial streaming velocity goes as $r^{-1/2}$. The calculated flux of matter onto the star is of the order

$$dm/dt \sim (2m)^2 \varrho_{\text{crit}}/\beta^3_{\text{crit}}$$

$$dm_{\text{conv}}/dt_{\text{conv}} \sim (2Gm_{\text{conv}}/c^2)^2 c^4 \varrho_{\text{crit conv}}/v^3_{\text{conv}}$$

Here ϱ_{crit} and β_{crit} are the density and the sound velocity out at the critical distance where the pull of the star first makes itself felt.

In the example we have for the sound velocity $\beta_{\text{crit}} \sim 0.4 \times 10^{-4}$. Take ϱ_{crit} to have the value 10^{-24} g/cm³ or $(0.74 \times 10^{-28}$ cm/g$) \times (10^{-24}$ g/cm³$) = 0.7 \times 10^{-52}$ cm^{-2} that one nominally adopts for interstellar matter. Note that $2m = 2 \times 1.47$ km $= 3 \times 10^5$ cm. Then we have $dm/dt \sim 10^{-28}$ (dimensionless) or in conventional units $\sim 10^{-15} M_\odot$/yr. If 0.1 of the incident rest mass is radiated, the corresponding output of energy is $\sim 3 \times 10^{30}$ erg/sec or $\sim 10^{-3}$ of the luminosity of the Sun. A nearby star will supply a much higher density of matter. For example, a star of the β-Lyrae type can give out $10^{-5} M_\odot$ of matter a year, corresponding at distances of the order of 10 solar radii to a mass density of the order $\varrho \sim 10^{-11}$ g/cm³. In this case the luminosity will be greatly increased.

The mechanism for disposition of the energy is compression, heating, and radiation. The converging matter arrives at supersonic velocities at a distance which in the example is calculated to be of the order of 10^{13} cm. On the basis of simplifying assumptions, which Zel'dovich and Novikov spell out and analyse in detail, Schwartzman concludes that "the temperature which develops during the adiabatic compression corresponds in order of magnitude to the kinetic energy of infall, and is thus approximately the temperature of the shock wave which would arise if the gas were to collide with an object at rest"—even though this gas falls without stopping. Temperatures achieved may run in the range 10^{10}, 10^{11} or 10^{12} °K. However, only a fraction of the radiation escapes because it comes from a region of high red shift, close to the Schwarzschild radius of the black hole. Zel'dovich, Novikov and Schwartzman conclude that an *isolated* interstellar black hole emits in the visible part of the spectrum, or in the X-ray and gamma ray region, according as the mass of the black hole is more or less than a certain critical figure. For the case of a *binary system*, material flowing from one star would have too much angular momentum to be able to fall directly onto a compact companion. It would instead form a spinning disc, in which the matter spirals inward at a rate determined by the efficiency of viscous dissipation. The energy liberated by the disc could again emerge as X-rays[98]. These circumstances emphasise the importance of X-ray astronomy, such as can only be done on board a satellite, in the search for black holes. In this connection one is especially eager to find out whether any one of the already known X-ray sources[101] can be related to a completely collapsed object. Of particular interest are the rapidly fluctuating (time-scales $\lesssim 0.1$ s) X-ray sources associated with binary stars; and especially those cases (e.g. Cygnus *X*-1) when the mass of the X-ray source, inferred from the dynamics of the binary system, seems too large for it to be a neutron star.

One can[102] speculate that (1) a significant fraction of all galaxies may become quasi-stellar objects for some periods in their histories; (2) a burned out quasi-stellar source may leave behind a black hole; (3) there may well be one or more massive ($M \gg M_{\odot}$) black holes in the nucleus of our own galaxy. This possibility increases the interest one has in looking for pulses of gravitational radiation and for X-rays associated with such an object.

CHAPTER 6

Quasi-Stellar Objects

GALAXIES VARY one from another in mass by more than an order of magnitude either way from the figure often quoted for a "typical" galaxy, $\sim 10^{11}$ $M_\odot$ (with $M_\odot = 1.989 \times 10^{33}$ g). From such a galaxy the output of electromagnetic radiation runs of the order of $10^{11} L_\odot$ (with $L_\odot = 4 \times 10^{33}$ erg/sec). By comparison a Seyfert galaxy may emit ~ 10 times as much radiant energy/sec, most of it from the concentrated nucleus of the galaxy; and a quasi-stellar object emits at ~ 100 times the rate from a region equally concentrated. Ups and downs of the intensity in the visible by 0.25 magnitude (ratio 1.26 : 1.00) over a time of one day[103] have made it seem reasonable to conclude that the region of emission is localised within one light day ($\sim 3 \times 10^{15}$ cm or 200 astronomical units), at least for a time of some days. There are at least two other ways to get at the size. Bahcall, Gunn and Schmidt[104,320] have found two quasi-stellar objects which are associated in direction with clusters of galaxies, whose red shifts agree with the average red shift of the associated clusters. This result makes it reasonable to believe that the quasi-stellar objects are at cosmological distances. The nature of quasar red shifts is still, however, the subject of lively controversy (see also Section 15.2). Intercontinental interferometry[105] has given as upper limit for the angular diameter of such objects 0.001″ (or 5×10^{-9} radian) as detected by radiotelescopes. For a distance of 2×10^9 light years this angular diameter implies a radio source of dimensions 10 lyr or less. Independently, one has detected variations in the radio emission with a time-scale of the order of a year[106], strongly suggesting that the radio source is less than 1 lyr across. That the limit on dimensions found by radio techniques is different from that found by optical means ties in with finding that the time variations of the two radiations are practically uncorrelated.

The power output of one of the best studied of quasi-stellar sources, 3C273 (red shift 0.158; estimated distance $\sim 2 \times 10^9$ lyr) has been evaluated to be approximately as follows

X-ray[107]	$\sim 7.3 \times 10^{45}$ erg/sec
ultraviolet	unobserved
visible[108]	$\sim 2 \times 10^{46}$ erg/sec

infrared[109]	$\sim 3 \times 10^{47}$ ergs/sec (excluding possible strong emission at wavelengths $\gtrsim 30\mu$)	
radio[110]	$\sim 9 \times 10^{43}$ ergs/sec	
total	$\sim 3 \times 10^{47}$ or more, or $\sim 10^{14} L_{\odot}$	

Some quasi-stellar objects emit more in the radio than 3C273; others have been detected only in the visible, not at all in the radio.

Besides *power* output, a remarkable feature of some of these objects is *directional energy* output of unprecedented magnitude. In 3C273 one detects both radio waves[110] and light coming from a jet of plasma of dimensions comparable to galactic ($\sim 10^5$ lyr) diameters. The spectrum indicates synchrotron radiation. One concludes that the charged particles and the magnetic lines of force about which they whirl were ejected from the galaxy in an explosion event $\sim 10^6$ years ago. The sum of particle energy and magnetic energy may be of the order of 10^{61} erg in some objects. Such an energy can be produced, so far as is known, only if (1) a significant fraction of the entire $\sim 10^{11} M_{\odot}$ in a galaxy undergoes thermonuclear combustion or (2) $\sim 10^7 M_{\odot}$ is converted almost completely into energy, or (3) some intermediate transformation takes place.

Several promising proposals have been made as to how the *power* output may arise: (1) compact cluster of $\sim 10^8$ stars at the center of the galaxy, with these stars colliding or otherwise growing to the point where they become supernovae; or (2) one giant star of $\gtrsim 10^5 M_{\odot}$, which not only vibrates or rotates but also converts a large fraction of its mass to energy; or (3) a giant black hole ($\sim 10^7 M_{\odot}$) which captures matter and in one or another of the ways already touched upon (discussed in an interesting recent paper of Lynden-Bell[102]) converts a substantial fraction of this matter to radiant energy; or (4) some complex combination of these mechanisms. It may well be that one of these mechanisms dominates in one kind of quasi-stellar source; another, in another. Current ideas strongly suggest that, whatever a quasar may be, the endpoint of its evolution is the formation of a massive black hole. In any case it is a central issue to uncover the evolutionary connection between quasi-stellar sources and Seyfert galactic nuclei and normal galactic nuclei.

For the *directional* output strong radio sources no such wealth of options is evident. One promising mechanism that comes to attention is the collapse of a magnetohydrodynamic system, with jetting of material out of one or both poles, a mechanism beautifully illustrated by the recent machine calculations of LeBlanc and Wilson[44] (see Figures 21 and 22).

Between the gravitational phenomena that go on in a quasi-stellar source and those in a single supernova there must exist a rich variety of phenomena intermediate in energy, which it will be of the greatest interest to explore.

From the value of the flux of 3C273 it is clear that the major part of the energy is emitted in the infrared wavelengths. In this region exists a very large absorption from the atmosphere (Figure 5); therefore an analysis of the energy output of quasars done from space platforms seems to be highly desirable. Moreover, the use of a space platform would allow a very long baseline for the radio interferometer analysis of quasars and consequently a very high resolution of their effective dimensions. One might then be able to discover whether the successive radio outbursts have the same location (as might be expected if a single supermassive object provided all the energy), or whether they correspond to explosions of separate stellar-mass bodies. These observations could also, as we discuss in Chapter 12, be of interest for cosmology.

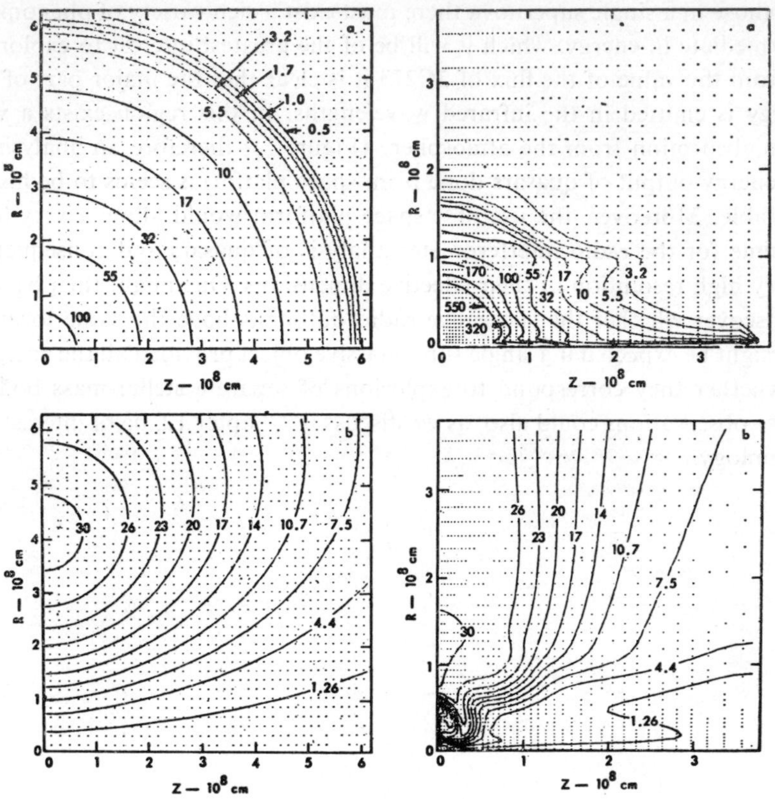

Figure 21

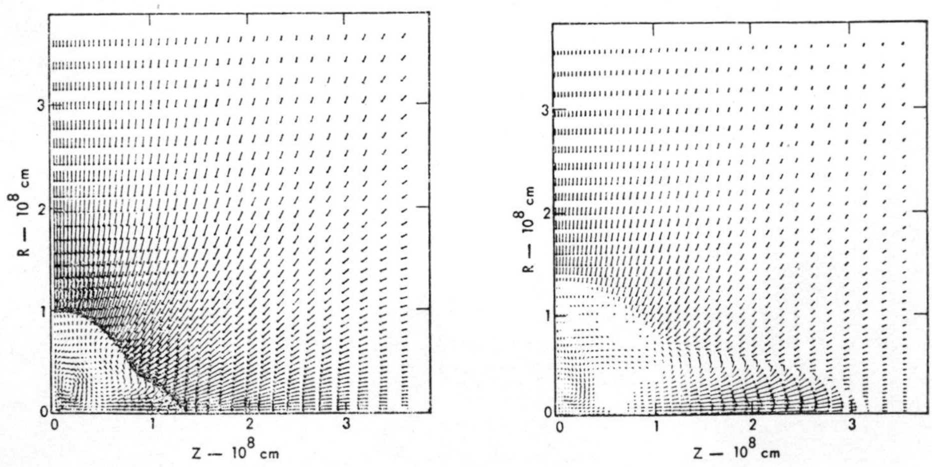

Figure 22

Figure 21 The LeBlanc-Wilson jet mechanism. Isodensity contours and magnetic flux contours parallel to the z axis for the gravitational collapse of a $7M_\odot$ rotating magnetic star as computed by J. M. LeBlanc and J. R. Wilson[44]. They have considered a star in which all the material has been burned to the end point of thermonuclear evolution. The star is also assumed to be initially in uniform rotation with an initial angular velocity of 0.7 rad/sec (angular momentum $\sim 4.6 \times 10^{50}$ g cm^2/sec) and to have a magnetic field parallel to the z-axis with an energy of the order of 0.025% the gravitational energy of the star. The computations are started from an equilibrium configuration. After a few seconds enough energy has been lost by neutrino emission that the star starts to collapse.

As the collapse proceeds the temperature of an interior region rises above the iron decomposition temperature and the collapse rapidly accelerates. During the collapse non-radial motions develop due to the increasing centrifugal force and the angular velocity approaches a vortex configuration. The shear in the velocity generates large magnetic fields along the axis of rotation. The multiplication of magnetic energy by this process is a hundred times larger than the one expected from simple compression. The collapse is stopped at a density of the order of 10^{11} g/cm^3. The combined effect of rotation and magnetic field produces a jet in the axial direction with a speed of the order of one-tenth of the speed of light. This jet carries a mass of the order of 2.1×10^{31} g and a kinetic energy of 1.6×10^{50} ergs, and with a considerable amount of magnetic energy ($\sim 3.5 \times 10^{49}$ ergs with fields of $\sim 10^{13}$ gauss). The two graphs in the upper part of the figure give the iso-density contours in units of 10^6 g/cm^3 at 0.72 sec and 2.67 sec after the beginning of gravitational collapse. With r is indicated the distance in the equatorial plane from the rotational axis (z-axis) in units of 10^8 cm.

The two graphs in the lower part indicate the magnetic flux contours parallel to the z-axis in units of 10^{22} gauss cm^2 again at 0.72 sec and at 2.67 sec.

Figure 22 Velocity vector in the jets developed in the collapse of a $7M_\odot$ rotating magnetic star (see caption of Figure 21). R gives the distance in the equatorial plane from the rotation axis (z direction) in units of 10^8 cm. The graph in the left hand illustration refers to a configuration 2.61 sec after $t = 0$ (start of gravitational collapse). The maximum velocity magnitude is 2.43×10^9 cm/sec. The graph in the right hand illustration refers to a configuration 2.64 sec after $t = 0.0$ with a maximum velocity of 4.38×10^9 cm/sec.

CHAPTER 7

Gravitational Radiation

THAT GEOMETRY can undulate and carry energy, Einstein showed in the very first days of general relativity[111]. Joseph Weber in the late 1950's was the first to launch a realistic effort to detect such "gravitational waves"[112]. His philosophy was pragmatic. We require a detector whether the waves come from a natural source or from a specially built artificial source. Therefore build the detector first. Moreover, a simple estimate suggests that any artificial source, if it is to be strong enough to be detected outside the near zone, will be expensive to build (Table IX).

Therefore emulate Hertz: explore with the detector for some unpredictable natural source of radiation before thinking about building an artificial source. Weber carried the economy consideration a step further at the beginning of his enterprise, not only hoping for a natural source but also relying on a natural detector, the natural quadrupole oscillation of the Earth itself. He did not find any clear evidence for energy being fed into this mode of vibration other than what could naturally be attributed to earthquakes and other noise. On this ground he was able to give the first observational upper limit ever for the intensity of gravitational radiation from outer space at a specified frequency,

$$\begin{aligned} dI/d\nu &= c^3 2\pi\, d\varrho_{\text{rad}}/d\omega < 27 \times 10^{30} \times 12 \times 10^{-25} \text{ g/cm}^3 \\ &= 3 \times 10^7 \text{ erg/cm}^2 \text{ sec Hz} \end{aligned}$$

at 3.1×10^{-4} Hz (vibration period 54 min)[114] (revised in our Table X to $< 2.6 \times 10^8$ erg/cm^2 sec Hz). Subsequent work of his has given, or allows one to estimate, upper limits at different frequencies.

Events on Weber's aluminium bar at College Park, Maryland: the bar extended or contracted in its principal longitudinal mode with three times thermal energy (k times 300°K) on a few dozen occasions a day and with five times thermal energy on less than a handful of times a day. Weber set up another bar at the Argonne National Laboratory near Chicago: it indicated a number of single events comparable to that of the first bar. Only a few events, of the order of one a day, were coincident. (There should in any case be a certain number of coincidences.) Events in the individual detectors coming at random times will occasionally overlap within the discrimination

Table IX Boundary r_b between near zone (gravitational field falling off with distance according to Newtonian predictions) and wave zone (true propagation with speed of light) for illustrative sources of gravitational radiation

Motion	Frequency	Near zone] [wave zone
Earth going round Sun	1/yr	3.8×10^{12} km
Quadrupole oscillation of Earth	1/54 min	3.9×10^{8} km
Quadrupole oscillation of Moon	1/15 min	1.1×10^{8} km
Two neutron stars or black holes, each of mass $M_\odot$, 100 km between centers, revolving in circle about center of gravity	41 sec^{-1}	2.9×10^{3} km
Oscillation of length of Weber's bar	$(10^4/2\pi)\ \text{sec}^{-1}$	74 km
Vibration of hot SiO molecule inside Earth	$3.7\times 10^{13}\ \text{sec}^{-1}$	3.2×10^{-4} cm

The value of r_b is estimated by comparing the actual tensorial wave with the scalar wave $u = R(r)\, Y(\theta, \varphi)$ ("potential" for the tensor wave) associated with a spherical harmonic $Y(\theta, \varphi)$ of order $l = 2$. The radial factor in this scalar wave satisfies the equation

$$d^2R/dr^2 + [k^2 - l(l+1)/r^2]\, R = 0$$

The boundary between the near zone (rapid fall off of R) and the wave zone (oscillation of R) occurs for $r = [l(l+1)]^{1/2}/k$ or, as better estimated by the JWKB "$(l+\frac{1}{2})$" approximation,[113] at

$$r = r_b = (l + \tfrac{1}{2})/k = 2.5/k = 2.5\lambdabar$$

Here $\lambdabar$ is the "reduced wave length" of the radiation (classically calculated distance of closest approach for a quantum with one unit $\hbar$ of angular momentum):

$$\lambdabar = 1/k = \lambda/2\pi = c/\omega = c/2\pi\nu.$$

time of the coincidence circuits (significant fraction of a second). Weber by statistical analysis and by counting coincidences when he introduces a time delay in one circuit, concludes that accidental coincidences are roughly one order of magnitude less frequent than the observed coincidences[115]. He has orally reported evidence that the coincidences are not associated with (*a*) solar flares (*b*) lightning (*c*) low-frequency electromagnetic waves used to signal from land to United States submarines (*d*) surges in the interstate electric power grid. He has also reported in the literature evidence that the coincidences are not associated with (*e*) an underground nuclear shot in November of 1968 or (*f*) seismic events, in any case implausible because of the ordinarily great difference in time of arrival of seismic waves at Chicago and Maryland, while a simple estimate shows that (*g*) the gravitational waves generated by an earthquake would be far too weak to actuate either bar and (*h*) the strains stored in the bar by reason of its past history of thermal cycles relieve themselves from time to time in "bar quakes". In such events crystallites in the aluminium change position like rocks in a rock pile. Pulses

Table X Upper limits to the intensity of the gravitational radiation arriving at the Earth from outer space at selected frequencies

	Earth	Weber's bar
Mode studied	Quadrupole vibration	Lowest mode of stretching vibration
Period	54 min	6.03×10^{-4} sec
Mass	5.98×10^{27} g	1.41×10^{6} g
Dimensions	Radius 6.27×10^{8} cm	153 cm long, 66 cm diameter
Q for this mode	400	$\sim 10^{5}$
Bandwidth $A_{\text{diss}} = \Delta\omega = \omega/Q = \tau_{1/e}^{-1}$	4.86×10^{-6} rad/sec	$\sim 10^{-1}$ rad/sec
$\int_{\text{resonance}} \sigma(\nu)\, d\nu$ (cf. text)	$4.7\ \text{cm}^2/\text{sec}$ $= 4.7\ \text{cm}^2$ Hz	$1.0 \times 10^{-21}\ \text{cm}^2/\text{sec}$ $= 1.0 \times 10^{-21}\ \text{cm}^2$ Hz
Information setting upper limit to continuous background gravitational radiation	Noise-power spectrum of normal mode excitation of earth during quiet periods, in the vicinity of 1 cycle/hr as measured by Weber and Larson[117]	Thermal noise of detector, at $T = 300°$K, if all attributed to gravitational radiation, implies rate of loss, $kT/\tau_{1/e} = kT\Delta\omega$
This limit, in its own natural units	$6.9 \times 10^{-14} \dfrac{(\text{cm/sec}^2)^2}{(\text{rad/sec})}$ [118]	4.1×10^{-15} erg/sec
This limit translated into terms of erg/sec uptake of energy from "gravitational waves"	1.2×10^{9} erg/sec (Note 1)	4.1×10^{-15} erg/sec
This divided by $\int \sigma\, d\nu$ gives upper limit to the flux at resonance	2.6×10^{8} erg/cm² sec Hz	4.1×10^{6} erg/cm² sec Hz
This multiplied by ν/c, gives upper limit to energy content for spectrum flat from $\nu = 0$ up to frequency of observation, ν	2.4×10^{-6} erg/cm³	2.2×10^{-1} erg/cm³
This translated into equivalent mass density	2.6×10^{-27} g/cm³	2.5×10^{-22} g/cm³
Order of magnitude of density required for closure of Universe based on best present values for age of Universe (13×10^{9} yr) and inverse Hubble constant (19×10^{9} yr)	$\sim 10^{-29}$ g/cm³	$\sim 10^{-29}$ g/cm³
Estimated upper limit to density compatible with homogeneous relativistic cosmology and maximum stretching of uncertainties in age and inverse Hubble constant	10^{-27} g/cm³	10^{-27} g/cm³

Table X (*continued*)

	Earth	Weber's bar
Do present upper limits on intensity of gravitational radiation improve on this cosmological limit?	No	No
Information setting upper limit to strength of arriving pulse of gravitational radiation	No useful upper limit. (Too many earthquakes! No way of discriminating between effects of earthquakes and effects of gravitational waves by coincidences—no easily reachable duplicate of the Earth!)	A pulse of $5kT$ (with $T = 300°K$) or more simultaneously in Argonne and Maryland detectors at most once in 81-day interval[116]
Energy uptake of detector in such a pulse	—	2×10^{-13} erg
This divided by $\int \sigma \, d\nu$ gives upper limit to the energy $I(\nu_0)$ of an incoming pulse per cm^2 of intercepted area and per unit frequency at resonance	—	2.0×10^{8} erg/cm^2 Hz
$I(\nu_0)$ multiplied by ν_0 gives upper limit to surface density of pulse energy at Earth for spectrum flat from $\nu = 0$ up to frequency of observation ν_0	—	3.3×10^{11} erg/cm^2
Upper limit to surface density of energy of pulse if it has same frequency as detector and ten times ($\Delta\omega = 10A_{diss}$) its band width (Note 2)	—	5×10^{7} erg/cm^2
—if $\Delta\omega = A_{diss}$	—	10^{7} erg/cm^2
—if $\Delta\omega \ll A_{diss}$	—	5×10^{6} erg/cm^2
Last multipled by $4\pi r^2$, with $r = 8.2$ kpc $= 2.5 \times 10^{22}$ cm, gives lower limit for output of gravitational wave energy from a source located at the galactic center if powerful enough to produce such a pulse	—	4.0×10^{52} erg
Amount of mass m, which would would have to be annihilated to produce such energy, assuming 50% maximum efficiency	—	$m \sim 0.04 M_\odot$

NOTE 1 The Weber and Larson figure for the noise-power spectrum of the Earth, as measured by gravimeter, is 6.9×10^{-14} (cm/sec^2)^2/(rad/sec) (mean square accelartion per unit interval of circular frequency, in the neighbourhood of the 54-min quadrupole mode). The corresponding circular frequency is $\omega_0 = 1.94 \times 10^{-3}$ rad/sec and the total damping rate is

$$A_{\text{damping}} = \Delta\omega = \omega_0/Q = \omega_0/400 = 4.9 \times 10^{-6} \text{ rad/sec}$$

All that part of the noise which lies in the interval of circular frequency given by the integral of the weighting factor $(\Delta\omega/2)^2/[(\omega - \omega_0)^2 + (\Delta\omega/2)^2]$ is attributed to the mode in question, namely, $(\pi/2)\,\Delta\omega$, or

$$6.9 \times 10^{-14} \times 7.7 \times 10^{-6} = 5.3 \times 10^{-19} \text{ (cm/sec}^2)^2$$

This mean square acceleration is divided by ω_0^2 to obtain the mean square up-and-down velocity of the surface of the Earth associated with this mode

$$1.4 \times 10^{-13} \text{ (cm/sec)}^2$$

There are five independent modes of quadrupole vibration, associated with the five spherical harmonics of order two; but all, except the mode in question, have zero amplitude at the point of observation, chosen to be the pole of the system of spherical coordinates. The kinetic energy for it, in the idealization of a globe of fluid of uniform density (cf. text below) is $(3/20)\ Ma^2\alpha^2 = (3/20)$ (mass of Earth) (vertical velocity at pole)^2; and average total energy in the mode in question is twice this amount, or

$$E = (3/10)\,(5.98 \times 10^{27} \text{ g})\,(1.4 \times 10^{-13} \text{ cm}^2/\text{sec}^2) = 2.5 \times 10^{14} \text{ erg}$$

The rate of dissipation of this energy is

$$EA_{\text{damping}} = 1.2 \times 10^{9} \text{ erg/sec}$$

The cited upper limit to the flux of gravitational radiation is obtained by taking this amount—which has to be continually replenished—as coming exclusively from gravitational radiation, and not at all from earthquakes, seiches, eruptions, etc., and dividing it by $\int \sigma\, d\nu = 4.7 \text{ cm}^2$ Hz. This gives the flux of energy at resonance. If we assume a spectrum flat from 0 to ν, we obtain the upper limit to the energy density multiplying the energy flux at resonance by ν/c.

NOTE 2 The uptake of energy, E, by Weber's bar detector is equated to the integral

$$E = \int I(\nu)\,\sigma(\nu)\,d\nu$$

where $I(\nu)$ is the spectrum of the assumed pulse of gravitational radiation (erg/cm^2 Hz). Five cases are of interest:

Broad spectrum, general shape	$I(\nu_0) = \dfrac{E}{\int_{\text{res}} \sigma(\nu)\,d\nu}$
Total energy (erg/cm^2) in pulse for case of a spectrum flat from $\nu = 0$ to ν just above the resonance frequency, ν_0	$\int I(\nu)\,d\nu = \nu_0 I(\nu_0)$
Total energy (erg/cm^2) in pulse for case of a spectrum of the form $I(\nu) = I_0(\Delta\omega/2)^2\,[(\omega-\omega_0)^2 + (\Delta\omega/2)^2]^{-1}$ with $\Delta\omega \gg A_{\text{diss}}$	$\int I(\nu)\,d\nu = (\Delta\omega/4)\,I(\nu_0)$
Total energy (erg/cm^2) in pulse for case of spectrum which depends upon frequency in the same way ($\Delta\omega = A_{\text{diss}}$) as does the cross-section itself	$\int I(\nu)\,d\nu = (A_{\text{diss}}/2)\,I(\nu_0)$
Total energy (erg/cm^2) in pulse for irradiation by a line much narrower ($\Delta\omega \ll A_{\text{diss}}$) than bandwith of receptor	$\int I(\nu)\,d\nu = (A_{\text{diss}}/4)\,I(\nu_0)$

NOTE 3 The Breit-Wigner formula contains the resonance denominator $(\omega - \omega_0)^2 + (A/2)^2$, whereas the standard formula for the response of damped simple harmonic oscillator contains the resonance denominator $(\omega^2 - \omega_0^2)^2 + 4\xi^2\omega_0^2\omega^2$. It is of no consequence for the discussion in the text that the one denominator far from resonance behaves very differently from the other. The integral of the cross-section comes almost exclusively from the immediate vicinity of the resonance, and there the two formulae agree in form, as one sees most easily by removing the factor $(\omega + \omega_0)^2$ from the latter resonance denominator and identiyfying ω in this factor with ω_0, as is justified in the vicinity of the resonance.

of such origin are roughly as numerous as other pulses in the first days after a 10°C change in temperature. There is no evident way for this mechanism to contribute to two detectors, except by accident. No one has yet come forward with a workable explanation for Weber's coincidences other than gravitational waves from outer space. However, the history of physics is rich with instances where supposedly new effects had to be attributed in the end to long familiar phenomena. Therefore it would seem difficult to rate the observed events as "battle tested". To achieve that confidence rating would seem to require confirmation with different equipment or under different circumstances, or both. In making this tentative assessment one can be excused for expressing at the same time the greatest admiration for the experimental ingenuity, energy and magnificent persistence shown by Joseph Weber in his search, over more than a decade, for the most elusive radiation on the books of physics. Moreover, he found no more than one coincidence in 81 days in which both pulses were five times thermal (300°K) or more[116], thereby allowing one to set (Table X) an upper limit to the incoming flux of gravitational radiation.

Approximately 150 Chicago-Maryland coincidence observed in a period of about six months correlated poorly with solar time, but significantly with galactic time[119]. The events were grouped in three bins. Included in the middle bin were all those events occurring in the four-hour period when the pair of detectors were most sensitive to gravitational waves coming from the center of the galaxy as well as those events occurring in the "quadrupole-symmetric" four-hour period twelve hours later, when the detectors were again maximally sensitive to gravitational waves from the galactic center. The "windows-in-time" for the other two bins were displaced four hours ahead and behind the "window-in-time" of the central bin. Interesting as it is that the count in the central bin was greater than that in either of the side bins, and by a significant margin, it is also noteworthy that natural statistical fluctuations have often given mistaken impressions of the reality of non-existent energy levels in nuclear physics and non-existent masses in elementary particle physics. Therefore it will be of interest to see how the frequency of coincidences depends upon galactic time in ensuing intervals of six months.

It is now appropriate to turn from the experimental search for gravitational radiation to the theory of gravitational radiation as it bears on potential sources of such radiation. The theory has undergone rich development, especially in the last two decades, thanks not least to the contributions of Professor Hermann Bondi and his colleagues. Doubts that hung over the subject in earlier days have now been dissipated. There was a time when even the reality of gravitational waves was doubted. Put a wave into the coordinate system. Let it propagate at whatever speed one will. This is all that a gravitational wave is—a paper phenomenon. It travels, not with the speed of light but with speed the of thought†. Moreover, any talk of a gravitational wave carrying energy is nonsense. There is no such thing as the local density of gravitational wave energy. The Einstein field equations themselves forbid. The source term $T_{\mu\nu}$ on the right-hand side of these equations contains every potential source of mass-energy *except* gravitation.

In time the true content of the theory could not be mistaken. Yes, arbitrary coordinate changes alter the metric coefficients. They also alter the connection coefficients $\Gamma^{a}_{\beta\gamma}$ obtained from the first derivatives of the metric coefficients. Non-zero Γ's may be altered to locally zero Γ's, and conversely (apparent gravitational field as sensed in a space ship that does or does not have its rocket engine turned on!). However, the tide-producing acceleration or Riemann curvature tensor, given by the second derivatives of the metric coefficients, if non-zero in one coordinate system is non-zero in all coordinate systems. It gives unambiguous evidence for the presence or passage of a gravitational disturbance. If it happens to be zero in a certain region of space for a certain interval of time, then there is no coordinate change that can produce for it any value but zero.

The local energy content of gravitational origin is the wrong thing to look for because curvature, the true physical effect, influences not one test particle at one location but the separation between two test particles at slightly separated locations. Global energy content, not local energy content of a packet of gravitational waves, is the quantity that has a simple meaning. To define this total mass-energy one wants to be dealing with a region of

† "Longitudinal-longitudinal [and] longitudinal-transverse ... gravitational waves ... have no fixed velocity ... They are not objective ... The are merely sinuosities in the coordinate system and the only speed of propagation relevant to them 'is the speed of thought'" Eddington.[120] What Eddington said to distinguish these fictitious (coordinate) waves from the real transverse-transverse gravitational waves was unfortunately misunderstood by certain other investigators and taken by them as an argument against the reality of all gravitational waves.

spacetime which is asymptotically flat. No one has ever succeeded in giving a well-defined meaning to total energy for a closed space, a finding natural to anyone who reflects that there is no platform on which to stand to measure the total attraction of the system! In an asymptotically flat geometry one determines the rate of approach to flatness (Schwarzschild mass-energy, m). For this purpose one measures the period of a planet in a Keplerian orbit about the center of attraction:

$$m\,(\text{cm}) = r^3[\omega(\text{cm}^{-1})]^2$$

This totalised mass energy lends itself to calculation by integration over the volume to which the undulation is confined. The resulting formula, qualitatively of the form[121]

$$m \sim (1/8\pi) \int \Gamma^2\, d^3x$$

reminds one of the familiar integral for electromagnetic mass-energy. However, there is an important difference between the integrals in the two cases. In the case of electrodynamics the local energy density has a well-defined meaning; in the case of geometrodynamics, it does not. In the gravitational case, reasonably enough, innumerable other integrals give altered values for the local density while giving always the same value for the total mass-energy[122].

Even electrodynamics teaches the need for caution about a purely local analysis of radiation energy. It says that

$$(E^2 + H^3)/8\pi$$

is the density of radiation energy and

$$(E \times H)/4\pi$$

the flux of radiation energy. However, put a highly charged conductor not too far from the N-pole of a bar magnet and select a point of observation where the electric field

$$E = 3000 \text{ volts/cm or } 10 \text{ es volts/cm (conventional units)}$$

or in geometric units (conversion factor $G^{1/2}/c^2 = 2.874 \times 10^{-25}$ cm^{-1}/gauss)

$$E = 3 \times 10^{-24}\ \text{cm}^{-1}$$

and where the magnetic field has the same magnitude

$$H_{\text{conv}} = 10 \text{ gauss}$$

$$H = 3 \times 10^{-24}\ \text{cm}^{-1}$$

and stands perpendicular to the electric field. The two Lorentz invariants $E^2 - H^2$ and $(E \times H)$, both vanishing in the present instance, serve to classify the electromagnetic field. Locally the field qualifies as a radiation field; globally it does not. The Riemann curvature tensor admits a similar local analysis[123], subject to similar caveats about its global relevance. Invariants of the Riemann curvature tensor, analogous to

$$E^2 - H^2 \quad \text{and} \quad (E \cdot H),$$

but more numerous, determine to which of the so-called "Petrov classes" the local tide-producing acceleration belongs. From local analysis one finds it useful to turn to the global picture for more insight into the way a gravitational wave propagates.

7.1 ANGULAR DISTRIBUTION OF GRAVITATIONAL RADIATION

The force of gravitation, like the electric force, falls off inversely as the square of the distance. Why then does the radiative component of the force fall only inversely with the first power of the distance? The answer has long been known in the case of electricity. The explanation goes back to the acceleration of the source. The disposition of the lines of force (Figure 23) is very different before and after the acceleration. The change from one field pattern to the other is confined to a shell of thickness δr equal to the acceleration time $\Delta\tau$. In the shell the lines of force that would otherwise have a length $\Delta\tau$ are stretched out to a length $r\Delta\beta_\perp$. Here $\Delta\beta_\perp$ is the component of the change in velocity of the particle perpendicular to the line of sight. This "stretching length" is greater the more distant the observer. Thus the normal field e/r^2 is augmented by the "stretching factor" $r\Delta\beta_\perp/\Delta\tau$ to the value

$$-(e/r^2)\,(r\Delta\beta_\perp/\Delta\tau) = -ea_\perp/r$$

Similar stretching occurs in the case of gravitational radiation. Here, however, the geometrical quantity stretched is a tensor, not a vector. Therefore the picture in this case is somewhat different from that shown in Figure 23. However, the qualitative result is the same: an effective field that falls off inversely as the first power of the distance.

When one comes to discuss angular dependence, one again has to reckon differently for gravitation and for electromagnetism because one is a tensor and the other is a vector. Figure 24 shows the receptor displaced by the polar angle θ from the line of motion of the source. The component of the acceleration perpendicular to the line of sight is

$$a_\perp = a \sin\theta$$

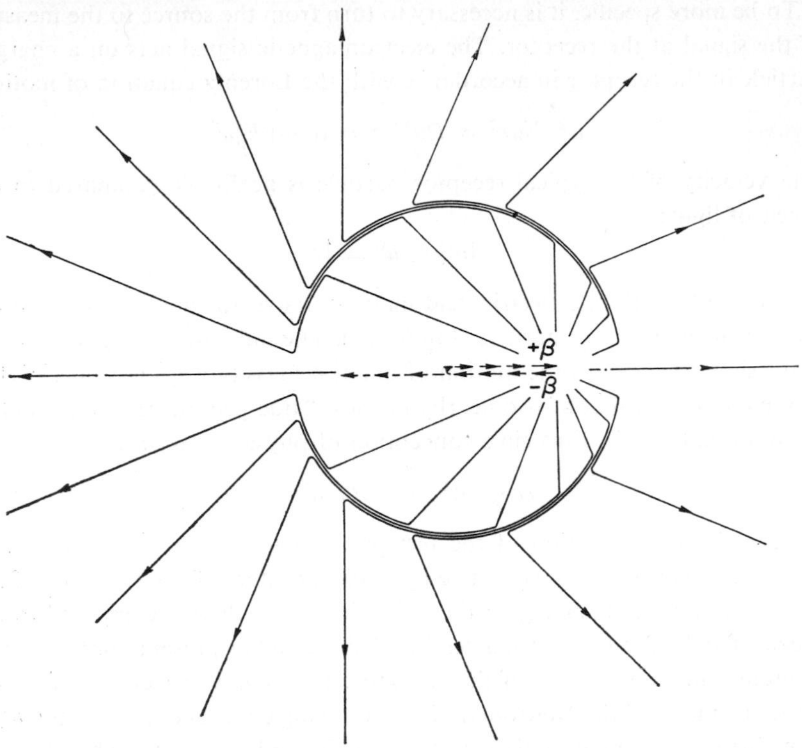

Figure 23 Explanation by J. J. Thomson of why the strength of an electromagnetic wave falls only as the inverse first power of distance r and why the amplitude of the wave varies (in the case of low velocities) as $\sin\theta$ (maximum in the plane perpendicular to the line of acceleration). The charge is moving to the left at uniform velocity. Far away from it the lines of force continue to move as if this uniform velocity were to continue forever (Coulomb field of point charge in slow motion). However, closer to the charge the field is that of a point charge moving to the right with uniform velocity ($1/r^2$ dpendence of strength upon distance). The change from the one field pattern to another is confined to a shell of thickness $\Delta\tau$ located at a distance r from the point of acceleration (amplification of field by "stretching factor" $r \sin\theta\, \Delta\beta/\Delta\tau$; see text).
We are indebted to C. Teitelboim for the construction of this diagram.

This quantity governs the strength of the electromagnetic field arriving at the receptor. In the gravitational case we deal with a tensor (2 index object). Consequently two factors of $\sin\theta$ enter into the calculation of the field at the receptor:

$$\text{signal amplitude} \propto \sin^2\theta$$

$$(\text{signal intensity} \propto \sin^4\theta)$$

To be more specific, it is necessary to turn from the source to the measure of the signal at the receptor. The electromagnetic signal acts on a charged particle in the receptor in accordance with the Lorentz equation of motion,

$$d^2x^\alpha/d\tau^2 = Du^\alpha/d\tau = (e/m)\, F^\alpha_{\ \beta}u^\beta$$

The velocity of the typical receptor particle is negligible compared to the speed of light:

$$|u^i| \ll u^0 \simeq 1$$

Consequently only the electric field matters. Its vector quality shows in the motion of the receptor particle (Figure 24). The influence of a gravitational wave shows up, not in the motion of one test mass (always a geodesic!) but in the relative motion of two nearby masses ("tide-producing component of gravitational field"; Einstein's conception of physics as *local*),

$$D^2\eta^\alpha/d\tau^2 = -R^\alpha_{\ \beta\gamma\delta}u^\beta\eta^\gamma u^\delta$$

Here η^α is the separation of the test particles (compare with two masses bound by a spring; or the effective dynamical centers of the two ends of the Weber aluminium bar). Again the velocities are negligible compared to the speed of light for the test masses in all detectors at present contemplated. Therefore no components of the curvature tensor have an effect except for what one may call the Newtonian tide-producing components; namely, $R^i_{\ ojo}$. Far from the source both electromagnetic and gravitational fields are "algebraically special". The magnetic field is equal in magnitude to the

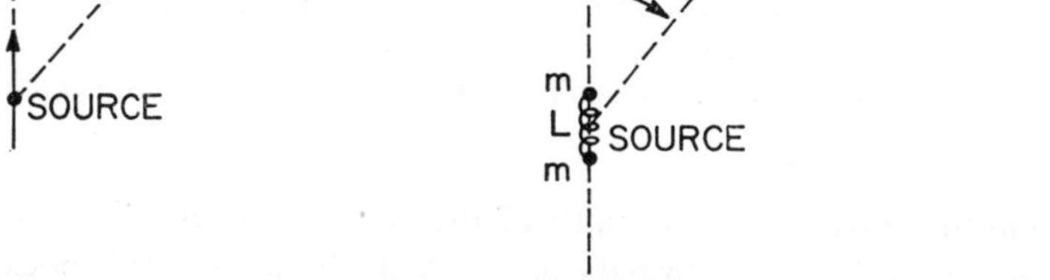

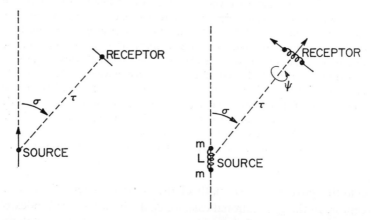

Figure 24 Orientation of receptor most favourable for the detection of waves from indicated source: left, electromagnetic radiation; right, gravitational radiation. In both cases the distance r and the polar angle θ of the receptor are considered to be prescribed, and only the orientation of the receptor is adjustable.

electric field, is perpendicular to it and both lie in the plane perpendicular to the direction of propagation. Similarly, the perturbation in the geometry is described by a transverse traceless tensor. Let the z-axis be pointed along the line of propagation, let the principal axes of polarization point in the x- and y-directions (right-hand column of Figure 25, upper two illustrations). Then the only non-zero Newtonian tide-producing components (Table XI) are

$$R^{y}_{oyo} = -R^{x}_{oxo} = (2/3)\,(\ddddot{Q}_{yy} - \ddddot{Q}_{xx}) \sin^2\theta/r$$

Table XI Magnitudes of potentials and fields for electromagnetic and gravitational radiation, compared and contrasted

	Electromagnetism	Gravitation
Potential	$A = e\beta_{\perp}/r \quad = e\beta \sin\theta/r$	$\delta g \sim \ddot{Q} \sin^2\theta/r$
		$\delta\Gamma \sim \dddot{Q} \sin^2\theta/r$
Field	$E = -ea_{\perp}/r = -ea \sin\theta/r$	$\delta R \sim \ddddot{Q} \sin^2\theta/r$
Meaning of field component	Magnitude of electric vector, this vector perpendicular to direction of propagation	Magnitude of tensor of Newtonian tide-producing force, this tide tensor transverse to direction of propagation and also traceless (zero net volume change)

in general the mass quadrupole moment is a traceless Cartesian tensor with the typical component

$$Q^{mn} = \iiint \varrho(3x^m x^n - \delta^{mn} x^s x_s)\, d^3x$$

For two localised masses, m, coupled by a spring of length L parallel to the y-axis, as in Figure 24, we have

$$Q_{xx} = -\int y^2 d\,(\text{mass}) = -I$$

$$Q_{yy} = 2\int y^2 d\,(\text{mass}) = 2I$$

$$Q_{zz} = -\int y^2 d\,(\text{mass}) = -I$$

where I is the moment of inertia,

$$I = 2m(L/2)^2$$

The receptor (Figure 24) operates best when oriented with the separation of the two test particles perpendicular to the line of sight and in the plane

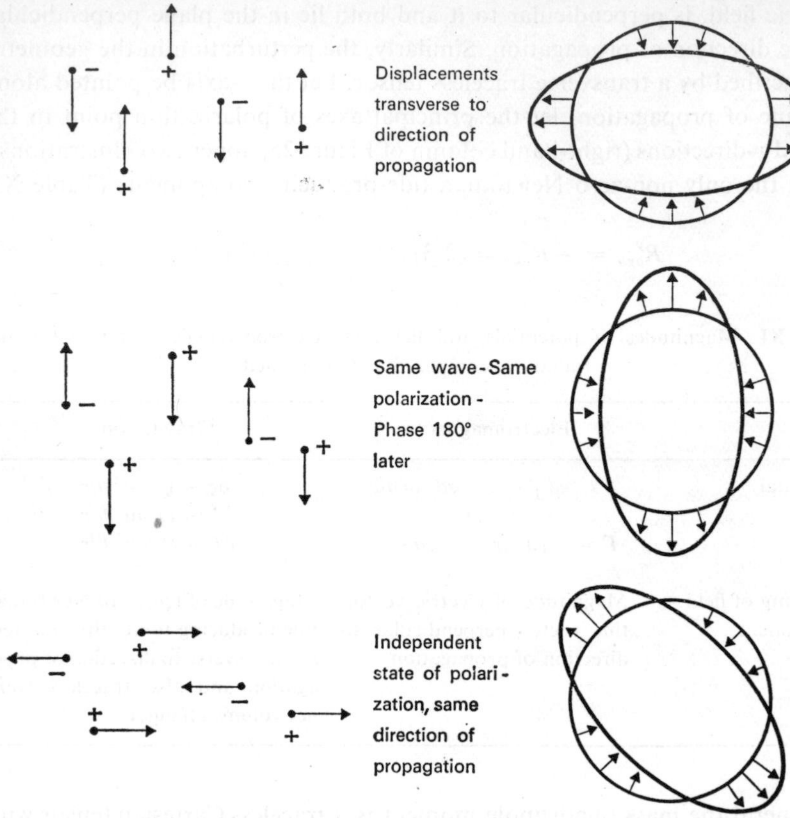

Figure 25 Polarization of electromagnetic waves and gravitational waves compared and contrasted. A gravitational wave travelling perpendicular to the plane of the paper and acting on a group of test particles spaced about the rim of a circle changes their separation as indicated by the ellipses. The two independent states of polarization are separated by 45 degrees, not by 90 degrees as for electromagnetic waves.

that contains the line of sight and the principal axis of vibration of the source. However, it gives an equally good response when rotated about the line of sight by the angle $\psi = \pm 90^\circ$. There is only one important difference between the signal received in this orientation and the signal received in the original orientation: a phase lag of 180°. (To produce a phase lag of 180° requires, one will recall, a rotation of 90° in the case of a transverse traceless tensor (spin 2); a rotation of 180° in the case of a transverse vector (spin 1); and a rotation of 360° for a field of spin $\frac{1}{2}$).

For any angle of rotation ψ around the line of sight intermediate between zero and 90° the magnitude of the tide-producing field that drives the detector is given by the standard law of tensor transformation,

$$R^{\bar{y}}_{o\bar{y}o} = \cos\psi\, R^{y}_{oyo} \cos\psi + \cos\psi\, R^{y}_{oxo} \sin\psi + \sin\psi\, R^{x}_{oyo} \cos\psi + \sin\psi\, R^{x}_{oxo} \sin\psi$$

$$= R^{y}_{oyo}(\cos^2\psi - \sin^2\psi) = (2/3)\,(\dddot{Q}_{yy} - \dddot{Q}_{xx})\,(\sin^2\theta/r)\cos 2\psi$$

We now turn to Figure 26 (note contrast with Figure 24; the vibration is now along the x-axis) and consider a receptor on a far-away star. We seek the magnitude of the signal or tide-producing acceleration produced at this star by an elementary time-changing quadrupole moment located at the point R on the surface of the Earth and oriented as shown in Figure 26. The transformation of the tensorial components is a little more complicated than that just considered. We carry it out in two steps. We note that Q_{ij} transforms like the product $V_i W_j$ of two vectors, the transformation law of which is a matter of elementary vector analysis. The steps in this transformation are listed here:

Laboratory Frame:

$Q_{\bar{\bar{x}}\bar{\bar{x}}} = 2I, \quad Q_{\bar{\bar{y}}\bar{\bar{y}}} = Q_{\bar{\bar{z}}\bar{\bar{z}}} = -I,$ all other components zero

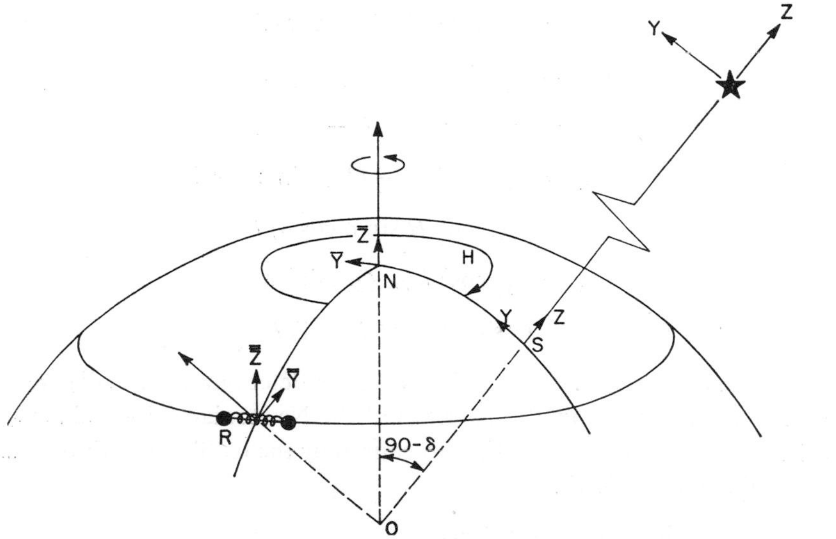

Figure 26 An idealized detector of gravitational waves, R, on the surface of the Earth, is driven by a source on a far-away star. The coupling between source and receptor is analysed in the text by using the principle of reciprocity (source at R, detector at star) and transforming tensorial components in turn from the laboratory frame (double barred coordinates) to a frame at the North Pole (barred coordinates) and then to a frame at S.

Turning Frame at North Pole:

$$Q_{\bar{x}\bar{x}} = Q_{\bar{\bar{x}}\bar{\bar{x}}} \cos^2 H + Q_{\bar{\bar{y}}\bar{\bar{y}}} \sin^2 H$$
$$= (3 \cos^2 H - 1)\, I$$
$$Q_{\bar{y}\bar{y}} = Q_{\bar{\bar{x}}\bar{\bar{x}}} \sin^2 \mathrm{H} + Q_{\bar{\bar{y}}\bar{\bar{y}}} \cos^2 H$$
$$= (2 - 3 \cos^2 H)\, I$$
$$Q_{\bar{x}\bar{y}} = Q_{\bar{\bar{x}}\bar{\bar{x}}} \sin H \cos H - Q_{\bar{\bar{y}}\bar{\bar{y}}} \sin H \cos H$$
$$= (3 \sin H \cos H)\, I$$
$$Q_{\bar{z}\bar{z}} = Q_{\bar{\bar{z}}\bar{\bar{z}}} = -I, \text{ all other components zero.}$$

Here H is the hour angle of the source (cf. Figure 26).

Frame at S:

$$Q_{xx} = Q_{\bar{x}\bar{x}} = (3 \cos^2 H - 1)\, I$$
$$Q_{yx} = Q_{xy} = Q_{\bar{x}\bar{y}} \sin \delta + Q_{\bar{x}\bar{z}} \cos \delta$$
$$= 3\, (\sin \delta \sin H \cos H)\, I$$
$$Q_{zx} = Q_{xz} = -Q_{\bar{x}\bar{y}} \cos \delta + Q_{\bar{x}\bar{z}} \sin \delta$$
$$= -3(\cos \delta \sin H \cos H)\, I$$
$$Q_{yy} = Q_{\bar{y}\bar{y}} \sin^2 \delta + Q_{\bar{z}\bar{z}} \cos^2 \delta$$
$$= (3 \sin^2 H \sin^2 \delta - 1)\, I$$
$$Q_{yz} = Q_{zy} = (Q_{\bar{z}\bar{z}} - Q_{\bar{y}\bar{y}}) \sin \delta \cos \delta$$
$$= -(3 \sin^2 H \sin \delta \cos \delta)\, I$$
$$Q_{zz} = Q_{\bar{y}\bar{y}} \cos^2 \delta + Q_{\bar{z}\bar{z}} \sin^2 \delta$$
$$= (3 \sin^2 H \cos^2 \delta - 1)\, I$$

The transverse traceless part of this tensor determines the radiative part of the perturbations in the metric, and the Newtonian tide-producing acceleration at the location of the star, caused by the oscillating pair of masses oscillating on the Earth:

$$\delta g_{xy} \equiv h_{xy} = (2/3r)\, \ddot{Q}_{xy}$$
$$\delta g_{xx} \equiv h_{xx} = -\, \delta g_{yy} \equiv -h_{yy} = (1/3r)\, (\ddot{Q}_{xx} - \ddot{Q}_{yy})$$

and

$$R_{\alpha\beta\gamma\delta} = \tfrac{1}{2}(\partial^2 h_{\alpha\delta}/\partial x^\beta \partial x^\gamma + \partial^2 h_{\beta\gamma}/\partial x^\alpha\, \partial x^\delta$$
$$-\partial^2 h_{\beta\delta}/\partial x^\alpha \partial x^\gamma - \partial^2 h_{\alpha\gamma}/\partial x^\beta\, \partial x^\delta)$$

$$
\begin{aligned}
R^x_{oxo} &= -\tfrac{1}{2}\partial^2 h_{xx}/\partial t^2 \\
&= -(1/6r)(\dddot{Q}_{xx} - \dddot{Q}_{yy}) = -R^y_{oyo} \\
&= (\sin^2 H \sin^2 \delta - \cos^2 H)(\dddot{I}/2r) \\
R^x_{oyo} &= R^y_{oxo} = -\tfrac{1}{2}\partial^2 h_{xy}/\partial t^2 = -(1/3r)\dddot{Q}_{xy} \\
&= -\sin\delta \sin 2H(\dddot{I}/2r)
\end{aligned}
$$

The flow of effective gravitational wave energy past the star (in the direction of increasing z) is given by the formula[124]†

$$
\begin{aligned}
(\text{energy flux}) &= (1/16\pi)[\dot{h}_{xy}^2 + (\tfrac{1}{2}\dot{h}_{xx} - \tfrac{1}{2}\dot{h}_{yy})^2] \\
&= (1/36\pi r^2)[\dddot{Q}_{xy}^2 + (\tfrac{1}{2}\dddot{Q}_{xx} - \tfrac{1}{2}\dddot{Q}_{yy})^2] \\
&= (\dddot{I}^2/16\pi r^2)[(\sin\delta \sin 2H)^2 + (\cos^2 H - \sin^2\delta \sin^2 H)^2]
\end{aligned}
$$

The total output of gravitational wave energy from the vibrator is

$$
\begin{aligned}
-dE/dt &= \int (\text{energy flux})\, r^2 \cos\delta\, d\delta\, dH \\
&= (2/15)\dddot{I}^2
\end{aligned}
$$

in geometrical units (where 1 cm of mass energy is

$1/(0.742 \times 10^{-28}\ \text{cm/g})$ of mass or $(9 \times 10^{20}\ \text{erg/g})/(0.742 \times 10^{-28}\text{cm/g})$

of energy, and where 1 cm of time is 1 cm/(3×10^{10} cm/sec)) or, in conventional units,

$$
-dE_{\text{conv}}/dt_{\text{conv}} = (2G/15c^5)(d^3 I_{\text{conv}}/dt^3_{\text{conv}})^2
$$

being identical with the more general result of Landau and Lifschitz[124],

$$
-dE_{\text{conv}}/dt_{\text{conv}} = (G/45c^5)\sum_{mn}(d^3 Q_{mn,\text{conv}}/dt^3_{\text{conv}})^2
$$

Now put the source at the star and let the oscillator on the Earth serve as the receptor. Evaluate its response from the principle of reciprocity. This principle connects the response of receptor B to source A with the response of A, employed as a receptor, to B serving as a source. We use this principle to study how a detector located on the Earth changes its response during the day to a gravitational wave coming from a star located at declination δ and hour angle H. For simplicity we will consider the case where the detector can be idealised as two point masses separated by a line

† (The coefficient in Eqs. (11-99) of Landau and Lifschitz should be changed from $(1/32\pi)$ to $(1/16\pi)$—the correct value being given in the French edition.)

perpendicular to the polar axis of the Earth and located at hour angle H (Figure 26). We look for the magnitude of the tide-producing acceleration R ((cm/sec^2)/cm in conventional units; cm^{-2} in the purely geometrical units employed throughout this analysis) that drives the receptor. It responds to the two independent components of polarization in the source at the distant star; thus

$$R^{\bar{\bar{x}}}_{o\bar{\bar{x}}o} \text{ (receptor on Earth)}$$

$$= (\dddot{Q}_{xx} - \dddot{Q}_{yy})_{\text{star}} (\sin^2 \delta \sin^2 H - \cos^2 H)/6r - (\dddot{Q}_{xy})_{\text{star}} \sin \delta \sin 2H/6r$$

An elementary oscillator oriented along the $\bar{\bar{y}}$-axis gives nothing new (mere change from H to $H \pm 90°$ to get $R^{\bar{\bar{y}}}_{o\bar{\bar{y}}o}$ from $R^{\bar{\bar{x}}}_{o\bar{\bar{x}}o}$!). When the detector is oriented along the $\bar{\bar{z}}$-axis (direction of polar axis!) the response will be independent of the rotation of the Earth:

$$R^{\bar{\bar{z}}}_{o\bar{\bar{z}}o} = (\dddot{Q}_{xx} - \dddot{Q}_{yy})_{\text{star}} \cos^2 \delta/6r$$

The flux of gravitational energy arriving at the Earth from the star, with the right polarization to be detected by an oscillator aligned along the $\bar{\bar{x}}$-axis, as indicated in Figure 26, will be

(flux of polarized gravitational wave energy at Earth)

$$= (1/144\pi r^2) [(\dddot{Q}_{xy})^2_{\text{star}} (\sin \delta \sin 2H)^2 + (\dddot{Q}_{xx} - \dddot{Q}_{yy})^2_{\text{star}} (\sin^2 \delta \sin^2 H - \cos^2 H)^2].$$

Averaged over sources with random orientation of polarization, located at hour angle H and declination δ, the flux is proportional to the "response factor" W (Figure 27 and Table XII)

$$W(H, \delta) = (\cos^2 H - \sin^2 \delta \sin^2 H)^2 + (\sin \delta \sin 2H)^2$$

Table XII Selected values of response factor $W(H, \delta)$ for detector oriented as in Figure 26

Declination, δ, of source	W_{peak} ($H = 0, \pm 180°$)	W_{min} ($H = \pm 90°, \pm 270°$)	W' (all H)
90° (polar axis)	1	1	0
60°	1	0.56	0.06
45°	1	0.25	0.25
30°	1	0.06	0.56
0° (equator)	1	0	1

Source, taken to have random polarization, is located at declination δ and hour angle H. Also given is the response factor $W'(\delta)$ for a detector arranged to oscillate parallel to the polar axis.

For an elementary detector oriented parallel to the polar axis the corresponding result is

$$W'(\delta) = \cos^4 \delta$$

(To normalise W or W' over the unit sphere, multiply by $15/32\pi$.)

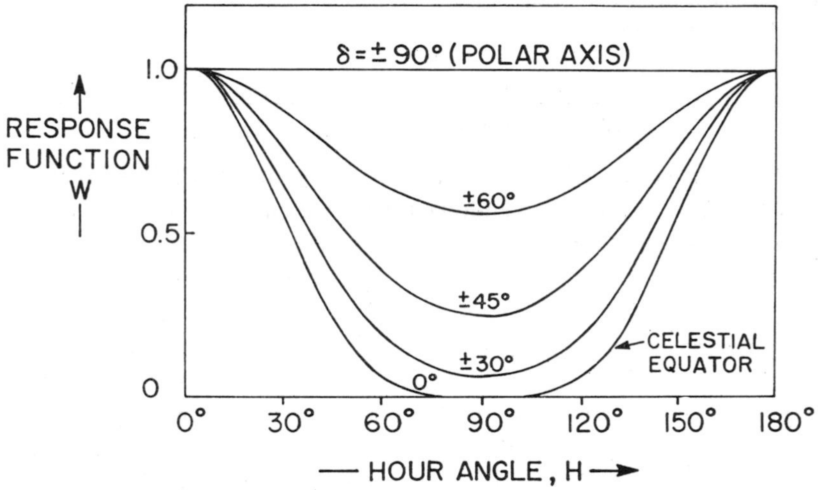

Figure 27 Response of a detector of gravitational radiation, aligned East-West as indicated in Figure 26, to sources of gravitational radiation of random polarization located at declination δ and hour angle H. The differential variation during the day will be greatest if the sources are located at the celestial equator ($\delta = 0$). The declination of the center of the Milky Way is $\delta = -29°$.

7.2 DETECTOR OF GRAVITATIONAL RADIATION

Considerations familiar in the absorption of light by atoms, in antenna design and in the physics of neutron capture, also govern the pick-up of gravitational radiation by an oscillator such as that illustrated in Figure 24 or Figure 26. In each of these instances it is appropriate to speak of an oscillator characterised by a natural or resonant circular frequency ω_0. This oscillator has a rate of damping

$$A_{\text{grav}} = \langle(-dE/dt)_{\text{grav}}\rangle_{\text{av}}/E$$

caused by radiation of gravitational waves, and a rate of damping caused by mechanical friction, electrical resistance, signal read-out, and all modes of dissipation other than gravitational radiation,

$$A_{\text{diss}} = \langle(-dE/dt_{\text{diss}})\rangle_{\text{av}}/E$$

The quantity A_{diss} is larger than A_{grav} by so many orders of magnitude in any detector so far imagined that it is reasonable to identify the total damping

$$A = A_{diss} + A_{grav}$$

with A_{diss} itself.

The cross-section for radiative capture of a neutron is described by the Breit-Wigner formula[125]

$$\sigma(\omega) = \pi \lambda\!\!\!^{-2} \frac{(2J+1)}{(2s+1)(2I+1)} \frac{A_{neut}A_{rad}}{(\omega - \omega_0)^2 + (\frac{1}{2}A_{neut} + \frac{1}{2}A_{rad})^2}$$

Here $\lambda\!\!\!^{-} = \lambda/2\pi$ is the reduced wavelength of the neutron (classical impact parameter $\hbar/mv$ for a neutron endowed with one quantum unit $\hbar$ of angular momentum), $\hbar\omega_0$ is the resonance energy of the compound nucleus, $\hbar\omega$ is the rest-plus-kinetic energy of the neutron plus the rest-plus-kinetic energy of the target nucleus in the center of mass frame of reference, and A_{neut} and A_{rad} ("line broadening", units sec^{-1}) are the decay rates of the resonance state of the compound nucleus due respectively to neutron re-emission and to gamma radiation. I is the spin of the target nucleus (random orientation in space!), $s = \frac{1}{2}$ is the spin of the neutron (unpolarized), and J is the spin of the resonance level of the compound nucleus. The cross-section integrated over resonance has the value

$$\text{"resonance integral"} = \int \sigma(\nu)\, d\nu = (1/2\pi) \int \sigma(\omega)\, d\omega$$

$$= \pi \lambda\!\!\!^{-2} \frac{(2J+1)}{(2s+1)(2I+1)} \frac{A_{neut}A_{rad}}{A_{neut} + A_{rad}}$$

When the neutron re-emission rate (sec^{-1}) is very low compared to the rate of gamma-ray emission, as is the case for slow neutron resonances, the last fraction in the expression for the resonance integral reduces to A_{neut}, independent of the precise broadening of the line by gamma-ray decay processes.

A similar formula applies to the absorption of light by an atom, except that here the fraction in question becomes

$$\frac{A_1(A_2 + A_3 + \cdots)}{A_1 + A_2 + A_3 + \cdots}$$

where A_1 is the rate of decay of the excited state by re-emission of a quantum of the original energy and A_2, A_3, ... are the Einstein rates for radiative transitions from the excited state to other lower states. Also, the statistical factor $(2s + 1)$ retains, even for light, with its spin 1, the value 2 corresponding to two independent states of polarization. For gravitational radiation the statistical factor $(2s + 1)$ again has the value 2 (two independent states of polarization) despite the fact that gravitons have spin 2. The cross-section near resonance has the value

$$\sigma = \pi \lambda\!\!\!^{-2} \frac{(2J+1)}{2(2I+1)} \frac{A_{\text{grav}} A}{(\omega - \omega_0)^2 + (A/2)^2}$$

with the

$$\text{"resonance integral"} = \int \sigma \, d\nu = \pi \lambda\!\!\!^{-2} \frac{2J+1}{2(2I+1)} A_{\text{grav}}$$

independent of damping (provided of course that the total damping is strong compared to the damping by gravitational radiation, a requirement all too easy to fulfil; and provided also that it is not so strong as to wash out the resonance (much less than critical damping!)).

The foregoing formulae for cross-section and resonance integral are appropriate for a quantum system that has random orientation and is responding to unpolarized gravitational radiation. For a hydrogen molecule "detector" it would be perfectly possible to consider in detail the individual vibration and vibration-rotation transitions and evaluate for each the relevant transition probabilities. However, no one has so far been able to imagine any source of the appropriate frequency with an intensity great enough to make such a molecular detector a device appropriate for consideration. Much more relevant is a system of macroscopic dimensions (quadrupole vibrations of Weber's aluminium bar, of the Earth and of the Moon). For such a system the orientation is well defined and the statistical factor $(2J + 1)/(2I + 1)$ is inappropriate. It is convenient to turn to the principle of detailed balance to evaluate the resonance integral.

In thermal equilibrium the detector will lose energy by radiation as rapidly as it gains energy by absorption at resonance:

$$\begin{pmatrix}\text{gravitational wave}\\ \text{energy absorbed per sec}\end{pmatrix} = \begin{pmatrix}\text{gravitational wave}\\ \text{energy radiated per sec}\end{pmatrix}$$

In detail, we have the equality

$$\begin{pmatrix}\text{energy incident per cm}^2\\ \text{and per sec and per Hz}\\ \text{at resonance frequency}\end{pmatrix} \begin{pmatrix}\text{"resonance integral",}\\ \int_{\text{res}} \sigma(\nu)\, d\nu\end{pmatrix}$$

$$= \begin{pmatrix}\text{energy of}\\ \text{oscillator}\end{pmatrix} \begin{pmatrix}\text{fraction of energy radiated}\\ \text{per sec in form of gravita-}\\ \text{tional waves}\end{pmatrix}$$

We evaluate the gravitational wave energy incident in thermal equilibrium from the Planck black-body spectrum (same considerations relevant as for electromagnetic radiation; again two independent states of polarization). Thus we have

$$c2(4\pi\nu^2/c^3)\, h\nu(e^{h\nu/kT} - 1)^{-1} \int_{\text{res}} \sigma(\nu)\, d\nu = A_{\text{grav}} h\nu(e^{h\nu/kT} - 1)^{-1}$$

or

$$\int_{\text{res}} \sigma(\nu)\, d\nu = (\pi/2)\, \lambda\!\!\!^{-2} A_{\text{grav}},$$

identical with what would follow from the Breit-Wigner formula on simply wiping out the factor $(2J+1)/(2I+1)$. This analysis refers to the case of radiation of random polarization incident from all directions. The response is greater when the radiation comes from a direction perpendicular to the line of the one-dimensional oscillator, and still greater when the radiation is polarized along one of the principal axes of dilatation. Specifically, taking $A_{\text{grav}} = (8/15)\, I\omega^4$ from the analysis of the oscillator (see later), we have

$$\int_{\text{res}} \sigma(\nu)\, d\nu = \begin{cases} (4\pi/15)\,(I/\lambda\!\!\!^{-2}) & \text{one-dimensional oscillator, random direction, random polarization} \\ (\pi/2)\sin^4\theta\,(I/\lambda\!\!\!^{-2}) & \text{one-dimensional oscillator, radiation incident at angle } \theta \text{ to line of oscillation, random polarization} \\ \pi(I/\lambda\!\!\!^{-2}) & \text{one-dimensional oscillator, radiation incident perpendicularly, most favourable polarization} \end{cases}$$

All these expressions are in geometrical units (I in cm of mass times cm^2 of distance; $I/\lambda\!\!\!^{-2}$ in cm; ν^{-1} in cm of light travel time). To use the three equations with conventional units for I and ν, insert on the right-hand side the factor $G/c = (0.742 \times 10^{-28}\ \text{cm/g})\,(3 \times 10^{10}\ \text{cm/sec}) = 2.22 \times 10^{-18}\ \text{cm}^2$ Hz/g. It will be noted that the energy picked up by the pair-of-masses-coupled-by-a-spring is proportional to the magnitude of those masses, provided that the unperturbed separation of the masses is kept fixed and provided that the spring is adjusted so as to keep the frequency unaltered. Thus the energy pick-up per unit of mass, and the amplitude of the resulting vibratory motion, is independent in this sense of the mass itself—a reflection of Einstein's principle of equivalence.

It may be of interest to compare the resonance integrals for gravitational radiation with the corresponding expressions for electromagnetic radiation (oscillator of charge e and mass m; c.g.s. units):

$$\int_{\text{res}} \sigma(\nu)\, d\nu = \begin{cases} \pi e^2/mc & \text{isotropic three-dimensional oscillator (the result familiar from the Kuhn-Reiche-Thomas sum rule}^{126}\text{)} \\ \pi e^2/3mc & \text{one-dimensional oscillator, random direction of incidence, random polarization} \\ (\pi/2)\sin^2\theta(e^2/mc) & \text{one-dimensional oscillator, radiation incident at angle } \theta \text{ to line of vibration, random polarization} \\ \pi e^2/mc & \text{one-dimensional oscillator, radiation incident perpendicularly, electric vector parallel to line of vibration} \end{cases}$$

The resonance integral is lower in the case of gravitational radiation by reason of two factors: first, the coupling constant, Gm^2, is small compared to e^2; second, the square of the dimension of the oscillator, I/m, is smaller than the square of the reduced wavelength, $\lambda\!\!\!^{-2}$, by a factor which in the simplest design is of the order (velocity of sound/velocity of light)^2.

An alternative derivation for the cross-section[127] and the resonance integral proceeds directly from the equation of motion for the displacement of the one charge

$$d^2x/dt^2_{\text{conv}} + \omega^2_{\text{conv}}x = eF_x(t)/m_{\text{conv}}$$

or the separation $\eta_x = L + \Delta L$ of the two masses,

$$c^{-2}(d^2\,\Delta L/dt^2_{\text{conv}} + \omega^2_{\text{conv}}\,\Delta L) = R^x_{oxo}(t)\,L_0$$

and gives the same results.

7.3 BAR AS MULTIPLE-MODE DETECTOR

Some interest attaches to a long bar as a detector of gravitational radiation endowed with a spectrum of characteristic modes. As pointed out to us by William Fairbank, it can sample the intensity of incoming gravitational radiation at several frequencies. We consider only the modes of longitudinal vibration. They have circular frequencies $\omega = vk = n\pi v/L$, where v is the speed of sound and L is the length of the bar (conventional units). The even-numbered modes give rise to no change in the quadrupole moment, so it is enough to consider the odd numbered ones, $n = 1, 3, 5, \ldots$ The displacement of the material at point x (x from—$L/2$ to $L/2$) from its normal equilibrium position is

$$\xi = \xi_0 \sin(n\pi x/L) \sin(\omega t)$$

This displacement changes the moment of inertia from $\int x^2 d$ (mass) to $\int (x + \xi)^2 d$ (mass); or to a first order in the displacement,

$$I = I_0 + (4M/Lk^2)\, \xi_0 \sin \omega t$$

The rate of emission of gravitational radiation is

$$\begin{aligned} \langle -dE/dt \rangle_{\text{av}} &= (2G/15c^5) \langle (d^3 I_{\text{conv}}/dt^3_{\text{conv}})^2 \rangle_{\text{av}} \\ &= (16/15)(GM^2\omega^2 v^4 \xi_0^2/c^5L^2) \\ &= (16\pi^2/15)(GM^2v^6/c^5L^4)\, \xi_0^2 n^2 \end{aligned}$$

The vibrational energy of the bar is

$$E = M\omega^2\xi_0^2/4$$

The gravitational damping coefficient

$$A_{\text{grav}} = \langle -dE/dt \rangle_{\text{av}}/E = (64/15)(G/c^5)(Mv^4/L^2)$$

is the same for all the odd-numbered modes of vibration. The resonance cross-section of any one of these modes for interception of gravitational radiation coming from random directions with random polarizations is

$$\int_{\text{res, random}} \sigma(\nu)\, d\nu = (\pi/2)\, \lambda\!\!\!^{-2} A_{\text{grav}} = (32/15\pi)(G/c)(v^2/c^2)(M/n^2)$$

with $n = 1, 3, 5, \ldots$ (falls as $1/n^2$ because proportional to $\lambda\!\!\!^{-2}$). As an example, consider the bar of Weber[112] with a mass of 1.4×10^6 g, composed of aluminium ($v = 6.42 \times 10^5$ cm/sec, $\beta = 2.14 \times 10^{-5}$) and operating in its lowest mode, $n = 1$:

$$\begin{aligned} \int_{\text{res, random}} \sigma(\nu)\, d\nu &= 0.679 \times (2.22 \times 10^{-18}\,\text{cm}^2\,\text{Hz/g}) \\ &\quad \times (1.4 \times 10^6\,\text{g}) \times (2.14 \times 10^{-5})^2 \\ &= 1.0 \times 10^{-21}\,\text{cm}^2\,\text{Hz} \end{aligned}$$

7.4 EARTH VIBRATIONS AS DETECTORS OF GRAVITATIONAL WAVES

Before considering the Earth itself or its mode of quadrupole oscillation as a detector of gravitational radiation, let us consider a globe of fluid of uniform density held in the shape of a sphere by gravitational forces alone (zero rigidity). Let the surface be displaced from $r = a$ to

$$r = a + a\alpha P_2(\cos\theta)$$

where θ is polar angle measured from the North Pole and α is the fractional elongation of the principal axis. The motion of lowest energy compatible with this change of shape is described by the velocity field

$$v_x = -\tfrac{1}{2}\dot{\alpha}x, \quad v_y = -\tfrac{1}{2}\dot{\alpha}y, \quad v_z = \dot{\alpha}z$$

(zero divergence, zero curl). The sum of the kinetic energy and the gravitational potential energy is

$$E = -(3/5)(GM^2/a)(1 - \alpha^2/5) + (3/20)\,Ma^2\dot{\alpha}^2$$

The circular frequency of free quadrupole vibration is

$$\omega = (16\pi/15)^{\frac{1}{2}}\,(G\varrho)^{\frac{1}{2}}$$

The quadrupole moments are

$$Q_{xx} = -I, \quad Q_{yy} = -I, \quad Q_{zz} = 2I$$

where I is an abbreviation for the quantity $I = (3/5)\,Ma^2\alpha$.
The rate of emission of gravitational energy, averaged over a period, is

$$\langle -dE/dt\rangle = (2G/15c^5)\,\tfrac{1}{2}(3Ma^2\dddot{\alpha}_{\text{peak}}/5)^2 = (3/125)(G/c^5)\,M^2a^4\omega^6\alpha^2_{\text{peak}}$$

whereas the energy available for emission is

$$E = (3/20)\,Ma^2\omega^2\alpha^2_{\text{peak}}$$

The ratio of these two quantities gives for the exponential rate of decay of energy by reason of gravitational wave damping, or "gravitational radiation line broadening",

$$A_{\text{grav}} = (4/25)(G/c^5)\,Ma^2\omega^4$$

The resonance integral of the absorption cross-section for radiation incident from random directions with random polarization is

$$\int_{\text{res, random}} \sigma(\nu)\,d\nu = (\pi/2)\,\lambda\!\!\!^{-2}A_{\text{grav}} = (2\pi/25)(G/c)\,Ma^2/\lambda\!\!\!^{-2}$$

This model of a globe of fluid of uniform density would imply for the Earth, with average density 5.517 g/cm^3, a quadrupole vibration period of 94 min, compared to the observed 54 min; and a moment of inertia $(2/5)Ma^2$ compared to the observed $0.33Ma^2$. Accordingly, in the expression for the resonance integral, we lower the inertial factor Ma^2 by the ratio 0.33/0.40 = 0.82, and insert for $\lambda\!\!\!^{-}$ the correct value for a 54-min gravitational wave,

$\lambda\!\!\!^{-} = 1.55 \times 10^{13}$ cm. In this way we arrive at the following estimate, that should be good to a factor of the order of two, for the resonance integral for the Earth,

$$\int_{\text{res, random}} \sigma(\nu)\, d\nu \simeq 0.251 \times (2.22 \times 10^{-18}\ \text{cm}^2\ \text{Hz/g}) \times (5.98 \times 10^{27}\ \text{g})$$
$$\times\, 0.82(6.37 \times 10^{8}\,\text{cm}/1.55 \times 10^{13}\,\text{cm})^2 \simeq 4.7\,\text{cm}^2\,\text{Hz}$$

A more precise evaluation would call for a detailed treatment of the quadrupole mode of vibration of the Earth, such as given for example by Pekeris[128], with allowance for the elasticity and distribution of density within the Earth.

7.5 SEISMIC RESPONSE OF EARTH TO GRAVITATIONAL RADIATION IN THE ONE-HERTZ REGION

Dyson has analysed the response of an elastic solid to an incident gravitational wave[129]. He shows that the response depends on irregularities in the elastic modulus for shear waves, and that it is strongest at a free surface. For the fraction of gravitational wave energy crossing a flat surface converted into energy of elastic motion of the solid he finds the expression

$$(\text{fraction}) = (8\pi G\varrho/\omega^2)\,(s/c)^3$$
$$\sin^2\theta|\cos\theta|^{-1}[1 + \cos^2\theta + (s/v)\sin^2\theta]$$

Here s and v are the velocities of shear waves and compressive waves and θ is the angle between the direction of propagation of the gravitational waves and the surface. Considering a flux of 2×10^{-5} erg/cm^2 sec incident horizontally ($\theta = \pi/2$; "divergent" factor $|\cos\theta|$ cancels out in calculation!) and taking s to be 4.5×10^5 cm/sec and ω to be 6 rad/sec he calculates that the 1 Hz horizontal displacement produced in the surface has an amplitude of

$$A \sim 2 \times 10^{-17}\ \text{cm}$$

which is too small by a factor of the order of 10^5 to be detected against background seismic noise. He points to the possibilities of improvements, especially via resonance (elastic waves reflected back and forth between two surfaces; Antarctic ice sheet).

7.6 CHANGES IN SOLAR SYSTEM DISTANCES NOT ADEQUATE AS DETECTORS OF GRAVITATIONAL WAVES

The distance from a laser on the Earth to a corner reflector on the Moon is as subject to change as the length of a bar in the laboratory under the influence of a passing gravitational wave. Likewise the distance from the Earth to a transponder in orbit around a planet or around the Sun varies with time. The length of Weber's 153-cm bar changes by about 3×10^{-14} cm when the principal mode receives an excitation of $kT = k\,300°K$, implying a fractional change in dimensions of 2×10^{-16}. A similar change in the Earth-Moon distance, 3.84×10^{10} cm, would amount to 8×10^{-6} cm. In contrast, laser technology is only gradually working up to the point of measuring this distance with a 10 cm accuracy (cf. chapter 10 on lunar motion). Therefore direct distance ranging today leaves much to be desired as a technique for searching for the effects of gravitational waves.

To be quantitative, denote by L the normal distance, a slowly varying function of time; and by ξ the small rapidly varying changes in this length; and by R^1_{010} the relevant component of the Riemann curvature tensor,

$$R^1_{010} = L^{-1}\, d^2\xi/dt^2 = L^{-1}c^{-2}\, d^2\xi/dt^2_{\mathrm{conv}}$$

Denote by $\lambda\!\!\!^{-} = \lambda/2\pi = 1/\omega = c/\omega_{\mathrm{conv}}$ the reduced wavelength of the gravitational radiation. For waves with $\lambda\!\!\!^{-}$ shorter than L, there will be a cancellation of plus and minus effects on the path from Earth to space station. It will be more appropriate to look for such waves (periods of the order of a second or less) by other means (seismometers, bars, etc.). Therefore we assume $\lambda\!\!\!^{-} \geqq L$, and write the minimum detectable curvature in the form

$$R^1_{010} \sim \Delta\xi/L\lambda\!\!\!^{-2}$$

where $\Delta\xi$ is the uncertainty in the distance measurement. The minimum detectable energy flow will be of the order

$$(\text{energy flow}) = (\lambda\!\!\!^{-2}/4\pi)\,\langle (R^1_{010})^2 \rangle \sim (1/4\pi)\,(\Delta\xi/L\lambda\!\!\!^{-})^2$$

With $\Delta\xi = 10$ cm and $\lambda\!\!\!^{-} = L = 3.8 \times 10^{10}$ cm we have for the minimum detectable energy flow 3.7×10^{-42} cm of energy per cm^2 of area and per cm of light travel time in geometrical units; or in conventional units

$$1.4 \times 10^{18}\ \mathrm{erg/cm^2\ sec}$$

a useless limit on the strength of incoming gravitational waves! Conversely, were there as much as 10^{-29} g/cm^3 of gravitational wave-energy in space with the relevant wavelength, the flux would be

$$(10^{-29}\ \mathrm{g/cm^3})\,(3 \times 10^{10}\ \mathrm{cm/sec})^3 = 270\ \mathrm{erg/cm^2\ sec}$$

and the precision of determination of the Earth-Moon distance should be

$$14 \times 10^{-8}\ \mathrm{cm} = 14\ \text{Å}$$

to pick up the effect of the radiation—a preposterous proposition!

The difficulties of alternative detection techniques emphasise the advantages of Weber's vibrating bar detector.

7.7 SOURCES OF GRAVITATIONAL RADIATION

7.7.1 Spinning rod

A rod spinning about an axis perpendicular to its length is one of the first sources of gravitational radiation ever to have been considered. The outgoing wave front, sectioned in the plane of the rotation, has the form of a spiral or rather of two inter-nested spirals (two equivalent wave crests per 360° rotation of the rod). Let I denote the moment of inertia of the rod about the axis of spin (z-axis). Then the quadrupole moments are

$$Q_{xx} = (3\cos^2\omega t - 1)\, I$$

$$Q_{yy} = (3\sin^2\omega t - 1)\, I$$

$$Q_{xy} = Q_{yx} = 3\sin\omega t\cos\omega t I$$

$$Q_{zz} = -I$$

The rate of emission of radiation is

$$-dE/dt = (32/5)\,(G/c^5)\, I^2\omega^6$$

Consider a rod of steel of radius $r = 100$ cm and length $L = 2000$ cm, of density 7.8 g/cm^3, of mass $M = 4.9 \times 10^8$ g (490 tons) and of tensile strength $T = 40{,}000$ pounds per square inch or 3×10^9 dyne/cm^2. The upper limit to the rate of rotation set by breaking of the rod is given by the formula

$$L\omega = (8T/\varrho)^{\frac{1}{2}} = 5.5 \times 10^4 \text{ cm/sec}$$

or

$$\omega = 28 \text{ rad/sec}, \quad \nu = 4.4 \text{ revolutions/sec}$$

The moment of inertia is

$$I = ML^2/12 = 1.6 \times 10^{14} \text{ g cm}^2$$

The rate of radiation is

$$-dE/dt = 2.2 \times 10^{-22} \text{ erg/sec}$$

Despite the smallness of the output, classical considerations are still appropriate. One quantum has energy

$$E_{\text{quantum}} = h2\nu = (6.62 \times 10^{-27} \text{ erg sec})\,(8.8 \text{ waves/sec}) = 5.8 \times 10^{-26} \text{ erg}$$

and per second 3.8×10^3 gravitons are emitted (correspondence principle limit of many quanta).

For a given tensile strength and given density the rate of emission goes up with the size of the system. Therefore it is appropriate to turn next to the consideration of a spinning star.

7.7.2 Spinning star

A spinning star will not give off gravitational radiation if it merely has a tidal bulge (axial symmetry around axis of rotation; no time-changing quadrupole moment). However, if it has a substantial departure from axial symmetry it can radiate powerfully. For example, Ferrari and Ruffini[32] consider as a possible model for the Crab pulsar a neutron star with these properties:

central density	6.0×10^{15} g/cm^3
mean radius	9.75 km
mass	$0.786 M_\odot$
a = one principal axis of figure in equatorial plane	1.0004×9.75 km
b = other principal axis of figure in equatorial plane	0.9996×9.75 km
$\varepsilon = (a-b)/(ab)^{1/2}$	8×10^{-4}
period of rotation	33 msec
circular frequency	190 rad/sec
rate of emission of gravitational wave energy	$-\dfrac{dE}{dt} = \dfrac{288 G I^2 \varepsilon^2 \omega^6}{45 c^5} = 1.18 \times 10^{38}$ erg/sec

Such a rate of emission would be sufficient by itself to account for the observed time rate of change of the period of the Crab pulsar (Table IV)

$$4.2 \times 10^{-13} \text{ sec/sec}$$

Of course the actual rate of emission of gravitational radiation could hardly be so extreme, because the production of the observed pulses, and the requirement that energetic X-ray-emitting electrons be continuously supplied to the nebula, implies that a sizeable part of the rotational energy be converted into electromagnetic energy.

The detailed treatment of the shape of a rapidly rotating neutron star is a difficult problem, involving hydrodynamics, gravitational potential, equation of state, rigidity of superdense matter, other issues of solid state physics, and even some corrections for general relativity. However, some insight into this problem for stars of quite different types can be gained by considering the equilibrium configurations of a rotating fluid mass in Newtonian theory[130], with an equation of state of the form $p = K\varrho^\gamma$ ("gamma law equation of state"). To permit equilibrium even without rotation the equation of state must be "harder" than $\gamma = 1.333$. The other extreme limit on γ is $\gamma = \infty$ (incompressible fluid). For any given value of γ, K and the mass of the star there is a critical angular momentum beyond which the system breaks up:

(*a*) for γ from 1.333 to 2.2 by shedding matter at the equator ("spin-off from rim")

(*b*) for γ from 2.2 to ∞ by binary (or conceivably multiple) fission.

In (*a*) there is no reason to anticipate any departure from axial symmetry nor any gravitational radiation. In (*b*) the possibilities are most vividly illustrated by the classical theory of figures of equilibrium of a rotating fluid mass, as summarised in Table XIII. The important point is the extended range of angular momenta for which the equilibrium figure is not axially symmetric and for which powerful gravitational radiation is thus allowed and expected. Moreover this range has the interesting property that the angular velocity increases as the angular momentum decreases (flattened figure relapsing towards sphericity!).

Table XIII Properties of the equilibrium configuration of a rotating mass of ideal incompressible fluid for angular momenta up to the maximum momentum allowable before fission

	$\frac{J}{(GM^3R)^{1/2}}$	$\frac{a}{R}$	$\frac{b}{R}$	$\frac{c}{R}$	$\frac{\omega}{(\pi G\varrho)^{1/2}}$	$\frac{I}{(2/5)MR^2}$	ε	$\frac{I\varepsilon}{(2/5)MR^2}$
MacLaurin spheroids	0.02539	1.00167	1.00167	0.99666	0.07308	1.00336	—	—
	0.05144	1.00683	1.00683	0.98648	0.14649	1.01368	—	—
	0.07882	1.01584	1.01584	0.96905	0.22050	1.03194	—	—
	0.10846	1.02949	1.02949	0.94354	0.29541	1.05984	—	—
	0.14163	1.04912	1.04912	0.90856	0.37147	1.10064	—	—
	0.18037	1.07721	1.07721	0.86177	0.44872	1.16040	—	—
	0.22834	1.11876	1.11876	0.79896	0.52663	1.25163	—	—
	0.29345	1.18563	1.18563	0.71138	0.60263	1.40572	—	—
	0.30375	1.19723	1.19723	0.69766	0.61174	1.43337	—	—
Jacobi ellipsoids	0.30747	1.24865	1.14875	0.64138	0.60803	1.43938	0.08340	0.12006
	0.31296	1.37864	1.04777	0.52613	0.60259	1.49924	0.27530	0.41274
	0.32192	1.46294	0.99480	0.46725	0.59384	1.56491	0.38806	0.60727
	0.33562	1.56624	0.93975	0.40764	0.58080	1.66812	0.51640	0.86141
	0.35594	1.69637	0.88211	0.34750	0.56213	1.82789	0.66569	1.21672
	0.36947	1.77523	0.85211	0.31731	0.55013	1.93878	0.75055	1.45516
	0.38980	1.88564	0.81503	0.28124	0.53294	2.10996	0.86360	1.82216

The circumstance that pulsar rotation frequencies diminish with time suggests that the neutron stars so far observed are not now in the Jacobi ellipsoid regime. It also leads one to believe that they are in the MacLaurin spheroid regime, or rather its equivalent for a compressible fluid. It is conceivable that at the time of formation the Crab or Vela pulsar rotated fast enough to become ellipsoidal. In that event the rate of emission of energy by gravitational radiation would have been great enough to remove the neutron star early in its history from the ellipsoidal to the spheroidal regime. It is not easy to see how to disprove this possibility by observations on a fresh pulsar, but one could imagine proving it by observing the intense gravitational radiation from the object in that early phase of its existence. The characteristic "signature" of such radiation would be its drift to higher frequencies followed by a sharp drop in intensity.

Whether the intensity drops all the way to zero depends on whether the star forgets its past; on whether it goes to a perfect MacLaurin spheroid, or goes to a spheroid with some residue of ellipsoidal deformation left in its crust. That there will be a crust with a depth from some metres to some kilometres seems inescapable according to the most elementary considerations of solid state physics, as argued persuasively by Ruderman[33], Pines *et al.*[40] and Smoluchowski[132], even if a full understanding of the physics of a neutron star seems not yet to have been achieved.

Certain though it is that an ideal incompressible fluid rotating sufficiently rapidly will take on an ellipsoidal shape, it is also certain that a sufficiently compressible rotating fluid mass will never become ellipsoidal. The critical degree of "hardness" of the gamma law equation of state required for the occurrence of ellipsoidal rotating figures of equilibrium is $\gamma = 2.2$, according to Jeans[131]. In contrast to the idealised fluid considered by Jeans, the material of a neutron star has an effective

$$\gamma = \frac{p + \varrho}{p} \frac{dp}{d\varrho}$$

which varies strongly from point to point (cf. Figure 3) and also, at the very highest densities ($\varrho > 10^{13}$ g/cm^3), varies strongly from model to model.

To Table XIII

Adapted and extended from Tables I and IV in Chandrasekhar[130] and Tables XVI and XVII in Jeans.[131] Here *a*, *b*, and *c* are the axes of the ellipsoid, $R = (abc)^{1/3}$ and $\varepsilon = (a - b)/\sqrt{ab}$. The term $I\varepsilon/(2/5)\, MR^2$ is the relevant quantity to evaluate the power emitted in gravitational waves (see text).

Configurations of higher angular momentum are unstable against fission. Last decimal not reliable.

In the Harrison-Wheeler equations of state γ exceeds 2.2 only in the outermost part of the crust, where the density ϱ has fallen to 400 g/cm³ and less. The thickness Δz of this layer of material depends upon the surface value of the acceleration of gravity, according to the formula

$$\Delta z = (1/g_{\text{conv}}) \int_{p=0}^{\varrho=400\text{g/cm}^3} dp_{\text{conv}}/\varrho_{\text{conv}}$$

$$= (4.9 \times 10^{13} \text{ cm}^2/\text{sec}^2)/g_{\text{conv}}$$

Table XIV Thickness of the surface layer of neutron star material for which the effective parameter exceeds the Jeans value $\gamma = 2.2$ (p_{conv} from 0 to 2.2×10^{16} dynes/cm², ϱ_{conv} from 7.8 g/cm³ to 400 g/cm³)

ϱ_0 (g/cm³)	$M/M_\odot$	R (km)	g (cm/sec²)	Δz
3.2×10^{15}	0.67	10.0	8.8×10^{13}	0.56 cm
2.6×10^{13}	0.18	323.0	2.2×10^{10}	22 m

For the specified central densities the properties of the star calculated in the second, third and fourth columns depend upon the equation of state at high densities which here for definiteness has been taken to be the H-W equation. Moreover, no correction has been made for any deformation of shape and change of the surface value of the effective acceleration of gravity which results from rotation (cf. for example Hartle and Thorne[7]). However, once the surface value of g (cm/sec²) has been determined in this or any other way, the thickness of the surface layer depends only on the equation of state at densities of 400 g/cm³ and less, a region where one has a good combination of experimental and theoretical information about the equation of state. (Other equations of state for neutron stars differ in this region from the H-W equation of state, not by reason of any alternative theoretical approach, but only because of earlier lack of interest in any detailed treatment of this outermost part of a neutron star.)

The thickness of this layer is evaluated for two neutron star configurations in Table XIV. The outer part of the star, with $\gamma > 2.2$, evidently constitutes a negligible fraction of the total

$$(\Delta z/R < 10^{-4}),$$

so that "γ-law stiffness" in this region should have no significant effect on the equilibrium configuration. Therefore, if only the immediately underlying region had to be considered, and it were to be idealised as a soft fluid ($\gamma < 2.2$), the neutron star might spin off material from an equatorial rim, but it would still retain rotational symmetry. Does not the freezing of the initially very hot neutron star ($T > 10^{10}$ °K) rigidify the system to a much greater depth (some kilometres) than the thickness of the surface layer just considered? And is not that rigidity then much more significant than the "γ-law stiffness" of the system? More significant in helping the system hold its shape, yes. Significant in determining what that shape will be, no. The effective γ is the important factor in determining the shape of a system spinning at critical speed; that this γ is deeper in the star than merely superficially is what really counts.

In the H-W equation of state γ never rises again as high as 2.2. If this equation of state is a reasonable approximation to the truth, a neutron star can hardly ever be expected to have an ellipsoidal shape. However, other authors have proposed an equation of state which is considerably stiffer in the region of supra-nuclear densities (Figures 28 and 29) ("hard core repulsions"). For example in the C-C-L-R equation of state, $\gamma = 2.2$ is reached and surpassed for densities of

$$\varrho = 4.4 \times 10^{13} \text{ g/cm}^3$$

and higher ("stiff region of core"). Even for this equation of state a star of sufficiently small mass will have no stiff core and the critical configuration will be one which spins off matter from an equatorial rim. However, stars

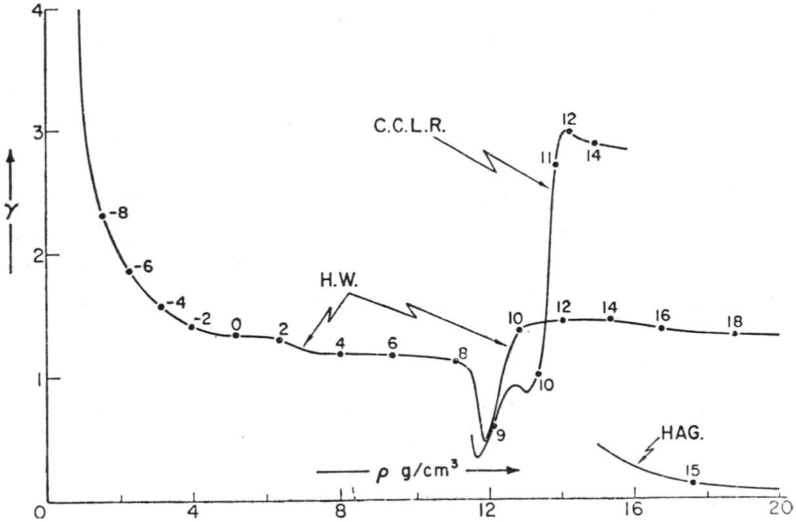

Figure 28 Values of $\gamma = \dfrac{p + \varrho}{p} \dfrac{dp}{d\varrho}$ are given as a function of the density ($\log_{10}\varrho$) for three different equations of state. H-W refers to the Harrison-Wheeler equation of state[7]; C-C-L-R refers to the Cameron-Cohen-Langer-Rosen equation of state[9]; HAG to the Hagedorn equation of state[14,133]. On the curves for selected values of the central density are given the corresponding values of the pressure ($\log_{10}p$).

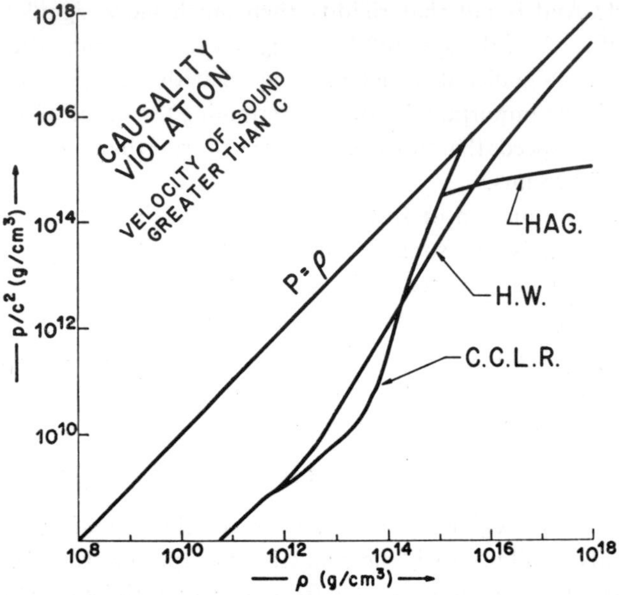

Figure 29 Value of the pressure ($\log_{10} p$) *vs.* density ($\log_{10} \varrho$) for three different equations of state. H-W refers to the Harrison-Wheeler equation of state[7]; C-C-L-R to the Cameron-Cohen-Langer-Rosen equation of state[9]; HAG to the Hagedorn equation of state[14,133]. The requirement that the velocity of sound be smaller than the velocity of light requires

$$\frac{dp}{d\varrho} \leqq 1.$$

with higher and higher central density will possess a stiff core which constitutes a larger and larger fraction of the whole (Table XV). Thus it may be that on this model the more massive neutron stars, turning at critical speed, will have ellipsoidal cores and be powerful emitters of gravitational radiation.

Table XV Size of "stiff region" ($\gamma \geqq 2.2$) at center of neutron star for selected values of central density, according to the C-C-L-R equation of state

ϱ_0 (g/cm³)	$M/M_\odot$	R (km)	$r_{\gamma=2.2}$ (km)	M (inside this radius)/$M_\odot$
1.6×10^{14}	0.13	17.7	7.9	0.11
2.5×10^{16}	2.40	10.9	10.2	2.36

For a given central density the calculated mass and mean radius will be somewhat altered from the values listed here if the star is rapidly rotating (cf. Hartle and Thorne[7] for the order of magnitude of the correction to be expected in typical cases), but no attempt is made to correct for the effects of rotation in this table.

7.7.3 **Double star system**

"It has been estimated that at least one-fifth of all the stars are binary systems," says Struve in the chapter, "The Origin and Development of Close Double Stars" in his book on stellar evolution (see page 171 of ref. 97). How it happens that double stars occur so frequently is a topic of astrophysics and hydrodynamics still in its beginnings. Binaries corresponding to even smaller separations could be formed by fission of a freshly formed, rapidly rotating neutron star or by fission of the collapsing core of a supernova even before the core has separated out as a neutron star. Table XVI lists a few representative examples of important types of double star systems and gives for each the rate of loss of energy by gravitational radiation calculated on the assumption of revolution in a circular orbit. The last two entries refer not to any known system, but to the idealised case of two compact objects (neutron star or black hole) of solar mass ($M_\odot = 1.98 \times 10^{33}$ g or 1.47 km) in revolution at one or other of two standard separations, 1000 km or 10,000 km (enormous differences in the calculated rate of emission!).

The separation of the components in any reasonable double star system is sufficiently large in comparison with the Schwarzschild radius that general relativity corrections can be disregarded in the theory of the orbit. Thus for the circular frequency of revolution of two stars of masses m_1 and m_2 in circular orbit about their common center of gravity we have the standard formula

$$\omega^2 = (m_1 + m_2)/r^3$$

(geometrical units for mass and time). General relativity shows most significantly in the fact that gravitational radiation can be emitted at all. The calculated rate of loss of energy by radiation is

$$-dE/dt = (32/5)\,[m_1 m_2/(m_1 + m_2)]^2\, r^4 \omega^6$$

For motion in an elliptic orbit of semi-major axis a and eccentricity ε Zel'dovich and Novikov[56] give

$$-dE/dt = (32/5)\, m_1^2 m_2^2 (m_1 + m_2)\, a^{-5} f(\varepsilon)$$

and supply a graph of $f(\varepsilon) = [1 + (73/24)\,\varepsilon^2 + (37/96)\,\varepsilon^4]/(1 - \varepsilon^2)^{7/2}$. These alternate ways of expressing the rate of loss of energy appear in Table XVII. Most striking is the proportionality of the radiated power to the tenth power of the velocity for a system of two identical stars of fixed mass, or fixed frequency of revolution, or fixed separation (first three entries in Table XVII). The last two entries in Table XVI illustrate the enormous difference in the calculated rate of emission between one "standard reference binary system" and another.

Table XVI Representative binary star systems and the calculated output of gravitation radiation from each

	Binary	Period	Masses $\begin{Bmatrix} M_A/M_\odot \\ M/M_\odot \end{Bmatrix}$	Distance from Earth (pc)	τ	$(-dE/dt)_{grav}$ erg/sec	Gravitational radiation at Earth (erg/cm² sec)
Resolved binaries	η Cas	480 yr	0.94, 0.58	5.9	3.8×10^{25} yr	5.6×10^{10}	1.4×10^{-29}
	ξ Boo	149.95	0.85, 0.75	6.7	1.5×10^{24}	3.6×10^{12}	6.7×10^{-28}
	Sirius	49.94	2.28, 0.98	2.6	2.9×10^{22}	1.1×10^{15}	1.3×10^{-24}
	Fu 46	13.12	0.31, 0.25	6.5	1.3×10^{22}	3.6×10^{14}	7.1×10^{-26}
Eclipsing binaries	β Lyr	12.925 day	19.48, 9.74	330	2.8×10^{12}	5.7×10^{28}	3.8×10^{-15}
	UWCMa	4.393	40.0, 31.0	1470	3.3×10^{10}	4.9×10^{31}	1.9×10^{-13}
	β Per	2.867	4.70, 0.94	30	1.3×10^{12}	1.4×10^{28}	1.3×10^{-13}
	WUMa	0.33	0.76, 0.57	110	2.5×10^{10}	4.7×10^{29}	3.2×10^{-13}
	WZSge	81 min	0.6, 0.03	100	4.9×10^{6}	3.5×10^{29}	2.9×10^{-13}
	10000 km binary	12.2 sec	1.0, 1.0	1000	13.0 yr	3.25×10^{41}	2.7×10^{-3}
	1000 km binary	0.39 sec	1.0, 1.0	1000	11.4 hr	3.25×10^{46}	2.7×10^{2}

First four entries—representative resolved binaries taken from the compilation of Van de Kamp.[134] Second four entries—representative eclipsing binaries, from the compilation of Gaposchkin.[134] Ninth entry—shortest period binary system yet observed (Kraft, Mathews and Greenstein[135]). Final two entries—calculations for idealised model of two neutron stars (or black holes), each of solar mass, separated by 1000 km and 10000 km, respectively. The fourth column gives distance from the Earth in pc (3.085×10^{18} cm); the fifth column the calculated characteristic time, $\tau = -E/(-dE/dt) = r/(-dr/dt) = (3/2)\,\omega/(d\omega/dt)$, for loss of energy by gravitational radiation.

In a close binary the rate of emission is so great that the period will increase with time at a significant rate (last entry in Table XVI). However, to observe such a change in period with time will not in itself constitute proof that gravitational radiation is responsible for the effect. Loss of mass from either component, or transfer of mass from one component to the other, will also cause the period to change (see pp. 205 and 237 of ref. 97). Moreover the very proximity that favours intense gravitational radiation would also seem to favour rapid mass transfer. To discriminate between the two mechanisms is not easy at any distance. However, if the two objects are as compact as a neutron star or black hole, mass loss and mass transfer would seem to drop out as significant factors. The problem is not then one

Table XVII Rate of emission of gravitational radiation from a double star system consisting of two identical stars of mass m, each in circular orbit, as a function of selected pairs of parameters (upper part of Table) and in a system consisting of a light mass m and a heavy mass M (lower part of Table)

	Give	Deduce		$-dE/dt$	τ
Twin Stars	β, m	$\omega = 4\beta^3/m$,	$r = m/2\beta^2$	$(2/5)\,(2\beta)^{10}$	$(5m/8)\,(2\beta)^{-8}$
	β, ω	$m = 4\beta^3/\omega$,	$r = 2\beta/\omega$	$(2/5)\,(2\beta)^{10}$	$(5/16\omega)\,(2\beta)^{-5}$
	β, r	$m = 2\beta^2 r$,	$\omega = 2\beta/r$	$(2/5)\,(2\beta)^{10}$	$(5r/16)\,(2\beta)^{-6}$
	m, ω	$\beta = (m\omega/4)^{1/3}$,	$r = (2m/\omega^2)^{1/3}$	$(2/5)\,(2m\omega)^{10/3}$	$(5/16)\,(2m)^{-5/3}\omega^{-8/3}$
	m, r	$\beta = (m/2r)^{1/2}$,	$\omega = (2m/r^3)^{1/2}$	$(2/5)\,(2m/r)^5$	$(5/128)\,(r^4/m^3)$
	ω, r	$\beta = \omega r/2$,	$m = \omega^2 r^3/2$	$(2/5)\,(\omega r)^5$	$(5/16)\,(1/\omega^6 r^5)$
Case $M \gg m$	β, M	$\omega = \beta^3/M$,	$r = M/\beta^2$	$(32/5)\,(m^2/M^2)\,\beta^{10}$	$(5/64)\,(M^2/m\beta^8)$
	β, ω	$M = \beta^3/\omega$,	$r = \beta/\omega$	$(32/5)\,m^2\beta^4\omega^2$	$(5/64)\,(1/m\beta^2\omega^2)$
	β, r	$M = \beta^2 r$,	$\omega = \beta/r$	$(32/5)\,m^2\beta^6/r^2$	$(5/64)\,(r^2/m\beta^4)$
	M, ω	$\beta = (M\omega)^{1/3}$,	$r = (M/\omega^2)^{1/3}$	$(32/5)\,m^2 M^{4/3}\omega^{10/3}$	$(5/64)\,(1/mM^{2/3}\omega^{8/3})$
	M, r	$\beta = (M/r)^{1/2}$,	$\omega = (M/r^3)^{1/2}$	$(32/5)\,m^2 M^3/r^5$	$(5/64)\,(r^4/mM^2)$
	ω, r	$\beta = \omega r$,	$M = \omega^2 r^3$	$(32/5)\,m^2\omega^6 r^4$	$(5/64)\,(1/m\omega^4 r^2)$

All quantities are in gravitational units. Conversion from conventional units via the factors 3×10^{10} cm/sec and 0.74×10^{-28} cm/g. The energy output, given in gravitational units (cm of mass energy emitted per cm of time, and therefore dimensionless) is translated to conventional units via the "standard power factor" $c^5/G = 3.6 \times 10^{59}$ erg/sec or $2.0 \times 10^5 M_\odot$/sec. $\omega = 2\pi\nu$ is the circular frequency, β is the velocity in orbit relative to the speed of light and E the energy of the system, $E = E_{\text{pot}} + E_{\text{kin}} = -m^2/r + m^2/2r = -m^2/2r$ (twin star system) or $E = -mM/2r$ (objects of very different mass). In the general case we have for the velocities of the two masses in their circular orbits

$$\beta_1 = m_2(m_1 + m_2)^{-1/2} r^{-1/2}$$

$$\beta_2 = m_1(m_1 + m_2)^{-1/2} r^{-1/2}$$

In terms of these velocities the rate of loss of energy by gravitational radiation is

$$-dE/dt = (32/5)\,\beta_1^2\beta_2^2(\beta_1 + \beta_2)^6$$

of identifying the mechanism of slowing down but of seeing the phenomenon at all. For this purpose no prospect seems more hopeful than to look for a pulsar, or pair of pulsars, with their period, or pair of periods, periodically modulated in frequency as the two stars revolve about their center of gravity.

7.7.4 **Pulsating neutron star**

A neutron star, freshly formed by whatever method, can hardly fail to be endowed with vibrational as well as rotational energy. The rotational energy can be kept for years if the star has axial symmetry (zero value of ε in Table XIII) or nearly so, as witness the pulsars; but the vibrational energy will normally be expected to be dissipated in gravitional waves, if not otherwise, in a time of the order of a few days or less, according to Zee and Wheeler[136]. They note that, "no gravitational radiation will come off... from the purely radial vibration of an ideal spherically symmetric distribution of mass†. However, the chances are overwhelming that the residual neutron star is endowed with a finite amount of angular momentum. Moreover, any natural amount of angular momentum, divided by the relatively very small moment of inertia of a neutron star, implies typically a very high angular velocity. In consequence, the neutron star will be expected to be more or less pancake shaped ($\sim$ oblate spheroid). This perturbation will couple the purely radial and the quadrupole modes of vibration. The admixture of quadrupole component, relative to the amplitude of the purely radial component, may be expected to be of the order of the eccentricity ε of the spheroid. Squaring amplitudes to get intensities, one concludes that the rate of damping of the radial oscillations by gravitational radiation should be of the order of ε^2 times the rate of damping of the quadrupole mode of vibration. For the exponent A in the formula

$$E(\text{quadrupole vibration}) = E_0(\text{quadrupole vibration})\, \varepsilon^{-At}$$

Z-W estimate a value of the order $A_{\text{quad}} \sim 1\ \text{sec}^{-1}$. Taking $\varepsilon \sim 0.01$ as an estimate of the eccentricity of a neutron star, which if anything would seem to be on the small side, one obtains

$$A_{\text{radial}} \sim \varepsilon^2 A_{\text{quad}} \sim 10^{-4}\ \text{sec}^{-1}$$

† A spherical mass, undergoing vibrations of spherical symmetry, will not be able to emit tensorial waves (Einstein theory; topological fixed point theorem) but could give off scalar waves if a scalar field existed (Jordan-Brans-Dicke scalar-tensor theory of gravity; damping time of order milliseconds, according to Morganstern and Chiu[137]). Thus the observation that the lowest radial mode has a damping time long in comparison to a minute would disprove the existence of a scalar field with anything like the proposed coupling constant.

Consequently it is difficult to see how the radial oscillations can last more than a few days".

It may be noted additionally that the Crab pulsar PSR 0531 rotating with a period of 33 ms, has a calculated rotation-induced eccentricity[32] of the order of $\varepsilon \sim 10^{-3}$, implying a damping of the order of 10^{-6} sec^{-1}, or a damping time of $\sim$ 10 days, for any radial mode of oscillation.

For energy contained in the rotation of the Crab pulsar to-day one estimates

$$E_{\text{rot}} \sim \tfrac{1}{2}I\omega^2 \sim \tfrac{1}{2}(4 \times 10^{44}\ \text{g/cm}^2)\ (190\ \text{rad/sec})^2 \sim 7 \times 10^{48}\ \text{erg}$$

The energy delivered into rotation at the time of formation of the neutron star could have been one or two orders of magnitude larger, say 10^{50} or 10^{51} erg. It is reasonable to think of a similar stockpile of energy going into vibration. Thus Zee and Wheeler, considering for a neutron star the idealised model of a sphere of incompressible fluid with

$$\text{density } 9.3 \times 10^{13}\ \text{g/cm}^3,\ \text{radius } 14\ \text{km},\ \text{mass } 1.1 \times 10^{33}\ \text{g} = 0.56 M_{\odot},$$

deformed at the moment of formation with a root mean squared departure from sphericity of

$$\xi_{\text{rms}} = \langle (r - R)^2/R^2 \rangle^{\frac{1}{2}} = 0.1$$

concluded that it would have an energy of deformation of the order of 3×10^{50} erg. What happens to this energy?

The energy of vibration of a neutron star immediately after formation, like the energy of the Earth immediately after an earthquake, is divided among the various independent modes of vibration in a way dependent upon the special features of the generating event. Easier to analyse than the reasons for the particular partition of the energy among the modes are the characteristic frequencies and rates of damping of the modes themselves. The spectrum of the oscillations of the Earth has been analysed in much detail, both observationally and theoretically[128]. The calculated frequencies and damping times for a neutron star are collected in Table XVIII.

The theory of the damping of the quadrupole modes is an interesting question of principle that has been analysed in papers of Thorne[140], Thorne and Campolattaro[141] and Price and Thorne[142]. They show how it is possible to obtain well-defined results for such a macroscopic quantity as the damping rate A_{grav} even though today one still does not know to express the microscopic forces at work by any simple gravitational analogue to the familiar formula $(2/3)\,(e^2/c^3)\dddot{x}$ for radiation reaction in electrodynamics.

The quadrupole vibration, superposed with a 90-degree phase difference upon the linearly independent mode of quadrupole oscillation turned 45°

Table XVIII Spectrum of excitations of a neutron star, with estimates of characteristic damping time.

(Equation of state) ϱ_0		H-W	3×10^{14} g/cm³	H-W	6×10^{15} g/cm³	L-S-T-C"V_γ"	5.15×10^{14} g/cm³	L-S-T-C"V_γ"	3×10^{15} g/cm³
M $2M/R$		$0.405M_\odot$	0.0574	$0.682M_\odot$	0.240	$0.677M_\odot$	0.159	$1.954M_\odot$	0.580
Mode		Frequency (Hz)	Damping time A_{grav}^{-1} (sec)	Frequency (Hz)	Damping time A_{grav}^{-1} (sec)	Frequency (Hz)	Damping time A_{grav}^{-1} (sec)	Frequency (Hz)	Damping time A_{grav}^{-1} (sec)
Radial	${}_0S_0$	588	5×10^5	4761	4×10^2	232	2×10^7	1978	7×10^4
	${}_1S_0$	769		5263		861		7874	
	${}_2S_0$	1111		8333		1660		12091	
Quadrupole	${}_0S_2$	835	13	3220	0.19	1430	1.7	2650	0.22
	${}_1S_2$			5840	0.28	4240	11	6430	1.6
	${}_2S_2$			8480	1.3			9750	2.6
	${}_3S_2$			10660	24				

In principle the first row in the Table (lowest frequency) should be the rotational mode; but this frequency depends upon energy, which differs from case to case (no eigenvalue!) and therefore is omitted from the compilation. Frequencies and damping times for quadrupole oscillations taken from Thorne[138]; frequencies for radial oscillations newly calculated by R. Ruffini; earlier calculations by Meltzer and Thorne[139] were not for the same masses and equations of state as those listed here. (H-W = Harrison-Wheeler equation of state as listed in updated form by Hartle and Thorne and in Table II of the present report; L-S-T-C "V_γ" = Levinger-Simmons-Tsuruta-Cameron "V_γ" equation of state also listed by Hartle and Thorne[7]. ϱ_0 denotes the central density; M, the mass ($M_\odot = 1.987 \times 10^{33}$ g); and $2M/R$, the standard factor in the expression $1-2M/R$ appearing in the Schwarzschild metric, evaluated on the surface of the star. Damping time (with respect to gravitational radiation) for radial oscillations infinite for ideal sphere, estimated in Table assuming a rotation-induced bulge with eccentricity $\varepsilon = 0.01$ from very rough formula

$$A_{grav}(\text{radial}) \sim (\omega_{rad}/\omega_{quad})^4 \varepsilon^2 A_{grav}(\text{quad}) \quad \text{(for lowest mode only!).}$$

A fuller compilation would include torsional and other oscillations of the crust and frequencies associated with the coupling of magnetic field with superfluid and with the crust[33,40,132].

from it in the equatorial plane, describes a rotatory excitation of fixed shape revolving at a fixed frequency. How does this mode of motion differ from rotation itself? One knows the answer from the theory of rotating fluid masses or from any one of the many applications of this theory such as the theory of nuclear vibrations and rotations[143]. The two modes of motion differ in no respect more sharply than that the frequency of the rotatory vibration is independent of energy (in the small amplitude approximation) whereas the frequency of rotation is proportional to the square root of the rotational energy (Table XIX).

Table XIX Rotation and rotatory vibration compared and contrasted (ideal incompressible fluid taken here as basis for simplified model of a neutron star).

	True rotation	Rotatory quadrupole vibration
Shape of surface	Prolate ellipsoid	Prolate ellipsoid
div v	Zero	Zero
curl v	2ω	Zero
Minimum energy of fluid inside compatible with motion of boundary?	No	Yes
Frequency is an eigenvalue?	No	Yes
Effective moment of inertia, J/ω	Value I for fluid as if frozen ("solid body value")	$I_{\text{effective}} \sim$ (deformation) I (goes to zero for small deformation)
Relation between energy and angular momentum	$E \simeq J^2/2I$	$E \simeq J\omega$
Relation between energy and frequency	$E \simeq \frac{1}{2}I\omega^2$	ω independent of E
These two modes of motion coupled at small rates of rotation?	No	No
Coupled at higher rates of rotation?	Yes	Yes
Energy in these two modes reasonably considered to be of same rough order of magnitude at time of formation of neutron star?	Yes	Yes

The numbers listed in Table XVIII give an impression of the rich diagnostics available if and when one succeeds in detecting gravitational radiation or any other effect correlated with vibration. Among the possibilities that claim attention are (1) determination of the mass of the star; (2) discrimination between one equation of state for superdense matter and another; (3) observation of frequency splitting (not shown in table; analogue of

Zeeman effect) induced in quadrupole vibrations by rotation of the system[128]; and (4) detection of elastic modes outside the framework of the idealised fluid model employed here[33,40,132] (solid state physics of the crust of the neutron star).

7.8 SPLASH OF GRAVITATIONAL RADIATION

If an electron in circular orbit about a nucleus is a familiar source of periodic electromagnetic radiation, equally familiar as a source of impulse radiation is an electron given high velocity in an X-ray tube, passing close to a nucleus in the copper target, and given a sudden transverse accleration as it flies by. The geometrodynamical analogue of such a source is a mass m flying with high velocity past another mass M, and thereby experiencing a sudden transverse change in velocity. Without such a change in velocity no radiation is to be expected, despite what might at first sight derive from the following line of reasoning: (1) The source of radiation is the sum of the squares of the moments $\dddot{Q}^{pq}$. (2) Each such moment is given by an expression to which each mass makes its contribution:

$$Q^{pq} = m(3x^p x^q - \delta^{pq}|x|^2) + M(3X^p X^q - \delta^{pq}|X|^2)$$

(case of two masses; generalisation obvious for case of many masses). (3) The coordinates are changing with time. (4) Therefore the Q^{pq} change with time. (5) Therefore the system emits gravitational radiation. This reasoning is wrong however. So long as the particles are envisaged as having uniform velocities the coordinates are linear functions of time. The quadrupole moments are therefore only second-degree functions of time. Their third time derivatives are identically zero. Interaction and deflections—or acceleration—are absolute requirements for the emission of gravitational radiation. To obtain a non-zero result one must allow for the interaction between the two masses. For our purpose it is enough to consider non-relativistic velocities and use the Newtonian expressions for accelerations; thus

$$m\ddot{x}^p = (mM/r^3)(X^p - x^p)$$

$$m\dddot{x}^p = (mM/r^3)(\dot{X}^p - \dot{x}^p) - (3mM\dot{r}/r^4)(X^p - x^p)$$

In this way one obtains for the third time rate of change of the mass quadrupole moment the expression

$$\begin{aligned}\dddot{Q}^{pq} &= m(3\dddot{x}^p x^q + 3x^p\dddot{x}^q - 2\delta^{pq}\dddot{x}^s x^s + 9\ddot{x}^p\dot{x}^q + 9\dot{x}^p\ddot{x}^q - 6\delta^{pq}\ddot{x}^s\dot{x}^s) \\ &\qquad + \text{similar term for the other mass} \\ &= (Mm/r^3)(2\delta^{pq}r\dot{r} + 18(\dot{r}/r)\,r^p r^q - 12\dot{r}^p r^q - 12r^p\dot{r}^q)\end{aligned}$$

The rate of loss of energy by gravitational radiation is given by the formula

$$-dE/dt = (1/45)\,\dddot{Q}^{pq}\dddot{Q}^{pq} = (8/15)\,(mM/r^2)^2\,(12\dot{\mathbf{r}}^2 - 11\dot{r}^2)$$

When the masses revolve about their center of gravity in circular orbits, we have $\dot{r}^2 = 0$ and $\dot{\mathbf{r}}^2 = \omega^2 r^2$ and we get back the results summarised for example in Table XVII. In the opposite extreme case where a small mass m flies past a large mass M, originally at rest, with impact parameter b so large and velocity β so great that the change in direction is small (straight line idealisation; parametric representation of motion $x = \beta t = b \tan\theta$, $y = b$, $r = b\sec\theta$, $\dot{r} = \beta\sin\theta$, $\dot{\mathbf{r}}^2 = \beta^2$) we have for the energy loss the expression

$$-\Delta E = \int (-dE/dt)\,dt = (37\pi/15)\,(m^2M^2\beta/b^3)$$

(Insert G^3/c^4 on right if switching from geometrical to conventional units.) Contrast this expression with the well-known formula for the loss of energy by electromagnetic energy when a particle of mass m and charge e flies past a larger mass of charge Q, again in nearly straight line motion:

$$-\Delta E = \int (2e^2/3c^3)\,\ddot{x}^2\,dt = \int (2e^2/3c^3)(Qe/mr^2)^2\,dt = (\pi/3)\,(Q^2e^4/m^2c^4b^3\beta)$$

(conventional units). The very different dependence on the velocity in the two cases reflects the difference between quadrupole and dipole radiation.

7.9 LOW-FREQUENCY PART OF SPLASH RADIATION

It is also interesting to look at the distribution in frequency of the outgoing radiation (Figure 30). Write

$$\dddot{Q}^{pq}(t) = (1/2\pi)^{\frac{1}{2}} \int_{-\infty}^{+\infty} \dddot{Q}^{pq}(\omega)\,e^{-i\omega t}\,d\omega$$

and assume

$$-\Delta E = (1/45) \int \dddot{Q}^{pq}(t)\,\dddot{Q}^{pq}(t)\,dt = (1/45) \int_{-\infty}^{+\infty} \dddot{Q}^{pq}(\omega)\,\bar{\dddot{Q}}^{pq}(\omega)\,d\omega$$

Here the Fourier amplitude for circular frequency ω is given by the expression

$$\dddot{Q}^{pq}(\omega) = (1/2\pi)^{\frac{1}{2}} \int_{-\infty}^{+\infty} \dddot{Q}^{pq}(t)\,e^{i\omega t}\,dt$$

The simplest case to consider, and the one of greatest interest, is that of low frequency. In this case the oscillatory factor drops out of the Fourier integral. The integration goes through directly, with no reference to any of the details of the processes of acceleration and deceleration, as well for the explosion of a bomb and the fall of a meteorite as for one star flying by another. We have

$$\ddot{Q}^{pq}(\omega \to 0) = (1/2\pi)^{\frac{1}{2}}\, \Delta\dot{Q}^{pq}(t)$$

with simple values for the $Q^{pq}(t)$ at times before and after any interaction takes place:

$\dot{Q}^{pq}(t) = 2m(3\dot{x}^p\dot{x}^q - \delta^{pq}\dot{x}^s\dot{x}^s)$ + a similar expression in M (and any other masses involved)

The intensity of the emission at low frequencies (ν small compared to the reciprocal of the interaction time) is given by the expression

$$\begin{aligned}
-d\,\Delta E/d\nu &= -2\pi d\,\Delta E/d\omega \\
&= (4\pi/45)\, \ddot{Q}^{pq}(\omega \to 0)\, \ddot{Q}^{pq}(\omega \to 0) \\
&= (2/45)\, \Delta\dot{Q}^{pq}(t)\, \Delta\dot{\bar{Q}}^{pq}(t) \\
&= (2/45)\, |2m(3\dot{x}^p\dot{x}^q - \delta^{pq}\dot{x}^s\dot{x}^s)_{\text{after}} \\
&\quad -2m(3\dot{x}^p\dot{x}^q - \delta^{pq}\dot{x}^s\dot{x}^t)_{\text{before}} + \text{similar terms in } M|^2
\end{aligned}$$

In the case of a mass m flying past a mass M with an impact parameter b, and these two objects exerting only gravitational forces on each other, we have for the velocities before and after the simple Newtonian results, m:

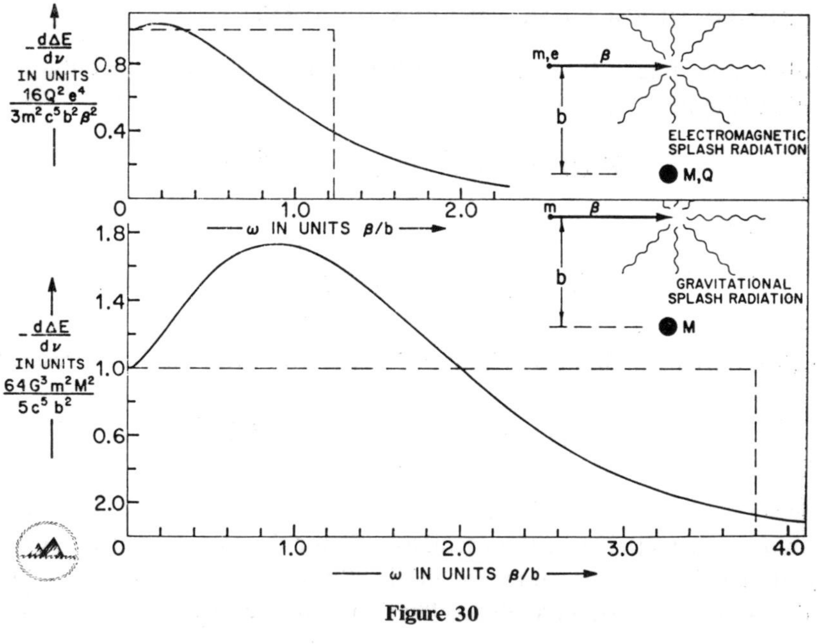

Figure 30

Figure 30 Electromagnetic and gravitational splash radiation compared. A particle of mass m, charge e, travelling at the non-relativistic speed $v = \beta c$, flies past a particle of much greater mass M, charge Q, originally at rest, with such a large impact parameter b that the motion is almost straight-line motion. The frequency distribution of the electromagnetic energy radiated is given by the formula

$$-d\Delta E/d\nu = (4e^2/3c^3)\left\{\left|\int_{-\infty}^{+\infty} \ddot{x}(t)\, e^{i\omega t}\, dt\right|^2 + \left|\int_{-\infty}^{\infty+} \ddot{y}(t)\, e^{i\omega t}\, dt\right|^2\right\}$$

with $(\ddot{x}, \ddot{y}) = (eQ/m)\,(vt, b)/[(vt)^2 + b^2]^{3/2}$; or

$$-d\Delta E/d\nu = (16Q^2e^4/3m^2c^5\beta^2)\,[u^2K_0^2(u) + u^2K_1^2(u)]$$

with $u = \omega b/v$ (conventional units) or $u = \omega b/\beta$ (geometric units) (formula given by N. Bohr[88]). The function of u in square brackets reduces to 1 at low frequencies and goes to $\pi u e^{-2u}$ at high frequencies, and, when integrated with respect to u, gives $\pi^2/8 = 1.234$ ("cut-off point of equivalent flat spectrum", as illustrated in diagram). Similarly, the distribution in frequency of the gravitational splash radiation is given by

$$-d\Delta E/d\nu = (2/45)\sum_{p,q}\left|\int_{-\infty}^{+\infty} \dddot{Q}^{pq}(t)\, e^{i\omega t}\, dt\right|^2$$

where in

$$\dddot{Q}^{pq} = (mM/r^3)\,(2\delta^{pq}\dot{r}/r + 18(\dot{r}/r)\,r^p r^q - 12\dot{r}^p r^q - 12 r^p \dot{r}^q)$$

we insert on the right expressions for the unperturbed straight line motion of m (now neutral!) relative to M (also uncharged). The first factor in $\dddot{Q}^{pq}$ has the value $mM\,(b^2 + \beta^2t^2)^{-3/2}$ and the second has the value $\beta^2 t$ times:

xx: $-4\beta^2 t - 18b^2\beta^2 t/(b^2 + \beta^2t^2)$

xy: $6b\beta - 18b^3\beta/(b^2 - \beta^2t^2)$

yy: $2\beta^2 t + 18b^2\beta^2 t/(b^2 + \beta^2t^2)$

zz: $2\beta^2 t$

The Fourier integrals $\int_{-\infty}^{+\infty} \dddot{Q}^{pq}(t)\, e^{i\omega t}\, dt$, relevant to the distribution in angle as well as to the distribution in frequency, are given by $4mM/b$ (no restriction on ratio of m to M) times:

xx: $-i(2uK_0 + 3u^2K_1)$

xy: $-3uK_1 - 3u^2K_0$

yy: iuK_0

zz: $i(uK_0 + 3u^2K_1)$

Thus

$$-d\Delta E/d\nu = (64m^2M^2/5b^2)\,[(u^2/3 + u^4)\,K_0^2 + 3u^3K_0K_1 + (u^2 + u^4)\,K_1^2]$$

(gravitational units; multiply by $G^3/c^5 = 1.23 \times 10^{-74}$ (erg/Hz) (cm^2/g^4) for conventional units). The factor in square brackets reduces to 1 for low $u = \omega b/\beta$, goes to $\pi u^3 e^{-2u}$ at high frequencies, and, when integrated with respect to u, gives $37\pi^2/96$ ("cut-off value for $\omega b/\beta$ for equivalent flat spectrum", as illustrated in the lower diagram).

$(0, 0, \beta) \to (0, 2M/b\beta, \beta)$ and M: $(0, 0, 0) \to (0, -2m/b\beta, 0)$; and thus

$$\Delta \ddot{Q}^{zy}(t) = \Delta \ddot{Q}^{yz}(t) = \ddot{Q}^{yz}_{\text{after}} - \ddot{Q}^{yz}_{\text{before}} = 12mM/b$$

All other components are either identically zero or of second order in the coupling factor mM. The intensity in the flat part of the spectrum is

$$(-d\Delta E/d\nu)_{\text{low}\,\nu} = (64/5)\,(m^2M^2/b^2)$$

One can define an effective cut-off frequency as that frequency which multiplied by the foregoing expression gives the total energy loss:

$$\nu_{\text{effective cut-off}} \equiv (-\Delta E)/(-d\Delta E/d\nu)_{\text{low}\,\nu} = \left(\frac{37\pi}{192}\right)\beta/b$$

$$\omega_{\text{cut-off}} = (37\pi^2/96)\,(\beta/b)$$

(geometric units). In comparison, the low-frequency electromagnetic radiation given off when a particle (mass m and charge e) flies past a massive center of attraction of charge Q has the intensity

$$\begin{aligned} -(d\Delta E/d\nu)_{\text{low}\,\nu} &= 2\pi(2e^2/3c^3)\,2\ddot{x}^2(\omega \to 0) \\ &= (4e^2/3c^3)\,(\Delta\dot{x})^2 \\ &= (4e^2/3c^3)\,(2Qe/mb\beta c)^2 \\ &= (16/3)\,(Q^2e^4/m^2c^5b^2\beta^2) \end{aligned}$$

(Same dependence on impact parameter as in the case of gravitational radiation; contrasting dependence upon velocity.) Dividing this energy per unit frequency into the total energy radiated, one has for effective cut-off frequency in this case

$$\nu_{\text{effective cut-off}} = \left(\frac{\pi}{16}\right)(c\beta)/b$$

with

$$\omega_{\text{cut-off}} = (\pi^2/8)\,(c\beta/b) \text{ (conventional units).}$$

As an illustration of the order of magnitude of gravitational splash radiation, consider a neutron star of one solar mass passing another neutron star, also of one solar mass, with an impact parameter of 100 km and with a velocity of $\beta = 0.01$ (3000 km/sec), the type of event that one can imagine taking place in the late stages of evolution of a dense galactic nucleus. The calculated magnitude of the gravitational splash radiation is

$$-\Delta E = (37\pi/15)\,(1.47\text{ km})^4\,(10^{-2})/(100\text{ km})^3 = 3.6 \times 10^{-7}\text{ km} = 0.036\text{ cm}$$

$$-\Delta E_{\text{conv}} = 0.036\text{cm}/(0.742 \times 10^{-28}\text{ cm/g}) = 4.8 \times 10^{26}\text{ g or } 4.4 \times 10^{47}\text{ erg,}$$

enough to decrease the relative velocity by 15 km/sec. Such a decrease in velocity, small though it is in absolute terms, can be enough to make the difference between two stars remaining free of each other's influence or becoming bound in closed orbits.

Zel'dovich and Novikov[45] have given a detailed treatment of the radiative capture from a hyperbolic orbit into an elliptic orbit (both reasonably enough idealised as Newtonian) by reason of gravitational radiation. They find for the cross-section for capture the expression

$$\sigma_{\substack{\text{capture into orbit by}\\ \text{gravitational radiation}}} \simeq \begin{cases} \pi(2M/\beta)^2\,(2m/M\beta^2)^{2/7} & \text{for} \quad m/M\beta^2 \gg 10 \\ 4\pi(2M/\beta)^2\,(1 + \exp(-20\beta^2 M/m)) & \text{for} \quad m/M\beta^2 \ll 10 \end{cases}$$

where m is the mass of the incoming object, M is the mass of the capturing center, and β is the relative velocity of approach at great distances.

7.10 RADIATION IN ELLIPTICAL ORBITS TREATED AS A SUCCESSION OF PULSES

When two stars move about their mutual center of gravity in long and extremely elliptical orbits, the radiation is most readily visualised as a succession of splashes, each given out in a close passage and each having a continuous spectrum. In reality the spectrum is discrete. For motion in an elliptic orbit of semi-major axis a and eccentricity ε both Peters and Mathews[144], Zel'dovich and Novikov[45] give for the power radiated in the n-th harmonic of the fundamental frequency the expression

$$-(dE/dt)_{\text{in nth harmonic}} = (32/5)\, m_1^2 m_2^2 (m_1 + m_2)\, a^{-5} g(n, \varepsilon)$$

with

$$\begin{aligned} g(n, \varepsilon) = (n^4/32)\, \{&[J_{n-2}(n\varepsilon) - 2eJ_{n-1}(n\varepsilon) + (2/n)\, J_n(n\varepsilon) + 2eJ_{n+1}(n\varepsilon) \\ &- J_{n+2}(n\varepsilon)]^2 \\ &+ (1 - \varepsilon^2)\, [J_{n-2}(n\varepsilon) - 2J_n(n\varepsilon) + J_{n+2}(n\varepsilon)]^2 \\ &+ (4/3n^2)\, [J_n(n\varepsilon)]^2\}. \end{aligned}$$

Although the spectrum is discrete and this formula gives the intensity of the individual lines, the spectrum looks continuous when viewed under poor resolution (Figure 31) because it consists of so many lines. As the ellipticity of the orbit decreases, the number of lines in the spectrum with appreciable intensity goes down. Only one line is left when the orbit becomes circular (analysis of previous section).

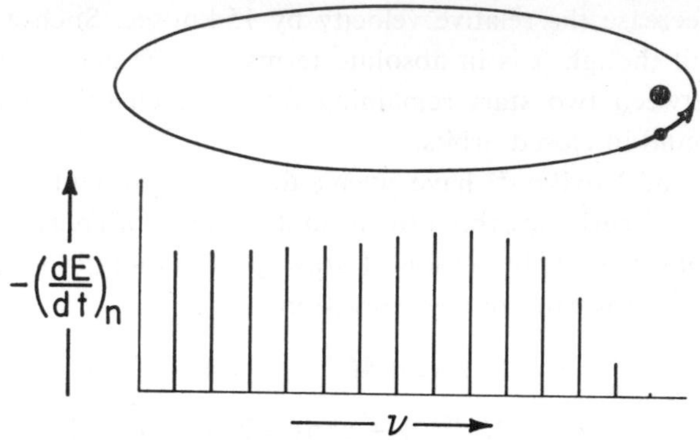

Figure 31 Qualitative form of the spectrum of gravitational radiation emitted by a pair of particles in narrow elliptical orbit about their mutual center of gravity (idealisation in which motion is treated as nearly Newtonian). The envelope gives the form of the spectrum that would be emitted in a single pass. The peaks come at integral multiples $\nu_n = n/T$ of the basic frequency of revolution in the elliptic orbit.

7.11 FALL INTO SCHWARZSCHILD BLACK HOLE

The opposite modification of the ellipticity carries attention, not to a circular orbit, but to the world line of one object falling straight towards another. If both objects have finite size, the acceleration on impact will be more important for the radiation emitted than the acceleration in flight. If the objects have indefinitely small size, the calculated radiation in Newtonian approximation

$$-dE/dt = (8/15)\,(mM\dot{r}/r^2)^2,$$

will rise to indefinitely high intensity. Actually a collapsed object of finite mass M will have a finite effective dimension, $r_{\text{Schwarzschild}} = 2M$. A much smaller object, of mass m, will not accelerate indefinitely as seen by the far-away observer. On the contrary, its approach to the Schwarzschild surface takes an infinite amount of Schwarzschild coordinate time (Figure 8; exponential approach $r \simeq 2M$ + constant $\exp(-t/2M)$). In consequence the gravitational radiation is perfectly finite in amount,

$$-\Delta E \sim 0.00246\, m^2/M$$

(value for $m \ll$ M and for case that mass m starts from rest, or essentially from rest, at infinity) and is peaked at a characteristic frequency of the order (see Figure 32)

$$\nu_{\text{peak}} \sim 0.024/M = 4.9 \times 10^3 \frac{M_\odot}{M}\ \text{Hz}$$

The foregoing figures for radiation upon capture by a black hole are only rough estimates, based upon two contradictory idealisations (1) a particle starting from rest at infinity and falling straight in according to the exact law for geodetic motion in the Schwarzschild geometry, given parametrically by the formulas

$$t = -4M(\eta^3/3 + \eta - \tfrac{1}{2}\ln(\eta + 1) + \tfrac{1}{2}\ln(\eta - 1))$$

$$r = 2M\eta^2$$

but (2) radiating as if it were moving in *flat* space,

$$-dE/dt = \frac{1}{45}\dddot{Q}^{pq}\dddot{Q}^{pq}$$

Actually the particle is moving through strongly curved space when it is radiating most strongly. Therefore a fuller treatment has to go back to first

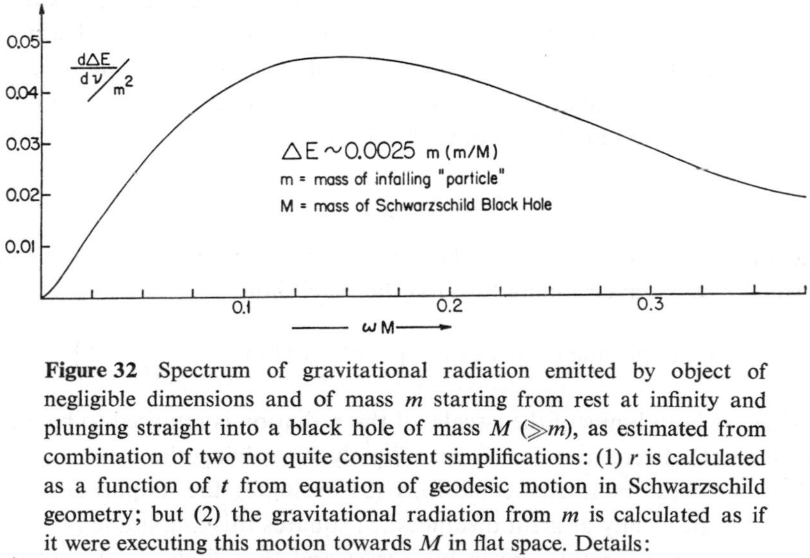

Figure 32 Spectrum of gravitational radiation emitted by object of negligible dimensions and of mass m starting from rest at infinity and plunging straight into a black hole of mass M ($\gg m$), as estimated from combination of two not quite consistent simplifications: (1) r is calculated as a function of t from equation of geodesic motion in Schwarzschild geometry; but (2) the gravitational radiation from m is calculated as if it were executing this motion towards M in flat space. Details:

$$-\frac{d\Delta E}{d\nu} = \frac{16m^2}{15}\left|\int_{-\infty}^{+\infty} (3\dot{z}\ddot{z} + z\dddot{z})\, e^{i\omega t}\, dt\right|$$

$$= \frac{16m^2}{15}\left|\int_1^{\infty} d\eta\left(\frac{3}{\eta^7} - \frac{1}{\eta^3}\right)\exp i4M\omega\left(\frac{\eta^3}{3} + \eta - \frac{1}{2}\ln\frac{\eta+1}{\eta-1}\right)\right|^2$$

$$-\Delta E = \frac{4m^2}{15M}\left(\frac{1}{7\cdot 9} - \frac{6}{11\cdot 13} + \frac{9}{15\cdot 17}\right)$$

principles. Zerilli[66] has given the foundations for such a treatment. He considered the incoming particle as making a small perturbation upon the background of the Schwarzschild geometry and analysed this perturbation in the geometry into tensorial spherical harmonics. For the radial factor in each harmonic he wrote down a second-order differential equation, a wave equation in curved space. On the right-hand side of each such equation appears a source term. The source represents the driving effect of the incoming particle. It contains as key factor the expression

$$m\delta(r - r(t))$$

The radial factor ("amplitude of specified harmonic") is obtained by solving the radial equation. The character of the solution is vitally affected by the choice of boundary conditions. In the present problem, as in almost every problem of radiation physics, the appropriate boundary condition at infinity is supplied by the requirement that there should be no incoming wave at infinity, only an outgoing wave (no "timing of sources at infinity in anticipation of the acceleration of the source"!). However, the same kind of causality requirement imposes still another demand on the radial wave: it may transport energy towards the black hole, but it must not transport energy out of the black hole. One requirement says that a typical Fourier component of the radial amplitude should behave at large r as an outgoing wave,

$$\sim \text{(amplitude factor)} \exp i\omega(r + 2M \ln r - t)$$

The other requirement says that the same Fourier component should behave at small r as a wave running towards the black hole,

$$\sim \text{(amplitude factor)} \exp i\omega\left(-r + 2M \ln \frac{2M}{r - 2M} - t\right)$$

Thus, at both limits of the r-scale all the radiation is required to be *escaping*. One (1) determines the wave uniquely by solving the wave equation subject to these boundary conditions (2) evaluates the amplitude of this wave asymptotically at great distances (3) squares this amplitude (4) inserts this square into formulas given by Zerilli and thus finds (5) the intensity of the outgoing radiation in units of energy per unit of frequency and per unit of solid angle and (6) the integrated intensity. Detailed calculations are not yet available†—hence the foregoing estimates based on a mixture of curved space and flat space arguments. It will be of great interest to carry through an analysis like Zerilli's for the radiation from an object of mass m falling with different impact parameters towards an extreme Kerr

† In this connection, see A.8.

black hole of mass M (angular momentum of the critical magnitude $J = M^2$), not least because of the greater possibilities in this case to get out a large fraction of the rest mass m in the form of radiation and the likelihood that the black holes formed in nature have nearly the critical amount of angular momentum.

7.12 RADIATION FROM GRAVITATIONAL COLLAPSE

If instead of one mass m falling towards a black hole of mass M one has a whole array of masses, all equal in magnitude, all marshalled with identical timing, all falling straight to the center of attraction (collapsing spherical shell), then all gravitational radiation is suppressed by destructive interference (no gravitational monopole radiation!). Likewise the collapse of a star endowed with spherical symmetry, barring development of instabilities and turbulence, will have no quadrupole moment and will produce no gravitational radiation. However, a star endowed with rotation will possess a quadrupole. Moreover, this quadrupole moment will change with time during collapse, as illustrated in Figure 33. Consequently a major burst of gravitational radiation will emerge. It will be followed by a succession of small and large pulses, with periodic gravitational radiation coming off between one pulse and the next, the details of the pulses and the discrete spectra depending upon the exact circumstances of the scenario (cf. caption, Figure 33). The first major pulse itself will be expected to have a continuous spectrum (cf. Figure 30) extending from $\nu = 0$ to $\nu_{\text{crit}} = \omega_{\text{crit}}/2\pi \sim (1/2\pi)$ (characteristic time, τ, associated with final stages of collapse to density $\varrho)^{-1} \sim (1/2\pi)\,(\pi G\varrho_{\text{conv}})^{-1/2}$, or in the example of Figure 33, $\nu_{\text{crit}} \sim (1/2\pi) \times (1/0.2\text{ ms}) \sim 10^3$ Hz). The energy emitted per unit frequency, at frequencies below ν_{crit}, will be of the order

$$-(d\,\Delta E/d\nu) = (2/45)\left(1 + \frac{1}{4} + \frac{1}{4}\right)(\Delta\ddot{Q}^{zz})^2 \sim (4/15)$$

$$\times \left[\int (3\beta_z^2 - \beta^2)\,\varrho\, d^3x\right]^2_{\text{stage of maximum inrush}}$$

or, in the example of Figure 33,

$$\begin{aligned}(1/15)\,(\Delta\ddot{Q}^{zz})^2 &\sim (1/15)\,(Q^{zz}_{\max}/\tau^2)^2 \sim (1/15)\,[0.15 \times 10^{20}\text{ cm}^3/(6 \times 10^6\text{cm})^2]^2\\ &\sim 10^{10}\text{ cm}^2 \text{ (geometrical units)}\\ &\sim 0.4 \text{ cm of mass energy per Hz (mixed units)}\\ &\sim 6 \times 10^{27}\text{ g of mass energy per Hz}\\ &\sim 5 \times 10^{48}\text{ erg/Hz (up to} \sim 1000\text{ Hz)}\end{aligned}$$

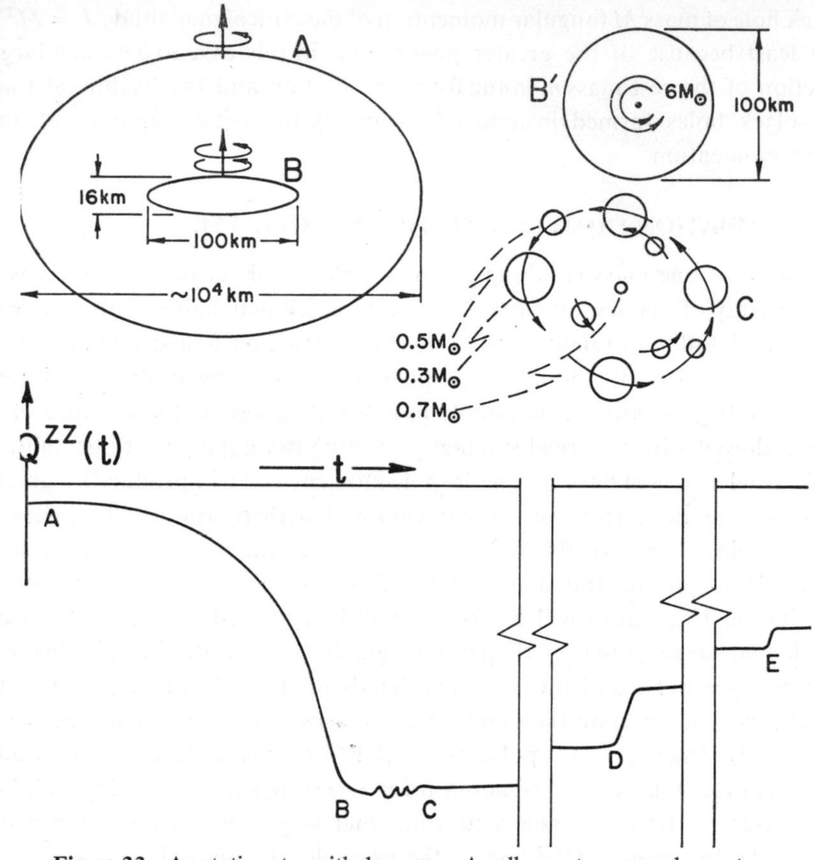

Figure 33 A rotating star with dense core A collapses to a pancake neutron star B; it fragments C; the fragments lose energy in periodic and splash gravitational radiation and recombine. The lower curve gives a schematic representation of the quadrupole moment as a function of time. Between B and C impulse radiation is created in the act of fragmentation not adequately described by the one indicated component of the quadrupole moment tensor. Between C and D multiply periodic radiation is given out until at D two fragments have lost enough angular momentum so that they combine with a splash of gravitational radiation; similarly at *E*, etc.

The total energy coming out in the splash of radiation will be of the order

$$-\Delta E \sim (1/30\pi)\,(Q^{zz}_{\max})^2/\tau^5$$

or in the example

$$-\Delta E \sim 5 \times 10^{51} \text{ erg}$$

The subsequent pulses (cf. Figure 33) will be expected to extend to higher values of ν_{crit} insofar as they arise from capture by, or amalgamation of, objects of density higher than the original density of the pancake.

7.13 EARTHQUAKES AND METEORITES

A Weber bar detector of gravitational radiation looking for events of the kind just discussed can be protected from seismic disturbances by soft enough mounting plus a seismometer in anti-coincidence. However, the seismometer only warns of disturbances that have arrived from an earthquake with seismic velocity. What of the gravitational wave disturbance that arrives from the earthquake (or a seismic event in the interior of the Earth) with the speed of light? A simple estimate shows that its effect is far too small to give even as much energy as $(1/10)(kT)$ (with $T = 300°\mathrm{K}$) to the bar. The same is true of an even more spectacular event—the impact of a giant meteorite ($\sim$ 1 km in diameter) on the Earth (S. African event of geological fame). The characteristic time, τ, for deceleration will be some small multiple of the time required for an acoustic wave to cross the object; thus,

$$\tau \sim 1\ \mathrm{km}/(5\ \mathrm{km/sec}) \sim (1/5)\ \mathrm{sec}$$

implying an upper limit to the important part of the spectrum of the order

$$\nu_{\mathrm{crit}} \sim \omega_{\mathrm{crit}}/2\pi \sim 1/2\pi\tau \sim 1\ \mathrm{Hz}$$

It is already clear from this one number that a natural mode of 1660 Hz will not respond to the event! In addition, the calculated output of energy

$$\begin{aligned}(-d\,\Delta E/d\nu) &\sim (4/15)\left[\int (3\beta_z^2 - \beta^2)\,\varrho\, d^3x\right]^2_{\text{stage of maximum inrush}} \\ &\sim (16/15)\,(m\beta^2)^2 \\ &\sim (4 \times 10^{15}\ \mathrm{g} \times 0.74 \times 10^{-28}\ \mathrm{cm/g})^2 \\ &\quad \times (10\ \mathrm{km\ sec^{-1}}/3 \times 10^5\ \mathrm{km\ sec^{-1}})^4 \\ &\sim 4 \times 10^{-5}\ \mathrm{erg/Hz}\end{aligned}$$

is many orders of magnitude too low to drive a detector 10,000 km away even if ν_{crit} did extend up to the frequency of the detector.

7.14 MICROSCOPIC PROCESSES

We do not treat here, because of their low intensity under almost all easily visualisable circumstances, the following otherwise very interesting sources of gravitational radiation: (1) quadrupole emission associated with molecular rotation and vibration (cf. especially Halpern[145]); (2) atomic transitions; (3) radiation given off in e^+e^- annihilation processes and in p^+p^- annihilation and in other elementary particle transformations[146].

Table XX Sources of gravitational radiation

Source	General features of the emitted gravitational radiation	Amount of gravitational radiation emitted for typical values of the parameters (at 100 pc the energy is spread over a sphere of surface 1.2×10^{42} cm^2)	Reasonable to detect
Spinning Rod	The type of spectrum to be expected is monochromatic with frequency falling as $\sim$const$/t^{1/4}$ and intensity falling as const$/t^{3/2}$. The flux of energy in the signal is given by $-dE/dt = (32/5)\, GI^2\omega^6/c^5$. The characteristic time to radiate a given amount E of rotational energy by emission of gravitational radiation is $$\tau = -E/(-dE/dt) = (5c^5/64GI\omega^4)$$ where $$c^5/G = 3.6 \times 10^{59} \text{ erg/sec}$$	We consider an iron cylinder with a diameter of 1 m, a length of 20 m, a mass of 4.9×10^8 g and rotating with an angular velocity $\omega = 28$ rad/sec (maximal rotation allowed by the tensile strength of the material). The flux of energy is given by $-dE/dt = 2.2 \times 10^{-22}$ erg/sec $= 3.8 \times 10^3$ gravitons/sec. At a distance from the source $d = 2.7 \times 10^9$ cm (inner boundary of wave zone; more than twice the diameter of the Earth) the flux at the detector will be 2.4×10^{-42} erg/cm^{2}sec or 4.7×10^{-17} gravitons/cm^{2}sec.	No
Revolving Double Star System (see also Tables XVI and XVII)	We indicate by m_1 and m_2 the masses of the stars and by r the distance between their centers. The spectrum of the radiation is monochromatic with frequency $\omega = [G(m_1 + m_2)/r^3]^{1/2}$ rising as const$/(t_0 - t)^{3/8}$ and intensity rising as const$/(t_0 - t)^{5/4}$. The flux of energy radiated away is given by $$-\frac{dE}{dt} = \frac{32G}{5c^5}\left(\frac{m_1 m_2}{m_1 + m_2}\right)^2 r^4\omega^6$$ and the characteristic time which the radiation will take place	In the case of Sirius the masses of the components are $m_1 = 0.98M_\odot$ and $m_2 = 2.28M_\odot$. The period of revolution $P \sim 50$ year. The flux of gravitational radiation emitted is $-dE/dt = 1.1 \times 10^{15}$ erg/sec. For a distance of 2.6 pc from the Earth the energy flux at the Earth's surface will be 1.3×10^{-24} erg/cm^{2}sec. For the system WZSge a simplified analysis gives for the masses of the components $m_1 = 6.46M_\odot$ and $m_2 = 6.46M_\odot$, the revolution period 81 min and the flux of energy $$dE/dt = -3.5 \times 10^{35} \text{ erg/sec.}$$	Impossible to detect directly the flux (too weak). Detection via optical observation of the gradual shortening of the period of rotation by reason of gravitational radiation would not be reliable due to transfer of mass from one component to other or competing effects.

	$\tau = -E/(-dE/dt) = r/(-dr/dt) = (2\omega/3)/(d\omega/dt) = 5c^5r^4/[64G^3m_1m_2(m_1+m_2)]$.	For a distance of 50–100 pc from the Earth, the energy flux at the Earth's surface could be as high as 1.2×10^{-6} erg/cm²sec.	
Spinning Neutron Star	We indicate by M the mass of the neutron star, by R its radius, by ω the angular velocity and by I the moment of inertia. We assume an equatorial eccentricity ε (accident of "geological" history!). The radiation emitted is expected to be monochromatic with frequency and intensity both falling with time. The flux of energy radiated away is given by $$-dE/dt = (288\, GI^2\varepsilon^2\omega^6)/45c^5$$ and the characteristic time in which the radiation will take place $$\tau = -E/(-dE/dt) = 45c^5/(576GI\varepsilon^2\omega^4).$$ The time $\Delta\tau$ required to sweep through the bandwidth $\Delta\omega$ of a detector is given approximately by $\Delta\tau = (\Delta\omega/\omega)\,\tau$.	For a neutron star of mass $m = 0.41 M_\odot$, radius $R = 21$ km, moment of inertia $I = 4.4 \times 10^{44}$ g cm², eccentricity $\varepsilon = 10^{-4}$, angular velocity $\omega = 10^3$ rad/sec, the power emitted in gravitational radiation is $$-dE/dt = 3.5 \times 10^{40} \text{ erg/sec}.$$ For a distance of the source from the Earth of 100 pc the calculated flux at the Earth's surface is 2.9×10^{-2} erg/cm²sec. For distance of the source of 1000 pc or 10000 pc the flux is 2.9×10^{-4} erg/cm²sec or 2.9×10^{-6} erg/cm²sec.	Yes—provided one has some means of observing the periodically changing luminosity (time modulated X-ray brightness; radio or optical pulses, etc.) and can thereby tune the gravitational radiation detector to a precise frequency.
Pulsating Neutron Star	The vibrations can be thought to be excited at the moment of formation of the neutron star or by subsequent impact of debris or starquakes. The spectrum is typical of the dilatation modes (mode "gravitational-wave-active" by quadrupole deformation induced by natural rotation) and modes of quadrupole vibrations. The frequency of the radiation is of the general order of magnitude $\omega \sim (\pi G\varrho)^{1/2}$. The rate of emission goes as the square of the relevant Fourier com-	For a neutron star of mass $M = 0.682 \times M_\odot$, radius $R = 8.4$ km, $\varrho_c = 6 \times 10^{15}$ g/cm³ the power emitted in gravitational radiation is $$-\frac{dE}{dt} \sim 3.0 \times 10^{53} \text{ erg/sec} \left\langle \left(\frac{\delta R}{R}\right)^2 \right\rangle$$ Here $\left\langle \left(\frac{\delta R}{R}\right)^2 \right\rangle$ = mean squared fractional departure of surface from sphericity $= \frac{\alpha^2}{5}$	Yes—for properly tuned narrow band "Weber band" detector and source 100 pc away. If source is 1000 pc away, only possible if the amplitude of the signal is as large as indicated and the bar is sufficiently cooled. No—if the source is 10000 pc away.

Table XX (*cont.*)

Source	General features of the emitted gravitational radiation	Amount of gravitational radiation emitted for typical values of the parameters (at 100 pc the energy is spread over a sphere of surface 1.2×10^{42} cm^2)	Reasonable to detect
	ponent of the quadrupole moment. The characteristic duration of the signal is $0.1 \lesssim \tau \lesssim 20$ sec (for details see K.S.Thorne[138] and Table XVIII).	$= 1/500$ if the dynamics of implosion cause, for example, a 10% reduction in length of symmetry axis (relative to sphere); in this case $-dE/dt = 6 \times 10^{50}$ erg/sec; period 0.31 msec; damping time 0.19 sec; $-\Delta E \sim 1 \times 10^{50}$ erg; energy per unit frequency near peak of spectrum $-d\Delta E/d\nu = 8 \times 10^{49}$ erg/Hz. For a source at 100 pc the calculated gravitational energy flux at the Earth is 5×10^8 erg/cm^2 sec (7×10^7 erg/cm^2/Hz). If instead distances of 1000 pc or 10000 pc ar attained the fluxes will be 5×10^6 erg/cm^2 sec (7×10^5 erg/cm^2 Hz) or 5×10^4 erg/cm^2 sec (7×10^3 erg/cm^2 Hz).	
Rapidly Spinning Neutron Star	If the effective γ of the matter constituting the neutron star is larger than 2.2 (see Sub-section 7.7.2) and if the angular momentum of the neutron star is large enough, the equilibrium configuration will be qualitatively like a Jacobi ellipsoid with the axis $a \neq b \neq c$. (if the effective γ is smaller than 2.2, shedding of matter from the equator of the star will take place and no equatorial eccentricity can develop.) The gravitational radiation emitted will be monochromatic, drifting toward *larger* values of the frequency. If we consider for simplicity the case in which one of the principal axes of inertia (z) coincides	Let us consider a neutron star of $M \sim 0.65 M_\odot$, $\bar{R} \sim 15$ km rotating with a period $P \sim 1.5$ msec. The expected eccentricity in the equatorial plane will be of the order of $\varepsilon \sim 0.87$ the power of gravitational radiation emitted will be $dE/dt \sim 0.6 \times 10^{51}$ erg/sec. The period starting from $P \sim 1.4$ msec (extreme Jacobi configuration) will decrease in a time $\tau \sim 0.4$ sec to a value of 1.3 msec. If the source is 100 pc away the flux at the Earth would be as high as 0.5×10^9 erg/cm^2 sec. If the source instead is 1000 or 10000 pc away the flux would be 0.5×10^7 erg/cm^{2}sec or 0.5×10^5 erg/cm^{2}sec.	Yes—this source should be clearly detectable with an antenna of the Weber type properly tuned at a period of ~ 1 msec. Moreover the signal with its characteristic drift towards higher values of the frequency would have a recognisable signature. Possible coincidence with associated electromagnetic emission could be important in the identification of the source.

with direction of the angular momentum, the amount of energy radiated will be given by

$$\frac{dE}{dt} = -\frac{32}{5}\frac{G}{c^5}(I_{11} - I_{22})^2 \Omega^6$$

where $\Omega \sim (\pi G \varrho)^{1/2}$. Assuming that the neutron star to be initially in an "extreme" Jacobi configuration (near the bifurcation point with the pear-shaped equilibrium configurations) the pulse of gravitational radiation will last for a time $\tau \sim 10^2(\overline{R}/(2Gm))^3 \overline{R} c^5$, where $\overline{R} = (a \cdot b \cdot c)^{1/3}$. In this time τ the neutron star will have drifted from a Jacobi configuration to an axially symmetric MacLaurin configuration, losing angular momentum by gravitational radiation and increasing its angular velocity.

"Particle" of mass m spiralling around a black hole of mass M

We consider the case in which the mass m is much smaller than the mass M of the black hole and dimensions of the "particle" are negligible with respect to the dimensions of the black hole—the effects of the tidal forces on the particle m are also considered to be negligible. If the particle is moving in a circular orbit the flux of gravitational radiation is

$$\frac{dE}{dt} = \frac{32\,G}{5c^5}\left(\frac{mM}{m+M}\right)^2 r^4 \Omega^6, \text{ with } \Omega^2 = G(m+M)/r^3.$$

We consider the case of a black hole of mass $M \sim 10^8 M_\odot$ with radius $R \sim 2.94 \times 10^8$ km and a "particle" (white dwarf star) of mass $m \sim 1.0 M_\odot$, radius $\sim 5.0 \times 10^3$ km. The particle is assumed to be in a circular orbit at a distance $r = 3R \sim 8.8 \times 10^8$ km from the center of the black hole. The power emitted in gravitational radiation is

$$dE/dt \sim 2.99 \times 10^{40} \text{ erg/sec}$$

with a period $P \sim 4.5 \times 10^4$ sec. Assuming the source to be 10000 pc away the flux of gravi-

No. The flux is too small to be detected at the Earth.

Table XX (*cont.*)

Source	General features of the emitted gravitational radiation	Amount of gravitational radiation emitted for typical values of the parameters (at 100 pc the energy is spread over a sphere of surface 1.2×10^{42} cm^2)	Reasonable to detect
	The radiation is monochromatic, with the frequency drifting toward higher values. The characteristic time in which the radiation will take place is $\tau = -E/-dE/dt = 5c^5r^4/[64G^3Mm/(m+M)]$.	tational radiation at the Earth's surface would be $\sim 2.4 \times 10^{-6}$ erg/sec	
	The *total* energy that can be emitted by the particle m before falling into the black hole is $0.0572 \times mc^2$ (binding of last stable circular orbit at $r = 6m$) in the case of a Schwarzschild black hole and $0.4235 \times mc^2$ in the case of a particle co-rotating in an extreme Kerr geometry (binding of last stable circular orbit at $r = m$). If the particle is in an elliptic orbit with eccentricity ε the emission will no longer be monochromatic but radiation will be also present in the n-th harmonic of the fundamental frequency (cf. Section 7.10). However, the maximum emission of radiation will take place at the moment of closest approach of the particle and the orbit will tend rapidly to become circular.	We take, as a limiting case, for the black hole a mass $M \sim 10M_\odot$ radius $R \sim 29.4$ km and for the "particle" (neutron star) a mass $m \sim 0.33M_\odot$ with radius ~ 26 km. The "particle" is assumed to be in a circular orbit at a distance $r = 3R \sim 88.2$ km from the center of the black hole. The power emitted in gravitational radiation is $-dE/dt \sim 3.3 \times 10^{53}$ erg/sec with a period $P = 4.5 \times 10^{-3}$ sec. Assuming the system to be 100 pc away the flux of gravitational energy at the Earth's surface would be 2.8×10^{11} erg/sec; for distances of 1000 pc and 10000 pc away there would be a flux of 2.8×10^9 erg/sec and 2.8×10^7 erg/sec.	Yes, but realistic computations should take into account the effects of tidal forces on the particle m (already in the extreme case under consideration the Roche limit is $R_{\text{limit}} \sim 2.45 \left(\frac{\varrho_{\text{B.H.}}}{\varrho_{\text{N.S.}}}\right) R_{\text{B.H.}} \sim 1.5 \times 10^3$ km!)

"Particle" m falling radially into a Schwarzschild black hole of mass M	We have analysed this problem in the framework of linearised theory, on the explicit assumptions that $m \ll M$, the dimensions of the particle are negligible with respect to the dimension of the black hole and the tidal effects on the particle are negligible. The total energy radiated in this process is $$\Delta E \sim 0.0025 mc^2 (m/M)$$ and the frequency has a peak at $\omega \sim 0.15/M$ (cf. Section 7.11).	Let us consider the emission of gravitational radiation due to a "particle" (neutron star) of mass $m \sim 0.68 M_{\odot}$ and radius $\sim$8.4 km into a black hole of mass $M \sim 10 M_{\odot}$ and radius $R \sim 29.4$ km. The total amount of gravitational radiation will take place in a time $\tau \sim 5 \times 10^{-4}$ sec and will be of the order of 2.3×10^{50} erg with a characteristic spectral distribution (see text) peaked around $P \sim 1.4 \times 10^{-4}$ sec.	Yes. However, our computations are valid only for an approximate order of magnitude and they should be extended to a complete relativistic regime.
		If we increase the mass of the black hole up to $M \sim 10^8 M_{\odot}$, radius $R \sim 2.94 \times 10^8$ km, the "particle" m (neutron star) of mass $m \sim 0.68 M_{\odot}$ and radius $\sim$8.4 km will emit a total amount of gravitational energy of the order of $\sim 2.3 \times 10^{43}$ erg in a time $\tau \sim 5 \times 10^3$ sec and frequency peaked around $P \sim 1.4 \times 10^3$ sec.	No! No realistic source of gravitational radiation can be expected for "particles" falling in a large mass black hole. The energy emitted is largely reduced by the factor $\dfrac{m}{M}$.

Negligible compared to the sources considered above are primordial gravitational radiation (cf. Section 7.15 for upper limits), atomic and molecular processes (cf. Section 7.14 for references), earthquakes, impact of a meteorite $\sim$1 km in diameter upon the Earth (cf. Section 7.13 for details), and solar flares (due to smallness of mass involved, $\sim 10^{26}$ g and the length of the time-scale of the dynamics, from minutes to hours).

Table XX summarises the sources of gravitational radiation considered so far, giving for each the frequency span and the order of magnitude of flux expected at the surface of the Earth. Valuable information on interesting astrophysical events flows by the Earth in a new channel waiting to be exploited. All these gravitational waves, of whatever wavelength, manifest their effects in the last analysis through changes of length. The longer the wavelength the longer is the base-line appropriate for the detection of that radiation. For some effects it is natural to consider a base-line located in space itself—the distance between two space stations—as the meter for the wave amplitude. The present-day accuracy of laser ranging (cf. Chapter 10) does not come close to rivalling the efficiency of Weber's laboratory bar for measuring fractional changes in length. However, it is not unreasonable to expect increases of many orders of magnitude in laser accuracy for future measurements of the distance between space stations. Hopes of such improvements and daily optical observations of pulsars that must have been born in a gigantic splash of gravitational radiation supply a double incentive for planning gravitational radiation studies.

The phenomena most readily detectable by equipment to Weber's are the "splashes" of kHz radiation from the collapse of stellar-mass objects in our Galaxy. The event rate expected on the basis of conventional astrophysical ideas is, however, only of the order of one per century—in striking contrast to the $\gtrsim 1$ pulse *per day* which Weber has reported!

CHAPTER 8

The Traditional Three Tests of Relativity

MEASUREMENTS OF the bending of light by the Sun, the gravitational red shift and the precession of the perihelion of Mercury as they stood in 1962 have been reviewed by Bertotti, Brill and Krotkov[147]. In the meantime two distinct California Institute of Technology groups[148,149]. working with different antennas and different frequencies, 9.6 GHz and 2.4 GHz, have measured the bending of radio waves by the Sun with improved accuracy. They took advantage of the pointlike character of the quasistellar sources 3C273 and 3C279 and the fact that the Sun passed close to them during the period 2-10 October 1969. The group operating at the shorter wavelength used an interferometer base line of the order of 1 km; the other, a base line of the order of 20 km. The Sun at closest approach came 4.5 solar radii from the nearer source (3C279), which in turn was 9.3° away from the other source (3C273). The bending of the radio waves is caused in part by the gravitational pull of the Sun (value predicted on the basis of standard general relativity at the solar limb 1.75″) and in part by refraction in the plasma of the Sun's corona. The effective refractive index (relative to an idealised flat space vacuum background) is

$$n(r) = 1 + \frac{2M_\odot}{r} - \frac{2\pi e^2}{m_e\omega^2} N_e(r)$$

Here the number density of electrons $N_e(r)$ can be represented for $r > 3R_\odot$ by the entirely empirical formula

$$N_e(r) = A/r^6 + B/r^{2.33}$$

with reasonable accuracy, as judged by the coronal light in solar eclipses, by space-probe measurements, by solar wind theory, and by observations of radio-astronomical scintillations. The coefficients A and B can vary by as much as a factor of 5 during the 11-yr solar cycle. Happily, the week of observation was "quiet" in terms of solar activity and there is no indication that A or B changed during that period. Consequently each group could decompose the observed deflections into a part attributable to gravitation and a part assignable to coronal refraction. The 9.6 GHz measurements gave 1.77″ ± 0.20″ (standard deviation) for the gravitational deflection as extrapolated to a ray passing the solar limb; i.e. 1.01 ± 0.11 times the Einstein value[148]. The 2.4 GHz observations yielded $1.04^{+0.15}_{-0.10}$ times the Einstein value[149]. The two groups, operating at frequencies differing by a factor of 4 so that the coronal refraction was in one case 16 times as much as in the other, nevertheless agreed as to the magnitudes of the gravitational bending and the coefficient in the coronal effect.

No decisive improvements have been made in knowledge of the gravitational redshift since the measurements of Pound and Rebka as cited in the review of Bertotti, Brill and Krotkov[147], and the experiment of Pound and Snider[150] in which they measured this effect with an accuracy of 1%. For the precession of the perihelion of Mercury, the old figure of 43.11″ ± 0.45″ per century, once regarded as in agreement with the relativistic figure of 43.03″, is no longer so regarded. In the meantime Dicke and Goldenberg[151] discovered that the Sun is oblate and have measured its oblateness. Dicke interprets the oblateness as caused by a solar core turning with a period of ~ 1 day, as compared with the familiar period of ~ 28 days in which the surface rotates. Regardless of all questi ons still outstanding as to the internal mechanism that produces the oblateness, there is little option about its external consequences *if* the observations indeed imply a solar quadrupole moment (for a contrary interpretation see Ingersoll and Spiegel[152]). The quadrupole moment inferred from the shope of the surface produces a $1/r^4$ component in the force additional to the $1/r^2$ force generated by a spherically symmetric center of attraction. This added component in the force is calculated to make a contribution of ~ 3″ to the precession of the perihelion of Mercury. Even before this last correction many other corrections had to be applied to get the "observed" 43″/century from the directly observed motion of Mercury, the largest among them being 5025.625″ ± 0.050″/century for the general precession of the equinox with respect to the distant stars and an amount of the same order of magnitude for the effect of other planets on Mercury. The oblateness of the Sun produces a new correction which brings the "old" (pre-Lincoln Laboratory) value of the precession down to ~ 40″/century, as compared to the Einstein value of 43″/century. It may be necessary to wait for the new Lincoln Laboratory reduction, at present under way (I. Shapiro and collaborators[153]), of all past observations, both telescopic observations of angle made over recent centuries, and radar determinations of velocities over recent years, before one can venture a judgment on the situation. They already find substantial alterations to be necessary in some of the important corrections that come into the motion of Mercury. However, their present tentative figure for the net precession agrees with the old figure and has about the same uncertainty.

CHAPTER 9

The Retardation of Light as it Passes by the Sun on its way to and Return from Venus

To THE classic three tests of relativity Einstein would surely have added a fourth, had radar ranging been an established part of technology in 1916: i.e. time delay of a radar pulse on its way from the Earth to Venus and back as it passes through the gravitational potential of the Sun. The time delay should be about 2×10^{-4} sec or 60 km of light travel time when the path of the ray goes close to the Sun. In order for the radar receptor on the Earth to detect the signal reflected from Venus free from interference from the radio emanations of the Sun, Venus must be a certain distance away from the solar limb ($\sim 1°$ for the Haystack, Massachusetts and Arecibo, Puerto Rico antennas). The superior conjunctions of Venus anyway normally bring that planet about 1° from the Sun (1° to 0.5° in the case of Mercury). The value calculated from Einstein's theory for the retardation is

$$\Delta t \simeq 4M_\odot \left[-\frac{3x_e + x_p}{2x_e} + \ln \frac{4x_e x_p}{b^2} \right]$$

(geometrical units; cm of light travel time; $M_\odot = 1.47$ km). Here b is the distance of the Earth $\rightleftarrows$ Venus beam from the center of the Sun at the point of closest approach ("impact parameter"; idealisation of flat space!); and from this point to the Earth and to the planet the distances are indicated respectively by x_e and x_p. The retardation is measured relative to the time of flight as it would be on the basis of standard Newtonian theory (planetary distance projected through region of conjunction on basis of classical interpolation; values of 23 astronomical parameters used in this Newtonian analysis based on more than 400 radar and 6000 optical observations). The solar corona was calculated to contribute less than 1 μs to the time delay measured by Shapiro and his Lincoln Laboratory colleagues[154,155]. They measured an effect equal to 0.9 ± 0.2 times the value predicted by Einstein's standard geometrodynamics.

The pulse signal power in these 1967 experiments was $\sim$ 300 kW. The echo signal returned from Venus was sometimes as low as 10^{-21} W, i.e. down by a factor of about 10^{27}. To improve the intensity, the Lincoln Laboratory group has recently proposed to transmit from Haystack, Massachusetts and receive with the much larger antenna at Goldstone, California.

CHAPTER 10

Relativistic Effects in Planetary and Lunar Motions

10.1 PLANETARY ORBITS

The energy received from a planet by reflection goes down as $1/r^4$. The energy received from an artificial pulsed source in orbit around the planet or sitting on the planet goes down as $1/r^2$. Therefore there is great appeal in the idea of tying a "transponder" in one way or another to a planet to receive a pulsed signal from the Earth and retransmit it, re-powered, to the Earth. Already with radar ranging one can determine distances to better than 15 m, and velocities to $\sim \frac{1}{2}$ mm/sec. With a transponder one should be able to reach out to greater distances and attain higher accuracy, $\sim$ 5 m. One is thus on the way to an order of magnitude improvement over traditional astronomy (angle measurements!) in the precision with which one knows the constants of the solar system, e.g. dimensions of orbits and masses of planets. In this way one should begin to see more and more relativistic effects emerging above the uncertainties of measurements.

From the passage of Mariner V past Venus and the resulting perturbation in its orbit the Jet Propulsion Laboratory group of Anderson and colleagues at Pasadena have already been able to determine[156,157] the mass of Venus with new precision.

$$GM_{\venus} = 324859.61 \pm 0.49 \text{ km}^3/\text{sec}^2$$

and at the same time (from the periodic monthly motion of the Earth towards Mariner V and away from it in the course of its long voyage) the ratio between the mass of the Earth and the mass of the Moon

$$M_{\oplus}/M = 81.3004 \pm 0.0007$$

Going beyond these classical Newtonian phenomena, general relativity predicts new effects in celestial mechanics. We owe the following precis of the literature to Anderson and Thorne of California Institute of Technology. In 1916 de Sitter[158] made a first analysis of some of the relevant effects. In 1928 Chazy published a two-volume book on the subject[159]. In 1963 Anderson and Lorell[157] gave the first order solution to the equations of motion in Schwarzschild geometry to the order e^2, where e is the eccentricity of the orbit. Reviews of the past work plus analysis of n-body effects and post-Newtonian corrections to the equations of motion have been given by Moyer[160] and Tausner[161] Thorne and Will give a 1969 survey of present status and future prospects and foresee the possibility ultimately of determining effects which in order of magnitude are one thousandth of the perihelion precession of Mercury.

The simplest non-cumulative effect is a periodic perturbation on the Schwarzschild coordinate given to first order in the eccentricity by

$$\delta r = -4em \cos M$$

Here M is the mean anomaly in the motion (standard orbit terminology; a monotonically increasing function of time), m is the mass of the center of attraction in geometrical units (1.47 km for the Sun, 0.047 cm for Mars) and e is the eccentricity of the orbit ($e = 0.093$ for Mars). The formula gives an amplitude of 550 m for the general relativistic effect in the orbit of Mars, but only a few mm for this effect in the path of Orbiter going past Mars.

Transponder in orbit around the planet (or the Sun)? Or sitting on the planet? In orbit the vehicle is subject to the solar wind, which rises and falls in strength, buffeting the spacecraft; and the pressure of solar radiation also pushes on it. Uncompensated on the way from the Earth to Mars, this force will push the object off schedule by kilometres, thereby destroying any chance of picking up the effects of relativity, unless compensated (dragfree satellite) or monitored. On the other hand, a transponder landed on a planet needs power, presumably from a battery, because no one has succeeded in operating a solar cell under a planetary atmosphere. It may turn out that a transponder orbiting Mars, for example, may prove a more practical method for keeping track of the orbit of this planet with high precision than a transponder landed on the planet. Whichever method is employed, one can look forward to the bringing to light of a host of new dynamical effects in the solar system, several of them of truly Einstein origin.

10.2 SEARCH VIA CORNER REFLECTOR ON THE MOON FOR RELATIVISTIC EFFECTS IN THE MOTION OF THE MOON PREDICTED BY BAIERLEIN

Already in 1962 Smullin and Fiocco[163] from M.I.T. had demonstrated that laser beams could be scattered from the surface of the Moon and detected back at the Earth. This experiment, though interesting, was of little use for testing general relativity or theories on the motion of the Earth-Moon system. The main reasons were the extremely weak return signal and the stretching in time of the return signal due to the curvature and inequalities of the reflecting surface on the Moon. One could imagine eliminating the first difficulty by using a more powerful laser than that originally built by Smullin and Fiocco. However, it was extremely difficult to overcome the second difficulty, namely to find a uniform and regular reflecting surface on the Moon.

After the 20 July 1969 lunar landing of Apollo 11 the situation changed drastically. An aluminium panel of 46 by 46 cm with 100 fused silica corner cubes each 3.8 cm in diameter was placed on the Moon (Figure 34). The reflectors are expected to have a lifetime in excess of ten years. The distance between the laser source on the Earth and the reflector in the Moon is now known with an accuracy of 15 cm, and will be followed systematically during the coming years.

The first observations were made at Lick Observatory (both source and receptor). The signal was sent at intervals of 30 seconds by a high-powered ruby laser. The spot of light on the Moon's surface was about 3.2 km in diameter. The return signal was received by the 100-inch telescope and detected by a photomultiplier. The photomultiplier and counting electronics was activated $\sim$ 2.5 sec after the pulse left the Earth for the Moon, approximately the time for the light to cover the distance Earth-Moon-Earth.

With such unprecedented accuracy it will be possible to study the orbital motion of the Moon, the vibrations of its surface, the change in the rotational motion of the Earth, and intercontinental distances and how they change with time. Baierlein as well as Krogh and Baierlein[164] have analysed general relativity effects in the Earth-Moon system. They conclude that there is no significant hope of observing general relativistic effects in the *light* propagation itself. However, general relativistic effects in lunar *motion* are quite significant, the major one being of the order

$$\sim 100 \text{ cm} \cos 2D$$

where D = mean longitude of Moon—mean longitude of Sun. This effect appears to be in reach of the present laser operation. Krogh and Baierlein, and also Nordtvedt[165] have analysed the difference between the Einstein predictions and those of the so-called scalar-tensor theory of gravitation. We should also mention the experiment, being prepared by a group at Stanford University,[166] to search for precession effects, predicted by relativity, in a gyroscope circling the earth in polar orbit. There are two effects: one, the geodetic precession (somewhat analogous to the Thomas effect) in the plane of the orbit, and the other, the Lense-Thirring precession induced by the spin of the Earth itself.

Figure 34 Corner reflector landed on the Moon by Apollo II in July 1969.

CHAPTER 11

The Expanding Universe

HUBBLE'S 1929 announcement[167] of an apparent proportionality between the redshift and distance of galaxies (which had been foreshadowed by the earlier work of Slipher, Wirtz and Lundmark) led to the concept of an *expanding* Universe. Subsequent developments have confirmed and strengthened Hubble's original argument and have also shown that this expansion is—on the largest observable scales—remarkably isotropic. This universal expansion really constitutes a far more striking test of relativity than the three traditional "crucial tests"—the bending of light, the gravitational redshift, and the presession of Mercury's perihelion. The observed isotropy allows us to describe the overall dynamics of the Universe by the simplest cosmological model which relativity theory permits. This model has faced three distinct waves of doubt: (i) "The Universe surely is not dynamic"; (ii) "the expansion must be speeding up (or else the stars would be older than the Universe!) whereas the theory predicts it should slow down"; and (iii) "there cannot be enough matter in the Universe to satisfy Einstein's argument for a closed Universe." These doubts have all now been allayed, vindicating relativistic cosmology. We therefore adopt this model as our basis for discussing the current evidence on the structure and evolution of our Universe. As the discussion progresses, it will become clear just how strong the evidence for large-scale homogeneity and isotropy really is.

11.1 METRIC AND FIELD EQUATIONS FOR HOMOGENEOUS ISOTROPIC UNIVERSE

As a first approximation ,we take a large-scale viewpoint, and treat the contents of the Universe (matter and radiation) as a homogeneous fluid. (A second approximation—see chapter 14—treats the observed structures on scales $\lesssim 10^8$ light years as perturbations on the large-scale background). *Homogeneity* of the Universe means that *through each event in the Universe there passes a spacelike hypersurface whose events are physically indistinguishable from each other* (i.e. the density, the pressure, the curvature of spacetime, etc. must be the same). The concept of *isotropy* can likewise be made more precise as follows: isotropy of the Universe means that *at any event an observer comoving with the cosmic fluid cannot distinguish any one of his space directions from the others by any local physical measurement.* It can be shown that strict isotropy in fact implies homogeneity. Also, the isotropy guarantees that the world lines of the fluid are *orthogonal* to the homogeneous hypersurfaces.

The natural way to set up a coordinate system is to choose space coordinates comoving with the fluid (i.e. which merely label the world lines of the fluid) and to choose the time coordinate so that t measures the proper time of any fluid element, and is constant over a given homogeneous hypersurface. The line element for spacetime then has the form

$$ds^2 = -dt^2 + g_{ij}\, dx^i\, dx^j \tag{11.1}$$

(i.e. $g_{tt} = 1$, because t measures time along lines of constant x^i). The isotropy and homogeneity are, however, such severe constraints that they allow no form of (11.1) other than the *Robertson-Walker* metric. There are only three possible versions of this metric, corresponding to universes whose space-like sections have positive, zero, or negative curvature respectively. This metric can be written as a natural generalisation from a 2-sphere of radius a (polar angles χ and φ) to a 3-sphere of radius a (polar angles χ, θ and φ); or, with the addition of time (metric locally Minkowskian!)

$$ds^2 = -dt^2 + a^2(t)\, [d\chi^2 + \sin^2 \chi(d\theta^2 + \sin^2 \theta\, d\varphi^2)] \tag{11.2}$$

If instead of using the polar angle χ one uses a parameter $u/2 = \tan \chi/2$ to measure relative distance from the "North Pole" (projection from the South Pole onto a plane tangential to the sphere at the North Pole) one has

$$ds^2 = -dt^2 + a^2(t) \frac{du^2 + u^2(d\theta^2 + \sin^2 \theta\, d\varphi^2)}{\left(1 + \left(\frac{k}{4}\right)u^2\right)} \tag{11.3}$$

with $k = 1$. The same formula applies to a spacetime whose space-like sections are flat ($k = 0$) and to a spacetime whose space-like sections have hyperbolic curvature ($k = -1$). (The metric in these two cases can be expressed in the form (11.2) if $\sin \chi$ is replaced by χ and $\sinh \chi$ respectively).

From (11.2) we see that the surface area of a 2-sphere centred on the origin and of coordinate radius χ is

$$4\pi(a^2(t)) \begin{cases} \sin^2 \chi \\ \chi^2 \\ \sinh^2 \chi \end{cases} \tag{11.4}$$

However the Robertson-Walker metric, derived from group-theoretical considerations alone, is not in itself enough to tell us the observable properties of a homogeneous isotropic Universe. To do this, we need in addition some dynamical equations which determine the function $a(t)$; and we must also specify the density $\varrho(t)$ and pressure $p(t)$ of the cosmological fluid. The Einstein equations then lead to two relations:

$$\left(\frac{\dot{a}}{a}\right)^2 - \frac{8}{3}\pi\varrho = -\frac{k}{a^2} \tag{11.5}$$

and

$$\dot{\varrho} + 3(p + \varrho)\frac{\dot{a}}{a} = 0 \tag{11.6}$$

The first of these has the structure of an "energy equation", being of the form (kinetic energy) + (potential energy) = total energy = constant, except for two circumstances: (i) It is divided through by the square of the "world radius" a because it deals with curvature. (ii) The quantity $(-k/a^2)$ that would normally be taken to represent the total energy, and would be adjustable, is *not* adjustable. There is no way of reaching in from outside to adjust it. The constant k has the value 1 for a closed universe ("negative apparent-energy"—insufficient energy to permit the system to fly apart; expansion always followed by recontraction) and -1 or 0 for "open" universes which continue to expand indefinitely. Specifically, standard geometrodynamics gives

$$\begin{pmatrix}\text{extrinsic}\\\text{curvature}\end{pmatrix} + \begin{pmatrix}\text{intrinsic}\\\text{curvature}\end{pmatrix} = 16\pi\begin{pmatrix}\text{energy}\\\text{density}\end{pmatrix}$$

or [inserting G and c explicitly in (11.6)].

$$6\left(\frac{\dot{a}}{a}\right)^2 + \frac{6kc^2}{a^2} = 16\pi G\varrho_{CGS} \tag{11.7}$$

Equation (11.6) is just a consequence of the first law of thermodynamics applied to any small element of the fluid with comoving boundaries. (Only knowledge of $p(t)$ and $\varrho(t)$ is required: the material need not be a perfect fluid with a well defined equation of state, nor in equilibrium.)

The solutions to (11.5) and (11.6) are easily derived in the interesting special cases of the "dust universe" ($p = 0$, and so, from (11.6), $\varrho \propto a^{-3}$) and the "radiation universe" ($p = \varrho/3$, $\varrho \propto a^{-4}$). These solutions—the Friedmann universes—are given in Table XXI, for the cases $k = -1, 0, +1$. Note that the initial behaviour (when the parameter η is $\ll 1$) is independent of k, but that the behaviour deviates when $\eta \simeq 1$ (i.e. when $a \simeq a_0$): the $k = +1$ models recontract, whereas for $k = 0$ or 1 the expansion continues indefinitely. All models always have the property that the expansion is *decelerating*—i.e. $a\ddot{a}/\dot{a}^2 < 0$. How this slowing down comes about is in no

Table XXI Radius a (cm) as a function of time t (cm) for homogeneous isotropic models of the Universe

k	p	Radius, a	Time, t	Hubble time, $H^{-1} = a/(da/dt)$
1	0	$(a_0/2)(1 - \cos\eta) \simeq a_0\eta^2/4$ for small η	$(a_0/2)(\eta - \sin\eta) \simeq a_0\eta^3/12$ for small η	$(a_0/2)(1 - \cos\eta)^2/\sin\eta$ $= t(1 - \cos\eta)^2/[\sin\eta(\eta - \sin\eta)] \geqq 1.5t$
0	0	$a_0\eta^2/4 = (9a_0t^2/4)^{1/3}$	$a_0\eta^3/12$	$1.5t$
-1	0	$(a_0/2)(\cosh\eta - 1) \simeq a_0\eta^2/4$ for small η	$(a_0/2)(\sinh\eta - \eta) \simeq a_0\eta^3/12$ for small η	$(a_0/2)(\cosh\eta - 1)^2/\sinh\eta$ $= t(\cosh\eta - 1)^2/[\sinh\eta(\sinh\eta - \eta] \leqq 1.5t$
1	$\varrho/3$	$a_0 \sin\eta \simeq a_0\eta$ for small η	$a_0(1 - \cos\eta) \simeq a_0\eta^2/2$ for small η	$a_0 \sin^2\eta/\cos\eta$ $= t \sin^2\eta/[\cos\eta(1 - \cos\eta)] \geqq 2t$
0	$\varrho/3$	$a_0\eta = (2a_0t)^{1/2}$	$a_0\eta^2/2$	$2t$
-1	$\varrho/3$	$a_0 \sinh\eta \simeq a_0\eta$ for small η	$a_0(\cosh\eta - 1) \simeq a_0\eta^2/2$ for small η	$a_0 \sinh^2\eta/\cosh\eta$ $= t \sinh^2\eta/[\cosh\eta(\cosh\eta - 1)] \leqq 2t$

($k = 1$, sphere; 0, flat space-like sections; -1 hyperbolic space-like sections; $p = 0$, dust; $p = \frac{1}{3}\varrho$, isotropic radiation.) In the equations the parameter η measures time in units of arc length travelled around the Universe: $d\eta = d$ (distance travelled)/radius $= dt/a(t)$. This parameter increases by 2π during the whole time of expansion and recontraction of the "dust-filled Universe" (time for photon barely to get around once) and in the radiation-filled Universe increases only by π (time only to get to antipodal point of Universe). The quantity a_0 is a constant which in the case $k = 1$ (3-sphere) measures the radius of the Universe at the phase of maximum expansion.

way more easily seen than by recalling that rocks driven apart into space by explosion of a planetoid are slowed down in their expansion by their mutual gravitational attraction. The same point shows up straightforwardly in the formal theory. (Indeed as Milne and McCrea[168] were the first to show, many of the results of Table XXI can be derived by applying Newtonian arguments to a small comoving sphere of matter, assuming the remainder of the Universe to exert no net effect on the local dynamics. General relativistic concepts are, however, essential when we come to consider light propagation over large distances.)

Introduction of the so-called "cosmological constant" into the field equations leads to an extra term $\Lambda/3$ on the right hand side of (11.5). This allows a wider range of cosmological models. If $\Lambda > 0$, it is possible for $a(t)$ to have a point of inflexion, and for the Universe to eventually expand in an *accelerating* fashion under the influence of the "cosmological repulsion". Of particular interest are the Lemaitre models, in which the Universe indulges in a long quasi-static "coasting phase" ($(\dot{a}/a) \ll (G\varrho)^{1/2}$) before resuming its expansion. These models have recently been discussed in connection with the problem of galaxy formation, and the distribution of quasar redshifts. There is, however, no observational evidence favouring a non-zero Λ. Einstein introduced the terms Λg_{ij} into his field equations only when he found that the equations otherwise would not permit a homogeneous isotropic Universe to be static. Except for its giving a static Universe, the new term was unreasonable, and had no correspondence with the rest of physics ("a force acting on everything but acted on by nothing"). When Hubble's work provided overwhelming evidence that the Universe really is expanding, Einstein abandoned the cosmological term, calling it "the biggest blunder of my life". We shall assume $\Lambda = 0$ throughout the following discussion.

The mean smoothed-out density of the matter within observed galaxies is *at least* 10^{-31} gm cm^{-3} (see section 15.1). On the other hand the mass-energy density of electromagnetic radiation in intergalactic space appears to be $\lesssim 10^{-33}$ gm cm^{-3} (see section 15.4 and chapter 16). Thus, unless there is a large background energy density in some undetected *relativistic* form (with $p \simeq \varrho/3$)—neutrinos or gravitational waves for instance—our *present* Universe is a "dust universe" in the sense that pressure is dynamically negligible. The existence of the apparently thermal microwave background radiation (see chapter 13) points towards a very different situation at early epochs when the Universe was vastly more compact than it is now ($a/a_{\text{now}} \lesssim 10^{-3}$); but we are probably justified in assuming $p = 0$ throughout the period when we observe discrete sources—galaxies, radio sources, quasars, etc.—directly ($a/a_{\text{now}} \gtrsim 0.2$). This means that except in its earliest phases, the Universe may be adequately described by one of the models in the top half of Table XXI. The basic questions then are:

1) Is the Universe "closed" ($k = +1$) or "open" ($k = 0$ or -1)?

2) What stage has the Universe reached in its evolution—in other words what is the present value of η (or, equivalently, of a_0/a_{now})?

The "classical" approach to the cosmological problem, pursued by Hubble and his successors, has entailed attempting to determine two observational parameters which are related to the "theoretical parameters" k, η and a_0. These are the *Hubble constant*

$$H = \dot{a}/a$$

and the *deceleration parameter*

$$q = -\frac{a\ddot{a}}{\dot{a}^2}$$

If the Universe is indeed closed, then q must exceed $\frac{1}{2}$. In recent years, the exploitation of new spectral windows has permitted astronomers to search for types of object (or diffuse material) which might be quite undetectable in the optical band. This allows us to compile a more complete inventory of the contents of the Universe (though some elusive forms of matter may still escape detection!) and could eventually answer the question—is the average density of mass-energy in space sufficient to curve up the Universe into closure? In such discussions it is convenient to define a dimensionless "density parameter"†

$$\Omega = \frac{8\pi G}{3H^2}\varrho$$

The "dust" models in Table XXI have the property that

$$\Omega = 2q$$

and

$$a_0/a_{\text{now}} = \frac{\Omega}{|\Omega - 1|}$$

Einstein's field equations by themselves do not require closure; but Einstein himself always reasoned that the theory was more than differential equations—it required boundary conditions too. He put forward arguments for considering closure to be the most natural of boundary conditions. If he was justified in these arguments the possibilities $k = 0$ and $k = -1$ are not allowed! In this case we would have to find $q > \frac{1}{2}$, $\Omega > 1$: the actual

† A different parameter, σ, is sometimes used in the literature. Our Ω is equal 2σ. In the following sections, Ω and H will denote the *present* values of these quantities (i.e. we omit the suffix "now").

values of these parameters would merely tell us how for the Universe has evolved along the cycle towards the epoch when the expansion will be replaced by recontraction to another singularity. (The curve $a(t)$ for $k = 1$ and $p = 0$ is a cycloid).

11.2 THE HUBBLE CONSTANT

If the Universe is indeed expanding uniformly, then the radiation from a source at distance r should be redshifted by an amount

$$\frac{\Delta\lambda}{\lambda} = z = Hr \tag{11.8}$$

This simple relation only holds for $z \ll 1$, when the definition of r is unambiguous; for larger z, r can only be defined in terms of a particular cosmology. For small z also, we can apply the ordinary inverse square law to infer that the apparent brightness l of a given type of object should vary as $l \propto z^{-2}$. The current evidence for the Hubble law is illustrated in Figure 35, which shows the corrected optical magnitude [$\propto -2.5 \log_{10}$ (apparent brightness)] for the brightest—or "first ranking"—member of clusters of

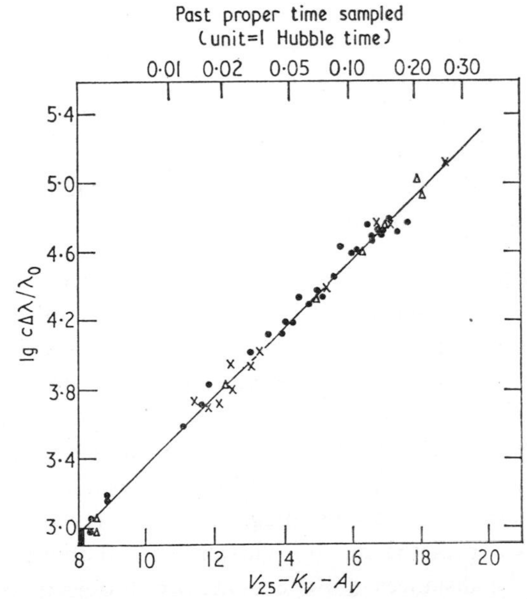

Figure 35 The Hubble diagram for brightest galaxies in clusters, taken from Sandage.[169,170] The quantity plotted on the horizontal scale is the visual magnitude, modified by the "k-correction" and a correction, depending on galactic latitude, for interstellar absorption.

galaxies. The accuracy with which the points lie on a line implies (*a*) the astrophysical contingency that the galaxies in the sample are indeed "standard candles", to $\lesssim 30$ per cent precision, *and* (*b*) that the Hubble law (11.8) is obeyed to better than $\sim$ 15 per cent. Hubble's original statement of his law was based only on relatively "local" objects with $z \lesssim 0.003$. (The possible deviations of the points from a line at the larger redshifts, which may at least in part reflect the *deceleration* of the expansion, is discussed later).

The data in Figure 35 provide firm evidence for the Hubble law, but they do not tell us the numerical value of H unless we have some independent measure of the distance (or intrinsic luminosity) of a typical bright galaxy. The determination of astronomical distances in a complex and uncertain procedure, which we shall not enter into in detail here. In brief, having determined the distance to some nearby stars kinematically, one then chooses a standard and easily recognisable class of star to gauge the distance of more remote clusters of stars. One can then pick a rarer, but intrinsically brighter, class of object to further extend the distance scale. The principal steps in the "cosmic distance ladder" are as follows.

1) *Kinematical methods* For the nearest stars, the trigonometrical parallax caused by the Earth's motion round the sun is detectable (by definition, a distance of 1 pc is that at which the semi-major axis of the Earth's orbit subtends one second of arc. One parsec is about 3.26 light years). This method is feasible out to $\sim$ 30 pc from the Sun. The method of "statistical parallax" and the so-called "moving cluster method" (which require information, derived from spectra, on the radial components of the velocities) may be applied out to $\sim$ 200 pc.

2) *Main sequence photometry* The kinematical methods allow direct determinations of the distances of enough stars to establish an empirical relation between absolute luminosity and spectral type (or colour) for "main sequence" (hydrogen burning) stars. One can then, in principle, determine the distance of any main sequence star whose *apparent* magnitude and spectral type are measured. In practice, it is more reliable to determine the distance of a cluster of stars by comparing the colour–*apparent* magnitude diagram for its member stars with the same diagram for the Hyades cluster, whose distance is known from the moving cluster method to be $\sim$ 40 pc. This method suffices for distances $\lesssim 10^5$ pc—i.e. for all objects within our own Galaxy.

3) *Variable stars* To extend the distance scale to nearby external galaxies, one needs distance indicators which are intrinsically brighter than

main sequence stars. The main objects used for this purpose are the Cepheid variables, for which there is a well-defined relation between the pulsation period and intrinsic luminosity. These can be detected out to ~ 4 Mpc. This distance is large enough to encompass the members of our own Local Group of galaxies (which includes the Andromeda galaxy M31).

4) *Globular clusters, HII regions, etc.* The key step in determining H is to obtain an estimate of the distance of the Virgo Cluster of galaxies. This is because (a) the Virgo Cluster is the nearest "rich"† cluster of galaxies, whose brightest member is likely to be typical of the galaxies used in the magnitude-redshift relation (Figure 35); and (b) the velocities of all galaxies closer than the Virgo Cluster are likely to be severely affected by local perturbations, whereas beyond the Virgo Cluster the universal recession velocity swamps the random component‡. It is therefore unfortunate that Cepheid variables are not bright enough to be visible in galaxies belonging to the Virgo Cluster. Among the intrinsically brighter objects which have been used as distance calibrators are *brightest stars*, *novae*, *supernovae*, *HII regions* and *globular clusters*. The last of these is perhaps the most reliable at present: by assuming that the brightest globular clusters in M87 are intrinsically similar to those in M31, Sandage has estimated the distance of M87 (NGC 4486) (which appears to be in the centre of the Virgo Cluster) as 17 Mpc.

An error in any rung of the "cosmic distance ladder" affects all higher rungs, so it is obvious that we still do not know the distance of the Virgo Cluster, nor the value of H, with any great confidence. Assuming that NGC 4472, in the Virgo Cluster, is typical of "brightest cluster member" (elliptical) galaxies, Sandage[169] finds

$$H = 75.3^{+19}_{-15} \text{ km s}^{-1} \text{ Mpc}^{-1}$$

However, de Vaucouleurs[171] finds evidence that the absolute magnitude of the brightest globular cluster in a galaxy is correlated with the luminosity of the galaxy. He also finds, as a result of recent observations indicating the presence of an extensive halo, that in fact NGC 4486 is the brightest galaxy in the Virgo Cluster. De Vaucouleurs obtains the value

$$H \simeq 50 \text{ km s}^{-1} \text{ Mpc}^{-1}$$

Estimates of H are likely to become firmer when other distance indicators can be calibrated. In particular, work by van den Bergh suggests that spiral

† The "richness" of a cluster is defined in terms of the number of galaxies in the cluster within two magnitudes of the third brightest member.

‡ See Chapter 17 for the evidence on this point from the isotropy of the background radiation.

galaxies can be classified, by form, in such a manner that each class is a good "standard candle". It should then be possible to plot a separate Hubble diagram for spirals instead of bright ellipticals. A recent attempt by Tammann and Sandage (see ref. 172) to measure the distance of the nearby spiral galaxy M101 (by using the various distance calibrators involved in "rung 4" of the distance ladder) suggest that this may lower the estimate of H to $\sim$ 50 km s^{-1} Mpc^{-1} or even less.

Van den Bergh[173] has combined nine different methods of estimating H, obtaining a weighted average

$$H = 95^{+15}_{-12} \text{ km s}^{-1} \text{ Mpc}^{-1}$$

It would be rash to regard H as known with better than $\sim$ 50 per cent precision. All than can safely be said is that it is unlikely to be far outside the range 50–100 km s^{-1} Mpc^{-1}, with recent work tending to favour the lower end of this range. There is reason to hope for improved precision in the foreseeable future. Sandage[172] remarks "I am also reasonably confident that programs now in progress [using various new methods to bridge the gap between "local" galaxies, where Cepheids, HII regions and brightest stars can be used, and the more distant galaxies where the redshift is a linear and single valued function of distance] should give definitive results within several years".

One recalls that Hubble's 1936 estimate of H was 530 km s^{-1} Mpc^{-1}. Among the main reasons for the drastic subsequent reduction are (a) the fact that Hubble apparently mistook some H II regions for bright stars [rung (4)] and (b) an error, discovered by Baade in 1952, in the estimated luminosity of Cepheids [rung (3)]. When a definite value of H is required, we shall hereafter take 75 km sec^{-1} Mpc^{-1}.

11.3 THE AGE OF THE UNIVERSE

The characteristic time-scale for the universal expansion is

$$\tau_H = H^{-1} = \frac{\text{(distance to typical galaxy)}}{\text{(recession velocity of that galaxy)}} = \begin{pmatrix}\text{time for zero separation} \\ \text{between galaxies, line-} \\ \text{arly extrapolating back} \\ \text{to start of expansion}\end{pmatrix}$$

$$= \left(\frac{\text{“radius of Universe”}}{\text{time rate of increase of “radius of Universe”}}\right)$$

Present estimates of H correspond to $\tau_H = (1\text{–}2) \times 10^{10}$ years. Because of the deceleration, the actual time since the singularity is less than τ_H. Provided that the dynamical effects of pressure are negligible almost all the time (as seems to be the case for the actual Universe), the time since the singularity is

$$\tau_H q(2q - 1)^{-3/2} [\cos^{-1}(q^{-1} - 1) - q^{-1}(2q - 1)^{1/2}] \quad (q \geqq \tfrac{1}{2})$$

or

$$\tau_H q(1 - 2q)^{-3/2} [q^{-1}(1 - 2q)^{1/2} - \cosh^{-1}(q^{-1} - 1)] \quad (q \leqq \tfrac{1}{2})$$

(this time is $< \frac{2}{3}\tau_H$ for closed models ($k = 1$) with $q > \frac{1}{2}$). Thus the age of the Universe, if it is closed and governed by Einstein's equations, can be up to $\sim 1.3 \times 10^{10}$ years (for $H \gtrsim 50$ km s^{-1} Mpc^{-1}). Current theories suggest that the oldest stars in globular clusters in our Galaxy are $\sim 10^{10}$ years old. By comparing the observed $^{235}U/^{238}U$ ratio, ~ 0.007, with that expected when the uranium was synthesised by rapid neutron capture (~ 1.65) one derives a *minimum* age of 6.6×10^9 years for the Galaxy. Other abundance ratios—in particular $^{187}Re/^{187}Os$—suggest an age slightly larger than 10^{10} years.[174] But there are no firm reasons—from stellar evolution or any other considerations—for doubting that our own Galaxy (or any other galaxy) could be younger than $\sim 1.3 \times 10^{10}$ years. These age considerations do however set an *upper limit* to Ω—if Ω exceeded ~ 10 (i.e. $q \gtrsim 5$) the inferred age of the Universe would be less than that of the Earth, even if H were as low as 50 km sec^{-1} Mpc^{-1}. This contrasts with the situation that prevailed in the 1940's, when τ_H was believed to be $\sim 1.8 \times 10^9$ years—less than half the Earth's estimated age of 4.5×10^9 years! This apparent inconsistency—along with general philosophical considerations—led Bondi, Gold and Hoyle,[175,176] in 1948, to propose replacing Einstein's theory not with an all-encompassing rival theory but with a simple prediction: a model universe that expands steadily ($a \propto e^{Ht}$) but nevertheless always keeps the same density ("steady state") by virtue of an assumed "continuous creation of matter". In this model, some galaxies could be much older than τ_H, even though the mean age of the matter present in the Universe would be only $\tau_H/3$, and the expansion would be accelerating ($q = -1$) (see Figure 36). The upward revisions in τ_H removed some of the motivation for the steady state model. But the theory was not generally abandoned until the discovery of the microwave background radiation (chapter 13) which can be understood as a relic of the cosmic fireball but finds no ready interpretation in a steady state universe.

11.4 DETERMINATION OF THE DECELERATION PARAMETER *q*

All cosmological models yield the same law (11.8) when $z \ll 1$, but at large distances the magnitude-redshift relation depends on the matter density (or, equivalently, on the deceleration parameter). To obtain the general

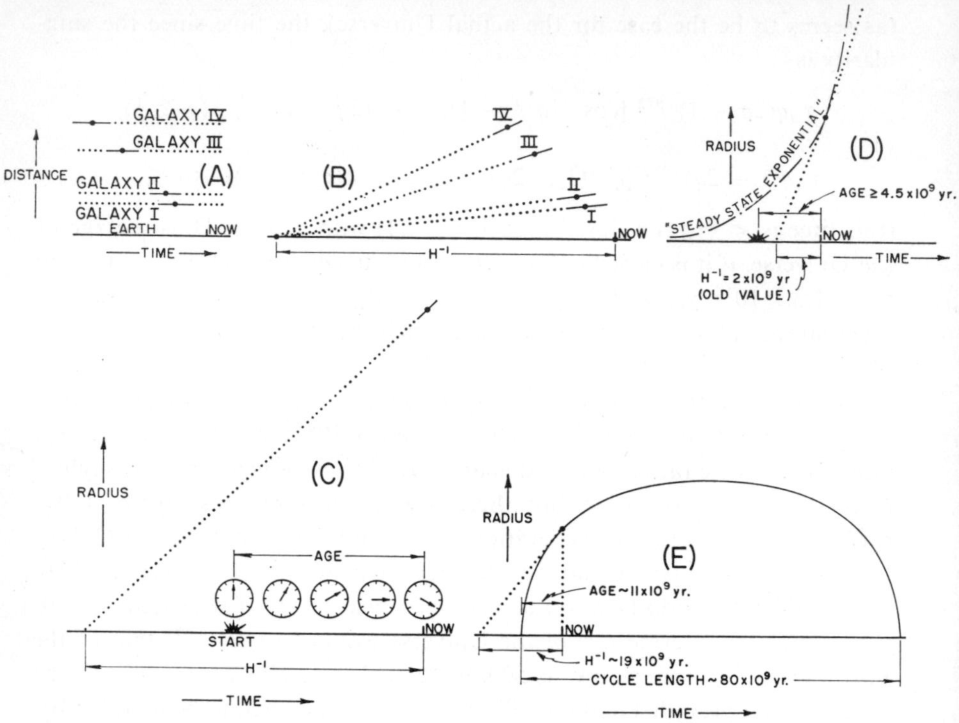

Figure 36 Expansion of Universe. A: Distances of galaxies from earth as they would be in the absence of expansion. Conception of Einstein in early days of this theory (a) before he learned that it predicted a dynamic universe and (b) before Hubble gave evidence for expansion. B: Doppler shifts reveal velocity of recession proportional to distance. H^{-1}, the inverse Hubble constant, is the time back to start of expansion as extrapolated linearly from present rates (same for all galaxies within limits of error and random galactic velocities). C: This time, common to all galaxies (linearly extrapolated or Hubble time) is to be contrasted with the actual time ("age") back to start of the expansion (symbolized in diagram by change in clock readings). D: The best values for the age and the Hubble time in the 1930's were inconsistent with the Einstein prediction of an expansion that slows down with time. They led Bondi, Gold and Hoyle to propose a "steady state" or exponential expansion of the universe with time, and a "continuous creation of matter", both of which concepts have subsequently generally been abandoned. E, Values for age and Hubble time reported by A. Sandage, $(13 \cdot 5 \pm 6) \times 10^9$ yrs[172] superposed on a diagram showing in the kind of expansion and recontraction of the universe expected on the basis of Einstein's theory for a closed universe (time from start to collapse, 80×10^9 yr, uncertain by a factor of two or more).

form of the Hubble law when z is not small, we must outline the derivation of formulae for observable quantities in isotropic cosmological models.

Consider a source at a radial coordinate χ_e (the observer being at $\chi = 0$). The radiation received by the observer at time t_{now} would have been emitted from the source at a time t_e given by

$$\chi_e = \int_{t_e}^{t_{\text{now}}} \frac{dt}{a(t)} \tag{11.9}$$

The redshift of the source is then given by

$$(1 + z) = \frac{a(t_{\text{now}})}{a(t_e)} \tag{11.10}$$

(i.e. wavelengths are "stretched" in proportion to the expansion factor between emission and reception). If the source has a (small) linear size d, then from (11.4) it subtends at the observer an angle

$$\Delta\theta = \frac{d}{a(t_e) \sin \chi_e} = \frac{d(1 + z)}{a(t_{\text{now}}) \sin \chi_e} \tag{11.11}$$

If the source (at time t_e) has a bolometric luminosity (i.e. a radiation output integrated over all frequencies) L, then the observed bolometric intensity is

$$l = \frac{L}{4\pi(1 + z)^2 \, (a(t_{\text{now}}) \sin \chi_e)^2} \tag{11.12}$$

The factor $(a(t_{\text{now}}) \sin \chi_e)$ is the radius of curvature of a 2-sphere centred on the galaxy and passing through the observer at the time of reception; the factor $(1 + z)^2$ appears because photons emitted in a time interval Δt are received during an interval $(1 + z)\,\Delta t$, and with energy reduced by $(1 + z)^{-1}$.

In practice what is generally observed, whether at radio or optical wavelengths, is not the bolometric intensity l but the "flux density" $S(\nu)$ per unit frequency at some frequency ν_0 (or in a band around ν_0). Because the radiation observed at ν_0 Hz was emitted at $\nu_0(1 + z)$ Hz, the source's *spectrum* is now relevant. If the power emitted between ν and $\nu + d\nu$ is $L(\nu)\, d\nu$, we have

$$S(\nu_0) = \frac{L(\nu_0)}{4\pi(1 + z)^2 \, (a(t_{\text{now}}) \sin \chi_e)^2} \left[(1 + z) \frac{S(\nu_0)}{S(\nu_0/(1 + z))} \right] \tag{11.13}$$

The extra factor in square brackets is known as the "k-correction". For power-law spectra $S(\nu) \propto \nu^{-\alpha}$ it reduces to $(1 + z)^{1-\alpha}$.

To compare (11.11)–(11.13) with observations, the quantity $(\sin \chi_e)^{-2}$ which appears in all the formulae†, and which is proportional to the solid angle subtended by the source at the observer, must be expressed in terms of z. From (11.9), (11.10) and (11.5) one easily finds

$$dz = -H(\Omega z + 1)^{1/2} (1 + z)^2 \, dt \tag{11.14}$$

and

$$a(t_{\text{now}}) \sin \chi_e = \frac{2c}{H\Omega^2(1 + z)} \{\Omega z + (\Omega - 2) [(\Omega z + 1)^{1/2} - 1]\} \tag{11.15}$$

Going to first order only in Ω $(= 2q)$, (11.11) then becomes

$$l = \frac{H^2 L}{4\pi z^2} (1 + (q - 1) z + \cdots) \tag{11.16}$$

This relation can alternatively be expressed in terms of the emitted power $L(\nu)$ at frequency ν and the flux density $S(\nu)$, if the "k-correction" is inserted on the right hand side. The deviations from the linear relation in Figure 35 thus yield, in principle, a measurement of the deceleration parameter q. The key question, of course, is whether q is $> \frac{1}{2}$ or $\leq \frac{1}{2}$; its precise value (if $\neq \frac{1}{2}$) would have no fundamental significance, being merely an indication of how far the Universe has evolved along its "cycle".‡

It is plain from Figure 35 that the available data are not really sufficient to determine q accurately, especially because, even for the galaxy with the largest redshift in the sample ($z = 0.461$), a change of order unity in q corresponds to only about a third of a magnitude change in l, which is no larger than the general scatter of the points in the diagram. The most recent estimate has been made by Sandage.[172] Taking the "k-correction" into account, but assuming that the 40 "first ranking" elliptical galaxies in his sample are indeed standard candles, he finds

$$q = 1.03^{+0.43}_{-0.26}$$

A similar analysis, but excluding eight of the more distant galaxies whose magnitudes were based on earlier work by Baum, yielded

$$q = 0.65^{+0.5}_{-0.3}$$

† For $k = 0$ and -1, the formulae still hold if $\sin \chi_e$ is replaced by χ_e or $\sinh \chi_e$ respectively.

‡ Note that in a radiation-dominated universe, the condition for closure is $q > 1$, not $q > \frac{1}{2}$.

Note that these estimates depend on assuming that the luminosities of the "first ranking" elliptical galaxies in the sample have no systematic z-dependence. (We recall that were it not for the remarkable fact that, at small z, the dispersion in the absolute magnitudes of these objects is less than 0.3 magnitudes, a well-defined magnitude-redshift relation would not be obtained even in the linear regime). Two possible systematic effects are the following.

"*Scott effect*" There is an inevitable tendency for progressively larger, richer and more exceptional clusters of galaxies to be detected as observations penetrate to larger redshifts.[177] If the brightness of the "first ranking" member is correlated with cluster size, this introduces a systematic effect which would lead us to *over*-estimate q. There is no understanding of what determines the luminosity function of the bright galaxies in clusters, so we cannot be sure how severe this effect might be.

Evolutionary corrections As we look out to larger redshifts, we see galaxies which are progressively younger. So, if the luminosity L of an elliptical galaxy changes significantly as it ages, an "evolutionary correction" must be applied to the estimates of q. Knowing the stellar content of elliptical galaxies, $L(t)$ can be estimated from the theory of stellar evolution. Spinrad[178] infers (from the colours and mass-to-light-ratios) that the luminosity function of the stars in ellipticals must be very steep, with a large number of faint stars on the lower main sequence, and that consequently L should slowly *increase* with t, as progressively larger numbers of stars evolve off the main sequence towards the giant branch. The consequent evolutionary correction is estimated, however, to be small. Tinsley[179], on the other hand, has obtained a very different result by following the evolution of a model elliptical galaxy, assuming various forms for the initial stellar luminosity function and birth rate. She proposes a *decrease* of L with t amounting to $\sim 10\%$ per 10^9 years which would be large enough to *lower* q by about 1 from the uncorrected estimates†.

Other uncertainties involved in determining q include (a) the possibility of absorption or scattering by intervening intergalactic matter, and (b) the effects of a non-uniform distribution of matter. (Dashevskii and Zeldovich[180] have pointed out that the usual simple relations between angular diameter—and therefore l—and redshift no longer hold if the matter distribution is irregular). We shall plainly have to await a better understanding of the many

† Note that the size of this correction depends on the value of H. In most other respects, however, the problems of determining H and q are entirely disjoint.

corrections involved, as well as the accumulation of much more data, before q can be determined directly from the magnitude-redshift relation for distant galaxies. Although the present evidence favours $q > \frac{1}{2}$, the "closure" of the Universe envisaged by Einstein is still not proven.† Sandage [172] comments, apropos of recent estimates of q, that "At present none of this must be considered very seriously. We need many more clusters whose redshifts are greater than $z = 0.2$ for a satisfactory solution. The only conclusion which I believe is currently warranted is that the steady state solution ($q = -1$ and no evolutionary correction) seems far from being probable".

† An alternative line of attack on the problem—searches for sufficient "missing mass" to make $\Omega > 1$—is discussed in Chapter 15.

CHAPTER 12

The Evolving Universe

THE ATTEMPTS to use ordinary bright galaxies to discriminate between different cosmological models are still somewhat inconclusive, basically because galaxies become invisibly faint (even with the aid of the 200 inch telescope) before one gets to the redshifts where substantial deviations from the linear Hubble law are expected. Ever since the source Cygnus A—the second strongest object in the radio sky—was identified twenty years ago as a radio galaxy with $z \simeq 0.05$, it has been evident that many of the unidentified sources revealed by radio surveys must have redshifts $z \gtrsim 0.5$, the galaxies which are their optical counterparts being too faint to be seen. The initial discovery of quasi-stellar objects (QSOs or "quasars") by Maarten Schmidt in 1963 has led to the measurement of redshifts as large as $z \simeq 3$ in some of these objects. Unfortunately, however, neither radio sources nor quasars have yet provided *any* useful estimate of q! The reasons for this regrettable situation are twofold.

First, there is a huge spread in the intrinsic luminosity of radio sources (and in the optical luminosities of quasars). Therefore no "standard candles" are available, so one must fall back on crude statistical tests.

Second, our physical understanding of these objects is so limited that we have no idea (as we *do*, to some extent at least, for normal galaxies) what evolutionary corrections to apply. It appears that quasars, N-galaxies, radio galaxies, and Seyfert galaxies are closely related phenomena, which though differing in scale and in their emission spectrum, are all manifestations of some kind of "violent activity" in galactic nuclei. Some non-stellar energy source—perhaps involving the gravitational collapse of either a massive object or many bodies of stellar mass—is probably involved, but none of the specific models so far proposed can claim to be more than speculative conjectures. (See Chapter 6.)

Despite their limited cosmological value, statistical studies of radio sources and quasars already tell us something which is surely crucially important for the astrophysics of compact objects and "violent events":

namely that, *whatever the value of q*, there must be strong secular evolution in the coordinate density of strong radio sources and quasars, in the sense that there were *many more* powerful emitters at earlier cosmic epochs.

12.1 RADIO (AND OPTICAL) COUNTS

Table XXII The luminosity function of extragalactic radio sources at the present epoch. Here P = radio luminosity per steradian = $L/4\pi$, and is measured at 178 MHz. The density $\varrho'(P)$ in the second column is defined by

$$\varrho'(P) = \int_{P/10^{1/5}}^{10^{1/5}P} \varrho(P)\, dP$$

i.e. $\varrho'(P)$ is the density of sources in equal logarithmic intervals of P

$P_{178}(WHz^{-1}sr^{-1})$	$\varrho'(P)\ (Mpc^{-3})$
2.5×10^{27}	3.5×10^{-11}
1.0×10^{27}	1.7×10^{-10}
4.0×10^{26}	9.0×10^{-10}
1.6×10^{26}	4.4×10^{-9}
6.3×10^{25}	2.4×10^{-8}
2.5×10^{25}	9.0×10^{-8}
1.0×10^{25}	3.2×10^{-7}
4.0×10^{24}	9.6×10^{-7}
1.6×10^{24}	3.3×10^{-6}
6.3×10^{23}	9.0×10^{-6}
2.5×10^{23}	9.4×10^{-5}
1.0×10^{23}	1.8×10^{-5}
4.0×10^{22}	4.7×10^{-5}
1.6×10^{22}	2.0×10^{-4}
6.3×10^{21}	8.3×10^{-4}
2.5×10^{21}	2.1×10^{-4}
1.0×10^{21}	6.7×10^{-3}
4.0×10^{20}	4.5×10^{-3}

The identified radio sources span an extremely broad range of both luminosity and linear dimensions, and no type of source has yet been found which serves as an adequate "standard candle" (also, of course, redshifts of radio sources cannot be determined unless optical counterparts can be identified). The radio power from normal spiral and elliptical galaxies is generally $\lesssim 10^{40}$ erg sec^{-1}. However a small fraction of all galaxies—often, as it happens, the brightest galaxies in clusters—are much more powerful radio emitters, with powers up to $\sim 10^{45}$ erg sec^{-1}. Some strong radio sources are associated with quasars. Because of their high optical luminosity, these can be identified out to redshifts exceeding 2. The estimated luminosity function of extragalactic radio sources is given in Table XXII (reproduced from ref. 181). The radio and optical properties of radio galaxies do not appear to be correlated in any obvious way. Nor are radio galaxies and quasars readily distinguishable on the basis of radio observations alone. In powerful sources, the radio flux typically emanates from two regions, one on each side of the optical galaxy (though in many cases there is a compact source in the galaxy itself). The nature of radio sources and the origin of their characteristic "dumb-bell" structure still defy agreed understanding, but this fascinating branch of "high-energy astrophysics" lies outside our present scope.

Because of this huge scatter in the radio properties of sources, the most fruitful cosmological test based entirely on radio data has been comparison of the relative numbers of strong and faint sources revealed by radio surveys. The conventional procedure is to plot, against flux density S, the number N of sources per steradian brighter than S. To get good statistics at high flux densities, where the number of sources is small, one needs a survey covering a large area of sky; however a smaller area suffices at lower flux levels (provided only that the Universe is sufficiently isotropic and homogeneous).

Theory

In a uniform static Euclidean universe, where the number of sources $N(r)$ within a radius r is proportional to r^3 and $S \propto r^{-2}$, we obviously would expect

$$N(> S) \propto S^{-3/2} \tag{12.1}$$

Since (12.1) holds for each class of source severally, this "three halves" law still holds even when the sources span a wide range in intrinsic luminosity $L(\nu)$.

The generalization of (12.1) to the case of a Friedmann cosmology is straightforward. Suppose that the Universe contains a class of sources with a range of luminosities $L(\nu)$, the density of sources with luminosities between $L(\nu)$ and $L(\nu) + dL(\nu)$ being $\varrho(L(\nu))\, dL(\nu)$ per unit *comoving* volume. If ϱ is independent of epoch ($\varrho = \varrho_{\text{now}}$)—so that the actual space density of objects varies as a^{-3}—we have a "source-conserving" model. Then the number of sources at coordinate distances $\chi - (\chi + d\chi)$ is

$$N(\chi)\, d\chi = 4\pi \varrho_{\text{now}} a_{\text{now}}^3 \sin^2 \chi\, d\chi \tag{12.2}$$

so

$$N(>S(\nu)) = \int_L \int_0^{\chi(L)} \varrho(L(\nu))\, a_{\text{now}}^3 \sin^2\chi\, d\chi\, dL(\nu) \tag{12.3}$$

where $\chi(L)$ is the solution of

$$L(\nu(1+z)) = 4\pi S(\nu)\,(1+z)\, a_{\text{now}}^2 \sin^2\chi \tag{12.4}$$

In computing the expected form of $N(S)$ in different model universes, it is convenient to work in terms of N/N_0, $N_0(S)$ being the 'Euclidean' prediction obtained by extrapolating the observed counts to very high flux density (where the sources would presumably be quite local) according to the law (12.1). From Eq. (12.2) the count for a particular luminosity class L in source-conserving cosmologies is

$$N(S, L) = \int_0^{\chi(L)} \varrho_{\text{now}}(L)\, a_{\text{now}}^3 \sin^2\chi\, d\chi$$

$$= \frac{\varrho_{\text{now}}(L)\, a_{\text{now}}^3}{4}(2\chi - \sin 2\chi) \tag{12.5}$$

when $\chi(S)$ is given by (12.4). In the Euclidean case however

$$N_0(L) = \frac{1}{24\pi^{3/2}}\, \varrho_{\text{now}}(L) \left(\frac{L}{S}\right)^{3/2}. \tag{12.6}$$

If the sources have spectra, over the relevant frequency range, such that $L(\nu) \propto \nu^{-\alpha}$, then we obtain an expression for N/N_0. We shall write this down for the three cases $k = 1, 0, -1$:

$$\frac{N(S)}{N_0(S)} = (1+z)^{-3/2(1+\alpha)} \begin{cases} \dfrac{3}{4} \cdot \dfrac{2\chi - \sin 2\chi}{\sin^3\chi} & (k = 1) \\ 1 & (k = 0) \\ \dfrac{3}{4} \cdot \dfrac{\sinh 2\chi - 2\chi}{\sinh^3\chi} & (k = -1) \end{cases} \tag{12.7}$$

where S and χ are related by

$$S(\nu) = \frac{L(\nu)}{4\pi a_{\text{now}}^2 (1+z)^{1+\alpha}} \begin{cases} \sin^{-2}\chi & (k = 1) \\ \chi^{-2} & (k = 0) \\ \sinh^{-2}\chi & (k = -1) \end{cases} \tag{12.8}$$

From these relations, and (11.15) which relates χ and z, one finds that in *open* models the terms involving χ in (12.7) are non-increasing, so if $\alpha > -1$

N/N_0 decreases monotonically as S decreases. In *closed* models one can show that, for $\alpha > 1$, N/N_0 is a decreasing function of distance (i.e. decreases as S decreases.) The radio sources detected in surveys typically have spectra falling towards higher frequencies, with $\alpha \simeq 0.75$, so we infer the important result that *in any source-conserving Friedmann model (whatever the luminosity function of the sources may be), the graph of* $\log N(> S)$ *against* $\log S$ *should have a slope flatter than* -1.5. The amount of flattening, being a cosmological effect, depends on the luminosity function $\varrho(L)$, and will only be significant at a given flux level S if a significant proportion of the sources with flux densities $> S$ have substantial redshifts.

Radio observations

Counting radio sources was recognised as a practicable cosmological test as soon as it was discovered that radio sources in directions away from the galactic plane are in general distant extragalactic objects. Source counts have now been performed by many different groups working independently, and when allowance is made for statistical uncertainty, and for the different frequencies at which the surveys were made, there is good agreement among all the results obtained. Figure 37 collects together the data from several different surveys.

The $N(S)$ relation presented by Pooley and Ryle[182] provides good statistics over the widest range of flux densities, and has formed the basis for many theoretical investigations. These counts (shown in Figure 38 in the form N/N_0) are based on the 5C2 survey, covering a few square degrees of sky down to a flux level at 408 MHz of $S_{408} \simeq 10^{-2}$ f. u.,† together with data for brighter sources in the earlier 3C and 4C surveys which had been carried out at 178 MHz‡.

The high flux (3C) part of the $\log N/\log S$ graph is approximately linear, with a slope $\beta \simeq -1.8$, but there is a gradual flattening towards the faint end, and at $S_{408} \simeq 0.01$ f.u., $\beta \simeq -0.8$ (of course the curve *must* eventually flatten to a slope > -1 to ensure convergence of the integrated radio background).

---

† One flux unit (f.u.) is 10^{-26} Wm^{-2} Hz^{-1}.

‡ If all the sources in the 3C and 4C surveys had power-law spectra, all with the same spectral index α, the earlier $N(S)$ relation published by Gower[183] could have been straightforwardly scaled from 178 MHz to 408 MHz and matched onto the 5C results. However, because there is a spread in the spectral indices, this simple procedure is illegitimate, and Pooley and Ryle[182] had to determine the 408 MHz flux for the brighter sources by interpolating (in the case of 3C sources whose spectra were already fairly well determined), and by a separate fan beam survey for a sample of 4C sources.

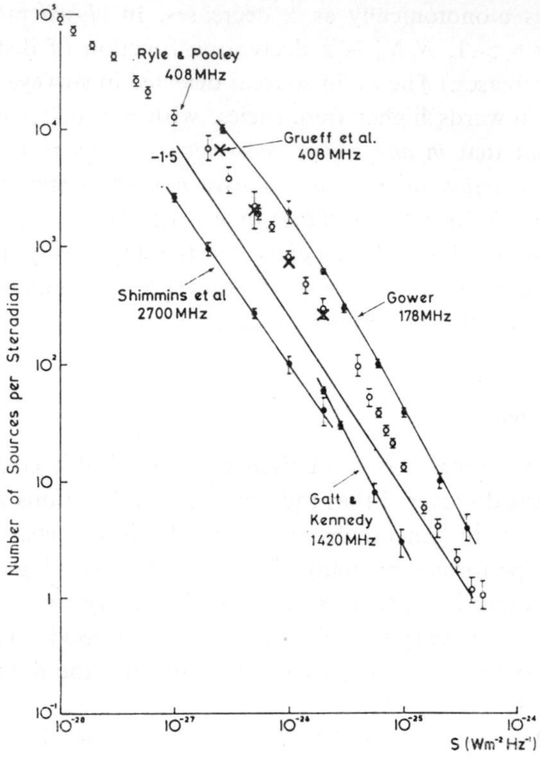

Figure 37
Counts of radio sources from various independent surveys.[182–186]

Independent surveys by Bolton *et al*[187]; MacLeod *et al*[188], Braccesi *et al*[189]; Grueff and Vigotti[184]; and others yield source counts entirely consistent with Pooley and Ryle's. (The first of these provided comforting assurances that the overall structure of the Universe appears more or less the same viewed from the Southern Hemisphere). More recently, several surveys have been carried out at much higher frequencies than 408 MHz. In comparing the results of these with the earlier work it is important to remember that, because there is a spread in the spectral indices of radio sources, one *should not expect* precisely the same $N(S)$ relation at all frequencies. High frequency surveys are weighted in favour of sources with flat or rising spectra, and there is no obvious reason why these should have the same spatial distribution as the type of source which predominates in the Cambridge surveys and in the other comparatively low frequency surveys mentioned above. In principle one can predict $N(S)$ at a frequency ν_2 in terms of the counts at a different frequency ν_1 if one knows the spectra of all the sources. However, if the spectral indices are widely spread, one needs to know $N(S)$, at ν_1, down to extremely low fluxes in order to predict even the bright end of the $N(S)$ relation at ν_2—for example some of the brightest sources in a 5 GHz survey have *rising* spectra, and would be below even the 5C threshold at 408 MHz.

In evaluating the reliability and significance of the $N(S)$ relation, a number of possible sources of error must be considered. (These matters are further discussed by Ryle[190], Scheuer[191], and Longair[181]; and in the original papers on the various surveys). Obviously it is essential that the flux estimates be reliable over the whole range of S, and that the surveys be complete. The effects of confusion tend to cause a systematic overestimate of S for the fainter sources; moreover *random* uncertainties in the individual fluxes, if they are proportionately larger for the fainter sources, cause *systematic* errors in the $N(S)$ relation. Another problem is that surveys made using interferometers fail to register the full flux of sources with large angular size. It is possible to estimate how important such effects are likely to be, and (to some extent) to correct for them in the final $N(S)$ curve. However all these effects contribute to the total errors, especially at low flux levels. Even if no errors or uncertainties of any kind entered into the observational procedure, there would still be a "formal error" attached to the $\log N/\log S$ curve, and to the quoted slope β, because of the finite numbers of sources involved. (We may envisage this "error" as a measure of the variations in the values of β that would be determined by an ensemble of observers scattered through space).

The crucial feature of the Pooley-Ryle counts is the steep slope $\beta \simeq -1.8$ of the 3C part of the $\log N/\log S$ relation. Assuming that the "error" results primarily from the finite number of sources, this slope differs from -1.5 at about the 3σ level†. This contradicts, in a highly significant manner, the predictions of *all* "source conserving" models, which would yield $|\beta| < 1.5$—indeed, a large fraction of the 3C sources are known to have substantial redshifts (this is true even if we exclude sources identified with quasars and consider only the radio galaxies whose redshifts are agreed to be "cosmological" in origin), and the source-conserving models predict a slope of only $\beta \simeq -1.3$ when these redshifts are allowed for.

---

† In estimating the slope and its uncertainty, it is important to remember that successive points on an *integral* $N(S)$ curve are *not* independent. There are consequently some advantages in plotting counts *differentially*. By fitting a 'maximum likelihood' straight line, Jauncey[192] derives a slope $-2.79^{+0.12}_{-0.11}$ for the *differential* 3C counts. (Note, however, that there is no reason to expect any part of the $\log N/\log S$ curve to be a precise straight line.)

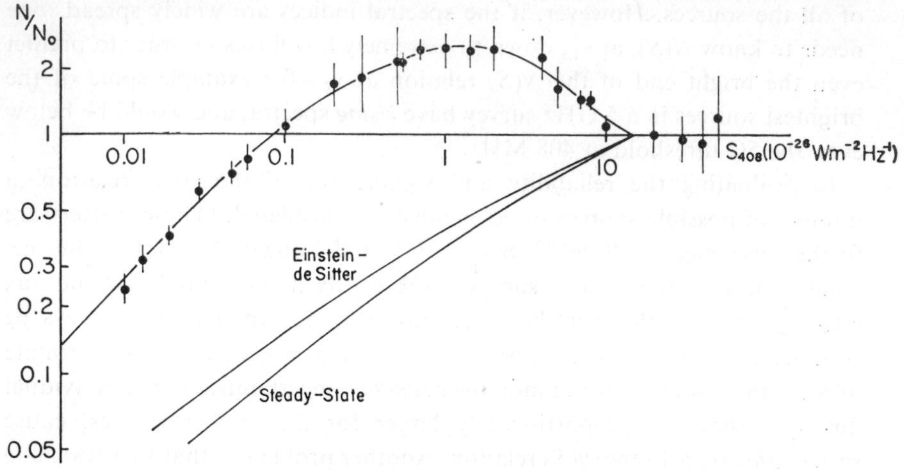

Figure 38 The radio source counts at 408 MHz from the Cambridge surveys,[182] plotted in the form N/N_0. The two lower curves are those expected if the comoving source density is independent of redshift, and the luminosity function has approximately the form given in Table XXII.

Figure 38, which plots Pooley and Ryle's data in the form N/N_0 [c.f. Eq. (12.7)], perhaps displays the problem posed by the counts in a clearer way. There is an apparent "excess" of intermediate-flux sources, compared with what one would expect in a Euclidean universe, amounting to $\sim$ 1000 sources per steradian. The "excess" is vastly greater if one compares with the predictions of a 'source-conserving' Friedmann model (also shown in the figure); however one can equally legitimately think in terms of a *deficit* of *bright* sources, in which case the numbers involved are at least $\sim$ 30 per steradian. The other radio surveys (see Table XXVI) reveal a similar form for N/N_0 (with the exception of the 2700 MHz counts presented by Shimmins *et al*,[185] though these were based on a survey of only 0.8 steradians of sky and therefore have rather poor statistical significance at the low values of N where the steep slope occurs).

The simplest explanation of the source counts is that the mean properties of radio sources have changed over cosmological timescales†, and that the luminosity function $\varrho(L)$ is a function of cosmic epoch. The "excess" shown in Figure 38 can then be attributed to a population of sources at

† Because this effect is not permitted in a strict steady state universe, the source counts can be taken as evidence against this cosmological theory. (Though, as Hoyle[193] has suggested, one cannot exclude on these grounds a "quasi-steady" state with fluctuations on time scale $\sim\tau_H$ and length scale comparable with the Hubble radius.)

some redshift z^* (say) whose nearby counterparts (which would tend to raise the high-S part of the curve, destroying the excess) are absent, or at least have a much lower density per comoving volume element. Obviously one can fit the observed peak in the N/N_0 diagram equally well by evolving $\varrho(L)$ with z in innumerable different ways. Most authors who have discussed this question have restricted themselves to simple functional forms for the evolution, but a great deal of freedom still remains. All the analyses, however, have these gross features in common:

1) *The evolution is probably a cosmological effect, occurring primarily at redshifts* $z^* \gtrsim 1$. It is clear from the identification that many sources (e.g. Cygnus A) have radio luminosities $\gtrsim 10^{44}$ erg sec^{-1}, and would still appear in radio surveys even if their redshifts were $\gtrsim 1$. If the excess occurs at $z^* \gtrsim 1$, the eventual fall-off in N/N_0 at low S can be attributed to the cosmological effects in (12.7) which become severe when z is *much* greater than unity. On the other hand, if the "excess" occurred at $z^* \ll 1$, we would have to postulate an intrinsic cut-off in the source population beyond z^*—or, in other words, assume that the excess sources lie in a "shell" around us—to explain the shape of the observed N/N_0, because the cosmological effects in (12.7) are unimportant at redshifts $\ll 1$.

2) *The evolution involves predominantly the more powerful sources.* Even if the "excess" sources occupied a narrow redshift range, the peak in the N/N_0 curve would be broadened by any spread in L. Indeed a luminosity function which had the same form at z^* as it has locally would neither reproduce the sharp maximum in the curve nor be compatible with the observed limits on the integrated background. The evolution cannot therefore consist of a simple scaling of the local $\varrho(L)$: instead, only the more powerful sources can be permitted to evolve strongly. Possible forms for this evolution are illustrated in Figure 39.

3) *The evolution must be very strong.* The comoving space density of powerful sources must be ~ 1000 times greater at $z \simeq 2$ than at the present epoch. Although the discrepancy in numbers between observed and "predicted" counts is only a factor ~ 5 (see Figure 38) the effects of evolution of an individual luminosity class is diluted by the broad spread in $\varrho(L)$.

Further discussion of this evolution and its implications (section 12.3) is deferred until we have mentioned the extra evidence provided by quasars and their redshifts.

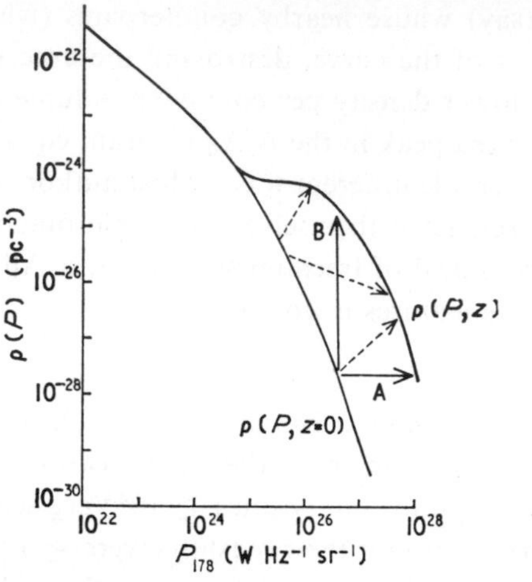

Figure 39 Illustrating various ways in which the evolution of the radio luminosity function can be described.[181]

(a) is called luminosity evolution,
(b) is called density evolution.

Optical counts

In the 1930's, Hubble[194] attempted to perform counts for *galaxies*, and showed that the number-magnitude relation down to 19th magnitude was compatible with a uniform distribution of galaxies in Euclidean space (the number going up by a factor 4 for each unit increase in limiting magnitude). There have been few extensions of Hubble's work, because the differences between the counts for different values of q are very small even at 21st magnitude. Also there are complications associated with obscuration by dust in our own Galaxy, and with clustering (and perhaps superclustering) of galaxies—problems which do not seem to affect radio counts so seriously.

Counts have recently been carried out of quasars† (and suspected quasars) in limited areas of the sky. Since these objects tend by virtue of their higher intrinsic luminosity to be be more distant than galaxies, cosmological effects might be expected to show up more clearly. The key advantage of optical over radio observations is that they lead to a determination of the redshifts. However one can perform optical counts, even without the redshifts, for objects which are suspected to be quasars on the basis of their colours.

† Schmidt[195] has given the following definition of a "quasar": "The class of objects of star-like appearance (or those containing a dominant star-like component) that exhibit redshifts much larger than those of ordinary stars in the Galaxy". For the purpose of the present discussion we assume that these redshifts are indeed due to the universal expansion.

Searches of selected fields for objects with ultraviolet excesses led Sandage and Luyten[196] to conclude that the number of quasars increased by a factor ~ 6 per magnitude. They find ~ 5 quasars per square degree with blue magnitude $B \lesssim 19.4$, and estimate ~ 100 per square degree brighter than $B = 21.4$. This is equivalent to an anomalously steep log N/log S relation ($\beta \simeq 2$). Bracessi and Formiggini[197] obtained a sample of 300 objects having both an ultraviolet *and* an infrared excess, a proven and efficient procedure for picking out candidate quasars and distinguishing them from ordinary blue stars in our Galaxy. They find a log N/log S slope of -1.74, again suggesting the same evolution as displayed by the radio sources (note that most of the quasars in these samples are "radio quiet"). Nevertheless the statistical reliability of these counts does not approach that achieved by the radio observations.

12.2 INFORMATION ADDED BY QUASAR REDSHIFTS: THE "LUMINOSITY-VOLUME" TEST

It turns out, perhaps surprisingly, that measurements of the redshifts of quasars has added very little to our knowledge of either the geometry and dynamics of the Universe, or the evolutionary properties of the objects themselves. The reason for this is manifest in Figure 40. The magnitude—redshift relation appears to be almost a pure "scatter diagram" (and a similar diagram is obtained if flux density is plotted against redshift for the radio quasars). Furthermore, no correlations between any parameters of quasars have been found which could allow us to pick out any "standard candles".

One would certainly not expect a well-defined relation between magnitude and redshift such as exists for the brightest elliptical galaxies in clusters, if for no other reason than that the luminosities of some individual quasars have varied by several magnitudes over the last few years. However Longair and Scheuer[199] were able to show that the lack of correlation between m and z cannot all be attributed to the breadth of the luminosity function—there must be some z—dependence in the mean luminosity of the objects as well. As was the case for radio sources, the data can be reconciled with *any* cosmological model by invoking appropriate evolution, so that main interest of these investigations lies in the light they shed on the evolution and properties of the quasars themselves.

Accepting, then, that the redshifts of quasars indicate their distance, (see 15.2 for discussion of this question) but that they are not "standard candles", what is the most efficient way to extract useful information from

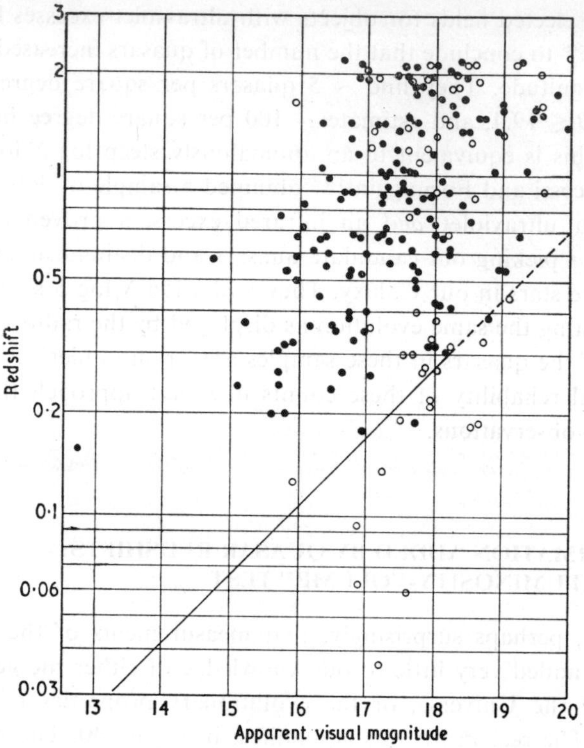

Figure 40 The magnitude-redshift relation for quasars.[198] Solid circles denote ratio sources and open circles radio quiet quasars. The straight line is the Hubble relation (extrapolated linearly) for brighest cluster galaxies.

the data? The most popular technique is the so-called "luminosity-volume" test. This test can be used even when the objects span a wide range of intrinsic luminosity, but to apply it rigorously one needs a sample which is "complete" in some well-defined sense. Adopting a particular cosmological model one then calculates, for each object, the volume V (in comoving coordinates) out to its redshift z. One also calculates the maximum redshift z_m which the same object could possess without becoming too faint to be included in the sample, and the corresponding volume V_m. (This stage in the procedure requires some information on the source's spectrum). If the assumed cosmological model were the true one, and if the sources were indeed uniformly distributed, then V/V_m should, on the average, be 0.5. If it turned out to be systematically greater than 0.5, implying that the sources tended to be towards the edge of their accessible volume, this would indicate an evolutionary effect.

Schmidt[200] applied this procedure to a sample of 33 identified quasars in the 3C Revised catalogue. These are believed to constitute a *complete* sample, over the area of sky covered, of quasars for which *both* $S_{178} \geqq 9 f.u.$ *and* the visual magnitude $m \leqq 18.4$ (in fact the optical cut-off is defined more precisely in terms of the flux density at 2500 Å). He found that for standard Friedmann models, $\langle V/V_m \rangle \simeq 0.70$, almost independently of q. This differs from 0.5 at a formal significance level of 0.01 per cent, thus providing strong evidence for cosmic evolution in these objects' properties. To estimate how strong this evolution had to be, Schmidt repeated the test in terms of a "fictitious volume" V' calculated by weighting each volume element in proportion to a mathematically convenient increasing function of z. He found that, for a weighting factor with roughly the form $(1 + z)^6$, $\langle V'/V_m \rangle$ was ~ 0.5. This indicates that the space distribution of the quasars in the sample is compatible with a simple density evolution $\propto (1 + z)^6$; though, as Schmidt stressed, no great weight should be attached to this particular functional form of z-dependence.

The main virtue of the luminosity-volume test is that it can be applied to a sample of sources defined by *more than one* selection criterion. Of Schmidt's 33 3CR quasars, for example, 22 were "radio limited" (in the sense that V_m was determined by the radio cut-off) and 11 optically limited. One cannot draw such significant conclusions from this small sample without taking both cut-offs into account[201]. (If one has a sample, with redshifts, limited by only one cut-off—as, for example, a sample of optically selected quasars—the luminosity-volume test is closely related to the source counts. Longair and Scheuer[202] have shown that in this simple case the counts, plus a knowledge of the *mean* redshift of the objects in the sample, yield essentially the same information as $\langle V/V_m \rangle$. Even in the more general case when there are two cut-offs, it is only the mean redshift of the objects in the sample which affects the result, and not the specific association of particular redshifts with particular objects.)

The method of the luminosity-volume test is in principle capable—given sufficient data—of establishing the luminosity function of quasars (or, more generally, the joint radio and optical luminosity function $\varrho(L_{\text{opt}}, L_{\text{rad}})$ and delineating how it evolves with z: for instance, is the evolution best fitted by a law of the form $(1 + z)^n$, or by some exponential function of cosmic time? The samples so far analysed are much too small to settle this question decisively, but Lynden-Bell[203] finds that the quasar evolution is rather better fitted by an *exponential* function. For a world model with $q = 1$, he derives

$$\varrho(L, t) \propto e^{11{,}8\left(1-\frac{t}{\tau_H}\right)} \tag{12.9}$$

This corresponds to a characteristic "decay time" of only 8.4×10^8 years (for $H = 75$ km s^{-1} Mpc^{-1}). Rowan-Robinson[204] finds that strong radio galaxies may evolve in a similar fashion.

In an attempt to derive the space distribution and luminosity function for quasars as a whole (and not just the radio quasars) Schmidt[195] has investigated a sample of 20 objects, selected from one of the fields studied by Sandage and Luyten, which is believed complete in the optical magnitude range 17.5–18.5. As already mentioned, counts of objects in the Sandage-Luyten fields show that the number per square degree rises by a factor ~ 6 per unit magnitude increment. This can be reproduced by an evolution $\varrho \propto (1 + z)^6$. By making this assumption for the objects in his sample, Schmidt is able to derive a luminosity function—it turns out that the co-ordinate density of objects (at a given z) increases by a factor 3 or 4 per magnitude with probably an even steeper cut-off at the bright end. Schmidt can then (again assuming the $(1 + z)^6$ law) predict the expected number and redshift distribution of quasars of all magnitudes. (See Table XXIII. In this table, f_{opt} denotes the observed flux density, in units of W m^{-2} Hz^{-1}, at an emitted wavelength of 2500 Å. In view of the small size of the observed sample, the entries in the table are not valid to more than one significant figure. The additional figures are given only for self-consistency.) This work leads to the conclusion that 24 per cent of 19th magnitude quasars should have z in the range 2.5–3. This is contrary to observation. It is therefore *not* consistent to apply the $(1 + z)^6$ law beyond $z = 2.5$, though one gets adequate agreement by supposing that at earlier epochs the quasar coordinate density stayed constant. (Perhaps the truncation of the power law evolution, which is also required in order to explain the convergence of the source counts at low S, is related to the epoch of galaxy formation).

A further very important point emerges when this sample of *all* quasars is compared with Schmidt's previous sample[200] of 3CR quasars: it is found that the 18th magnitude objects in the two samples have *more or less the same* distribution of redshifts (see Table XXIV). This is highly significant for the following reason. If the radio and optical luminosities were independent, the 3CR quasars should tend to have *smaller* redshifts than the others, because at large z only those quasars which were exceptionally bright radio-wise would appear in the radio catalogue, whereas at small z even quite (intrinsically) weak radio sources would be included. Schmidt explains this result by proposing that the radio and optical luminosity functions are not independent, and tentatively suggests that the combined radio-optical luminosity function can be written

$$\varrho(L_{\text{opt}}, L_{\text{rad}}) = \varrho(L_{\text{opt}})\, \psi(L_{\text{rad}}/L_{\text{opt}}) \tag{12.10}$$

Table XXIII (From Schmidt[195]) Theoretical redshift-magnitude table for quasi-stellar objects with density $\propto(1+z)^6$ in a cosmological model $q_0 = 1$

	Δv	$(1+z^6)\,\Delta v$	$\log f_{\rm opt}$										
$\log z$	(Gpc^3)	(Gpc^3)	–27.8(13)	–28.2(14)	–28.6(15)	–29.0(16)	–29.4(17)	–29.8(18)	–30.2(19)	–30.6(20)	–31.0(21)	–31.4(22)	–31.8(23)
+0.5	24	(148000)	—	—	—	—	—	—	(35×10^3)	(480×10^3)	(1.9×10^6)	(8×10^6)	(15×10^6)
+0.3	21	17100	—	—	—	—	—	4×10^3	56×10^3	217×10^3	1.0×10^6	2×10^6	9×10^6
+0.1	14	2150	—	—	—	—	503	7×10^3	27×10^3	124×10^3	0.2×10^6	1×10^6	—
−0.1	8.5	315	—	—	—	74	1026	4×10^3	18×10^3	32×10^3	0.2×10^6	—	—
−0.3	4.2	52	—	—	12	169	660	3×10^3	5×10^3	27×10^3	—	—	
−0.5	1.75	9.7	—	2	32	123	560	1×10^3	5×10^3	—	—	—	—
−0.7	0.63	1.96	0	6	25	113	202	1×10^3	—	—	—	—	—
−0.9	0.21	0.43	1	6	25	44	219	—	—	—	—	—	—
−1.1	0.061	0.098	1	6	10	50	—	—	—	—	—	—	—
−1.3	0.017	0.023	1	2	12	—	—	—	—	—	—	—	—
−1.5	0.0047	0.0056	1	3	—	—	—	—	—	—	—	—	—
−1.7	0.0012	0.0014	1	—	—	—	—	—	—	—	—	—	—
$n_{QSO}(\log f_{\rm opt})$			5	25	116	573	3170	20×10^3	111×10^3	400×10^3	1.4×10^6	3×10^6	9×10^6

NOTE Numbers in parentheses following $\log f_{\rm opt}$ are the approximate visual magnitudes.

This would mean that the ratio of the *apparent* brightness of quasars at radio and optical wavelengths—which can be thought of as a generalized "colour"—was distributed according to a universal function independent of optical luminosity and (apart from a slight dependence on the difference between optical and radio spectral indices) of redshift. In table XXV (pp. 189–193) we reproduce data from a recent catalogue[206] of quasars with known redshifts. The columns list respectively (1) the designation of the object; (2) and (3) its right ascension and declination; (4) its approximate visual magnitude; and (5) its emission line redshift (an asterisk denotes the presence of absorption lines).

Table XXIV Distribution of Redshifts of Sandage-Luyten (radio-quiet) and 3*CR* quasars of approximate optical magnitude 18, according to data of Schmidt[195,200]

$\log_{10} z$	Radio-quiet	3*CR*
0.2 to 0.4	5	3
0.0 to 0.2	7	6
−0.2 to 0.0	2	6
−0.4 to −0.2	3	2
−0.6 to −0.4	0	1
−0.8 to −0.6	2	0
Unknown	1	1
Total	20	19

12.3 THE PHYSICAL NATURE OF THE EVOLUTION

The most dramatic inference from all this work is that the density of powerful sources—both radio galaxies and quasars—must increase by a factor 10^2–10^3 per comoving volume (i.e. over and above the $(1 + z)^3$ dilution factor) between the present epoch and the epoch corresponding to $z \simeq (2–3)$. Drastic though this effect undoubtedly is, we recall that the most remote known sources emitted the radiation we now detect when the Universe was ~ 20 per cent of its present age. Perhaps we should not be suprised when we infer gross differences between such ancient epochs and the present day.

Since we can infer nothing about the geometry of the universe from these observations, it is more interesting to invert the problem and consider instead how much can be learnt about the epoch-dependence of the source parameters. All the evidence indicates that radio quasars and radio quiet quasars both display the *same* evolution as had been inferred for strong radio sources as a whole from analyses of the radio source counts. This suggests that the evolution is predominantly due to a drastic secular decrease in the propensity of galactic nuclei to undergo "violent events"†. The similarity of the optical and radio evolution is, however, rather surprising. The optical measurements refer to a compact object, with dimensions $\lesssim 1$ pc, whereas the radio properties refer to structures on scales up to 100 Kpc, and the relevant physical processes are quite distinct. In particular, most models of extended radio sources invoke an intergalactic gas to confine the source components. The higher density of this gas at earlier epochs might be expected to cause some variation of radio properties with z quite independent of the evolution of the optical properties. Ideally, one would like to have enough observations to be able to subdivide the data and determine the evolutionary behaviour for different classes of object separately. At present such procedures result in rather small numbers of objects in each category, and also introduce extra selection effects; so this program must await the identification of many more quasars and radio galaxies (and accurate magnitudes of the latter out to the largest possible redshift). At present everything is consistent with a similar evolution law for all powerful radio sources and quasars‡—either an exponential evolution of the form (12.9), or else a power law $(1 + z)^n$ with $n \simeq 6$ and a truncation at $z \simeq 2.5$.

A feature common to almost all proposed astrophysical models is that the active lifetime of a strong radio source or quasar is probably no more than 10^6–10^7 years. This is much shorter than the Hubble time τ_H at all relevant redshifts. At any cosmic epoch, we must therefore be observing sources on every part of their evolutionary track, the relative number of sources (per unit volume) at each particular stage of development being inversely related to the rate at which they evolve through that stage. Indeed, much of the scatter in the observed linear sizes and luminosities of radio sources can be interpreted as different stages in the development of a single type of object, the compact sources growing into extended doubles.

---

† At first sight one might be tempted to adduce the higher rate of violent events at early epochs as evidence that they are associated with the *formation* of galaxies, rather than with the eventual collapse and "death" of galactic nuclei. This inference is however unwarranted. Just as the human infant mortality rate decreases more or less monotonically with age, so it is possible that many galaxies form with dense nuclei poised on the verge of unstability (soon therefore to collapse) and that those surviving to the present epoch are almost all so stable that the current death rate is low.

‡ On the basis of radio properties alone, there is no clear-cut way of distinguishing a radio galaxy from a quasar. In both types of object, one finds compact (and sometimes variable) radio components $\lesssim 1$ pc in size, as well as very extended components.

Consequently observations of quasars with large redshifts do not reveal objects which are individually younger—in the sense of having experienced a shorter active lifetime—than similar nearby objects. The inferred epoch-dependence of $\varrho(L)$ (illustrated schematically in Figure 39) must therefore reflect either a change in the evolutionary track traced out by each source, or a change in the source production rate (or, more probably, in both these things).

The foregoing discussion of source counts and the luminosity-volume test has served at least to emphasize the necessity of a firmer understanding of the physics of extragalactic objects, and of their astrophysical evolution, before we can use these techniques to infer the geometry of the Universe. (All that can be said at this time is that the observations are incompatible with a strict steady state universe, where no secular dependence of mean source parameters is allowed.) As mentioned later, however (section 15.3), quasars are of indirect importance for cosmology, because searches for absorption effects, dispersion, etc. in their radiation allows us to deduce physical conditions in the intervening space.

12.4 ANGULAR DIAMETER–REDSHIFT TESTS: THE UNIVERSE AS A LENS

Equation (11.11) tells us how the observed angular size of a "rigid rod" depends on redshift. The precise dependence is a function of q. In all Friedmann models (open as well as closed) there is a certain critical redshift beyond which the angular size starts *increasing* (the increase being always $\propto (1 + z)$ for sufficiently large z). In effect, curved space acts as a lens of great focal length. The effect is attributable to the matter along the line of sight, which tends to focus any beam of radiation from a source. This curving of light rays has little effect on the apparent size of nearby objects. However, in a closed universe the angular diameters attain a minimum at a redshift of only $z \simeq 1$, and distant galaxies—galaxies from a quarter to a half the way round the Universe—may appear greatly magnified (see Figure 41 and Figure 42).

"Dumb-bell" radio-sources suggest themselves as a class of objects, detectable even at large redshifts, to which this procedure can be applied. However they certainly are not "standard rigid rods" any more than they are "standard candles" so one can, at best, hope to discern some trend in a relation involving much scatter. Miley[207] finds a correlation between fringe-visibility and redshift for radio quasars with steep spectra (flat spectrum quasars probably being so small that they are unresolved even at smallest redshifts). This is one of the few pieces of fairly direct evidence that quasar redshifts are indeed a measure of distance, but (accepting this) one cannot safely infer anything else because, owing to observational selection, the only large redshift objects studied are those with high L, which might differ systematically from the typical lower redshift source in the sample. Legg[208] plotted the angular size of 57 double radio sources against redshift. On the assumption that there is a well-defined (*and redshift independent*) upper limit to the linear size of strong sources, he found that the data agreed better with a de Sitter (steady state type) model than with a Friedmann universe. However there *may* be an epoch-dependence in the typical size of double sources—for example the components may be braked more rapidly at earlier epochs because they are moving through a denser external medium—and in our present state of ignorance these effects are just as unpredictable as the evolution in the luminosity function. Thus, as Legg points out, even if we were sure that there were no selection effects operating, this test cannot be used to estimate q.

Longair and Pooley[209] have performed a different kind of angular diameter test which involves *no* knowledge of the redshifts. Using the data on the size distribution of 3C sources they have calculated, assuming a Friedmann model and neglecting evolutionary effects, that there should only be ~ 2 faint sources in the 5C2 and 5C3 surveys with angular size $\gtrsim 70''$: in fact 9 such sources were found. This is probably just an evolutionary effect—a $(1 + z)^3$ z-dependence of the source density would be sufficient, which is less drastic than the evolution implied by the source counts, though the latter refers only to more powerful sources.

Observations of angular diameters of ordinary galaxies and clusters, even though presently restricted to $z \lesssim 0.5$, may also involve uncertain evolutionary corrections. We are still too ignorant of cluster formation and evolution to know whether rich clusters of galaxies are adequate "rigid rods". Attempts to measure the angular sizes of individual galaxies are bedevilled by another problem. A galaxy does not have a sharp edge—the angular diameter that is measured refers to the part of a galaxy whose *surface brightness* exceeds a certain limiting value. Thus smaller physical diameters are measured when the galaxy has a large redshift. These diameters are called *isophotal diameters*, and Sandage[205] has calculated that the isophotal angular diameter of a typical galaxy would decrease monotonically with increasing z, the dependence being insensitive to q. Recently, however, Baum[210] claims to have overcome this difficulty, and to be able to measure the "metric diameter" of galaxies by studying their optical images. His provisional estimate of q is 0.3.

Will one be able in future to find any characteristic distance that will serve as a natural standard of length not only at very great distances ($z = 2$

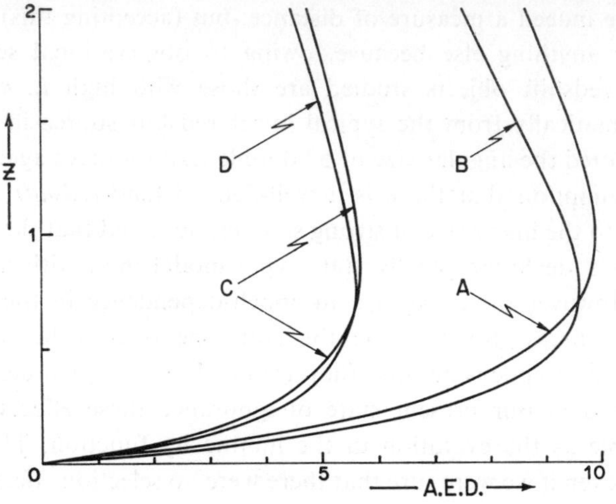

Figure 41

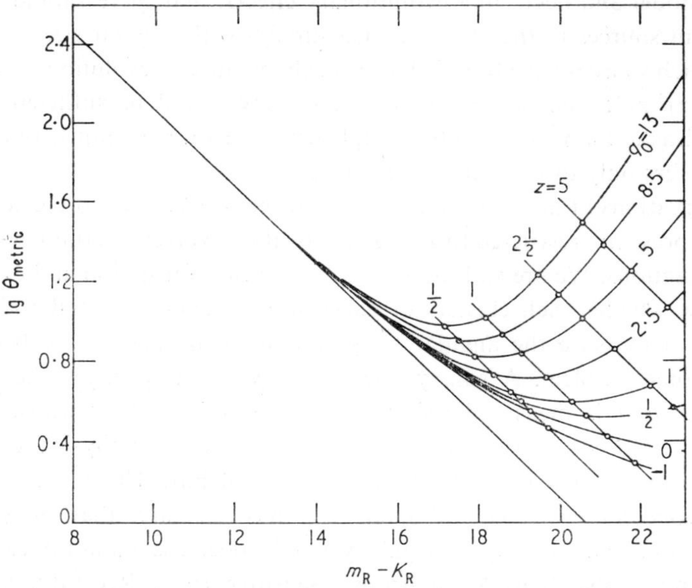

Figure 42 Diagram, due to Sandage,[206] of metric angular diameters of galaxies as a function of corrected magnitude and redshift. Curves are plotted for different values of the deceleration parameter ($q_0 = -1$ denotes the steady state model). Note that for all Friedmann models the angular diameters pass through a minimum, which corresponds to the maximum in the "angular effective distance" plotted in Figure 41.

Figure 41 Angle effective distance (= actual transverse dimension of object/angle subtended by it) as a function of red shift for the following cases

	Specified parameters					Derived quantities		
	Curve	p/ϱ	t (in 10^9 yr)	H^{-1} (in 10^9 yr)	η	a (in 10^9 l yr)	a_0 (in 10^9 l yr)	ϱ (in 10^{-30} g/cm^3)
closed space ($k = 1$)	A	0 (dust)	11	19	1.511	40.323	85.726	2.316
	B	0 (dust)	11	25	2.221	24.775	30.858	3.599
	C	$\frac{1}{3}$ (radiation)	11	25	0.667	31.751	51.324	2.845
	D	$\frac{1}{3}$ (radiation)	11	30	0.953	21.299	26.122	4.795

Table XXVI Counts of radio sources. The total number of sources in a survey is frequently greater than the number counted because some surveys are incomplete at the lowest flux densities and others exclude regions close to the galactic plane. All flux densities are in units of 10^{-26} W m^{-2} Hz^{-1} = 1 flux unit (f.u.). The counts are described by $N(S) \propto S^{-\beta}$

Observatory	Frequency (MHz)	Limiting flux density	Number of sources in survey	Flux density range of counts	β	References
Bologna						
BI	408	1.0	654	≥ 0.6	1.58 ± 0.1	a
BIS		0.6	38			
	408	0.15	328	0.25–2.0	1.5 ± 0.1	b
Cambridge						
revised 3C	178	9	328	≥ 9	1.8	c } reviewed in f
4C	178	2	~5000	2–10	1.66	d } reviewed in f
North Polar Survey	178	0.25	87	0.25–2	1.3	e } reviewed in f
5C2	408	0.0115	207	0.01–0.1	0.8	g, h
5C3	408	0.012	213	0.012–0.1	0.8	i
5C4	408	0.016	189	0.016–0.1	0.8	j
Dominion Radio Astronomy Observatory (DA)	1420	1	615	2	1.88 ± 0.07	k
Illinois	610.5	0.8	239		1.8	
NRAO	1400	2	235	≥ 2	1.90 ± 0.12	m
		0.5	200	0.5–2	1.50 ± 0.08	m
	5000	0.6	275	≥ 0.6	1.76 ± 0.11	n
		0.6–3	262	0.6–3	1.66 ± 0.14	n
		0.067–0.6	97	0.067–0.6	1.72 ± 0.21	n
Ohio	1415	0.2	2100	0.5–1.5	1.57	o
	1415	0.16	8100	≥ 1	1.7	p
				< 0.4	0.8	p

Parkes						
$-20° > \delta > -60°$	408	4	297	$\geqq 4$	1.85	q
$20° > \delta > 0°$	408	1	564	$\geqq 3.5$	1.8 ± 0.15	r
$0° > \delta > -20°$	408	1	628	$\geqq 3$	1.89 ± 0.10	s
$27° > \delta > 20°$	408	1.5	397	$\geqq 1.5$	1.86 ± 0.10	t
$\left\{ \begin{array}{l} -60° > \delta > -75° \\ -75° > \delta > -90° \end{array} \right.$	408	$\left. \begin{array}{c} (\approx 1 \text{ at } 1410 \text{ MHz}) \\ 0.5 \end{array} \right\}$	247	$S_{1410} \geqq 1$	1.7 ± 0.3	u
	2700	0.4 0.08	135 210	$\geqq 0.08$	$1.38^{+0.12}_{-0.08}$	v

a Braccesi, A., Ceccarelli, M., Fanti, R., Gelato, G., Giovannini, C., Harris, D., Rosatelli, C., Sinigaglia, G., and Volders, L., *Nuovo Cimento* **40B**, 267 (1965); Braccesi, A., Ceccarelli, M., Fanti, R., and Giovannini, C., *Nuovo Cimento* **41B**, 92 (1966).

b Grueff, G., and Vigotti, M., *Astrophys. Letters* **2**, 113 (1968).

c Bennett, A. S., *Mem. R. Astr. Soc.* **67**, 163 (1962).

d Pilkington, J. D. H., and Scott, P. F., *Mem. R. Astr. Soc.* **69**, 183 (1965); Gower, J. F. R., Scott, P. F., and Wills, D., *Mem. R. Astr. Soc.* **71**, 49 (1967).

e Ryle, M., and Neville, A. C., *Mon. Not. R. Astr. Soc.* **125**, 39 (1962).

f Gower, J. F. R., *Mon. Not. R. Astr. Soc.* **133**, 151 (1966).

g Pooley, G. C., and Ryle, M., *Mon. Not. R. Astr. Soc.* **139**, 515 (1968).

h Pilkington, J. D. H., Scott, P. F., *Mem. R. Astr. Soc.* **69**, 183 (1965); Gower, J. F. R., Scott, P. F., and Wills, D., *Mem. R. Astr. Soc.* **71**, 49 (1967).

i Pooley, G. C., *Mon. Not. R. Astr. Soc.* **144**, 101 (1969).

j Willson, M. A. G., *Mon. Not. R. Astr. Soc.* **151**, 1 (1970).

k Galt, J. A., and Kennedy, J. E. D., *Astron. J.* **73**, 135 (1968).

l Macleod, J. M., Swenson, G. W., Yang, K. S., Dickel, J. R., *Astron. J.* **70**, 756 (1965); McVittie, G. C., and Schusterman, L., *Astron. J.* **71**, 137 (1967).

m Bridle, A. H., Davies, M. M., Fomalont, E. B., and Lequeux, J., in press.

n Kellermann, K. I., Pauliny-Toth, I. I. K., and Davies, M. M., *Astrophys. Letters* **2**, 105 (1968); Kellermann, K. I., Davies, M. M., and Pauliny-Toth, I. I. K., *Astrophys. J.* **170**, L. 1 (1971).

o Fitch, L. T., Dixon, R. S., and Kraus, J. D., *Astron. J.* **74**, 612 (1969).

p Harris, B. J., and Kraus, J. D., *Nature* **227**, 785 (1970).

q Bolton, J. G., Gardner, F., and Mackay, M. B., *Aust. J. Phys.* **17**, 340 (1964).

r Day, G. A., Shimmins, A. J., Ekers, R. D., and Cole, D. J., *Aust. J. Phys.* **19**, 35 (1966).

s Shimmins, A. J., Day, D. A., Ekers, R. D., and Cole, D. J., *Aust. J. Phys.* **19**, 837 (1966).

t Shimmins, A. J., and Day, G. A., *Aust. J. Phys.* **21**, 377 (1968).

u Price, R. M., and Milne, D. K., *Aust. J. Phys.* **18**, 329 (1965).

v Shimmins, A. J., Bolton, J. G., and Wall, J. V., *Nature* **217**, 818 (1968).

or 3), but in galaxies closer at hand? Perhaps not, in which case the angular diameter test will never help to tell us whether the universe is closed. However it would seem unwise to discount this possibility, with all the advantages it would bring, in view of the remarkable development in very long baseline interferometry (paired radio telescopes at intercontinental distances, yielding angular resolutions better than $10^{-3''}$), and especially in view of the demonstrated ability of skilled observers to find regularities where one had no right to expect them in advance.

12.5 SUMMARY

Cosmic time and redshift are related, in dust-filled Friedmann models, by the parametric equations of Table XXI. Thus, by observing discrete sources out to redshifts $z \simeq 3$, we can in effect probe $\sim$ 80 per cent of cosmic history. A hypothetical astronomer, observing only $\sim 2 \times 10^9$ years after the initial singularity, would see a vastly more active and dramatic cosmic scene. Whereas our nearest bright quasar is 3C273 ($\sim$ 600 Mpc distant), he would be likely to find a similar object only $\sim$ 10 Mpc away ($\sim$ 50 times closer), and as bright as a fourth magnitude star. He would observe correspondingly greater activity in the radio sky, the integrated flux at meter wavelengths being perhaps 100 times higher than it is today.

This is the message of the radio source counts, reiterated by the distribution of quasars, and it is obviously crucially important to our understanding of the astrophysical evolution of the contents of the Universe. It is, however, disappointing that the discovery of objects with very large redshifts has not led to any progress in "classical" or "geometrical" cosmology. It is still studies of normal bright galaxies (despite the smaller redshifts involved) which are likely to yield the most significant limits on q.

Table XXV Catalog of quasars

OBJECT	α(1950)	δ(1950)	V MAGNITUDE	Z
PHL 658	0 3 25.4	15 53 10	16.40	.450
3C2	0 3 49.0	-1 21 7	19.35	1.037
3C9	0 17 49.8	15 24 16	18.21	2.012
3C13	0 27 3.0	39 32 12	19.00	
4C 42.01	0 32 23.0	42 21 48	18.30	1.588*
PHL 923	0 56 31.7	-0 9 16	17.33	.717
PHL 938	0 58 12.0	1 56 0	17.16	1.930*
PHL 957	1 0 36.0	13 0 0	16.60	2.720*
PKS 0106+01	1 6 4.5	1 19 1	18.39	2.107
PKS 0115+02	1 15 42.8	2 42 32	17.50	.672
PKS 0119-04	1 19 55.9	-4 37 7	16.88	1.955*
PKS 0122-00	1 22 55.8	-0 21 34	16.70	1.070
4C 25.05	1 23 56.0	25 43 54	17.50	2.360*
PHL 3375	1 28 24.0	7 28 0	18.02	.390
PHL 1027	1 30 30.0	3 22 0	17.04	.363
PHL 3424	1 31 12.0	5 32 0	18.25	1.847
3C 47	1 33 40.3	20 42 16	18.10	.425
3C48	1 34 49.8	32 54 20	16.20	.367
PHL 1078	1 35 29.0	-5 42 6	18.25	.308
PHL 1093	1 37 23.0	1 16 18	17.07	.258
PHL 3632	1 39 54.0	6 10 0	18.15	1.479
4C 33.03	1 41 18.0	33 57 0	17.50	1.455
PHL 1127	1 41 30.0	5 14 0	18.29	1.990*
PHL 1186	1 47 36.0	9 1 0	18.60	.270
PHL 1194	1 48 42.0	9 2 0	17.50	.298
PHL 1222	1 51 12.0	4 48 0	17.63	1.910*
PHL 1226	1 51 48.0	4 34 0	18.20	.404
PKS 0155-10	1 55 15.0	-10 58 0	17.09	.616
3C57	1 59 30.4	-11 47 0	16.40	.680
DW 0202+31	2 2 10.0	31 57 18	18.00	1.466
PKS 0202-17	2 2 34.0	-17 15 37	18.00	1.740
PKS 0214+10	2 14 26.9	10 50 24	17.00	.408
PKS0225-014	2 25 35.0	-1 29 12	18.00	.685
PHL 1305	2 26 21.0	-3 54 0	16.96	2.064
PKS 0229+13	2 29 2.3	13 9 41	17.71	2.065*
PKS0231+022	2 31 14.6	2 16 18	18.00	.322
PHL 1377	2 32 36.6	-4 15 5	16.46	1.434*
PKS 0237-23	2 37 52.6	-23 22 6	16.63	2.224*
NRAO 140	3 33 22.3	32 8 36	17.50	1.258
3C93.1/113	3 48 37.0	33 4 54	18.50	

PKS 0349-14	3	49	9.5	-14	38	7	16.22	.614
3C94	3	50	4.1	-7	19	50	16.49	.962
PKS 0403-13	4	3	14.0	-13	16	18	17.17	.571
PKS 0405-12	4	5	27.4	-12	19	34	17.07	.574
PKS 0424-13	4	24	48.0	-13	9	36	17.50	2.165*
3C138	5	18	16.5	16	35	27	18.84	.759
3C147	5	38	43.5	49	49	43	17.80	.545
3C175	7	10	15.3	11	51	30	16.60	.768
3C181	7	25	20.1	14	43	47	18.92	1.382
3C184.1/140	7	29	22.0	81	52	36	17.50	1.022
PKS								
PKS 0736+01	7	36	42.4	1	43	57	17.47	.191
3C186	7	40	56.7	38	0	32	17.60	1.063
3C191	8	2	3.8	10	23	58	18.40	1.952*
4C 05.34	8	5	20.0	4	41	24	18.00	2.877*
3C196	8	9	59.4	48	22	8	17.79	.871
PKS 0812+02	8	12	47.2	2	4	11	18.50	.402*
4C 37.24	8	27	55.0	37	53	42	18.11	.914
3C204	8	33	18.2	65	24	6	18.21	1.112
3C205	8	35	10.0	54	4	42	17.62	1.534
4C 19.31	8	36	15.0	19	33	0	17.60	1.691
PKS 0837	8	37	28.0	-12	3	54	15.76	.200
3C207	8	38	1.7	13	23	5	18.15	.684
LB 8707	8	46	1.0	14	32	0	18.10	
LB 8755	8	48	5.0	15	33	30	17.70	2.010
3C208	8	50	22.8	14	3	58	17.42	1.109
4C 17.46	8	56	3.0	17	3	12	17.40	1.449
PKS 0859-14	8	59	55.0	-14	3	37	17.80	1.327
3C215	9	3	44.2	16	58	16	18.27	.411
PKS 0922+14	9	22	22.3	14	57	26	17.96	.896
PKS0922+005	9	22	35.7	0	32	6	18.07	1.720
4C 39.25	9	23	55.0	39	15	18	17.86	.698
PKS 0932+02	9	32	42.6	2	17	42	17.39	.659*
AD 0952+1	9	52	11.8	17	57	44	17.23	1.472
3C232	9	55	25.4	32	38	23	15.78	.530
PKS 0957+00	9	57	43.8	0	19	50	17.57	.907
PKS 1004+13	10	4	44.0	13	3	30	15.15	.240
TON 490	10	11	6.0	25	6	0	15.40	1.631
4C 48.28	10	12	49.0	48	53	12	19.00	.385
3C245	10	40	6.1	12	19	15	17.29	1.029
PKS 1049-09	10	48	59.5	-9	2	12	16.79	.344
5C 02.10	10	49	41.0	48	55	54	18.00	.478
5C 02.56	10	55	18.0	49	55	36	20.00	2.380
PKS 1055+20	10	55	36.9	20	8	18	17.07	1.110
3C249.1	11	0	27.4	77	15	8	15.72	.311

Figure 43 Radiometers used by R. A. Stokes, R. B. Partridge and D. T. Wilkinson at White Mountain (Calif.), Summer 1967 (from front to back radiometers for wavelengths of 3.2 cm, 1.6 cm and 0.86 cm). [Photograph kindly supplied by R. B. Partridge.]

QS1108+285	11 8 26.5	28 57 56	20.00	2.192
3C254	11 11 53.4	40 53 42	17.98	.734
PKS 1116+12	11 16 20.8	12 51 6	19.25	2.118*
PKS 1127-14	11 27 35.6	-14 32 54	16.90	1.187
3C261	11 32 16.3	30 22 1	18.24	.614
PKS 1136-13	11 36 38.6	-13 34 9	17.80	.554
3C263	11 37 9.0	66 4 28	16.32	.652*
PKS 1148-00	11 48 10.2	-0 7 13	17.60	1.982
4C 49.22	11 50 48.0	49 47 48	16.10	.334*
4C 31.38	11 53 44.0	31 44 47	18.96	1.557
4C 29.45	11 56 58.0	29 33 24	15.60	.729
3C268.4	12 6 41.7	43 56 5	18.42	1.400
PKS 1217+02	12 17 38.3	2 20 21	16.53	.240
3C270.1	12 18 4.0	33 59 50	18.61	1.519*
4C 21.35	12 22 23.0	21 40 0	17.50	.435
TON 1530	12 23 12.0	22 51 0	17.00	2.051*
4C 25.40	12 23 12.0	25 14 30	16.00	.268
3C273	12 26 33.3	2 19 42	12.80	.158
PKS 1299.02	12 29 25.9	-2 7 31	16.75	.388*
TON 1542	12 29 48.0	20 25 0	15.30	.064
PKS 1233-24	12 32 59.4	-24 55 46	17.20	.355
3C275.1	12 41 27.7	16 39 19	19.00	.557
B 19	12 45 2.7	34 31 29	17.94	2.065
BSO 1	12 46 29.0	37 46 25	16.98	1.241*
B 46	12 46 29.2	34 40 49	17.83	.271
BSO 2	12 48 17.7	33 47 4	18.64	.186
B 86	12 49 40.6	33 54 42	17.58	1.431
3C277.1	12 50 15.3	56 50 37	17.93	.320
PKS 1252+11	12 52 7.7	11 57 21	16.64	.870
B 114	12 52 57.7	35 55 26	17.92	.221
3C279	12 53 35.8	-5 31 8	17.75	.536
B 142	12 54 55.0	37 3 29	17.84	.280
B 154	12 55 2.1	35 21 21	18.56	.183
B 185	12 55 40.2	37 15 19	18.12	1.530
B 194	12 56 7.8	35 44 56	17.96	1.864*
B 189	12 56 51.1	36 48 10	19.22	2.075
B 201	12 57 26.6	34 39 34	16.79	1.375
3C280.1	12 58 14.2	40 25 15	19.44	1.659
B 246	12 58 31.0	34 4 48	18.18	.690
B 196	12 58 42.0	35 38 48	18.28	.323
B 471	12 58 49.0	34 22 42	17.66	.774

B 228	12	59	21.4	36	46	22	17.83	1.194
BSO 6	12	59	30.5	34	27	15	17.87	1.956
B 264	12	59	30.9	32	21	58	16.89	.095
B 234	13	0	43.2	36	7	48	17.52	.060
B 272	13	1	35.0	37	30	48	17.25	.036
B 286	13	1	43.0	35	49	32	18.65	.330
B 288	13	2	18.0	35	45	22	18.39	1.293
B 340	13	4	47.1	34	40	39	16.97	.184
B 312	13	4	53.1	37	29	33	19.08	.450
3C281	13	5	22.5	6	58	16	17.02	.599
B 330	13	5	26.0	36	25	54	18.01	.920
B 337	13	5	30.2	35	17	50	17.62	.300
B 382	13	6	53.0	35	1	32	17.55	.194
B 360	13	8	17.6	38	14	17	17.56	2.090
BSO 8	13	9	15.0	34	3	8	17.43	1.750
BSO 11	13	11	22.1	36	16	30	18.41	2.084*
PKS 1317-00	13	17	4.5	-0	34	21	17.32	.890*
TON 153	13	17	24.0	27	45	0	15.30	
TON 156	13	18	54.0	29	6	0	16.00	
TON 157	13	21	0.0	29	27	0	16.00	.960
PKS 1327-21	13	27	23.2	-21	26	34	16.74	.528
3C287	13	28	15.9	25	24	38	17.67	1.055
3C286	13	28	49.7	30	45	58	17.25	.846
RS 12	13	31	11.0	28	8	24	18.61	
RS 13	13	31	30.0	27	45	48	17.94	1.287
4C 55.27	13	32	17.0	55	15	48	16.00	.249
RS 23	13	33	55.0	28	40	18	18.74	1.908*
PKS1335+023	13	35	7.3	2	22	6	18.00	.610
MSH13-011	13	35	31.3	-6	11	57	17.68	.625
RS 32	13	36	3.0	26	29	6	18.91	.341
3C288.1	13	40	30.4	60	36	55	18.12	.961
DW 1349+02	13	49	58.0	2	47	42	19.00	
PKS 1354+19	13	54	42.3	19	33	41	16.02	.720
3C298	14	16	38.8	6	42	21	16.79	1.439*
4C 20.33	14	22	37.0	20	13	49	17.86	.871
TON 202	14	25	18.0	26	46	0	15.68	.366
MSH14-121	14	53	12.2	-10	56	40	17.37	.938
PKS 1454-06	14	54	2.7	-6	5	45	18.00	1.249
3C309.1	14	58	56.7	71	52	11	16.78	.905*
3C311	15	2	58.0	60	12	36	18.00P	1.022

PKS 1510-08	15 10 8.9	-8 54 47	16.52	.361*
4C 37.43	15 12 46.0	37 2 30	15.50	.370
3C323.1	15 45 31.1	21 1 33	16.69	.264
DA 406	16 11 50.0	34 20 0	17.50	1.401
TON 256	16 12 8.7	26 13 0	15.41	.131
3C334	16 18 7.5	17 43 29	16.41	.555
3C336	16 22 32.0	23 52 1	17.47	.927
PKS 1634+26	16 34 21.4	26 54 18	17.75	.561
3C345	16 41 17.6	39 54 11	15.96	.594
4C 29.50	17 2 11.0	29 50 0	19.14	1.927
3C351	17 4 3.7	60 48 29	15.28	.371
PKS 1801+01	18 1 44.0	1 1 18	19.00	1.522
3C380	18 28 13.5	48 42 40	16.81	.691
3C407	20 5 46.0	-4 27 18	18.00	1.864
PKS 2115-30	21 15 11.0	-30 31 46	16.47	.980
3C432	21 20 25.4	16 51 56	17.96	1.805
PKS 2128-12	21 28 52.6	-12 20 21	15.99	.501
AO 2128+08	21 28 54.0	8 59 42		
PKS2134+004	21 34 3.7	0 28 12	18.00	1.930
PKS 2135-14	21 35 .8	-14 46 27	15.53	.200
PKS 2144-17	21 44 17.7	-17 54 5	19.50	.684
PKS 2145+06	21 45 36.1	6 43 41	16.47	.367
PKS 2146-13	21 46 46.1	-13 18 24	20.00	1.800*
PKS2153-204	21 53 48.0	-20 26 30	17.50	1.310
PKS 2216-03	22 16 16.3	-3 50 41	16.38	.901
3C446	22 23 11.0	-5 12 17	18.39	1.404
PHL 5200	22 25 50.6	-5 34 0	18.00	1.981*
CTA 102	22 30 7.8	11 28 23	17.33	1.037
3C454	22 49 7.6	18 32 47	18.40	1.757
3C454.3	22 51 29.5	15 52 54	16.10	.859
PKS 2251+11	22 51 40.6	11 20 39	15.82	.323
PKS2254+024	22 54 43.9	2 27 32	18.00	2.090
CID 141	23 25 43.0	29 20 o	17.30	1.015
OZ-252	23 31 18.0	-24 0 14	17.00	
PKS 2344+09	23 44 3.8	9 14 4	15.97	.677
PKS 2345-16	23 45 27.6	-16 47 52	18.00	.600
PKS 2354+14	23 54 44.7	14 29 26	18.18	1.810

CHAPTER 13

The Microwave Background: The Primordial Fireball

ONE CAN in principle learn something about sources with $z > 3$, if they exist, by investigating their integrated contribution to the background in the infrared, optical and x-ray bands (we return to this topic in Chapter 16); and it is conceivable that individual sources whose redshifts substantially exceed 3 may sometime be discovered. But such measurements tell us nothing about the period before sources were formed! Current knowledge of pre-galactic epochs depends almost exclusively on the *microwave background.* It is the apparently thermal spectrum of this radiation, and its high isotropy, which provides the most compelling evidence for the Universe's hot dense origins; for the validity of Friedmann models as a description of all but (perhaps) the *very* earliest stages in the expansion; and for physical conditions at those epochs. Unless the commonly accepted interpretation of the microwave background is entirely misconceived, the detection of this radiation—in 1965—must rank as the most important advance in cosmology since Hubble discovered the universal expansion.

13.1 OBSERVATIONS AND INTERPRETATIONS OF THE MICROWAVE BACKGROUND

Penzias and Wilson[211] of Bell Telephone Laboratories, measuring the effective noise temperature of a 20 foot horn reflector antenna at a wavelength of 7.4 cm, discovered an excess temperature of $3.5 \pm 1^{\circ}$K. They also noticed that the excess temperature was, over a period of the order of one year, isotropic, unpolarized and free from detectable variations. The horn-shaped antenna they used was originally designed to receive signals reflected by Telstar satellites. This inferred radiation field was $\gtrsim 100$ times more intense than the expected background (extrapolated from longer wavelengths) from non-thermal radio sources. In the same issue of Astrophysical Journal as Penzias and Wilson's announcement of "A Measurement of Excess Antenna Temperature at 4080 MHz", Dicke *et al*[212] published an interpretation of this background as "primordial cosmic fireball radiation" surviving from an early epoch in which the Universe was enormously hotter and denser. Gamow[213] had, fifteen years earlier, analysed the physics of a homogeneous Universe expanding in accordance with Einstein's equations, and had predicted that such "relic" radiation might exist. In 1956 he suggested[214] a present temperature of 6°K—within a factor ~ 2 of that found by Penzias and Wilson.

The simple "primordial fireball" hypothesis (see section 13.2) predicts that the background should be black body radiation, but this one measurement, referring to a single frequency, provided no spectral information. The hypothesis therefore received strong support when Roll and Wilkinson—who were already in 1964 constructing a specially conceived radiometer† to detect cosmic background radiation—reported[215] a new measurement giving a temperature 3.0 ± 0.5°K at a wavelength of 3.2 cm. Subsequently a series of measurements between 3.3 mm and 74 cm have all yielded results compatible with a black body spectrum. The best estimate of the temperature is 2.7°K. Most of these measurements refer to the Rayleigh-Jeans part of the spectrum ($h\nu \ll kT$ and intensity $\propto \nu^2$), but two of them—those by Boynton *et al*[216] at 8.6 mm (carried out at the White Mountain Research Station, California, whose altitude exceeds 12,000 feet, using the bolometer shown in Figure 43) and Millea *et al*[217] at 3 mm—are claimed to be sensitive enough to discriminate between a "grey body" (ν^2) law and a true black body, and to favour the latter.

A 2.7°K black body spectrum peaks at ~ 2 mm and falls off exponentially at shorter wavelengths. An important test of Gamow's hypothesis would therefore be to make measurements at millimeter wavelengths and see whether the background spectrum peaks and cuts off as a black body should. Unfortunately the atmosphere is not sufficiently transparent, mainly because of O_2 and water vapor, for these crucial observations to be possible from ground level (even at 3 mm, where there is an atmospheric "window" and apparently successful measurements *have* been made, the emission from the atmosphere exceeds the expected background by a factor ~ 10). Evidence on the form of the background spectrum at millimeter wavelengths comes instead from (a) observations of interstellar molecules and (b) rockets and balloons.

a) Studies of the molecular absorption lines seen in some stellar spectra tell us what fraction of interstellar molecules, whose low-lying rotational

---

† The radiometer employed Dicke's switching technique, in which the receiver is switched at 100 cps between the sky and a liquid helium reference source. The 100 cps oscillations in the receiver output then provide a sensitive measure of the difference of these two temperatures.

levels can be excited by an ambient millimeter radiation field, are in excited states. In fact in 1941 McKellar[218] had discovered that cyanogen (CN) in an interstellar cloud between us and the star ζ Ophiuchi was absorbing not only in the $R(0)$ [$J = 0 \rightarrow J = 1$; λ3874.608 Å] transition, but also in the $R(1)$ [$J = 1 \rightarrow J = 2$; λ 3873.998 Å] transition. From the respective line strengths could be inferred the proportion of CN radicals in the $J = 1$ and $J = 0$ rotational state of the ground electronic configuration. The corresponding excitation temperature was $\sim$ 2.3°K but no satisfactory explanation of the phenomenon was given. In Herzberg's classic book on diatomic molecules[219], this result is quoted with the comment that the temperature "has of course only a very restricted meaning". It did not appear that collisions with ions, electrons, or atoms could maintain this excitation at the density appropriate to the interstellar medium, even in dense interstellar clouds, but after the discovery of the microwave background Woolf, and Field and Hitchcock[220], independently realized that these observations could be understood if the CN was bathed in radiation with the appropriate temperature, and found the same effect in a second star ζ Persei. The method has been developed further by Thaddeus and his associates[221], who have studied CN absorption in the optical spectra of 11 different stars, finding similar excitation conditions in all cases. This suggests an excitation mechanism which is uniform, at least throughout our part of the Galaxy. The wavelength corresponding to this transition is 2.62 mm, and the radiation temperature inferred from these observations is $\sim$ 2.6°K. The problem of determining intensities in the millimeter region is thus transformed from a difficult direct operation to a more straightforward procedure carried out in the optical band where the atmosphere is no longer opaque.

No absorption lines at 3873.369 Å, corresponding to the $R(2)$ [$J = 2 \rightarrow J = 3$] transition in CN, have been seen. This sets an upper limit to the population in the $J = 2$ state of the ground electronic configuration, and hence to the radiation temperature at 1.31 mm. Limits have similarly been placed on the temperatures at 0.56 mm and 0.36 mm, from studies of CH and CH^+ absorption respectively. These limits lie above a 2.7°K black body, but fall significantly below a grey body ($\propto \nu^2$) extrapolation of the Rayleigh-Jeans segment of the spectrum.

The molecular results, together with all the results of direct ground-based measurements, are plotted in Figure 44.

b) Ideally, to find out the real shape of the background radiation in the millimeter band, one should make a direct measurement of intensity above the atmosphere. In 1968 a joint NRL—Cornell group[231] flew a rocket containing a liquid helium cooled detector sensitive to radiation in the

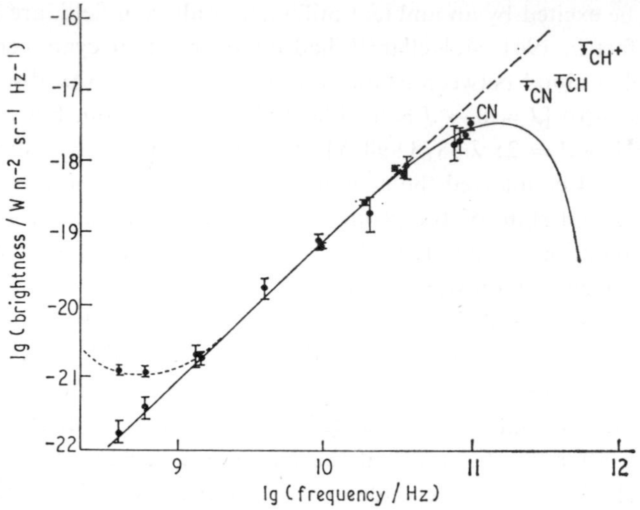

Figure 44 The direct ground based measurements of the microwave background radiation, together with the results inferred from interstellar molecules. In order of increasing frequency (decreasing wavelength) the observations are by:

Frequency (GHz)	Wavelength (cm)	Reference
0.408	73.5	Howell and Shakeshaft[222]
0.610	49	
1.415	21.2	Penzias and Wilson[223]
1.45	20.7	Howell and Shakeshaft[224]
4.08	7.35	Penzias and Wilson[211]
9.4	3.2	Roll and Wilkinson[215]
9.4	3.2	Stokes, Partridge and Wilkinson[225]
19	1.58	
20	1.5	Welch, Keachie, Thornton and Wrixon[226]
32.4	0.924	Ewing, Burke and Staelin[227]
35	0.856	Wilkinson[228]
36.6	0.82	Puzhano, Salomonovich, and Stankevich[229]
84	0.358	Kislyakov, Chernushev, Lebsky, Maltsev, and Serov[230]
91	0.33	Boynton, Stokes, and Wilkinson[216]
91	0.33	Millea, McColl, Pederson, and Vernon[217]

The CN point at 2.6 mm and the upper limits at 1.32, 0.56 and 0.36 mm are due to Bortolot, Clauser and Thaddeus.[221] Below 10^9 Hz, the isotropic background can be split into a component having spectrum $I \propto \nu^2$ and another having $I \propto \nu^{-0.8}$, the latter being attributable to discrete extragalactic sources.[222]

0.4–1.3 mm band, which was exposed at an altitude of > 100 km. Extraordinarily strong radiation, with an estimated energy density $\sim$ 20 eV cm^{-3}, was recorded. There was no evidence for any anisotropy, nor of any obvious discrete sources for this radiation. This experiment was subsequently repeated, with similar results. (Later recalibrations, however, led the experimenters to reduce[232] the estimated intensity by a factor $\sim$ 2 from the 20 eV cm^{-3} originally quoted). The detectors flown in these rocket experiments gave no spectral information, but Muehlner and Weiss[233], who detected a similar excess flux at balloon altitudes, achieved a certain amount of spectral resolution by using three different windows. Their data were claimed to be consistent with the radiation being concentrated in the band between 0.8 mm and 1 mm. If this were in fact so, these results would not conflict with the molecular upper limits, even if the intense sub-millimeter flux pervaded the whole Galaxy. However, observations with high spectral resolution made at mountain altitudes reveal no line features rising more than $\sim$ 10°K above the continuum[234], showing that this flux cannot be concentrated in fewer than five lines with $\Delta\lambda/\lambda \lesssim 0.05$ unless these happen to coincide with lines of the H_2O spectrum where atmospheric absorption is especially heavy).

In view of the difficulty in carrying out these millimeter measurements, the present data should perhaps be treated with a certain caution. Indeed, a recent rocket flight by Blair *et al*[235] found no evidence for any excess radiation at wavelengths $\gtrsim$ 0.8 mm, in contradiction to the result of Muehlner and Weiss[233]. These observations certainly cannot claim the same accuracy as the radiometer measurements at centimeter wavelengths. Even the reality of this excess radiation is presently questionable, and its universal nature is certainly not yet established—it may originate in our own Galaxy, or even the Solar System or upper atmosphere. If it is real, and indeed cosmological, it must originate within a narrow range or redshifts, or else the lines would be so broadened as to conflict with the molecular upper limits. It is thus perhaps premature to analyse the millimeter background in a cosmological context. We can only urge that these vital observations be repeated, soon and often, since so much hangs on them. (Though, of course, repetition is equally desirable for *all* the microwave background measurements—those that accord with common preconceptions as well as those that not. We must be on guard against "bandwagon effects".)

We defer to section 13.2 a discussion of the cosmological implications of the microwave background. But it may be appropriate to summarize at this stage the difficulties of accounting for this background in other ways, since it is only because these difficulties seem so overwhelming that the "hot big bang" has gained such widespread acceptance.

Any theory of the microwave background is constrained by the following requirements:

i) *Energy* The energy density of a 2.7 °K black body radiation field is $\sim$ 0.25 eV cm^{-3}; and the millimeter "excess" could be a factor $\sim$ 40 larger that this. If this radiation were generated, from the matter, at a redshift z^*, then rest energy would have to be released, and converted into the appropriate form, with an efficiency

$$\varepsilon \simeq 3 \times 10^{-5}\, b\Omega^{-1}(1 + z^*)$$

where b is unity if the black body energy alone is considered, but could be as much as $\sim$ 40 if the millimeter excess is real, and is included as well.

ii) *Spectrum* Even though the millimeter spectrum is ill-determined, there is no doubt that a Rayleigh-Jeans spectrum is closely obeyed from 8.5 mm down to 74 cm (i.e. $\nu^{2\pm 0.05}$).

iii) *Isotropy* Conklin and Bracewell[236] placed an upper limit $\lesssim 0.2\%$ to the intensity fluctuations on angular scales comparable with their telescopic beam width ($\sim$ 10 arc minutes). Other workers[237-240] have derived similar limits, and no evidence has been found for any small-scale "graininess" in the background (limits of comparable precision have also been placed on large-scale anisotropies[237,238,242], but these constitute constraints on the cosmological model rather than on the actual emission mechanism). The energetic requirements are not in themselves prohibitive—indeed, many authors have noted the coincidence that, if $b = 1$, the required energy could be produced by transmuting $\sim$ 30 per cent of the matter in galaxies, with $\Omega_{gal} \simeq 0.02$ (see section 15.1), into helium at $z^* \lesssim 1$. Two possibilities have been explored as alternative to the "hot big bang" model: that the background is due to discrete sources (e.g. radio sources with self-absorbed spectra yielding a Rayleigh-Jeans law); or that it results from emission by dust at a temperature of $\sim 3(1 + z^*)$°K. We discuss the snags of these two theories in turn.

Discrete sources

The problem with this type of model is that the microwave background at centimeter wavelengths is hundreds of times stronger than the estimated integrated contribution from known classes of radio sources. Worse still, very few observed discrete sources, even among those revealed by high frequency surveys, actually have rising spectra resembling a Rayleigh-Jeans law. One is consequently forced to postulate a new population of sources with ν^2 spectra at centimeter wavelengths. If such sources had a sufficiently low luminosity $L(\nu)$, but sufficiently high space density ϱ, they could in principle dominate the integrated background (contribution $\propto \varrho L$) even though, because they would only be individually detectable out to small distances, they would be relatively inconspicuous in surveys (contribution $\propto \varrho L^{3/2}$). Sciama[243] and Hazard and Salpeter[244] found that the absence of observed sources which could conceivably belong to this hypothetical population implies, if the sources are uniformly distributed, that they are at least $\sim 10^4$ times more numerous than galaxies. The only possible candidates would presumably be intergalactic objects in some way related to globular clusters.

This argument setting a lower limit to the source density could be relaxed if no member of the population existed in our locality, either because of a "local hole" or because the dominant contribution came from large redshifts. In such circumstances no individual sources would appear in radio surveys. However a stringent constraint is still provided by the remarkable small-scale isotropy which the background displays (see chapter 17), and many authors[244–248] have considered this aspect of the integrated source model. Even making the most favourable assumptions about the z-dependence and spectrum, Smith and Partridge[248] find that the sources must still be as numerous as galaxies, unless they are restricted to redshifts so large that the radiation is all scattered and isotropised by intervening intergalactic gas. (In making all these estimates, the sources have been assumed randomly distributed throughout space. If the sources were actually clustered in the same manner as galaxies, this assumption would certainly lead to an *under*-estimate of the fluctuations on scales up to $\sim$ 30 Mpc).

Energetic considerations do not provide an unsurmountable objection to the hypothesis. The spectrum of the sources is entirely hypothetical, and could be supposed to have the same form as the observed background spectrum. Alternatively, as discussed by Wolfe and Burbidge[247], each source could have a "δ-function" spectrum, the observed background continuum being the integrated contribution from a very wide range of redshifts. In this latter case, the isotropy data of Penzias *et al*[241] may be more crucial than the (somewhat more sensitive) limits provided by Conklin and Bracewell[236,237]. This is because the former refer to shorter wavelengths (3.3 mm) and thus, in this form of the discrete source model, to sources closer to us.

Note that in these source models the intensity of the centimeter-wavelength background and the wavelength at which the spectrum turns over (which must be at a few millimeters) are independent parameters, and it is purely coincidental that the values are consistent with a black body spectrum (for which, of course, the one determines the other.)

Dust models

The possibility that thermal emission by grains might be the cause of the centimeter background has also been considered. Narlikar and Wickramasinghe[249] envisaged large numbers of grains, either spread uniformly through intergalactic space or concentrated in localized sources, which emitted a line at a particular frequency. Integrating over all redshifts one obtains a continuum, as in the discrete "δ-function" source model discussed by Wolfe and Burbidge[247]. Alternatively[250], there might be sufficient intergalactic solid hydrogen to make the Universe optically thick at centimeter wavelengths out to $z \simeq 1$. If the grain temperature were $\sim 3\,°\mathrm{K}$ (which vapour pressure data indicate would necessarily be the case for solid hydrogen in near-vacuum conditions) one might hope to obtain a Rayleigh-Jeans spectrum. A serious difficulty with this idea, and also with Narlikar and Wickramasinghe's suggestion, is that grains radiate very inefficiently at wavelengths much larger than their own dimensions. Field[251] has shown that, whatever the grains are made of, the average value of Q (the ratio of the absorption—and emission—cross-section to the geometrical cross-section) over any waveband with $\Delta\lambda \simeq \lambda$ satisfies $\langle Q \rangle \lesssim 4\pi^2 d/2$. Thus the minimum necessary column density out to $z \simeq 1$ is *independent* of the grain size d; it does, however, depend somewhat on the *shape* of the grains. Even if all the material in the Universe were in the form of solid hydrogen it is only marginally possible to obtain an optical depth $\gtrsim 1$ at wavelengths as long as ~ 10 cm.

A further constraint is that the dust must not cause too much reddening or extinction in the visible. The only circumstance in which Q_{optical} would not greatly exceed Q microwave would be if $d \simeq 10$ cm (i.e. "bricks" or "snow", not grains). The only other possibility is that the grains, though individually small, may be concentrated into large clouds which are, as a whole, completely opaque to visible light. However the microwave background isotropy then sets further non-trivial constants on the properties of these discrete clouds.

The above work was largely motivated by the wish to reconcile the observed cosmic background with a steady state cosmology, but dust models have also been considered by Layzer[252], a proponent of the so-called "cold universe" which expanded from a singularity but initially had zero temperature. Layzer's suggestion is that the microwave background results from radiation emitted at $z^* \simeq 10$ that has been thermalized by dust. In some respects, this suggestion is not faced by such severe problems as arise in the steady state theory—the column density varies roughly as $(1 + z^*)^{3/2}$, and one gains a further factor $(1 + z^*)$ because Field's inequality must now be applied at a shorter wavelength. On the other hand, as the grain temperature would have to exceed $2.7(1 + z^*)$°K, one must rely on heavy elements, since solid hydrogen grains could not exist.

Further information on the spectrum and isotropy of the millimeter flux should help towards a better understanding of its origin. If the spectrum turned out to have precisely the form of a black body, this would constitute utterly compelling evidence for the canonical "big bang" first envisaged by Gamow. The rival theories that have been proposed would lose whatever plausibility they ever possessed if—in addition to satisfying the constraints (i)–(iii)—they were also required to reproduce, by pure coincidence, an exact black body spectrum. To some extent, therefore, confirmation of an "excess" universal millimeter flux would weaken the case for the hot big bang vis a vis the alternatives. However the discrete source and dust models are hard pressed to satisfy the constraints (ii) and (iii) (the ν^2 spectrum at centimeter wavelengths and the isotropy) which apply *irrespective* of the form of the millimeter spectrum. So, at least provided that this spectrum is moderately smooth, an interpretation in terms of the primordial fireball may still be the most plausible one available. It may, however, be necessary to drop the assumption that the early universe was strictly uniform and homogeneous, and to attribute the deviations from a pure black body spectrum to heat injection associated with the dissipation of primordial *in*homogeneities (see chapter 17). The introduction of inhomogeneities obviously, in a certain sense, detracts from the elegance of thc canonical big bang concepts, but it is unclear that the assumption of complete initial uniformity is anything more than a convenient mathematical simplification.

On the other hand, one may retain the homogeneous big bang as an interpretation for the centimetric background, and attribute the millimeter excess, if real, to some unrelated process (which, of course, is not constrained by the present data to display especially precise isotropy). The nuclei of Seyfert galaxies are suspected to emit most of their energy at $\sim 100\mu$, and Setti and Woltjer[253] propose that redshifted far infrared emission from Seyfert galaxies at $z \simeq 2$ is responsible for the millimeter background. If the excess has a spectrum with sharp features in it, it would be hard to devise any plausible alternative to a galactic origin. (Note that a millimeter background temperature *below* 2.7°K would however be a serious embarrassment for the "primordial fireball" theory).

It is probably because no satisfactory alternative theory has suggested itself that even the "conservative element" in astrophysics has been prepared to accept the exceedingly radical idea that the microwave background photons are "primordial", and have been propagating through space since an epoch long before any galaxies or other discrete sources existed.

The microwave background thus seems to provide us with extraordinarily direct evidence on the very early universe, and, not surprisingly, its discovery has stimulated much work on the *physical conditions* prevailing at those epochs (and not just the overall dynamics and geometry). In the "canonical" hot big bang model, it is supposed that the thermal microwave background has pervaded the Universe ever since $t = 0$, and inhomogeneities and anisotropies are ignored. Though doubtless a gross oversimplification of the true "world picture", this is a much more acceptable model than any specific alternative so far proposed. It should therefore, in the opinion of the present authors, be adopted as a basis for calculation until either some glaring contradiction emerges, or else a radically new and improved concept is devised. We now summarize current ideas on the evolution of this model universe, and describe some of the processes that occur at various stages in the expansion.

13.2 THE CANONICAL "HOT BIG BANG" COSMOLOGY

The mass-energy density of 2.7°K black body radiation is $\varrho_\gamma \simeq 4.4 \times 10^{-34}$ gm cm^{-3}. This corresponds to $\Omega_\gamma \simeq 4 \times 10^{-5}$ (definition in section 11.1; the precise value depends on the Hubble constant H, which we here take as 75 km sec^{-1} Mpc^{-1}). Thus the microwave background is dynamically negligible at the present epoch, since even the material in galaxies—very much a lower limit to the total matter content ϱ_m in space—contributes $\Omega_{\text{gal}} \simeq 2 \times 10^{-2}$. This conclusion still stands even if "excess" millimeter radiation is all included in Ω_γ. During the expansion, however, conservation of particles in a comoving volume implies

$$\varrho_m \propto a^{-3}; \tag{13.1}$$

whereas if the radiation is primordial and its energy is unaffected by interaction with the matter (the latter assumption will be justified later)

$$\varrho_\gamma \propto a^{-4}, \tag{13.2}$$

since, though photon numbers are conserved, photon energies vary as a^{-1}. These relations tell us that the radiation would have been dynamically dominant during the early phases of the expansion. In fact we find that $\varrho_\gamma/\varrho_m > 1$ for

$$\left(\frac{a}{a_{\text{now}}}\right)^{-1} \gtrsim \left(\frac{a_c}{a_{\text{now}}}\right)^{-1} = (1 + z_c) = 2.5 \times 10^4\,\Omega \tag{13.3}$$

The relations (13.1) and (13.2), plus knowledge of the present values of ϱ_m and ϱ_γ allow us to determine the general behaviour of $a(t)$ from Eq. (11.5) and (11.6). During the "radiation-dominated era", when $a < a_c$, the expansion proceeds according to the law

$$a \propto t^{\frac{1}{2}} \tag{13.4}$$

(This corresponds to the small-η limit of the $p = \frac{1}{3}\varrho$ solutions in Table XXI. Even when ϱ_m has its lowest allowable present value, equivalent to $\Omega = \Omega_{\text{gal}}$, the curvature term in (11.5) is never significant in these isotropic models when $a < a_c$). When $a > a_c$, only the matter is dynamically important,† so, provided that the pressure exerted by the matter is negligible, the expansion will obey one of the $p = 0$ solutions in Table XXI. When the curvature term is negligible the expansion obeys

$$a \propto t^{2/3} \tag{13.5}$$

(13.5) is always strictly true if $\Omega = 1$ ($k = 0$). If $\Omega \ll 1$, it holds for $a \lesssim a_0 \simeq \Omega a_{\text{now}}$, but for $a \gtrsim \Omega a_{\text{now}}$ the behaviour changes to the law $a \propto t$, representing undecelerated expansion. If $\Omega > 1$, the deceleration during the later stages is greater than implied by (13.5).

During the expansion, the black body temperature varies as $T \propto \varrho_\gamma^{\frac{1}{4}} \propto a^{-1}$, and it is more convenient to label the various stages in the expansion by the value of T than by the value of a. During the "radiation-dominated" phases there is, provided approximate thermodynamic equilibrium prevails, an essentially unique relation between temperature and energy density. This implies the following temperature-time dependence:

$$t = f(T_{10})\, T_{10}^{-2} \text{ seconds} \tag{13.6}$$

when T_{10} is the photon temperature in units of 10^{10} °K. The function f depends on the species of particles (neutrinos, etc.) present as well as photons, and is *of order unity* at least for $T_{10} \lesssim 100$. We give its form more precisely later.

Relations (13.1) and (13.2) shows that the ratio ϱ_γ/ϱ_m decreases monotonically during the expansion. However the ratio of the *number* of photons to the number of particles (or, equivalently, the entropy per baryon) would be a strict constant in the absence of interactions between matter and radiation. The photon density in a black body radiation field of temperature T is $20T^3$ cm^{-3}; and so for a present temperature of 2.7°K this ratio is now

$$\frac{n_\gamma}{n_b} \simeq 6 \times 10^7\, \Omega^{-1} \tag{13.7}$$

---

† This assumes that there is not a large energy density in neutrinos or other relativistic form. In the "canonical" primordial fireball, the neutrinos contribute an energy density of $0.45\varrho_\gamma$, which should, strictly speaking, be taken into account in determining a_c.

The fact that this is a very *large* number implies that the heat capacity of the matter is only $\sim 10^{-7}$ times that of the radiation. This guarantees that the entropy per baryon *will* be almost precisely constant throughout most of the expansion, even when matter and radiation are coupled, and also justifies our earlier assumption that changes in the thermal properties of the matter cannot significantly invalidate (13.2).

The initial state hypothesized by the canonical big bang theory consists of a uniform radiation field with a "small" but non-zero baryon number. Obviously conventional physics will be incomplete, or perhaps even utterly inapplicable, before some small time $t_{\min} > 0$, even if we rule out complications such as anisotropies. Some cosmologists have considered the epoch $t \simeq 10^{-23}$ sec (horizon size $\simeq$ muon Compton wavelength) and even $t \simeq 10^{-44}$ sec (horizon size $\simeq$ Planck length $(G\hbar/c^3)^{1/2}$). We shall however concentrate attention on epochs when the material is below nuclear density—i.e. $T \lesssim 10^{13}$ °K and $t \gtrsim 10^{-6}$ sec. Thus we are only excluding the events of the first 10^{-6} seconds (though this omission looks more serious if one views time logarithmically!). In discussing the evolution of the contents of the Universe, it is convenient to distinguish four main phases, the dividing line between them being associated with important changes in the properties of the material.

Note: It is worth remembering that the physical history of the matter in an isotropic universe is exactly the same as the behaviour of matter inside a small insulating box, with reflecting walls, whose linear dimensions expand in proportion to a. (Indeed, in the more general case of anisotropic models it is still legitimate to limit attention to a box, with comoving walls, whose shape varies appropriately during the expansion). The *time-dependence* of a, however, also involves global and relativistic considerations. (It is often helpful to visualize the behaviour of *any* isotropic background radiation field in this way, even at epochs when the universe is transparent).

The four "eras" in the history of a canonical hot big bang universe, in chronological order, are as follows:

1) *The hadron era:* $t_{\min} < t \lesssim 10^{-5}$ *sec* At sufficiently early stages of the expansion, there would be complete thermodynamic equilibrium between photons, electrons, positrons, neutrons, protons, neutrons and various hyperons. Thus during the earliest phases, if $T \gtrsim 10^{13}$ °K or 1 Bev ($t \lesssim 10^{-6}$ sec) proton-antiproton pairs and other types of heavy particles will be present in roughly the same numbers as photons. Relation (13.7) then implies that there will be almost exactly equal numbers of particles and antiparticles (the disparity being only 1 part in $\sim 10^7$). The physics during the "hadron era" is still controversial: see the reviews by Hage-

dorn[245], Kundt[255] and Omnes[256] for a fuller treatment of this topic. We recall here only the important issue of whether there is a "limiting temperature" at which the specific heat of equilibrium radiation becomes infinite owing to the multiplicity of species of particles.

Since the equation of state is uncertain during the hadron era, we do not know whether the expansion then obeys $a \propto t^{1/2}$ or not; nor of course do we know the time-dependence of T. Fortunately, however, no aspects of the later development of the canonical big bang universe are sensitive to the details of the hadron era (though a proposed "phase transition" occurring during this period, when $T \simeq 2 \times 10^{12}$ °K, plays a crucial role in separating matter and anitmatter in the "zero baryon number" cosmology explored by Omnes and his associates).

2) *The lepton era: 10^{-5} sec $\lesssim t \lesssim$ 10 sec* When the temperature drops so that kT is far below the rest energy of a proton, then the proton antiproton pairs and other hadrons annihilate (unless, as Omnes has argued, they are already separated on a macroscopic scale.) The annihilation cross-sections are so large that essentially no antiprotons or antineutrons would survive, and we would be left with only the primordial excess of 1 baryon in 10^7. (If there were no initial baryon excess then only one baryon would survive for every $\sim 10^{18}$ photons, unless macroscopic matter-antimatter separation had occurred during the hadron era.) We then enter the "lepton era", where the energy is initially shared between photons, electrons and positrons, muons, and both types of neutrinos and antineutrinos, all species being sufficiently well coupled by electromagnetic and weak interactions to ensure that they keep at equal temperature. (The protons and neutrons are both dynamically and thermally negligible, but each of the four types of neutrinos contributes an energy density $7/8aT^4$, as do the ultra-relativistic electrons and positrons).

At $t \simeq 10^{-4}$ seconds, muon pairs decay. Most of these decays occur while the photons are still well enough coupled to the other types of particle (including the ν_μ) that the muon annihilation energy is shared between all the remaining species. The ν_μ and $\bar{\nu}_\mu$ decouple when the muons have disappeared, but the electron neutrinos may remain in equlibrium somewhat longer by $e^- + e^+ \leftrightarrow \nu_e + \bar{\nu}_e$ and electron-neutrino scattering.

The lepton era continues until, when $T \simeq 5 \times 10^9$ °K ($kT \simeq 0.5$ MeV $\simeq m_e c^2$) most of the electron-positron pairs annihilate. By this stage however all the neutrinos are decoupled, so that the $e^+ - e^-$ annihilation energy goes exclusively into photons. This means that thereafter the photons are hotter than the neutrinos, by a factor which is straigthforwardly calculated to be $(11/4)^{1/3}$. After $e^+ - e^-$ annihilation (i.e. for $T \lesssim 5 \times 10^9$ °K) the factor f in (13.6) has the value 1.92; for 10^{12} °K $\gtrsim T \gtrsim 5 \times 10^9$ °K its value is about 1.09; and above 10^{12} °K it is uncertain (in Hagedorn's model $f \to 0$ as T tends to the limiting temperature.)

When $T \gg 10^{10}$ °K, kT greatly exceeds the proton-neutron mass difference, so protons and neutrons are almost equally abundant. But even before the time when $e^+ - e^-$ annihilation occurs, the Boltzmann factor starts to favour protons, so n_n/n_p decreases below unity. This is the main phenomenon determining the primordial helium abundance, and we return to it in section 13.3.

3) *The plasma era: 10 sec $\lesssim t \lesssim 10^{12}$ sec* Even after the electron-positron pairs have all annihilated, repeated Compton scattering of the photons by the remaining unpaired electrons† provides enough coupling (a) to ensure that the matter and radiation behave more or less like a single fluid, in which the radiation provides essentially all the pressure; and (b) to keep the matter and radiation at almost the same temperature, in spite of the tendency for the matter (with $\gamma = 5/3$; (energy per particle) $\propto$ (momentum per particle)^2 $\propto$ (de Broglie wavelength)$^{-2}$ $\propto a^{-2}$) to cool adiabatically twice as fast as the radiation (for which $\gamma = 4/3$; (energy per photon) $\propto$ (momentum per photon) $\propto a^{-1}$).

The "plasma era" continues until the temperature is too low to maintain a high level of hydrogen ionization. This stage is reached at $T \simeq 4000$°K (the primeval helium having recombined somewhat sooner). The fractional ionization then falls rapidly, the net recombination rate being determined by the two-photon emission rate from the 2s state of the hydrogen atom, since this is the only way an atom can reach the ground state without producing a photon which immediately photoionizes another atom. At T 1000°K only 1 electron in $\sim 10^4$ remains free[257,258]. Sunyaev and Zeldovich[259] give a simple formula for the degree of ionization $x(z) = n_p/(n_p + n_H)$ during the recombination period:

$$x(z) = \frac{A}{\Omega^{\frac{1}{2}} z} e^{-B/z} \qquad (13.8)$$

where $A = 6 \times 10^6$ and $B = 1.458 \times 10^4$. This formula holds for $1500 \gtrsim z \gtrsim 900$. The age of the universe at this epoch depends somewhat on the present overall matter density, because if $\Omega \gtrsim 0.05$ the radiation stops being *dynamic-*

† Note that the assumption of macroscopic charge neutrality requires us to postulate a primeval electron excess equal to the proton excess. One conventionally assumes that the other conserved lepton numbers are zero. However it is conceivable that, for example, the excess of ν_e over $\bar{\nu}_e$ or vice versa, is comparable with the photon density, and this would modify primordial nucleosynthesis (see Chapter 17).

ally dominant (at $a = a_c$ and a time $t = t_c$) before the recombination occurs†, and the expansion law then switches from (13.4) [the constant of proportionality being given by (13.6)] to (13.5). Throughout most of the "plasma era", growth of perturbations of galactic scale is inhibited by the radiation field. Some quantitative details of the effects, and of the damping processes which result from radiative diffusion and viscosity, are described in Chapter 14.

4) *The post-recombination era:* $t \gtrsim t_{rec} \simeq 10^{12}$ *sec* When the free electrons have almost all recombined, thermal coupling between matter and radiation is lost. Figure 45 shows the changing transparency of the Universe through the recombination period taking account of both Thomson and Rayleigh scattering. The temperature of the non-relativistic matter then drops as a^{-2}, and *if* there had been no subsequent heat input, the gas at the present epoch, when the radiation temperature is $\sim 2.7°$K, would have

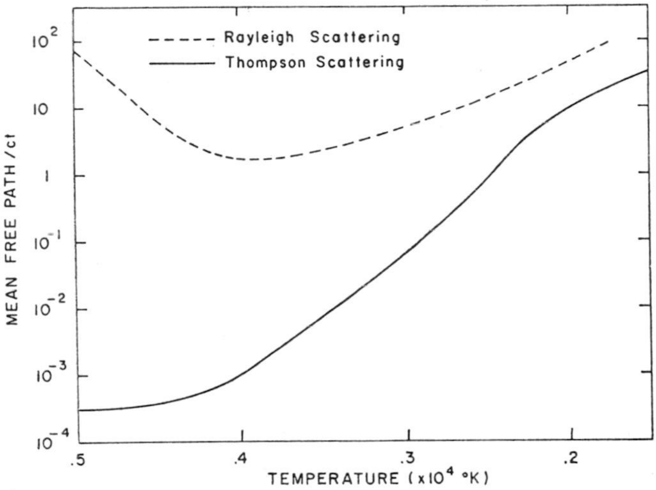

Figure 45 The mean free path of photons, in units of *ct*, through the period of plasma recombination for a Friedmann universe with $\Omega = 1$. Rayleigh scattering is always less important than Thomson scattering. [Reproduced from Yu and Peebles.[260]]

† It is important not to confuse the epoch t_c when radiation ceases to be *dynamically* dominant with the recombination epoch discussed here. The latter epoch signals the breakdown of *local* coupling between matter and radiation. In the canonical big bang these two times coincide, at least to within an order of magnitude or so. However, t_c occurs at a temperature $\propto n_\gamma/n_b$, whereas the recombination occurs at a *fixed* temperature, and it is purely coincidental that the entropy per baryon (an arbitrary parameter) has a value which happens to make these two times comparable.

cooled to $\sim 10^{-2}$ °K. The neutral matter would be transparent to the black body radiation†. This means that radiation pressure no longer hinders the growth of density perturbations, so that all masses $\gtrsim 10^5 M_\odot$—below which scale gradients in the *gas* pressure alone $[(3/2)\, nkT_{gas}]$ exert a significant restoring force—are unstable. Another consequence is that the microwave background photons which we now observe could have propagated freely, without scattering, since the termination of the recombination epoch ($T \simeq 2000$°K), and thus provide direct information about physical conditions on a "last scattering surface" at $z \simeq 1000$. If the gas were reionized at a subsequent epoch, however, this conclusion may be altered (see Chapter 17).

Detailed computations have shown that, even in a strictly homogeneous universe, the primeval black body radiation becomes somewhat distorted during the epoch of plasma recombination. This is because the recombination photons produce an excess background on the high frequency (exponential) part of the black body curve. The energy content of this extra contribution, however, is undetectably small. When $z \lesssim 900$ Eq. (13.8) breaks down, because the free remaining electrons are unable to pair up with protons within the expansion time scale. Thus a small residual ionization ($x \sim 10^{-4}$ to 10^{-5}) "freezes out". These free electrons do not provide a significant opacity at $z \ll 1000$, but (especially when $\Omega \ll 1$) scattering of the microwave photons can keep the gas as hot as the radiation until $z \lesssim 200$. (ref. 258).

The "post-recombination era" occupies $\gtrsim 99.99$ per cent of the history of the big bang universe. During the later part of this era ($t \gtrsim 10^9$ years?) galaxy formation must have occurred. The deviations from homogeneity, and the possibility of a heat input which raises the uncondensed gas to a high temperature (see section 15.3), will complicate the picture at redshifts $z \lesssim 3$, even if it was an adequate approximation at earlier stages.

Figure 46 illustrates and encapsulates the density and temperature of the universe, as functions of t, for the four "eras" of the canonical big bang model.

13.3 NUCLEOSYNTHESIS IN THE PRIMORDIAL FIREBALL

After the termination of the hadron era ($t \simeq 10^{-5}$ sec) the canonical big bang predicts a unique relationship between time, temperature, and (if the present value of Ω is given) particle density. One can therefore ask: what

† The primordial fireball can be compared to the confined part of a hydrogen bomb explosion with radiation and matter in equilibrium. Only some seconds (in the case of the bomb) or $\sim 10^5$ years (for the Universe) after the initial explosion does the material open up enough to let the optical radiation escape.

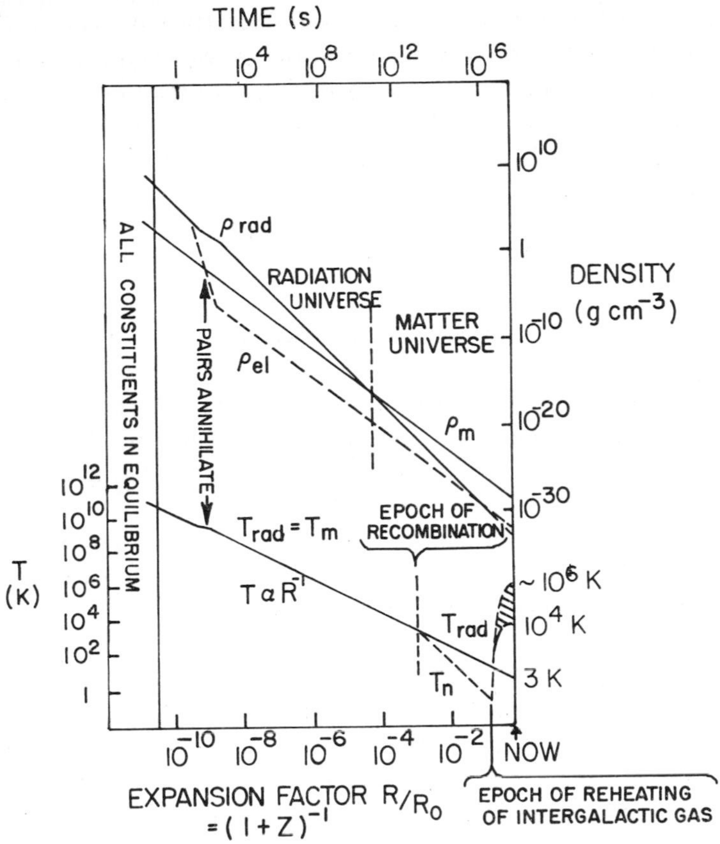

Figure 46 The evolution of a "canonical" primordial fireball. The densities of radiation, baryons and electrons are shown as a function of time. Also shown are the matter and radiation temperatures. After recombination, thermal contact between the matter and radiation is lost. Consequently T_m drops faster than T_{rad} during the adiabatic expansion. As discussed in the text (section 15.3) the gas must have been reheated at a redshift $z \gtrsim 3$. Two possible thermal histories are illustrated.

nuclear processes would have occurred during the hot early phases, and what composition does this cosmological model predict for the material from which, at a much later stage, galaxies first formed?

During the lepton era, protons and neutrons can transform into one another, via the weak interactions

$$p + \bar{\nu}_e \leftrightarrow n + e^+$$

$$p + e^- \leftrightarrow n + \nu_e$$

The timescale for these reactions is $\sim T_{10}^{-5}$ sec (two powers of T being attributable to the energy-dependence of the cross-sections, and the other three to the change in particle density during the expansion). Comparison with the expansion time-scale (13.6) shows that the proton/neutron ratio is maintained in thermal equilibrium until T has dropped to $\sim 10^{10}$ °K. This equilibrium ratio is

$$\frac{n_n}{n_p} = e^{\frac{-(\Delta m)c^2}{kT}} \simeq e^{\frac{-1.5}{T_{10}}}, \qquad (13.9)$$

Δm being the proton-neutron mass difference, and at $T \simeq 10^{10}$ °K its value is ~ 0.22.

At lower temperatures, the expansion time-scale ($\propto T^{-2}$) becomes shorter than the equilibrium time-scale (and of course the latter increases even more drastically when the $e^+ - e^-$ pairs annihilate). Thus the neutron/proton ratio effectively "freezes out" at $\sim 10^{10}$ °K.

The neutrons would, of course, eventually decay into protons with the laboratory half-life of ~ 10 minutes. However the proton density in the canonical fireball would be high enough for the reaction

$$n + p \rightarrow D + \gamma \qquad (13.10)$$

to proceed much faster. While T exceeds 10^9 °K ($t \lesssim 200$ sec), the deuterium is rapidly photodisintegrated. At lower temperatures, however, when the black body radiation is softer, the equlibrium deuterium abundance becomes large enough for a series of further reactions to occur, which lead to the synthesis of helium. For instance:

$$D + D \begin{array}{l} \nearrow {}^3He + n \rightarrow {}^3H + p \\ \searrow H^3 + p \end{array} \qquad (13.11)$$

$$ {}^3H + D \rightarrow {}^4He + n$$

These reactions are sufficiently rapid for essentially all the deuterium that is produced (and survives photodisintegration) to go towards the formation of He. Moreover, if the proton and neutron density in the early universe corresponds to a *present* density $\gtrsim 10^{-32}$ gm cm^{-3}, i.e. $\Omega \gtrsim 10^{-3}$ (as seem certainly to be the case), then (13.10) proceeds more or less to completion in the time-scale given by (13.6). If *all* the "frozen-out" neutrons get incorporated into helium nuclei, then our earlier estimate of the neutron/proton ratio leads to a helium abundance of 36 per cent *by mass*. More detailed calculations[261,262] which take account of all the relevant reactions (and allow for spontaneous neutron decay, etc.) yield estimates of $\sim 25\%$. These calculations confirm that the primordial helium abundance is insensitive to the present matter density, at least as long as Ω lies within the range $0.025 \lesssim \Omega \lesssim 5$ permitted by other observations. Note that the matter-density at the helium formation epoch is only $10^{-3}\,\Omega$ gm cm^{-3}—low by terrestrial standards. The radiation mass-energy density at $T \simeq 10^9$ °K is, however, ~ 50 gm cm^{-3}.

The evidence pertaining to the helium abundance—both direct spectroscopic observations of stars and nebulae and inferences from the theory of stellar structure and evolution—have recently been reviewed by Tayler[263] and Danziger[264]. Although the data are somewhat fragmentary, they favour a universal helium abundance of 25 per cent. Danziger notes, however, that the observations of population II B stars, which should in principle provide the best test of the helium abundance at the birth of the galaxy, are discordant with the other evidence. Also the low fluxes found by Davis's solar neutrino experiment[265] are more easily reconciled with theory if the solar helium abundance is below 25 per cent. If our galaxy had maintained its present luminosity for $\sim 10^{10}$ years, it could have transformed only 2–3 per cent of its hydrogen into helium, and no generally acceptable way has been suggested for producing 25 per cent helium via the type of nuclear processes currently occurring, without at the same time processing too much material into heavier elements. Thus it is tempting to invoke a cosmological origin for the bulk of helium we observe, and it is widely considered one of the "triumphs" of the hot big bang theory that, in its simplest form, it naturally predicts ~ 25 per cent helium. (We return in Chapter 17 to consider how the predicted abundance would be altered if conditions in the early universe had differed in various ways from the canonical picture).

The most elaborate computations of nucleosynthesis in the primordial fireball are those made by Wagoner, Fowler and Hoyle[262] (see Figure 47), who obtained a helium abundance of 25–27 per cent by mass (the higher value corresponding to a higher Ω). This value was somewhat higher than the abundance derived from observations. However, since Wagoner *et al* performed their computation there has been an upward revision in the weak interaction coupling constant (a consequence of a 10 per cent reduction in the neutron lifetime[266] from the previously accepted experimental value). This causes the neutron/proton ratio to freeze out *later*, and therefore a *lower* neutron abundance. This correction would lower the helium abundance by ~ 10 percent of the value calculated earlier[267,268].

Wagoner *et al*[262] included in their computations a total of 144 reactions, involving heavy nuclei up to Mg. Their work showed, however, that it was not possible (within the limits of the canonical parameters) to synthesize significant quantities of any elements heavier than ^{4}He, except possibly ^{7}Li, the main "hangup" being the non-existence of stable nuclei with atomic

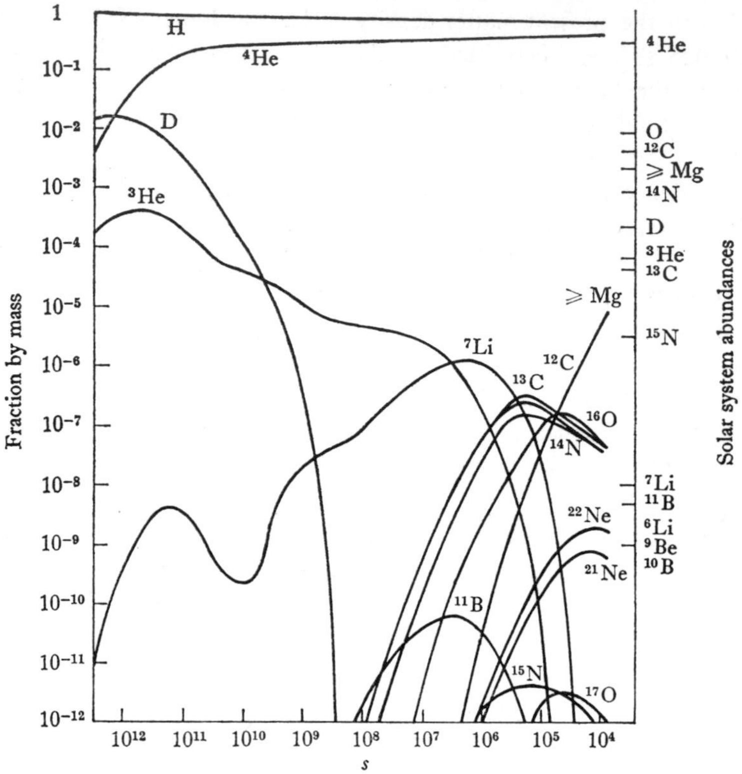

Figure 47 The composition of matter emerging from the primordial fireball according to calculations by Wagoner *et al.*[262] The horizontal scale is the entropy per baryon.

weights 5 and 8, which impedes synthesis via $p + {}^4\text{He}$, $n + {}^4\text{He}$ or ${}^4\text{He} + {}^4\text{He}$ collisions. (Also, the high Coulomb barrier in the reactions ${}^4\text{He} + {}^3\text{H} \rightarrow {}^7\text{Li} + \gamma$ and ${}^4\text{He} + {}^3\text{He} \rightarrow {}^7\text{Be} + \gamma$ prevents them from competing effectively with the reactions which yield ^{4}He). Although Gamow and his associates had originally hoped that all elements might have been synthesized (via repeated neutron capture) in the fireball, hopefully emerging with their "cosmological abundances", studies of stellar evolution and nucleosynthesis have, over the last 15 years, persuaded most astrophysicists that it is feasible for all the heavy elements to have been formed during the history of the Galaxy. The correlations of heavy element abundance with age and with location in the Galaxy—contrasting with the apparant *lack* of dependence of helium abundance on these parameters—favours this view, and it would in fact be an embarrassment for modern advocates of the hot big bang if this theory entailed a substantial primordial heavy element abundance. (It is interesting, however, that the relative abundances of elements heavier than carbon can best be explained *not* by steady combustion in stars, but instead by a process of "explosive nucleosynthesis"[269 270], where nuclei such as ^{12}C, ^{16}O and ^{28}Si are heated to temperatures $\sim 2 \times 10^9$ °K, and then burn for a brief period limited by the adiabatic cooling rate)†. Thus it is only the helium, and perhaps the deuterium, in the galaxy to which most workers would wish to ascribe a primordial origin. Note that the expected abundance of deuterium, unlike that of helium, is sensitive to the value of Ω.

† Exploration of the properties of an *oscillating* universe first led Dicke and his Princeton colleagues to the concept of the primordial fireball. They realized that it was impossible to destroy the heavy elements synthesised in one cycle, and thus permit a 'new deal' in the next cycle, unless the temperature attained during the 'bounce' was $\gtrsim 10^{10}$ °K.

CHAPTER 14

The Fate of Fluctuations: Galaxy Formation

THE MANIFEST non-uniformity of the actual Universe—evidenced by the existence of galaxies, stars and ourselves—does not render the Friedmann models inapplicable. They give a valid overall description provided that the typical fluctuations $\Delta\varrho$ of the mean density over length scales d satisfy

$$G(\Delta\varrho)\, d^2 \ll c^2 \tag{14.1}$$

for all d. This ensures that the fractional departures from the Robertson-Walker metric (11.2) are everywhere $\ll 1$, and that the peculiar velocities induced by the inhomogeneous mass distribution are $\ll c$. Both for galaxies and clusters of galaxies, (14.1) is fulfilled by a margin of at least $\sim 10^4$, and the isotropy of the background radiation strongly suggests (see Chapter 17) that it holds on all larger scales up to the "Hubble radius" $\sim c\tau_H$.

The mean density of matter within galaxies is, very roughly, in the range 10^{-1}–10^3 particles cm^{-3}. Obviously galaxies—and, a fortiori, clusters of galaxies—could not have existed in their present form when the *mean* density of the universe (now $\sim 10^{-5}\,\Omega$ particles cm^{-3}) was higher than this. Protogalaxies may then have been merely regions of slightly enhanced density, whose expansion was retarded, and eventually halted, by their excess gravitational field. Although gravitational forces promote instability, pressure gradients (especially radiation pressure in the "plasma era") and viscous dissipation tend to counteract this growth. We now describe the interaction between these processes, and the consequent fate of various types of perturbation in the "primordial fireball". The main aim of recent work on this problem has been to try to understand the occurrence of structure with the characteristic observed scales without artificially choosing the initial conditions with this end in view; and (more ambitiously) to understand the morphology and angular momentum distribution of galaxies. No very spectacular results have yet been achieved, but the fact that these investigations can even be contemplated illustrates how the discovery of the microwave background has enlarged the range of phenomena amenable to quantitative discussion.

The view that present structural features evolved from small perturbations in an initially almost homogeneous universe is, however, no more than a working hypothesis: there is no firm evidence that, on the scale of galaxies and clusters, the Universe was ever smoother than it is now. Ambartsumian, for instance, has long been voicing the opinion that matter "emerges" locally from a high density singular state, but this idea cannot be developed quantitatively in our state of knowledge (c.f. Ne'eman and Tauber's "lagging core" theory[271], and the concept of "white holes"). A further incentive for pursuing studies of the fate of small initial perturbations is that only in this approach is it consistent to assume a homogeneous background cosmology: if galaxies had always had the same contrast density $\Delta\varrho/\varrho$, then (14.1) would be violated when ϱ was large, so the mathematically tractable Friedmann models would be irrelevant to the dense early stages.

14.1 GRAVITATIONAL INSTABILITY OF PERFECT FLUID FRIEDMANN MODELS

The first full relativistic treatment of perturbed Friedmann models was given by Lifshitz[272] in 1946. Zeldovich[273], Harrison[274] and Field[275], among others, have reviewed subsequent work on this topic.

Dust universe

It is easy to see that a dust-filled Friedmann model is unstable to the growth of density perturbations. The dynamics of the matter within any comoving sphere are unaffected by the external matter. Conversely, the behaviour of the external universe would be unaltered if the region within the sphere were replaced by a Schwarzschild metric whose mass is chosen so that it joins smoothly onto the unperturbed Friedmann model at the boundary. The central part of the Schwarzschild metric can then in turn be replaced by an expanding homogeneous sphere whose dynamics mimic a section of a *different* Friedmann model. Envisaging many such spherical holes, we have the "swiss cheese" universe discussed by Einstein and Strauss[276]. Any perturbed spherical region thus expands just like a Friedmann universe of different curvature (different a_0 in Table XXI). All Friedmann dust models start off (when $\eta \ll 1$) with $a \propto t^{2/3}$ and the same constant of proportionality, but the behaviour diverges as a approaches a_0. When η is small, the values of a attained by different models at a given t differ by a term of order η^2. This means that, while the perturbation is still small,

$$\left|\frac{\Delta a}{a}\right| \propto \frac{\Delta\varrho}{\varrho} \propto a \propto t^{2/3} \tag{14.2}$$

This shows that an expanding dust universe is unstable to the growth of those perturbations which can be regarded as fluctuations in the local total energy, or in k/a_0^2 [Eq. (11.7)]. And of course any density irregularity—not merely a uniform sphere—would exhibit the same general type of growth. (One can also imagine initial irregularities in which the energy, or a_0, is unperturbed, but the expansion starts at different times. This type of fluctuation also emerges from a more elaborate mathematical treatment of the problem but is of little interest because it *diminishes* in amplitude in proportion to $a^{-3/2}$ during the expansion.)

A useful concept in this subject is a hypothetical observer's "particle horizon", which consists of those points at an arc distance $\eta(t)$ from his location on the 3-sphere, where

$$\eta(t) = \int_0^t \frac{dt}{a(t)}.$$

Objects at the particle horizon have infinite redshift, and in all Friedmann (decelerating) models the horizon grows to encompass more and more matter as t increases.

Provided $\Delta\varrho/\varrho$ is interpreted as the fractional deviation from the background density which different co-moving observers would measure *at the same proper time t*, (14.2) still holds even when the scale of the fluctuation is large compared to the particle horizon.† Of course, this is true for *any* scale at sufficiently early times, because in dust models with $a \propto t^{2/3}$ the mass M_H within the particle horizon ($\sim \varrho(ct)^3$) varies as

$$M_H \propto a^{3/2} \propto t \tag{14.3}$$

Relation (14.2) obviously ceases to apply when the fluctuations attain order unity. However the behaviour can still be easily visualized if one thinks of a section of a universe with a larger k/a^2 than the background cosmology. In a $k = 0$ background, any region of enhanced density will stop expanding (at a time when it has $9\pi^2/16$ times the background density) and will then start to collapse. Qualitatitively similar considerations hold in any model, provided that $\Delta\varrho/\varrho$ attains order unity when a/a_0 (for the unperturbed

† When the perturbed sphere is *small* compared with the particle horizon, and its boundary expands with a speed $v \ll c$, we can derive (14.2) by the following elementary Newtonian argument. A change in energy corresponds to a given change $\Delta(v^2)$ in v^2 which is independent of a. For an Einstein-de Sitter ("zero energy") model, or any Friedman model when $a \ll a_0$, $v \propto a^{-1/2}$, so $\Delta v/v \propto a$. This means that the fractional differences in the times taken by the perturbed an unperturbed spheres to attain a given radius (or, equivently, the fractional density perturbation at a given time) increases in proportion to a.

model) is still $\ll 1$. In a closed ($k = 1$) model a fluctuation whose amplitude is still small when $a \simeq a_0$ will not separate out until the whole background universe is in a state of collapse. Any fluctuations which start to recontract while still smaller than M_H would not really constitute a part of our universe. If $k = -1$, a perturbation which still has $\Delta\varrho/\varrho \ll 1$ when $a \simeq a_0$ will retain almost *constant* amplitude thereafter. This is because self-gravitation of the perturbed region would then become unimportant, and the fluctuation would continue, like the unperturbed universe, in undecelerated expansion. These simple results for gravitational instability of spherical irregularities can be straightforwardly extended to more general perturbations, but the results are not qualitatively different.

If p is non-zero, the Friedmann dust solution still describes the overall dynamics if $p \ll \varrho$, but pressure gradients can then stabilize fluctuations on sufficiently small scales.† They will not grow, but instead oscillate like sound waves. The minimum length scale over which gravitational forces overwhelm pressure is the so-called "Jeans length", first considered by Sir James Jeans in 1902:

$$\lambda_J = c_s\left(\frac{\pi}{G\varrho}\right)^{\frac{1}{2}} \tag{14.4}$$

where c_s is the sound speed. The corresponding "Jeans mass" is

$$M_J \simeq \frac{c_s^3}{G^{3/2}\varrho^{1/2}}. \tag{14.5}$$

In a cosmological model where curvature effects can be neglected (i.e. $|k/a^2| \ll 1$) this mass is such that the differential expansion velocity across it is $\sim c_s$. The effects of pressure are negligible for all fluctuations with $M \gg M_J$, so in such cases the above discussion of dust models is applicable.

Radiation universe

One can also derive simple approximations for the behaviour of perturbations in a "radiation universe" containing a perfect fluid with $p = \frac{1}{3}\varrho$. If the radiation mean free path is small enough to confer fluid-like properties on the material—as is the case (on all relevant scales) during the "plasma era" of the canonical big bang—then the effective sound speed is $c_s = c/\sqrt{3}$. Pressure therefore stabilizes all masses substantially smaller then the particle horizon (in other words $M_J \simeq M_H$). However this universe *is* unstable to

† When the gas is capable of cooling by emission of radiation, the pressure and density are no longer related by a simple equation of state, and there is a possibility of *non-gravitational thermal instabilities*. These may occur after the reheating of intergalactic gas.

perturbations with scales $\gg M_H$ and we find (by comparing $a(t)$ for different models, as we did earlier for the dust case) that

$$\frac{\Delta\varrho}{\varrho} \propto a^2 \propto t \tag{14.6}$$

Because M_H increases with t, a perturbation of given scale will start off by growing according to (14.6) but almost as soon as it comes within the horizon pressure gradients become important and it starts oscillating.

14.2 PERTURBATIONS DURING THE "PLASMA ERA"

The original studies of gravitational instability were motivated by the hope that galaxies and clusters might have condensed from the random $\sqrt{N}/N$ fluctuations which would naturally be expected in a universe composed of discrete atoms. For a galactic mass of $\sim 10^{11} M_\odot$, however, the statistical fluctuations are only $\sim 10^{-34}$, and these would not have condensed out by the present epoch unless one assumes that the growth was initiated at a stage when the particle horizon encompassed only a few atoms. It is not clear that much physical significance attaches to this result. The problem is that the overall expansion, which has a time scale equal to the "*e*-folding" time of the Jeans instability, transforms the growth rate of the density contrast from an *exponential* to a (much slower) *power law*, which means that one must either suppose that the fluctuations came into being at an exceedingly early epoch, or else assume larger initial amplitudes. For example, the existence of galaxies could be accounted for in a "dust" universe if there were perturbations with $\Delta\varrho/\varrho \simeq 10^{-7}$ at the time when such a mass first came within the horizon. Though much larger than the "statistical" amplitude, these are still "small" perturbations in the sense that they would cause the geometry of the universe to deviate only negligibly from the strictly homogeneous case.

However, this type of hypothesis is unsatisfactory because it really explains nothing. The initial amplitude must be chosen in a purely ad hoc manner: if it were too small, galaxies would not yet have condensed; but if it were too large they would have stopped expanding at an early epoch, and would have higher mean densities than are observed. In addition, when $p = 0$, perturbations of all scales grow at the same rate, given by (14.2). This means that there is no preferred scale of instability, but that the postulated initial spectrum must be weighted in favour of whatever masses are required. As T. Gold has put it: "Things are as they are because they were as they were". It is true that nonzero pressure could stabilize scales which were $< M_J$ during certain stages of the expansion, with the result that the growth of large scales would be favoured. However, this effect does not seem capable of impressing any prominent features on an initially smooth mass spectrum of fluctuations.

The detection of the thermal microwave background, and its interpretation as "primeval" radiation, stimulated renewed interest in this approach to the problem of galaxy formation. There are several reasons why this should have been so:

i) This radiation constituted the first really compelling evidence for a hot, dense, opaque "fireball phase" in the evolution of the Universe.

ii) The isotropy of the background, established to $\lesssim 0.1$ per cent—unprecedented precision for a cosmological measurement—provided the first firm evidence that simple Friedmann models may provide an adequate dynamical description of the early universe.

iii) The hot big bang model allows us to predict a definite equation of state, except perhaps during the "hadron era".

iv) The fate of perturbations in the "fireball" is *not*, as in perfect fluid models, governed solely by competition between pressure gradients and gravitation: the coupling between matter and radiation is not perfect, and this leads to dissipative processes. The resulting damping acts preferentially on certain scales (and certain types) of perturbation. It would be gratifying if the irregularities most likely to survive and amplify bore some relation to the characteristic scales of observed structures in the universe.

Radiative damping of small amplitude oscillations

In the "post-recombination era" ($t \gtrsim t_{rec}$) the universe is essentially transparent, and motions of the matter are essentially unaffected by the black body radiation. During the "plasma era", on the other hand, each photon is scattered many times by free electrons. Thus the electrons must share the mean motion of the photon gas; the nucleons, being electrostatically coupled to the electrons, are therefore dragged along as well. When the mean photon energy ($\sim 3kT$) is $\ll m_e c^2$, the Thomson cross-section $\sigma_T = 6.6 \times 10^{-25}$ cm^2 is appropriate. The mass $M_{\tau=1}$ of matter within a sphere of optical depth unity would be $\lambda^3_{\tau=1}\varrho_m$, where $\lambda_{\tau=1} = (n_e\sigma_T)^{-1}$ is the photon mean free path between Thomson scatterings. At t_{rec}, when $n_e \simeq 10^4\,\Omega$ cm^{-3}, this mass is

$$M_{\tau=1} \simeq 10^7\,\Omega^{-2}\,M_\odot, \tag{14.7}$$

and it diminishes rapidly ($\propto T^{-6}$) as we extrapolate further back into the past. Thus, at least for perturbations on galactic scales ($\sim 10^{11}M_\odot$), the matter and radiation behave as a single fluid for $t \lesssim t_{rec}$. The sound speed in this composite fluid (evaluated taking the inertia of both matter and radiation into account, but neglecting the pressure of the matter) is

$$c_s = \frac{c}{\sqrt{3}}\left(1 + \frac{3}{4}\frac{\varrho_m}{\varrho_\gamma}\right)^{-\frac{1}{2}} \tag{14.8}$$

Before t_c (when $\varrho_\gamma > \varrho_m$), $c_s \simeq c/\sqrt{3}$ and $M_J \simeq M_H$. Between t_c and t_{rec}, M_J remains constant, at a value

$$M_1 \simeq 3 \times 10^{15}\,\Omega^{-2}\,M_\odot; \tag{14.9}$$

After t_{rec}, M_J drops drastically because the thermal pressure of the particles is only $\sim 10^{-7}\,\Omega$ of the radiation pressure, so M_1 is the largest mass which can ever be stabilized by pressure forces.†

Non-uniformities in ϱ_γ on scales $\ll M_J$ would, in effect, behave like sound waves. The effective sound speed for wavelengths λ is actually $c_s(1 - \lambda^2/\lambda_J^2)^{1/2}$, the effect of gravity being to "soften" the fluid, by an amount which becomes negligible as $\lambda/\lambda_J \to 0$. If the matter and radiation were perfectly coupled (no viscosity or damping) we could readily infer the behaviour of the amplitude $\Delta\varrho_\gamma/\varrho_\gamma$ by applying the usual theory of adiabatic invariants to "standing waves"—the energy within one wavelength varies inversely as the period $\mathcal{N}$ of the oscillation. For perturbations with a given comoving wavelength λ ($\ll \lambda_J$), the oscillation energy varies during the expansion as

$$(\text{amplitude})^2 \times (\text{sound speed})^2 \times \begin{pmatrix}\text{mass-energy within a cubic}\\ \text{co-moving wavelength}\end{pmatrix};$$

i.e. it is proportional to

$$\left(\frac{\Delta\varrho_\gamma}{\varrho_\gamma}\right)^2 c_s^2[(\varrho_m + \varrho_\gamma)\,a^3] \tag{14.10}$$

The period of oscillation ((wavelength)/(sound speed)) varies roughly as

$$\left.\begin{aligned} N &\propto a \qquad (t \lesssim t_c)\\ N &\propto a^{3/2} \quad (t_c \lesssim t \lesssim t_{rec}) \end{aligned}\right\} \tag{14.11}$$

so, approximately,

$$\frac{\Delta\varrho}{\varrho} \propto \begin{cases}\text{constant} & (t \lesssim t_c)\\ a^{-1/4} \propto t^{1/6} & (t \gtrsim t_c)\end{cases} \tag{14.12}$$

The physical interpretation of the behaviour when $t \lesssim t_c$ is that the inertial mass within a comoving volume varies as a^{-1}, so the reduction in energy

† If $\Omega \lesssim 0.05$, $t_c \gtrsim t_{rec}$. (14.9) is then inapplicable, and the largest scale which can be stabilized by pressure then contains a particle mass of $\sim 3 \times 10^{19}\Omega M_\odot$.

($\propto a^{-1}$) required by the adiabatic invariant can be absorbed by the decrease in mass, without necessitating any decrease in the amplitude.

The photon mean-free-paths, however, are not completely negligible, and a more accurate treatment allowing for photon diffusion and viscosity reveals that the oscillations would be damped[277-280]. For length scales $\ll \lambda_J$ one can treat small amplitude oscillations as acoustic waves, neglecting general relativistic effects. The damping can be attributed to two effects:

i) The photons tend to leak out of compressed regions, and there is a consequent phase lag between the matter oscillations and the radiation fluid (i.e. pressure) oscillations, leading to attenuation of the waves.

ii) The radiation field becomes slightly anisotropic in sheared regions, and this leads to viscous dissipation. Taking both these effects into account, a linear analysis by Weinberg[280] shows that oscillations of wavelength λ are damped at a rate

$$\tau_{\text{damping}}^{-1} = \frac{2\pi^2 c}{3} \frac{\lambda_{\tau=1}}{\lambda^2} \frac{\varrho_\gamma}{\left(\varrho_m + \frac{4}{3}\varrho_\gamma\right)} \left\{\frac{16}{15} + \frac{\varrho_m^2}{\varrho_\gamma\left(\varrho_m + \frac{4}{3}\varrho_\gamma\right)}\right\} \tag{14.13}$$

The two terms in the braces represent the contributions from effects i) and ii) respectively. When $\varrho_m \ll \varrho_\gamma$ (i.e. $t \ll t_c$) the viscosity term dominates, but when $\varrho_m \gg \varrho_\gamma$ ($t \gg t_c$) the "phase lag" effect is more important. In all cases, however, τ_{damping} is comparable with the time a photon takes to "random walk" a distance λ. The damping is more important on smaller scales, and at any epoch t there will be a *minimum undamped lengthscale* λ_D for which $\tau_{\text{damping}} \simeq t$. In fact $\lambda_D \simeq (\lambda_{\tau=1}\lambda_H)^{1/2}$. By the end of the "plasma era", oscillatory motion involving all masses up to a certain mass M_2 would have been severely attenuated. We find

$$M_2 \simeq \begin{cases} 6 \times 10^{12}\,\Omega^{-5/4}\,M_\odot & (\Omega \gtrsim 0.03) \\ 1.5 \times 10^{13}\,\Omega^{-1}\,M_\odot & (\Omega \lesssim 0.03) \end{cases} \tag{14.14}$$

Figure 48 illustrates how various characteristic masses behave during the "plasma era", in a universe with $\Omega = 1$. The mass M_H of the *matter* within the particle horizon is plotted (we observe that a whole galactic mass first comes within the horizon when $t \simeq 1$ year). Also plotted are M_J, $M_{\tau=1}$ and the minimum undamped mass $M_D \simeq \varrho_m \lambda_D^3$. A region in which the photon density is perturbed therefore behaves according to (14.6) until the baryon mass M associated with it falls below M_H. If $M < M_1$ [relation (14.9)], the perturbation comes within the horizon while the universe is still radiation dominated (and $M_J \simeq M_H$), so it then starts to oscillate according to (14.12). The oscillations will be attenuated if $M \lesssim M_D$. So for all mass scales with $M > M_2$ [relation (14.14)], $\Delta\varrho_\gamma/\varrho_\gamma$ will be reduced exponentially by the time the plasma recombines. (Fluctuations with mass $M > M_1$ would never, of course, oscillate).

The relevant parameter for galaxy formation is $\Delta\varrho_m/\varrho_m$, and not $\Delta\varrho_\gamma/\varrho_\gamma$, and it is only when the initial perturbations preserved the photon/baryon

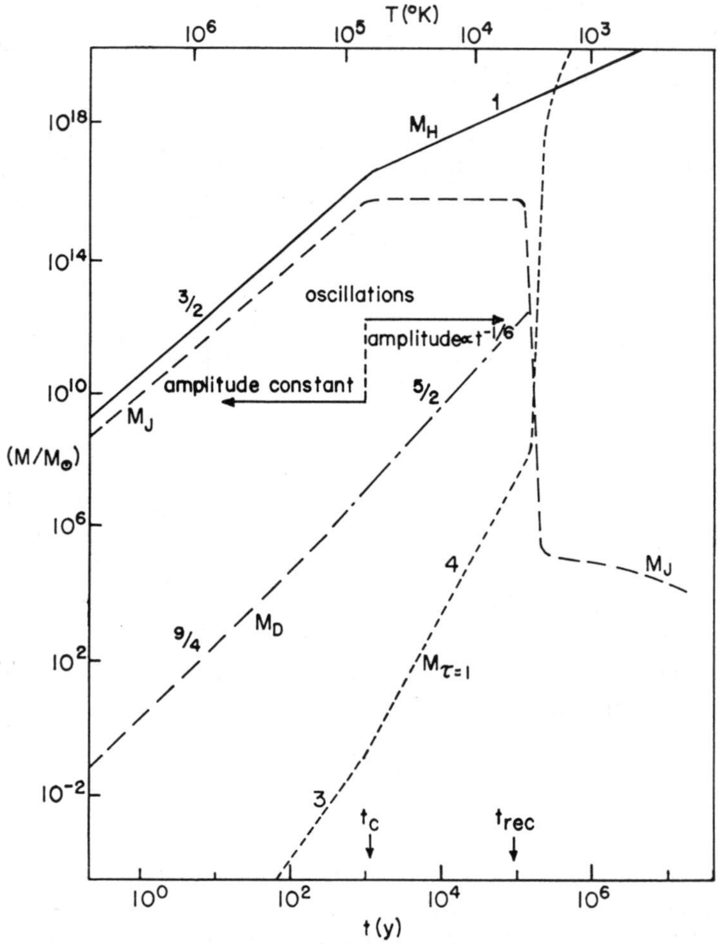

Figure 48 Some characteristic masses for a cosmology with $\Omega \simeq 1$ and present temperature 3°K. M_H denotes the mass of particles within the particle horizon (note that a galaxy comes within the horizon after ~1 year). Adiabatic perturbations on scales $< M_D$ are severely damped. During the expansion, the various masses increase as the powers of t which are given alongside the curves.

ratio (i.e. $\Delta\varrho_\gamma/\varrho_\gamma = (4/3)\,\Delta\varrho_m/\varrho_m$) that the damping of the radiation distribution reduces $\Delta\varrho_m$ to zero also. These *adiabatic perturbations* are very much a special case: in the more general situation where $\Delta\varrho_\gamma/\varrho_\gamma$ and $\Delta\varrho_m/\varrho_m$ are independent (and the entropy per baryon is itself perturbed), a residual *isothermal perturbation*

$$\frac{\Delta'\varrho_m}{\varrho_m} = \left(\frac{\Delta\varrho_m}{\varrho_m} - \frac{3}{4}\,\frac{\Delta\varrho_\gamma}{\varrho_\gamma}\right)$$

will remain even after the radiation distribution has smoothed out. Isothermal perturbations on scales $\ll M_J$ are in effect "frozen in" throughout the "plasma era" because radiative drag constrains the matter to expand at the same rate as the photon gas. A perturbation with $M \ll M_1$ which was isothermal when it first came within the horizon would never oscillate, but would maintain the same amplitude until t_{rec}.

There is at present no special motive for postulating one type of initial perturbation rather than another. Precisely adiabatic perturbations would occur if, for some reason, the entropy per baryon were strictly constant. One process leading to almost purely isothermal fluctuations has been discussed by Harrison[281]. He points out that, in a universe possessing overall symmetry between matter and antimatter, small non-uniformities (1 part in $\sim 10^7$) in the matter-antimatter ratio in the hadron era could yield isothermal fluctuations of order unity at later times.

Primordial vorticity

One can also envisage initial perturbations involving *vorticity*. If the associated peculiar velocities are $\Delta v \ll c_s$ and the scales $\ll M_H$, then it may be valid to apply the theory of *in*compressible turbulence because the fractional induced density inhomogeneities would be $\sim (\Delta v/c_s)^2$. The characteristic rotational velocities V_{rot} on a given comoving scale vary as

$$V_{rot} \propto \begin{cases} \text{constant} & (t \lesssim t_c) \\ a^{-1} & (t \gtrsim t_c) \end{cases} \tag{14.15}$$

For motions on scales $\ll M_H$, (14.15) follows immediately from conservation of angular momentum ($\propto a^5\omega(\varrho_m + \varrho_\gamma) \propto a^4 V_{rot}(\varrho_m + \varrho_\gamma)$) within a comoving volume. Photon viscosity would however damp out small eddies. In a linear situation where there is no energy input from larger eddies, the smallest eddy to survive viscous damping until t_{rec} has a mass[274,282]

$$M_2' \simeq 10^{10}\,\Omega^{-11/4}\,M_\odot \tag{14.16}$$

The reason why, for $\Omega \simeq 1$, this is much less than M_2, is that when $t > t_c$ the momentum is carried mainly by the nucleons, whereas the viscosity involves only the photons. (The calculation of M_2 assumes that the eddy is roughly spherical: damping of flattened eddies would be important on larger scales.)

Other linear processes

Other processes which affect irregularities before the recombination epoch include:

i) Neutrino viscosity, which is maximally effective when the ν_e mean free path is $\sim ct$ (i.e. at $t \simeq 1$ sec, $T \simeq 10^{10}$ °K) but has no effect on galactic-scale irregularities[283] because M_H is then $\lesssim 10^{-4} M_\odot$. As Misner[284] was the first to emphasize, however, neutrino viscosity may play a key role in reducing any initial *anisotropy* which the Universe might have possessed.

ii) Adiabatic oscillations can in principle be damped if the *bulk viscosity* is non-zero. The bulk viscosity is most significant when the electron-positron pairs are no longer ultrarelativistic but have not yet all annihilated—i.e. when $T \simeq 5 \times 10^9$ °K. Even then, the effect is unlikely to be important because the electron collision rates are much shorter than the oscillation frequencies of interesting scales.

iii) Field[285] has discussed a non-gravitational instability, first pointed out by Gamow, which arises because the electron temperature is always slightly below T_γ (owing to the expansion) and each particle tends to shield its neighbours from the intense ambient radiation. The resulting "mock gravity"—an inverse square law attraction—does not seem strong enough to be of interest.

The development of small-amplitude irregularities on the scales which seem directly relevant to galaxy formation is thus governed, when $t \lesssim t_{\rm rec}$, by gravitation, radiation pressure, and radiative viscosity. If the amplitudes are small enough for a linear analysis to be applicable, we can therefore relate the amplitude at $t_{\rm rec}$ as a function of mass to the "initial" spectrum. There is no a priori reason for postulating any particular form for the initial spectrum. Since there appears to be no effective mechanism for damping the largest scales—and indeed all masses $> M_1$ [Eq. (14.9)] grow continuously in amplitude—the apparent large-scale uniformity of the universe at the present time forces us to assume that the corresponding initial amplitudes were very small. The least artificial spectrum is one in which, at a given instant, the "power" is concentrated in smaller scales (or higher wavenumbers). Many authors have adopted a spectrum in which the amplitude of adiabatic fluctuations, measured at an epoch before any relevant scales have been stabilized by pressure or attenuated by damping, is proportional to (wave-number)$^{2/3}$ or equivalently, $\propto M^{-2/9}$. This choice of spectrum has the property that all scales M have equal amplitudes at the time when $M \simeq M_H$ (the smaller masses come within the horizon first). The necessary amplitude must be fixed a posteriori, but the possibility that some perturbations may be selectively damped or amplified raises the hope that certain preferred scales of irregularity may be especially prominent when they emerge from the fireball. Figure 49, taken from Sunyaev and Zeldovich[286], shows how the amplitude of adiabatic perturbations on various scales evolves before t_{rec}. Yu and Peebles[280], and Michie[279], have followed such fluctuations through to the completion of recombination, but find no appreciable increase in the minimum undamped mass above M_2.

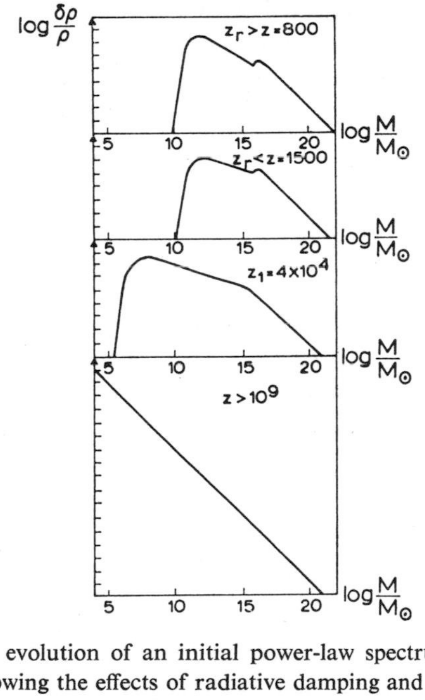

Figure 49 The evolution of an initial power-law spectrum of adiabatic fluctuations, showing the effects of radiative damping and pressure stabilization of the smaller scales, for a universe with $\Omega = 1$. [From Sunyaev and Zeldovich[286]].

Non-linear effects

Granted that some non-infinitesimal initial perturbations must be present to account for the present-day structure of the Universe, there is no reason why their amplitude should not be so large that a linear analysis is inadequate.

Recently several authors have made more ambitious investigations of the non-linear interactions between different scales of irregularity in the plasma era.

Adiabatic oscillations of large amplitude would develop shock waves (oscillations of wavelength λ develop harmonics of wavelength $\lambda/2$ which...). Peebles[287] shows that this process would reduce the amplitude of adiabatic oscillations with mass M to

$$\frac{\Delta\varrho}{\varrho} \lesssim \frac{\left(1+\frac{3}{4}\frac{\varrho_m}{\varrho_\gamma}\right)^2}{\left(1+\frac{\varrho_m}{\varrho_\gamma}\right)\left(1+\frac{21}{16}\frac{\varrho_m}{\varrho_\gamma}\right)} \left(\frac{M}{M_H}\right)^{1/6} \tag{14.17}$$

This process would therefore affect fluctuations of masses $> M_2$ if their amplitudes before recombination were sufficiently large.

If the velocities V_{rot} associated with rotational perturbations on scale λ are such that $V_{\text{rot}}t \gtrsim \lambda$, transfer of energy between eddies of different scale will modify the initial spectrum. Ozernoi and his coworkers[282,288,289] have shown that initial *rotational* perturbations could establish a Kolmogorov turbulence spectrum over a range of eddy scales corresponding to masses up to

$$5 \times 10^{15} \left(\frac{V_{\text{rot}}}{c}\right)^3 \Omega^{-2} M_\odot \tag{14.18}$$

This spectrum (for which the rotational velocities $\propto M^{1/9}$) would continue down to a mass *less* than M_2' [Eq. (14.16)] because energy fed in from larger scales will compensate for viscous dissipation. Even though the character of the primordial irregularities is unspecified and "ad hoc", these non-linear processes tend, irrespective of the initial spectrum, to establish a definite form for the spectrum at t_{rec} even on the range of scales unaffected by exponential damping. It is this spectrum which, albeit perhaps indirectly, determines the mass distribution of galaxies and clusters and their angular momentum.

14.3 PROCESSES AFTER RECOMBINATION

The work described above shows that all purely adiabatic fluctuations with masses up to those of the largest galaxies, or even somewhat larger, would have been attenuated by radiative viscosity at $t \lesssim t_{\text{rec}}$. However, *isothermal* fluctuations would survive with their initial amplitude, as would the isothermal component of any more general initial perturbation. Also, primordial turbulence—even though, for $V_{rot} \ll c_s$, it would have been essentially incompressible before t_{rec}— would quickly generate density irregularities as soon as the recombination had reduced the effective sound speed.

Any density fluctuations whose amplitude just after t_{rec} was already $\gtrsim 1$ would form condensations with density $\gg 10^4\,\Omega$ atoms cm^{-3}. Therefore any agglomerations of galactic mass which are destined to form ordinary galaxies (typical densities 10^{-1}–10^3 atoms cm^{-3}) probably had $(\Delta\varrho/\varrho) \ll 1$ just after decoupling. This provides some reassurance that a linear treatment may be appropriate before t_{rec}; and may also, as discussed by Peebles[290], raise problems for theories of primeval turbulence[289,291–293].

After t_{rec}—or at least by the time T has fallen to $\sim$ 1000°K—the primordial plasma would have recombined sufficiently that there would be no dynamical coupling between radiation and matter, as far as fluctuations with $M \lesssim M_H$ are concerned. (Both the radiation drag acting on optically thin fluctuations and the pressure due to the high frequency tail of the radiation spectrum in the Lyman lines or Lyman continuum are easily shown to be indeed negligible). Unless magnetic fields are dynamically significant, gravitational instability after t_{rec} is thus inhibited by gas pressure alone, the corresponding Jeans mass being only

$$M_3 \simeq 3 \times 10^5\,\Omega^{-1/2}\,M_\odot \qquad (14.19)$$

Masses in the range that interests us can therefore condense unimpeded by pressure. Their behaviour is therefore given by (14.2).†

Since all perturbations $\gg M_J$ grow at essentially the same rate, the scale of the first condensations to separate out would be that for which, just after t_{rec}, $\Delta\varrho_m/\varrho_m$ was greatest. If the initial fluctuations were strictly adiabatic and irrotational the smallest scales surviving at t_{rec} would have mass $\sim M_2$. If the amplitude decreased towards larger scales, the first condensations to form would be of mass $\sim M_2$, and it is tempting to identify them with large galaxies. If we knew the maximum radius attained by protogalaxies, we could infer the epoch at which they stopped expanding. Studies by Eggen, Lynden-Bell and Sandage[294] of stars in our Galaxy with highly eccentric orbits—stars whose low heavy element content identify them as among the oldest in the Galaxy—suggest that these stars formed during a free-fall collapse from a radius of $\sim$ 50 kpc. The mean density corresponding to the maximum radius is $\sim 10^{-2}$ particles cm^{-3}, and one easily calculates that if $\Omega \simeq 1$ the

† But note that a non-interacting background of, for example, gravitons or degenerate neutrinos which dominated the mass-energy density would reduce the overall expansion timescale ($\sim (G\varrho_{total})^{-1/2}$). Density perturbations in the matter (growing on a timescale $\sim (G\varrho_m)^{-1/2} > (G\varrho_{total})^{-1/2}$) then maintain almost constant amplitude instead of growing according to (14.2).

expansion of our Galaxy must have halted at a redshift $z \simeq 5$. The collapse would then have been completed by $z \simeq 2.5$. Relation (14.2) then tells us that, at t_{rec}, $\Delta\varrho/\varrho$ must have been ~ 0.5 per cent.

If the original perturbations were *not* precisely adiabatic and the initial spectrum favoured smaller masses, we would expect condensations of $\sim 10^6 M_\odot$ to develop first. (Peebles[295] has shown that, for various assumptions about the initial spectrum of the perturbations, a typical condensation would be within an order of magnitude of M_3). Peebles and Dicke[296] suggest that some of these might form globular clusters. This hypothesis accounts for the remarkably "standardised" properties of these objects, and predicts that there should be large numbers of them in intergalactic space. On this picture, galaxies would result from inelastic collisions between clouds of $\sim 10^6 M_\odot$ occurring before they had condensed into stars.

Doroshkevich, Zeldovich and Novikov[297] propose a more complex chain of events to explain galaxies. They suggest that the "primary" condensations of $\sim 10^6 M_\odot$ do not develop in the manner envisaged by Peebles and Dicke, but instead form unstable "superstars" which explode and heat up their surroundings. Even if only one part in $\sim 10^4$ of the material in the Universe exploded in this fashion, enough heat would be generated to raise the temperature of all the remaining gas to $\sim 10^6$ °K. If the redshift at which this happened was $z \lesssim 20$—as would be the case if $\Delta\varrho/\varrho$ were $\lesssim 2$ per cent at t_{rec}—the gas would not have time to cool down again. The consequent increase in the Jeans mass acts like a thermostat and inhibits the formation of further superstars. The irregular heating would, however, create inhomogeneities on scales 10^9–$10^{12} M_\odot$ (corresponding to the "sphere of influence" of each superstar) even if such scales were completely absent from the initial spectrum of perturbations. These could then condense to form galaxies or quasars.

In the "primordial turbulence" model, the perturbations before recombination involve primarily the *velocities* and not the density. However at t_{rec} the effective speed of sound drops by a factor $\sim 3 \times 10^3\, \Omega^{-1/2}$ so the motions (previously subsonic) become supersonic and generate density irregularities. Ozernoi and Chibisov[289] have discussed the subsequent supersonic turbulence in general semi-quantitative terms. Their work suggests that density irregularities are, after recombination, stabilized by an effective pressure ($\gg$ thermal gas pressure) contributed by turbulent eddies on scales smaller than the given irregularity, and that proto-galaxies would not isolate themselves from the expanding background until $z \simeq 100$. It is argued that all the angular momentum of galaxies can, in this approach, be ascribed to the primordial vorticity. For initial perturbations that are irrotational (whether adiabatic or isothermal), the present spin of galaxies must be attributed to tidal interactions between neighbouring proto-galaxies. There has been some recent debate in the literature about whether this process can be effective enough[291,298].

At the moment, the main firm achievement of these studies of perturbations in the "fireball" has been to show that if the initial fluctuation spectrum is "smooth", but decreases in amplitude as the scale increases, three characteristic scales are favoured:

M_1: the minimum mass whose growth is never interrupted by pressure forces ($\sim 10^{15} M_\odot$).

M_2: the mass of the smallest adiabatic perturbation which escapes severe damping during its oscillatory phase before t_{rec} ($\sim 10^{12} M_\odot$).

M_3: the minimum mass able to condense freely after t_{rec} ($\sim 10^6 M_\odot$).

The nature and spectrum of the initial irregularities is conjectural, but, if the amplitudes are large enough, non-linear processes may set up a calculable spectrum that is insensitive to the initial conditions. It should then be feasible to calculate the mass-density relation (and perhaps also the mass-angular momentum dependence).

At present, however, it is little more than speculation even to associate M_1, M_2 and M_3 with observed scales of agglomeration (clusters of galaxies? galaxies? globular clusters?) in the Universe. Moreover there is still no understanding of the basic distinction between spiral galaxies and ellipticals (though this probably reflects the angular momentum per unit mass in the protogalaxy); still less can we expect to interpret the morphology of galaxies in fuller detail. However these pioneering studies of perturbed cosmological models may well turn out to provide a valid and fruitful insight into the fundamental problem of the origin of the observed structures in the Universe.

An obvious prerequisite for any serious future comparison of theory and observation is fuller information on the actual mass spectrum and morphology of galaxies and clusters—and superclusters, if they exist (see section 15.1). There are doubtless many omissions from our present catalogue of the contents of the Universe. Indeed the material in galaxies apparently contributes only 1–5 per cent of the density required by a closed universe. But there is no reason to believe that all the matter in the Universe shines, so optical astronomy may give an impression of the characteristic scales of "clumpiness" that is biased as well as exceedingly incomplete.

CHAPTER 15

The Mystery of the Missing Mass: The Contents of the Universe

SINCE DIRECT measurements of the deceleration parameter q are still inconclusive, one may wonder whether astronomical observations of the contents of extragalactic space can yield a firmer answer to the question: is the combined mass-energy density of all forms of matter sufficient to close the universe? (Note that the density corresponding to $\Omega = 1$ is 1.1×10^{-29} g cm^{-3} for $H = 75$ km sec^{-1} Mpc^{-1}, and scales as H^2). This approach turns out to be no more satisfactory, since many forms of mass-energy could have evaded detection by all present techniques. However, because such information is the foundation for any understanding of the astrophysical evolution of the universe, we summarize what is now known about the various forms of cosmic matter. We also estimate their respective contributions to the overall density parameter Ω.

15.1 GALAXIES AND CLUSTERS OF GALAXIES

The detailed morphology of galaxies, based on their optical appearance, and on their colours and spectroscopic properties, is beyond our present scope. The main categories are "irregulars", spirals and ellipticals—the latter, together with $S\,0$ galaxies (disk-shaped galaxies devoid of apparent spiral structure) predominate within rich clusters of galaxies. Hubble classified galaxies into a morphological scheme, which was later elaborated by Sandage and de Vaucouleurs (see "The Hubble Atlas of Galaxies" for superb illustrations depicting all these types). It was originally believed that galaxies evolved from one type to another—spirals and barred spirals transforming into ellipticals. But there is no evidence for any systematic age difference between spirals and ellipticals; nor is it clear how one type can evolve into the other in $\lesssim 10^{10}$ years, especially as spirals are now known to possess systematically more angular momentum per unit mass than ellipticals. Many galaxies do not fit into Hubble's classification scheme. Examples are illustrated in Arp's "Atlas of Peculiar Galaxies".[299] These pathological objects may provide valuable clues to galactic structure in general. As Arp comments in the introduction to his atlas: "If we could analyze a galaxy in the laboratory, we would deform it, shock it, probe it in order to discover its properties. The peculiarities ... represent perturbations, deformations and interactions which should enable us to analyze the nature of the real galaxies which we observe and which are too remote to experiment on directly". In recent years, the various classes of galaxy which display "violent activity"—Seyfert galaxies (which have a small nucleus characterized by a spectrum with strong, broad emission lines), *N*-galaxies (i.e. "nucleated" galaxies), and "compact" galaxies—have received increasing attention. They are discussed at length in Burbidge's excellent review article[300], and in the proceedings of the 1970 Vatican Semaine d'Étude on "Nuclei of Galaxies". Such objects, however, comprise only a small proportion of all galaxies.

The masses of individual galaxies are best estimated from rotation curves (for spirals especially), and from the velocity dispersion of the stars (for nearby ellipticals). Statistical estimates of masses come also from dynamical studies of "double galaxies" which are apparently orbiting around each other (assuming randomly oriented orbits)[301]. The mass-to-light ratio (M/L) can then be estimated. This is normally expressed in solar units—i.e. as a multiple of $M_\odot/L_\odot$—and is ~ 1 for irregulars, 1–10 for spirals, but substantially larger (perhaps ~ 30) for ellipticals. The mean space density of galaxies of different absolute luminosities can also be estimated. Abell's luminosity function[302] for galaxies in clusters is shown in Figure 50. A smooth extrapolation of this function indicates that the contribution of very faint galaxies to the total mass would be insignificant unless they had abnormally high M/L ratios. Oort[303] in 1958 attempted to use all the available data to infer the smoothed-out density of galaxies. He obtained

$$\Omega_{\text{gal}} \simeq 0.03\dagger$$

This corresponds to a density of ~ 0.03 "mean galaxies" per cubic Mpc (a "mean galaxy" having a mass $\sim 10^{11} M_\odot$ and absolute magnitude ~ -19.5). Recently Shapiro[304] has made an independent estimate, using more modern M/L ratios, and finds $\Omega_{\text{gal}} \simeq 0.01$. Note, however, that these estimates refer only to the region closer than ~ 20 Mpc. Since the individual galactic masses are estimated dynamically, these estimates of Ω_{gal} include most forms of invisible matter within the galaxies themselves, such as, for instance, massive black holes at the centres of elliptical galaxies, neutron stars, or molecular hydrogen. (The only exception would be a massive

† In fact Oort[303] assumed a Hubble constant $H = 75$ km sec^{-1} Mpc^{-1}, and obtained $\varrho_{\text{gal}} \simeq 3 \times 10^{-31}$ gm cm^{-3}. However ϱ_{gal} would scale as H^2, so Ω_{gal} is actually independent of H.

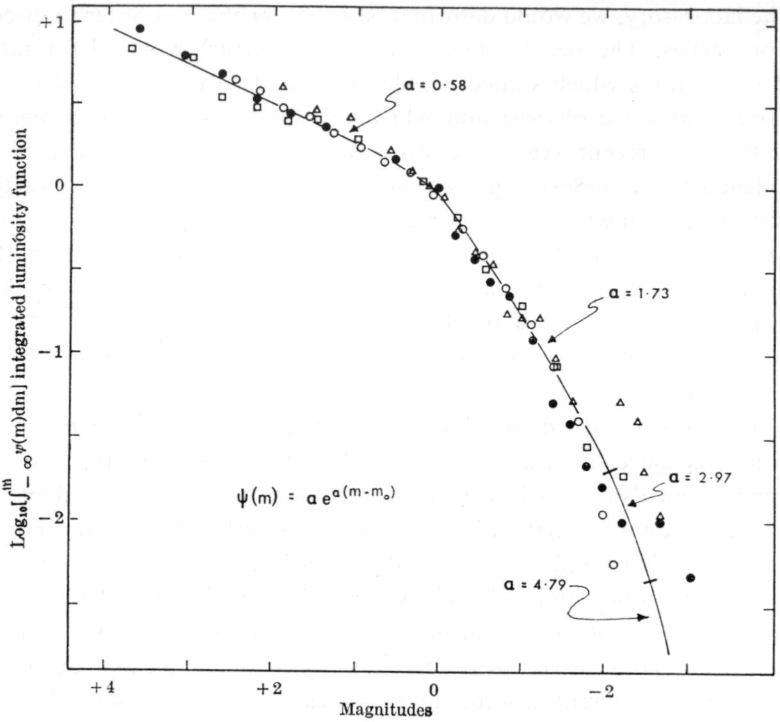

Figure 50 The luminosity function for galaxies in clusters, according to Abell. Vertical axis shows $\log_{10}\left[\int_{-\infty}^{m} \psi(m)\, dm\right]$, the integrated luminosity function.

extended spherical halo, which would obviously not affect the interior dynamics. This could arise from invisible remnants of a first generation of stars which formed when the galaxies were initially collapsing. By taking very "deep" photographs sensitive to low surface brightness, Arp and Bertola[305] have recently found that elliptical galaxies extend to much larger radii than was previously apparent).

Galaxies are concentrated into clusters, varying from "rich" clusters like that in Coma, consisting largely of elliptical and SO galaxies, to small groups like the Local Group, of which the Andromeda galaxy and our own Galaxy are the dominant members. The claim by Abell and de Vaucouleurs that the clusters are themselves grouped into "superclusters" with scales up to ~ 100 Mpc remains a controversial issue, and the question of a heirarchy of clustering on even larger scales is even more uncertain. As discussed in Chapter 17, the microwave background isotropy restricts the permissible amplitude of fluctuations on very large scales. Radio surveys reveal no evidence that sources are clustered on large scales. When larger samples of quasars have been identified, it may be feasible to use them to check the homogeneity on scales of $\sim$ 1000 Mpc.

To relate the occurrence of clustering to the initial fluctuation spectrum in the fireball, as already discussed in chapter 14, one obviously requires a full autocorrelation function for the density distribution of galaxies. This is a formidable observational task. Another key question is whether there indeed exists a characteristic, or preferred, galactic mass, which we could perhaps identify with the mass M_2 [Eq. (14.14)]. Although the cluster luminosity function shown in Figure 50 displays a broad range, most of the integrated light comes from galaxies with luminosities near the "kink", $\sim$ 2 magnitudes fainter than the brightest member. Unless the mass-to-light ratio depends strongly on mass, this implies that these galaxies contain most of the mass. It is not, however, clear whether field galaxies have a similar luminosity function. Arp[306] has emphasized that many types of intrinsically faint galaxies may have escaped recognition altogether: if their linear dimensions were too large, their low surface brightness would render them undetectable above the night sky; whereas if they were very compact, they would be mistaken for stars.

Of the reality of large clusters (e.g. the Coma Cluster, the Perseus Cluster and the Virgo Cluster) there can be no doubt, and it is the dynamics of these systems that suggests most insistently the presence of "missing mass" in some form other than normal galaxies.

Rich clusters of galaxies (Coma being the prime example) have the appearance of dynamically stable systems. If they were not gravitationally bound, the velocity dispersions among their members ($\sim$ 1000 km sec^{-1}) are large enough that they would disrupt in $\sim$ 20 per cent of the Hubble time. But straightforward application of the scalar virial theorem [(kinetic energy) = $\frac{1}{2}$(gravitational binding energy)] to these clusters reveals that the "visible mass" in the form of galaxies falls far short of that required to bind them, if one adopts the usual estimates for the mass-to-light ratios of individual galaxies. The discrepancy amounts to a factor $\gtrsim$ 50 in the Virgo and Perseus clusters, and perhaps $\sim$ 10 in the case of Coma. If these rich clusters are typical, then the smoothed-out density of clusters, including not just the galaxies themselves but the (much larger) contribution from the "missing mass", could contribute $\Omega_{\text{cluster}} \simeq 1$—in other words, sufficient gravitating matter to close the universe.

What could be the form of this "missing matter" in clusters?

Gas Uncondensed gas is one obvious possibility, but a sufficiently high gas density (an average of $\sim 3 \times 10^{-3}$ particles cm^{-3} in Coma and perhaps ten times higher in the cluster center) is hard to reconcile with the various observational limits. In both Coma and Virgo, 21 cm observations rule out this amount of *neutral* gas[307,308]. In the Coma cluster, the absence of $H\beta$ emission excludes an ionized gas with $T_e \lesssim 10^4$ °K; and the X-ray limits would be exceeded if $T_e \gtrsim 10^6$ °K (ref. 309). As Turnrose and Rood[310] have discussed, it is hard to see what could maintain the gas in the temperature range 10^4 °K $\lesssim T_e \lesssim 10^6$ °K and prevent it from settling into the centre of the cluster. Thus even though there are indications of *some* intergalactic gas in Coma—both from X-ray astronomy[311] and from the radio evidence that material is streaming from a radio galaxy to one of its neighbours[312]—there is probably not enough to bind the cluster.

Dust Zwicky[313] claims that fewer faint (background) clusters are observed in the fields of relatively nearby clusters than would be expected from a random distribution, and attributes this effect to obscuration by dust in the foreground clusters. However the amount of mass required to cause this obscuration is a quite negligible fraction of the "missing mass" unless the "dust" consists of grains hundreds of times larger than those in interstellar space.

Faint stars, "dead galaxies", black holes If the intergalactic space in the Coma cluster were pervaded by stars, their integrated light would not be detectable if their masses were $\lesssim 0.2M_\odot$ (for main sequence stars $M/L \propto M^{-2}$). A rather less plausible alternative is that clusters may contain invisible "dead galaxies"[314], in which all the stars had already completed their evolution. This would require stellar masses $\gtrsim 2M_\odot$, and such galaxies would have been very luminous when they were young. Sunyaev and Zeldovich[315] point out that the integrated emission from such objects would be in danger of exceeding observational limits on the extragalactic sky brightness (see Chapter 16). Dead galaxies and black holes would manifest themselves gravitationally, and van den Bergh[316] argues that the absence of tidally distorted spiral galaxies in the Virgo cluster shows that, in that case at least, the "missing mass" cannot be in the form of dark objects whose individual masses are of galactic order. However, objects with smaller individual mass (e.g. collapsed intergalactic globular clusters?) would be undetectable, even if they collectively contributed enough mass to bind the cluster.

The dynamics of clusters is thus an outstanding unsolved problem, but there is no reason at the moment to abandon the hypothesis that "missing matter", in some as-yet-undiscovered form, is present, in which case Ω_{cluster}

$\simeq 1$. We recall, however, the radical views of Ambartsumian, Arp and others to the effect that clusters may in fact by flying apart†. These views seem especially cogent, however, when applied to small groups of galaxies where the virial discrepancy is sometimes even larger than in the great clusters and the problem of hiding the missing mass even more severe. In some systems—Seyfert's sextet, a very compact system with both spiral and irregular members, being perhaps the most dramatic example[318]—the crossing time is only $\sim 10^8$ years, and it seems unlikely that, even if bound, such a compact system could have survived for a Hubble time without its members colliding and coalescing. Such systems are hard to incorporate in the scheme of galaxy formation outlined in the last section, and point towards the need for some more original thinking.

15.2 QUASARS

If quasars are indeed "hyper-active" galactic nuclei (as we assumed in chapter 12), then they involve only one galaxy in $\sim 10^5$ at the present epoch, and make no significant contribution to Ω. On the other hand if, as has been frequently suggested, their redshifts were not cosmological but gravitational, the situation would be very different. The known quasars could then be much closer to us (with a correspondingly higher number density) and could each be very much more massive than galaxies.

The local theory was first proposed in order to reduce the overall energetic requirements, and to alleviate some problems associated with the *concentration* of energy implied by the rapid variability ($\gtrsim 10^{46}$ erg sec^{-1} from a region $\lesssim 1$ light day across!) However, interpreting the redshift as a gravitational effect also leads to difficulties: no satisfactory *stable* model has been given which yields a potential well sufficiently deep (z up to ~ 3) and "flat bottomed" ($\Delta\lambda/\lambda \lesssim 0.01$ for the emission lines and $\lesssim 0.001$ for the absorption lines). The fact that forbidden lines are seen in quasar spectra limits the plasma density n_e in the emitting region to $\lesssim 10^6$ cm^{-3}. This sets a lower limit on the volume V of gas required, because the luminosity $L \propto n_e^2 V$. The volume within the potential well $\propto M^3$, so $L \propto M^3$ (assuming that n_e has the maximum allowed value). In the local theory the distance d to a typical quasar, and hence L, is unknown; but the *apparent* brightness *is*

---

† Such views are in principle testable by detailed analysis of the velocity distribution across a cluster. These studies can also determine whether the "missing mass" could be a few *very* massive black holes in the centre of the cluster (by analogy with Wolfe and Burbidge's studies[317] of individual elliptical galaxies).

known, so we have $L \propto d^2$. Therefore $M \propto d^{2/3}$ and so $\Omega_{\text{quasar}} \propto Md^{-3} \propto d^{-7/3}$. This means that if the observed quasars were too close, then (assuming them to have a uniform spatial distribution), $\Omega_{\text{quasar}} \gg 1$, which would be incompatible with the limits on q! Putting in the numerical coefficients, one finds $d \gtrsim 40$ Mpc and $M \gtrsim 10^{13} M_{\odot}$. A similar lower limit on d (but one that is independent of any specific theory of the redshifts) comes from applying an Olbers-type argument to the radio background from unresolved quasars. Thus the quasars cannot be *very* local unless our Galaxy occupies a specially priveleged position. However, if there is a substantial non-cosmological contribution to their redshifts, Ω_{quasar} may be significant (and indeed seems *bound* to be if the redshifts are gravitational in origin). In the speculative theory of Verontsov-Velyaminov[319], where quasars are *clusters* of galaxies in the process of explosive formation (along the lines advocated by Ambartzumian), as well as in Ne'eman and Tauber's model[271], which interprets quasars as "lagging cores"—regions left over from the "big bang" and only now expanding out of the initial singularity—Ω_{quasar} might be substantial.

There seems to us, however, no strong reason for invoking a special interpretation of the redshifts in quasars. All the alleged theoretical difficulties with the "cosmological" interpretation apply with almost equal strength to radio sources, *N*-galaxies and related objects. No one has ever *proved* that any modification of physical law is required to understand the operation of a quasar. However when one recalls the "scatter diagram" (Figure 40) revealed by plotting the redshifts and magnitudes of quasars, and the hypothetical "evolution" which is invoked to cancel out the expected trend (Chapter 12), one cannot avoid the suspicion that this whole procedure is very much a bootstrapping operation, and its self-consistency is no guarantee of truth. In no way could doubts on quasar redshifts be more decisively set to rest than by discovering a quasar and an ordinary galaxy indubitably physically associated, and determining the redshift of each. At the time of writing, Gunn[320] has found a moderately convincing case of a quasar (PKS 2251 $\pm$ 11) which has the same redshift ($z = 0.323$) as a small cluster of galaxies apparently associated with it; on the other hand, Arp has given several examples (see, for instance, ref. 321) where quasars and other compact objects appear to be physically linked by "bridges" to galaxies with smaller redshift. The statistical significance of the latter type of evidence is hard to quantify (as in some other astronomical contexts, objects are *pre-selected* for study by virtue of their peculiar appearance); one fervently hopes that this work, which affects not only the value of Ω_{quasar} but also all other cosmological evidence involving quasars, will be urgently pursued.

15.3 DIFFUSE INTERCLUSTER GAS

There is no evidence for the presence of any intergalactic gas in the space outside clusters of galaxies. However any matter in this form is exceptionally elusive, and any constraints or limits on its properties are cosmologically important, because of its possibly large contribution to the density parameter Ω. If the "hot big bang" picture indeed resembles the actual universe, and if galaxy formation proceeded in the general way outlined in chapter 14, it would actually be highly surprising for all the material to have condensed into clusters of galaxies, leaving nothing behind. More probably, a fraction of the matter would condense into bound systems (proto-galaxies? proto-globular clusters? massive objects?) and the power output from these objects would heat up the remaining material at some redshift z_{reheat}. Having been reionized and perhaps heated to a high temperature this material would subsequently be less prone to form condensations and so might still remain as a diffuse gas. A possible schematic temperature-history for the uncondensed intergalactic gas is illustrated in Figure 46. If $H = 75$ km sec^{-1} Mpc^{-1} and protons and helium nuclei occur in the ratio 10 : 1 (72 per cent H and 28 per cent He by mass), the hydrogen density is $n_H = 4.5 \times 10^{-6}\, \Omega_{IGG}$ cm^{-3}, and the electron density $n_e = 5.4 \times 10^{-6}\, \Omega_{IGG}$ cm^{-3}.

We shall now give a list of the observations which constrain the properties of the gas. These rely on searches for absorption (or other effects such as dispersion) in the radiation from discrete sources, or on searches for emission from the gas itself. The latter class of observations, especially, are cosmology-dependent, and set limits to the thermal history of the gas at redshifts larger than those of any known discrete sources. (See the review by Field[308] for a thorough and up-to-date treatment of these topics.)

1. Intergalactic neutral gas

a) *21 cm measurements* Evidence on the amount of intergalactic neutral hydrogen at recent epochs comes from measurements of the 21 cm line. One may search for 21 cm *emission* from the gas (which would contribute to the radio background at all wavelengths *longer* than 21 cm, and give rise to a "step" in the background temperature at 21 cm) or alternatively search for *absorption* in the spectrum of a strong extragalactic radio source of redshift z, where one might expect a trough extending from 21 cm to $21(1 + z)$ cm, because all radiation received in this wavelength range would have had precisely the 21 cm wavelength with respect to the ambient gas at some point along its path to us. In this trough, the optical depth at a frequency ν is

$$\tau(\nu) = \left(\frac{a}{\dot{a}}\right) c n_{HI} \left\{1 - e^{-\frac{h\nu_{21\text{cm}}}{kT_s}}\right\} \frac{1}{\nu_{21\text{cm}}} \int \sigma(\nu)\, d\nu \qquad (15.1)$$

where a, $\dot{a}$, n_{HI} and the spin temperature T_s are all evaluated at the epoch when $a/a_{\text{now}} = \nu/\nu_{21\,\text{cm}}$, and the integral is carried out over the line width. Since under all reasonable conditions $h\nu_{21\text{cm}} \ll kT_s$, the term in braces can be approximated as $h\nu_{21\text{cm}}/kT_s$.

The *emission* measurements have the advantage of being independent of the spin temperature T_s (the step being $\propto T_s\tau$ when, from (15.1), $\tau \propto T_s^{-1}$) provided that this substantially exceeds the continuum background temperature of $\sim$ 3°K (otherwise the predicted height of the step, for a given n_{HI}, involves an extra factor $(T_s - T_{\text{continuum}})/T_s$); whereas the absorption measurements, though they involve T_s directly, allow one to point one's telescope at a bright source like Cygnus A so that even a very small optical depth can yield a detectable decrease in antenna temperature.

The best limit from the emission measurements[322] corresponds to a present-day neutral hydrogen density of $\lesssim 3 \times 10^{-6}[T_s/(T_s - T_{\text{continuum}})]$ cm^{-3} (i.e. $\Omega_{HI} \lesssim 0.3$ if $T_s \gg 3$°K). The absorption limits correspond to $\Omega_{HI} \lesssim 3 \times 10^{-3} T_s$. Allen[323] has shown, by combining the emission and absorption limits, that the limit $\Omega_{HI} \lesssim 0.3$ cannot be evaded by having a spin temperature $\sim$ 3°K, and that, if the density of hydrogen in intergalactic space exceeds $\sim 10^{-6}$ cm^{-3}, then it must be more than 70 per cent ionized regardless of its kinetic temperature. It is unlikely, however, that T_s would exceed $\sim$ 30°K even if the kinetic temperature were high enough to render the gas predominately ionized. If $T_s \simeq 10$°K, the 21 cm absorption measurements would imply $\Omega_{HI} \lesssim 0.03$.

b) *Lyman α absorption* A more stringent limit on the intergalactic HI density—though it refers to redshifts $z \simeq 2$ rather than the current epoch—is set by the absence of detectable absorption on the "red" side of the redshifted Lyman α wavelength (i.e. $\lambda > 1216\ (1 + z)$ Å) in quasar spectra. The scattering cross section for Lyman α photons is very large, and the absence of detectable absorption ($\lesssim$ 5 per cent, say) in any quasar implies that[324 325] $n_{HI} \lesssim 6 \times 10^{-12}$ cm^{-3} at $z \simeq 2$†, or $\Omega_{HI} \lesssim 2 \times 10^{-7}$. It hardly seems conceivable that the total intergalactic gas density would be as low as this limit, so it would appear more natural to suppose that most of the gas is ionized‡. This would indicate that reheating of the gas took place at

† The opacity is given by an expression similar to (15.1) but with the simplification that the exponential (stimulated emission) term is negligible. The precise limit on n_H depends on the Hubble constant at $z \approx 2$, and therefore on the assumed cosmological model as well as on H.

‡ A limit which is ~200 times *less* stringent can similarly be set to the amount of intergalactic *molecular* hydrogen at $z \approx 2$ from the absence of detectable absorption due to the Lyman or Werner bands.[326]

a redshift $z_{\text{reheat}} \gtrsim 3$ (since the largest redshift of any quasar known at the time of writing is 2.89).

The validity of these Lyman α limits depends crucially on the assumption that quasar redshifts are cosmological. It should soon be possible to use telescopes on space platforms to search for the Lyman α absorption trough, at ultraviolet wavelengths, in "bona fide" galaxies. This could set equally tight limits on the neutral hydrogen density at small redshifts, and might also help to settle the vexed question of the nature of quasar redshifts.

iii) *Other tests for neutral gas and heavy ions* Ultraviolet transitions in other ions could also cause absorption in the spectra of distant objects. Because the intergalactic densities are so low, essentially all the bound electrons are in their ground states or in very low lying states that can be excited by the microwave background. Therefore only resonance lines can be utilized for this test. Shklovski[327] first applied the method to the resonance doublet of MgII (λ2798) and many other transitions have subsequently been considered in the same way[328,329]. There is no evidence that intergalactic absorption attributable to any of these ions occurs, but the interpretation is unclear because our ignorance of the ionization level means that we do not know what lines to expect. In any case, there is no reason to expect heavy elements to be present in intergalactic space with anything approaching their solar-system abundances—the primordial abundances would presumably be very low, and contamination by material ejected from galaxies need not be substantial. It is unfortunate that the resonance lines of He(584Å) and He^+(304 Å) are too far into the ultraviolet to be observable from the ground, even in the spectra of the quasars with the largest known redshifts.

If the excess soft X-ray background (see section 16.7) comes from cosmological distances, the condition that the intergalactic medium be transparent to the X-rays requires $\Omega_{\text{He+He}^+} \lesssim 0.1$ (ref. 315). This means that if Ω_{IGG} is very high the helium is probably mainly doubly ionized.

It has often been suggested that the absorption lines seen in some quasar spectra, with several different redshifts $z_{\text{abs}} \lesssim$ the emission line redshift, are due to clouds of material along the line of sight, with *cosmological* redshifts z_{abs}. Alternatively, the absorbing material may have been ejected from the quasar itself at speeds up to $\sim c/2$. The former possibility would imply that spectra of all quasars with similar emission line redshifts should display comparable numbers of absorption redshifts. It can thus be tested when more spectra of uniform quality become available[330]. If the absorption lines (or at least some of them) do indeed turn out to be intergalactic, the study of quasar spectra will obviously teach us something about physical conditions and elemental abundances in intergalactic space.

2. Intergalactic ionized gas

The absence of a Lyman α absorption "step" in quasar spectra indicates that, if there is any appreciable amount of intergalactic gas, it must be largely ionized for $z \lesssim z_{\text{reheat}}$ (where $z_{\text{reheat}} \gtrsim 3$). The most sensitive and interesting tests for the presence of hot ionized gas are searches for thermal bremsstrahlung in the X-ray band, and possible distortions (due to bremsstrahlung at centimeter wavelengths and to Compton scattering) of the primordial microwave background spectrum. These effects depend mainly on the thermal history of the gas at redshifts z much greater than 3, and we shall first deal with some effects which could, in principle, tell us about conditions at more recent epochs.

a) *Electron scattering* The electron scattering optical depth, back to a redshift z, due to ionized intergalactic gas is

$$\tau_{es} = \int_{\frac{a_{\text{now}}}{(1+z)}}^{a_{\text{now}}} n_e \sigma_T c(da/\dot{a})$$

If the intergalactic gas is ionized, and its density varies as $a^{-3} \propto (1+z)^3$, then

$$\tau_{es} \simeq 0.04\left(\frac{\Omega_{IGG}}{\Omega}\right)\{(1+\Omega z)^{1/2}(3\Omega + \Omega z - 2) - (3\Omega - 2)\}. \qquad (15.2)$$

Ω is the *total* density parameter (which governs the dynamics of the universe), Ω_{IGG} (assumed constant in (15.2), and $< \Omega$) is the contribution from intergalactic gas, and we assume $\Lambda = 0$. Even if $\Omega_{IGG} = 1$, we have to go back to $z = 7$ to reach unit optical depth, so this effect is not large even for quasars (τ_{es} varies as $(1+z)^{3/2}$ in the high redshift limit). Moreover, since the attenuation is the same for all frequencies for which $h\nu \ll m_e c^2$, it would be difficult to recognize†. It is interesting that this redshift-dependent effect would significantly bias the determination of q from the $M - z$ relation[324,331], even though redshifts of only $\lesssim 0.5$ are involved. Its neglect would lead to an underestimate of q.

b) *Low frequency radio absorption* There is some indication that the non-thermal radio background, due to the integrated effect of extragalactic sources, cuts off below $\sim$ 2 MHz, and this could be due to free-free absorp-

† For hard X-rays ($h\nu \simeq m_e c^2$) the effect depends on ν, since the relevant cross-section then drops below the Thomson limit. In this case the scattering occurs even if the electrons are bound.

tion by intergalactic plasma. We can confirm that this is quantitatively plausible by evaluating the optical depth of a uniform slab of length $c\tau_H$. If the gas temperature is T_e this is

$$\tau_0 \simeq 0.2\Omega_{IGG}^2 \left(\frac{T_e}{10^4}\right)^{-3/2} \nu_{\mathrm{MHz}}^{-2} \tag{15.3}$$

A proper calculation of the free-free opacity back to a large z requires some estimate of the redshift-dependence of T_e. However if T_e were *constant* the optical depth back to a redshift z would [using similar notation to Eq. (15.2)] be[191]

$$\tau_{ff} = \frac{2\tau_0}{\Omega^3}\left\{(1 + \Omega z)^{1/2}\left(\frac{1}{5}(1 + \Omega z)^2 + \frac{2}{3}(\Omega - 1)(1 + \Omega z) + (\Omega - 1)^2\right)\right.$$
$$\left. - \Omega^2 + \frac{4}{3}\Omega - \frac{8}{15}\right\} \tag{15.4}$$

which varies as $(1 + z)^{5/2}$ in the limit of large redshift.

Bridle[332] and Clark *et al*[333] report a cut-off at $\sim$ 2 MHz in the isotropic radio background, but it is difficult to reconcile the parameters needed to ex-plaint his as a genuinely intergalactic effect ($\Omega_{IGG} \simeq$ (1–10) and $T_e \lesssim 10^4$ °K) with other evidence on the intergalactic gas.

c) *Dispersion of radio waves* Waves of frequency ν propagate through plasma with a group velocity of $c\,(1 - \nu_p^2/\nu^2)^{1/2}$, where

$$\nu_p = \left(\frac{n_e e^2}{\pi m_e}\right)^{1/2} \simeq 9 \times 10^3 n_e^{1/2}\ \mathrm{Hz}$$

is the plasma frequency.

Thus a sharp radio pulse emitted from a cosmologically distant source will be smeared out at reception, the lower frequencies arriving later than the high frequencies. The time delay in a signal from a redshift z is[334]

$$\Delta t \simeq 10^7 \Omega_{IGG}\, \nu_{MHz}^{-2}((1 + \Omega z)^{1/2} - 1)\ \text{seconds} \tag{15.5}$$

The only presently known variations in extragalactic sources occur at high radio frequencies ($\nu \gtrsim$ 1000 MHz) and have timescales of the order of months. If the radiation is incoherent synchrotron emission, the regions responsible for the emission at $\lesssim$ 100 MHz must, typically, be at least 10 pc across in order that their surface brightness temperature should not exceed the kinetic temperature of the relativistic electrons, and so rapid fluctuations would be unlikely. Thus it does not seem feasible to search for intergalactic dispersions in *known* radio variables (especially because there may be complicated frequency-dependent time delays within the sources themselves). However it is not inconceivable that a scaled-up version of the coherent emission mechanism operating in pulsars may occur in some extragalactic objects. Such a mechanism could then lead to low frequency variability on timescales of hours, minutes, or even less. If such phenomena were ever discovered, intergalactic dispersion would be readily detected. Because the z-dependence of (15.5) is a function of Ω, comparison of the dispersion in different objects would provide a new cosmological test[335].

d) *Thermal bremsstrahlung X-rays* A hot dilute gas emits bremsstrahlung (or "free-free" radiation) with a characteristic flat spectrum up to a frequency $\nu \simeq kT_e/h$, above which the emission decreases exponentially. If the intergalactic gas is now hotter than $\sim 10^6$ °K (or, more generally, was hotter than $\sim 10^6(1 + z)$°K at a redshift z) we would expect its bremsstrahlung emission to extend into the X-ray band, and many authors have described how the observed X-ray background can be used to set upper limits to the temperature and density of the gas. It is clear that the exact spectrum depends upon the temperature history $T_e(z)$, but in all realistic models the characteristic exponential cut-off is expected†. Since the observations (see section 16.7) strongly suggest that the X-ray background is *not* exponential but is of a power-law nature, it is normally assumed that the background at energies $\gtrsim 1$ keV is primarily *non-thermal* (and that its intensity merely sets upper limits to the bremsstrahlung contribution from hot intergalactic gas). Many authors have attributed the soft X-ray ($\lesssim 1$ keV) "excess" background to a diffuse intergalactic medium, but (as described further in chapter 16) the present data by no means warrant any firm statement.

Weymann[336] and Bergeron[337], in particular, have addressed themselves to the following question. Is it possible for an intergalactic medium with the "critical density" ($\Omega_{IGG} \simeq 1$) to be hot enough to yield the high degree of ionization at $z \simeq 2$ inferred from the quasar optical spectra (only 1 atom in $\sim 10^7$ neutral!) without its bremsstrahlung emission *exceeding* the X-ray background limits? If the ionization is due primarily to collisional ionization by thermal electrons, the value of T_e at $z \simeq 2$ must be several million degrees. (The nature of the heat input is not specified, but could derive, for example,

† Note, however, that for a very hot, low density gas the collisional relaxation timescale of $\sim 10^{-2} n_e^{-1} T_e^{3/2}$ seconds might exceed the Hubble time. There would then be no reason to expect a Maxwellian distribution with a well defined temperature, and there would not be any clear-cut distinction between "thermal" and "non-thermal" particles.

from dissipation of primordial turbulence, motions of galaxies, or low energy cosmic rays escaping from sources). Weymann[336] showed that the data were just compatible with a model in which the gas was heated at a redshift of 2 or 3, and the heat input subsequently switched off, allowing the gas to cool adiabatically after the epoch corresponding to $z \simeq 2$. One of Weymann's models in fact yielded a 0.25 keV photon flux comparable with the measured soft X-ray background. However this agreement should not be regarded as more than fortuitous, because the heating laws which Weymann chose were computationally simple and perhaps not especially realistic. Also, for a given Ω_{IGG} the minimum ionization level required at $z \simeq 2$ is proportional to H (or τ_H^{-1}) and the X-ray background intensity at a given T_e is proportional to H^3. Thus the uncertainty in the Hubble constant in any case precludes any detailed estimate of the X-ray flux from these considerations. It is apparent, however, that the permissible thermal histories of an intergalactic gas with $\Omega_{IGG} = 1$ are very closely constrained by the data (though the constraints would be less severe if $H \lesssim 50$ km s^{-1} Mpc^{-1}). It seems rather unlikely that so much gas actually exists in intergalactic space, especially when we remember that is it likely to be "clumpy", and the emission ($\propto \langle n_e^2 \rangle$) consequently enhanced.

If the gas at $z \simeq 2$ is heated by ultraviolet radiation, essentially complete ionization can occur even for electron temperatures $T_e \simeq 10^4$ °K (as in an ordinary HII region), and the gas would not get much hotter than this, however intense the ultraviolet radiation field was. The X-ray limits provide no constraint on Ω_{IGG} in this case. However, estimates of the likely ultraviolet background contributed by quasars and other non-thermal sources (even allowing for the effects of evolution) suggests that it is unlikely to be strong enough to ionize the gas unless[338,339] $\Omega_{IGG} \lesssim 0.1$ (An ultraviolet flux sufficient to photoionize a denser intergalactic medium at $z \simeq 2$ would probably conflict with the available ultraviolet background limits at wavelengths longward of the Lyman limit[340]. Recombination radiation from the gas itself—stronger when $T_e \simeq 10^4$ °K than when $T_e \simeq 10^6$ °K—would also be a problem). An intergalactic medium *photoionized* at $z \simeq 2$ could have partially recombined by the present epoch. Thus the lack of Lyman-α absorption in quasar spectra does not mean that we should necessarily expect no 21 cm emission or absorption due to HI at smaller redshifts. If the intergalactic gas were non-uniformly distributed, the constraints set by the background radiation becomes more severe, since the depends on the mean-square density. The Lyman-α absorption constraint is, however, eased, but Peebles[241] has shown that even if the gas were all concentrated in pressure-supported neutral self-gravitating clouds it could not contribute more than $\Omega_{HI} \simeq 0.03$ without having detectable effects on quasar spectra.

e) *Indirect tests* There are a number of other observations which are relevant to the intergalactic medium, but which do not permit such direct inferences because their interpretation involves other poorly known quantities.

Most of the observed Faraday rotation of extragalactic radio sources is probably attributable to our own galaxy, since there is a strong correlation between the rotation measure (a measure of $\int n_e B_{II}\, dl$ along line of sight) and the galactic latitude of sources. Some of the rotation measure may be intrinsic to the source. An upper limit to the intergalactic Faraday rotation of sources with $z \simeq 1$ would set a limit to the product $(n_e B_{II})$. Were there a *uniform* intergalactic field, there should be a direction-dependent contribution to the rotation correlated with the redshifts of the sources. Recently, several authors[342-344] have claimed such an effect, which implies $\Omega_{IGG} \simeq 10^9 B^{-1}$, where B is the strength, in gauss, of the (uniform intergalactic field. (If the field were not uniform over the whole universe, but had a correlation length L, it would have to be stronger by $\sim(c\tau_H/L)^{1/2}$ to produce the same net rotation measure in a distant source). This "positive" result is so far based on limited statistics, but more extensive future observations of distant radio sources should indicate whether the effect is genuine or not. Although Harrison[345] has shown that primordial vorticity can lead to the generation of a weak "seed" magnetic field, any ordered field strong enough to contribute significantly to the Faraday rotation would have to be part of the initial conditions. The existence of extended radio sources itself supports the view that intergalactic space cannot be completely empty.

The limits on the intensity of an isotropic γ-ray background (see section 16.8) imply $\Omega_{IGG}\varepsilon_{EG}/\varepsilon_G < 0.1$, where ε_G and ε_{EG} are the galactic and extragalactic cosmic ray energy densities respectively—otherwise the flux of γ-rays from the decay of collisionally produced π_0 would conflict with the observational limits[308, 346].

3. Limits on the density and "thermal history" for $z \gtrsim 3$

The quasar observations have told us that the intergalactic gas, even if its density is very low, must have been reionized at some redshift $z_{\text{reheat}} \gtrsim 3$. We now consider the observational constraints on the condition of the gas at $z \gg 3$. As no discrete sources are known with such large redshifts, we must obviously rely on background measurements.

Since the bremsstrahlung emissivity per unit comoving volume varies as $(1 + z)^3$, the fact that even at $z \simeq 2$ the X-ray background limit is a serious constraint means that we must have $T_e(1 + z) \lesssim 10^6$ °K at all substantially earlier epochs. What other constraints are there on $T_e(z)$?

One important constraint arises because hot electrons tend (statistically) to lose energy when they scatter soft photons[347,348]. The average fractional energy *gain* per photon each time it is scattered is $\sim kT_e/m_ec^2$. Thus the cumulative effect of many scatterings would be to distort the primordial black body radiation towards a "grey body" spectrum (or, more precisely a Bose distribution with $I(\nu) = \nu^3/\{\exp[(h\nu/kT_e) + \mu] - 1\}$ where $\mu > 0$) which contains the same photon number density as the original background, but a higher average energy per photon. The magnitude of these cumulative distortions obviously depends on the parameter

$$y = \int_0^{\tau_{es}(z_{\text{reheat}})} [kT_e(z)/m_ec^2]\, d\tau_e \tag{15.6}$$

where τ_{es} is given by (15.2)

Zeldovich and Sunyaev[348] show that for small values of y, the effective temperature of the Rayleigh-Jeans part of the background spectrum is decreased from its undistorted value T_0 to $T_{R-J} = T_0e^{-2y}$. The amount of energy under the distorted spectrum is increased to $aT_0^4e^{4y}$. Expressing this in terms of T_{R-J}, which is what is actually observed, we find that the energy density is $a\,T_{R-J}^4e^{12y}$. Zeldovich and Sunyaev argue that the molecular limits (see chapter 13) imply

$$y \lesssim 0.15. \tag{15.7}$$

If the intergalactic gas was reheated at a large redshift, its bremsstrahlung emission would enhance the black body background at *long* (radio) wavelengths. The observations of Howell and Shakeshaft[222] imply that this contribution to the background must amount to $\lesssim 3\,°\text{K}$ at wavelengths $\gtrsim 80$ cm, and this means that

$$\Omega^2 \int_0^{z_{\text{reheat}}} \frac{(1+z)}{(1+\Omega z)^{1/2}} T_e^{-1/2}(z)\, dz \lesssim 1.8 \tag{15.8}$$

Note that, for a given value of z_{reheat}, (15.7) and (15.8) set *upper* and *lower* limits respectively to T_e. Taken together they imply that if $\Omega_{IGG} \simeq 1$, $z_{\text{reheat}} \lesssim 200$ (and of course at these redshifts it is unlikely that bound systems have separated out, so essentially all the material in the universe must, in this context, be included in Ω_{IGG} even if this does not remain true at more modest redshifts). Indeed, these considerations show that a period when the hydrogen is predominantly neutral is inevitable in the canonical hot big bang model unless $\Omega \lesssim 0.1$†.

† It can easily be checked that for $z \lesssim 1000$, the production of photons via bremsstraht lung (leading to the constraint (15.8)) is not copious enough to invalidate the argument that led to (15.7), which assumed a constant comoving-density of photons.

It is unfortunately not possible to pin down the value of z_{reheat}, even when $\Omega_{IGG} \simeq 1$, within any better limits than $200 \gtrsim z_{\text{reheat}} \gtrsim 3$. It is tempting to associate the reheating with the epoch at which galaxies (or perhaps, more generally, the first generation of bound systems) condensed out from the expanding universe. This epoch, however, is not well determined either (see chapter 14). As discussed by Doroshkevich and Sunyaev[349], the power per comoving volume required to maintain the gas ionized at a fixed temperature increases with z: bremsstrahlung losses $\propto (1 + z)^3 T_e^{1/2}$; inverse Compton losses $\propto (1 + z)^4 T_e$, and adiabatic losses (assuming a cosmology with $\Omega \simeq 1$) $\propto (1 + z)^{3/2} T_e$. However in view of our uncertainties of the form which this energy input takes, and of its origin, we cannot use these arguments about energy to restrict z_{reheat}.

15.4 ELECTROMAGNETIC RADIATION

Evidence on the intergalactic background radiation, in all bands, is summarized schematically in Figure 51 (see also Table XXVII). The largest known energy density occurs in the microwave band (or at millimeter wavelengths, if the strong background revealed by some rocket and balloon measurements is genuinely universal). However, even though observations are

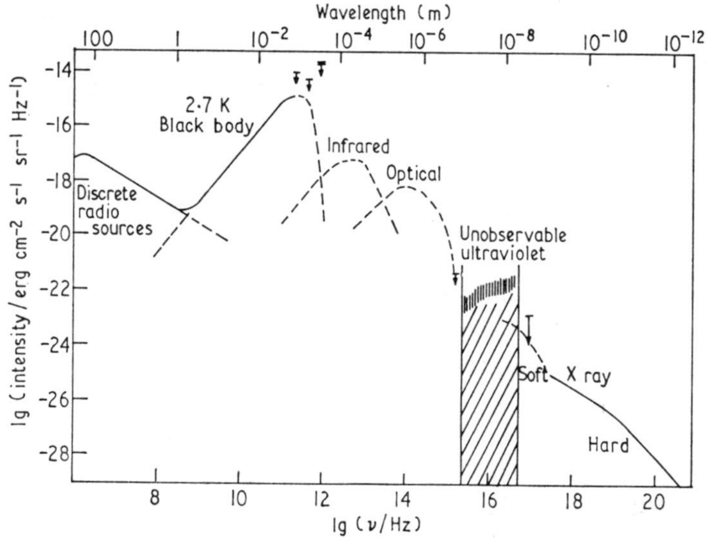

Figure 51 Schematic spectrum of the background radiation in all bands reproduced from reference 350. Directly observed portions of the curve are shown as solid lines. The associated energy densities in different bands are given in Table XXVII.

sparse in the infrared band, it seems very unlikely that the *total* radiation density integrated over all bands exceeds ~ 10 eV cm^{-3} (corresponding to $\Omega_{rad} \simeq 10^{-3}$). Thus electromagnetic background radiation is dynamically and gravitationally negligible at the present epoch, and the same is probably ture of cosmic ray particles. Nevertheless (as is clear from our discussion of the canonical hot big bang) primordial radiation could be dynamically dominant when

$$a/a_{now} \ll 1.$$

The background radiation is nevertheless of the greatest interest for astrophysical cosmology, and we return to it in the next chapter.

Table XXVII Energy density and number density of photons of the isotropic background radiation

Wavelength range	Energy density of radiation (eV cm^{-3})	Number density of photons (cm^{-3})
Radio	10^{-7}	1
Microwave background radiation	0.25	400
Infrared	$\sim 10^{-2}$ [a]	~ 1 [a]
Optical	$\sim 3\times 10^{-3}$ [a]	$\sim 10^{-3}$ [a]
Soft X-rays ($\varepsilon < 1$ keV)	10^{-4}–10^{-5}	3×10^{-7}–3×10^{-8}
Hard X-rays ($\varepsilon > 1$ keV)	10^{-4}	3×10^{-9}
Soft γ-rays ($1 < \varepsilon < 10$ MeV)	3×10^{-5}	10^{-11}
Hard γ-rays ($\varepsilon > 10$ MeV)	$\leqq 10^{-5}$	$\leqq 10^{-12}$

[a] Estimates (see Section 16.3).

15.5 NEUTRINOS, etc.

According to the canonical big bang theory we would expect a background of ν_e and ν_μ with temperature $\sim 2°\mathrm{K}(= (4/11)^{1/3}\, T_\gamma)$. However an extra background "sea" of low energy neutrinos sufficient to yield $\Omega_\nu \simeq 1$ would correspond to a Fermi level of $\sim 10^{-2}$ eV and be quite undetectable at the present day. The best available limit[350] on the level of this "Fermi sea" is $\lesssim 2$ eV (i.e. $\Omega_\nu \lesssim 10^9$!) This follows from the occurrence of protons in the cosmic radiation with Lorentz factors as high as $\gamma \simeq 10^9$. In the rest frame of such particles, 1 eV neutrinos would appear to have BeV energies, and inverse β-decays would reduce a proton's momentum on a timescale which would be $\lesssim 5\times 10^7$ years if the Fermi level exceeded ~ 2 eV. Laboratory measurements of the electron spectrum near the endpoint of the β-decay

$H^3 \rightarrow He^3 + e^- + \bar{\nu}_e$ reveal no anomalies attributable to a background of degenerate low energy neutrinos; but these are even less sensitive, telling us merely that the Fermi level is $\lesssim 60$ eV.

Even though the direct present-day limits are far from being able to exclude $\Omega_\nu \simeq 1$, a neutrino density of this magnitude drastically affects the expansion rate of the early universe, and the helium production (see chapter 17).

It is interesting that, if one accepts the "primordial fireball" as a valid description of the universe, at least back to the lepton era, one can place a limit of $\lesssim 100$ eV (4 orders of magnitude below current laboratory limits) on the *rest mass* of ν_μ.[352] Otherwise these particles, with a present density $\sim 100\ \text{cm}^{-3}$, would make a contribution to the mass-energy density of the universe exceeding the permissible limit ($\Omega \simeq 5$). Zeldovich[353] has discussed how similar arguments may be devised to set limits on the properties of quarks and other potential constituents of the primordial fireball.

15.6 GRAVITATIONAL WAVES

A consequence of the canonical big bang theory is that *thermal* gravitational radiation should pervade the cosmos[354]. Only during the very earliest phases ($t \ll 10^{-6}$ sec) would these gravitational waves be coupled to the other constituents. But even after they decouple the wavelengths continue to stretch in proportion to the scale factor a, the energy density decreasing as a^{-4}. This is the same as the behaviour of the *electromagnetic* energy density. Thus, the present gravitation energy density would be comparable with that of the 3°K microwave background (in fact it would be somewhat smaller because the electromagnetic background is augmented by the energy of species annihilating after the gravitons have decoupled). These gravitational waves—frequencies $\sim 10^{11}$ Hz—would have no observational consequences. Indeed, the mass-energy density of such waves could certainly reach $\sim 10^{-29}\ \text{gm cm}^{-3}$ without their being detectable by any feasible technique (other than by their contribution to the deceleration of the Universe).

The same is true of the acoustic frequency (~ 1 kHz) waves which one might infer, on the basis of Weber's claims, to be emanating profusely from galactic nuclei.

Of possibly greater interest are gravitational waves of $\gtrsim$ Mpc wavelength ($\lesssim 10^{-14}$ Hz) associated with the primordial fluctuations that give rise to galaxies and clusters. Our brief discussion of the genesis of galaxies (chapter 14) showed that the early universe cannot have been perfectly homogeneous and isotropic, and the perturbations in density and velocity which must be invoked to explain galaxies can be visualized as potential sources of gravitational radiation.

Taking $z = 3(a/a_{\text{now}} \simeq 0.25)$ as a characteristic redshift of galaxy formation, we can make the following crude estimate of how strong these waves may be. The amplitude of the perturbation in the metric at that time (at wavelengths of the order of the then current galactic separations) can be estimated by evaluating the gravitational redshift between the centre and outside of a galaxy:

$$\delta g \simeq \frac{(\text{mass})}{(\text{radius})} \simeq \frac{10^{11} M_\odot}{15 \text{ kpc}} \simeq \frac{1.5 \times 10^{11} \text{ km}}{4.5 \times 10^{17} \text{ km}}$$

This quantity measures the strength of the semi-static, adiabatically changing perturbations in the geometry. The gravitational wave component of such long wavelength perturbations is likely to be much smaller, because they are governed by the time rate of change of large-scale quadrupole moments. Nevertheless for a crude estimate we use the entire $\delta g \simeq 3 \times 10^{-7}$. The waves in question have reduced wavelength $\lambda\!\!\!^{-} = \lambda/2\pi \simeq (1/2\pi)$ (then current galactic separations) $\simeq (1/2\pi) \times (1/4) \times$ (present separations of $\sim 10^{25}$ cm). The effective energy density is

$$\varrho_{\text{radiation}} \simeq \frac{\Gamma^2}{4\pi} < \left(\frac{\delta g}{\lambda\!\!\!^{-}}\right)^2 \bigg/ 4\pi \simeq \left(\frac{3 \times 10^{-7}}{4 \times 10^{23} \text{ cm}}\right)^2 \bigg/ 4\pi \simeq 4 \times 10^{-62} \text{ cm}^{-2}$$

($\sim 6 \times 10^{-34}$ gm cm^{-3} in conventional units, conversion factor 0.742×10^{-28} cm gm^{-1}). The expansion of the Universe between then and now will have lowered the radiation density according to

$$\begin{pmatrix}\text{density of} \\ \text{radiation energy}\end{pmatrix} = \begin{pmatrix}\text{number density} \\ \text{of gravitations}\end{pmatrix} \begin{pmatrix}\text{energy per} \\ \text{graviton}\end{pmatrix} \propto a^{-4} \qquad (15.9)$$

so this estimate for the present-day energy density of these waves is $\sim 3 \times 10^{-36}$ gm cm^{-3}. This is much too small to contribute significantly to the effective density of mass-energy. But it has to be emphasized that ways of generating such long wavelength gravitational waves, other than those considered here, cannot be excluded. Indeed if the early universe were exceedingly "chaotic", such waves could even perhaps have a mass-energy density of 10^{-29} gm cm^{-3}. There might then be interesting effects on the appearance of clusters and groups of galaxies—the waves would induce apparent velocity gradients comparable with the differential Hubble recession over length scales $\lesssim$ (wavelength)[355]. In fact a mass-energy density $\sim 10^{-29}$ gm cm^{-3} in waves of $\gg 10$ Mpc wavelength would actually be incompatible with the accurate linearity of the Hubble law!

It is of interest to learn from the work of Misner[284,356] that there is one mode of gravitational radiation to which the formula (15.9) does not apply. This is the mode of oscillation of the geometry having the longest wavelength that can fit into a closed universe. This so-called "mixmaster" mode of oscillation leaves a homogeneous universe homogeneous, but changes an isotropic universe into one with different radii of curvature in the three principal directions of curvature at each point. The effective density of energy associated with this mode of excitation is

$$\begin{pmatrix} \text{effective density} \\ \text{of energy associated} \\ \text{with mixmaster} \\ \text{mode of oscillation} \end{pmatrix} \propto a^{-6} \tag{15.10}$$

a being defined so that a comoving volume element $\propto a^3$. We shall return in chapter 17 to the "mixmaster universe" in connection with the evidence on the present isotropy of our Universe.

15.7 BLACK HOLES IN INTERGALACTIC SPACE

The prospects of detecting isolated (non-accreting) black holes—argued already in section 5.8 to be exceedingly poor—become still worse when the search shifts from the vicinity of our own Galaxy to incomparably more remote regions of intergalactic space. Even a collapsed object of galactic mass ($\sim 10^{11} M_\odot$) has a Schwarzschild radius smaller than 10^{-2} parsecs. Formation of massive black holes could well be an important feature of the galaxy formation scenario outlined in chapter 14. Indeed, any region whose expansion was halted before t_{rec} would appear inexorably fated to collapse (like any supermassive star supported primarily by radiation pressure) unless rotation can stabilise it. And it is not only very large masses which could thus form: even masses $M \ll M_\odot$ could have collapsed at very early stages when $M_H \simeq M$, if the initial fluctuations had large enough amplitudes. Zeldovich and Novikov[357] have however emphasized the likelihood that any "seed" black hole in the primordial fireball would accrete rapaciously from its surroundings, to emerge at t_{rec} with a mass $M_1 \simeq 10^{15} M_\odot$ [Eq. (14.9)]. For this case the Schwarzschild radius is still only ~ 100 parsecs. Even if such objects contributed all the "missing mass", the separation between neighbours would be ~ 10 Mpc, and there is only a very small chance of sufficiently precise alignment with a background source to yield an efficient "gravitational lens" with large magnification. Since any lens preserves surface brightness, and the image angular diameter is

$$\sim (\text{Schwarzschild radius of lens})^{1/2}/(\text{distance of source})^{1/2}$$

$$\propto (\text{lens mass})^{1/2},$$

only sources whose unmagnified angular diameter is much less than this can be appreciably magnified, even for perfect alignment. But the probability that a given background source (assumed sufficiently compact) *should* be magnified by this mechanism is actually dependent only on $\Omega_{\text{lens}} \propto \varrho_{\text{lens}} M_{\text{lens}}$, and not on M_{lens} itself. If $\Omega_{\text{lens}} \simeq 1$, Refsdal[358] has shown that the apparent magnitude of a set of point-source "standard candles" at $z \gtrsim 1$ would be subjected to a random scatter. The probability of a magnification factor $f \gg 1$ is $\sim f^{-2}$. If the lens masses were $\sim 10^{11} M_\odot$, Refsdal's calculations would apply to quasars, whose radiative output is concentrated in a region at most a few parsecs in size, but normal galaxies would be too diffuse to be affected. However if the lens masses were $\gtrsim 10^{14} M_\odot$ the apparent brightness of galaxies would also be affected, and the scatter would introduce systematic errors into the determination of q.

Thus, even if intergalactic black holes do not manifest themselves as a result of accretion processes, they are in principle detectable—statistically if not individually—via the gravitational lens effect, if their masses exceed $10^{11} M_\odot$. However we cannot at present rule out a contribution $\Omega \simeq 1$ from such objects; and there seems no practical way at all of estimating the mass-energy density contributed by collapsed objects of mass $\ll 10^{11} M_\odot$ in intergalactic space, unless we were fortunate enough to find "lens images" of compact sources which could be resolved by very long baseline radio interferometers.

CHAPTER 16

Cosmic Background Radiation

DIFFUSE BACKGROUND radiation has been detected over almost 16 decades of frequency—from a few MHz up to $\sim 3 \times 10^{16}$ MHz (100 MeV)—and upper limits extend even beyond this range. Generally an important contribution comes from the Galactic Disc, but in some wavebands it has proved possible to isolate a truly cosmic isotropic component originating beyond our own Galaxy. Indeed in two bands—the x-ray band, and also, as we have seen (Chapter 13), the microwave part of the spectrum—the cosmic background is so strong that it swamps the emission from the Galaxy.

If one observes the sky with primitive equipment and sufficiently poor resolution (in any waveband), one will fail to resolve any sources, and so will detect nothing but background. But investigations of background radiation retain importance, especially for the cosmologist, even when one can also study discrete sources. There are two general reasons for this:

i) In some wavebands, most of the emission may not come from discrete sources at all. Other possibilities are:

a) emission from diffuse intergalactic material or cosmic ray particles;
b) emission from *pre*-galactic gas at large redshifts;

or, of course

c) "primordial" radiation, which is just as much a part of the initial conditions of the universe as the baryon content.

Much of the matter in the Universe may take the form of diffuse gas in metagalactic space (see Section 15.3), so attempts to determine the density and other properties of this gas by studying the radiation it absorbs or emits are of obvious cosmological importance. Also, even if the background is actually all attributable to discrete sources, these sources may be so intrinsically faint that they do not show up individually, even in surveys with "sophisticated" equipment. Only their integrated emission can then be studied. As a corollary, limits on the intensity of background radiation set important constraints on the luminosity function of faint sources.

ii) Whether it is contributed by discrete sources or not, any investigations of isotropic (and therefore extragalactic) background radiation are relevant to cosmology. This is because simple arguments of the type used in discussing Olbers' paradox shows that the bulk of any extragalactic radiation field comes at least from redshifts of order unity: and if evolutionary effects are important, the bulk of the observed background may have survived from $z \gg 1$.

We shall now summarize the observations of background radiation, recounting some proposed interpretations—of differing plausibility, but in no case definitive—starting at the longest wavelengths. Figure 51 summarizes all the data pertaining to the background in a schematic way.

16.1 RADIO BAND: $10^6 \lesssim \nu \lesssim 10^9$ Hz

Throughout this waveband the galactic background dominates, and it is impossible to isolate an extragalactic contribution without first *assuming* a spectrum. However the data are consistent with there being an extragalactic component in the 10^7–10^9 Hz band with a spectral index α in the range 0.7–0.9, and a brightness temperature at 178 MHz of 30 ± 7°K (if $\alpha = 0.7$) or 15 ± 3°K (if $\alpha = 0.9$)[332]. This radiation is probably synchrotron emission from discrete radio sources—in fact the integrated contribution from sources whose flux density exceeds the limit of the 5C surveys ($S_{408} \geqq 10^{-2}$ f.u) would produce about half of it, and have the right spectrum; and a plausible extrapolation of the $N(S)$ curve to still lower fluxes could account for it all. The fact that it is not stronger constrains the permitted evolution of intrinsically weak radio sources (see section 12.1).

Below 10 MHz there is evidence that the background spectrum flattens, or even cuts off[333,359]. This could reflect an intrinsic property of the source; but it could alternatively be due to free-free absorption. If the absorption occurred in intergalactic space, the latter would require a gas temperature $T \lesssim 10^4$ °K and a very high density—ie. $\Omega_{IGG} \gtrsim 1$. (If the gas were very clumpy, however, less would be needed).

16.2 MICROWAVE AND MILLIMETER BAND: $10^9 \lesssim \nu \lesssim 3 \times 10^{11}$ Hz

The strong isotropic background in this band, and its fundamental cosmological implications, were discussed in chapter 13. (See also Chapter 17 for a description of the isotropy data and their interpretation). We recall here only the important role this radiation plays in many high energy astrophysical processes by virtue of its high photon density. One such process—the production of background x-rays—is discussed in section 16.7. Another—the absorption of γ-rays by pair production — is mentioned again in Section 16.9.

16.3 INFRARED BAND: $3\times10^{11} \lesssim \nu \lesssim 3\times10^{14}$ Hz

With the exception of a tentative measurement at 100μ (3×10^{12} Hz) due to Houck and Harwit[360], only upper limits[361–363] to the extragalactic infrared background have so far been claimed (Figure 52). These observations must be made from balloons, rockets or satellites, because of severe absorption in the upper atmosphere due principally to water vapour, oxygen, ozone and carbon dioxide. This waveband is nevertheless, likely to be of great importance for high energy astrophysics, because some external galaxies (especially Seyfert galaxies) and quasars appear to emit most of their power in the infrared[109,373], having spectra which probably peak somewhere around 100μ. Figure 53 shows the observed spectra of several such objects. (There is no agreed interpretation of this phenomenon. Possibilities include: synchroton or inverse Compton radiation with absorption at wavelengths $\gtrsim$ 100μ; plasma oscillations, perhaps of matter falling onto collapsed objects; or stimulated molecular line emission from dense turbulent gas clouds in galactic nuclei).

There may also be a significant infrared background from interstellar grains, and from fine structure transitions in interstellar gas atoms in normal galaxies[374].

Measurements in the nearer infrared (1–10μ) could help to pin down the epoch of galaxy formation. When galaxies form, much of their gravitational binding energy must be dissipated and radiated in a free-fall timescale of $\sim 10^8$ years. For this reason (as well as because a large proportion of the first generation of stars will be 0 and B stars, which have low mass-to-light ratios and also evolve rapidly to the supernova stage) young galaxies should be very bright, radiating most of their energy in the near ultraviolet. If galaxy formation occurred at a redshift $\lesssim$ 10, most of this energy would be received in the near infrared.[375,376]

16.4 OPTICAL BAND: $3\times10^{14} \lesssim \nu \lesssim 10^{15}$ Hz

The extragalactic component of the background optical emission is swamped by the starlight from the Milky Way. The detection problem is aggravated by emission from the Earth's atmosphere, and the Zodiacal Light (solar radiation scattered by interplanetary dust). For these reasons, only upper limits to the extragalactic optical background are so far available. Roach and Smith[368] set a limit of $\lesssim$ 1 per cent of the total observed sky brightness at ground level. (The actual limit they quote refers to 5300 Å and corresponds, at this wavelength, to 5 tenth-magnitude stars per square degree). More recently, Lillie[369] has set a limit, from rocket observations, to the isotropic background at 4100 Å. If the extragalactic background spectrum is the same as that from our own Galaxy, this latter limit is stronger by a factor ~ 2 than Roach and Smith's (see Figure 52).

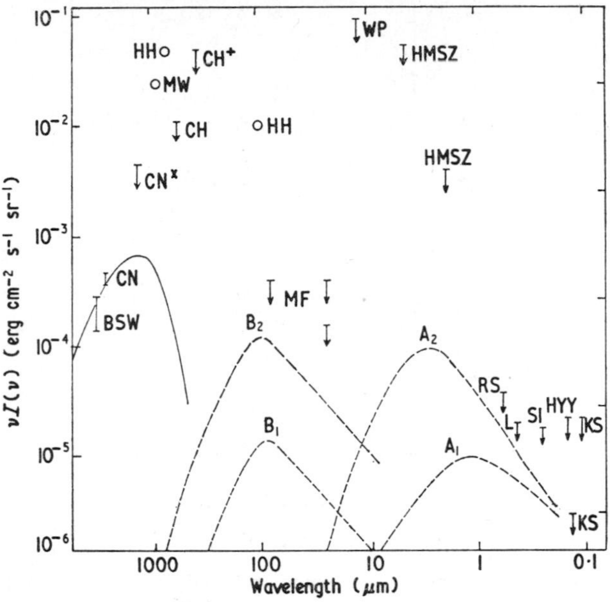

Figure 52 Upper limits to the infrared and optical background [from Peebles,[364] with additional points in the ultraviolet]. Vertical axis shows $\nu I(\nu)$ in (erg cm^{-2} s^{-1} st^{-1}).

A_1 estimate by Peebles and Partridge[365] of optical background due to galaxies.

A_2 as above, but including strong evolution.

B_1 infrared background due to infrared galaxies, without evolution, as estimated by Low and Tucker.[366]

B_2 as above, but with evolution.

The observations are due to:

BSW Boynton, Stokes and Wilkinson[216]

CN, CN*, CH, CH⁺ Bortolot, Clauser and Thaddeus[221]

HH Houck and Harwit[360,367]

MW Muehlner and Weiss[233]

MF McNutt and Feldman[361]

WP Walker and Price[362]

HMSZ Harwit, McNutt, Shivanandan, and Zajac[363]

RS Roach and Smith[368]

L Lillie[369]

SI Sudbury and Ingham[370]

HYY Hayawaka, Yamashito and Yoshioka[371]

KS Kurt and Sunyaev[372,340] (The later value, referring to wavelength shortward of Lyman α, yielding the *higher* limit.)

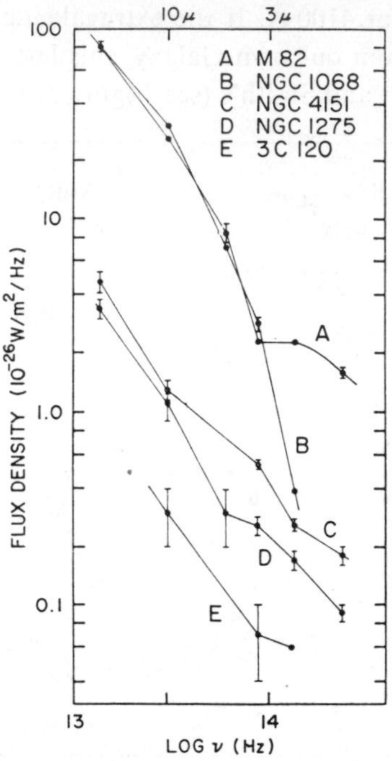

Figure 53 Infrared spectra of some external galaxies, according to Kleinmann and Low.[109]

These upper limits have been used by Peebles and Partridge[365] to set a *lower* limit to the mean mass-to-light ratio of the "missing mass" in the Universe. They find that, if the mass has a smoothed-out density $\sim 10^{-29}$ gm cm^{-3} ($\Omega \simeq 1$), this must exceed 80 solar units. (Of course all estimates of the integrated background from sources depend not only on the assumed evolution but also on the deceleration parameter, but the latter does not in practice introduce any large uncertainty into the estimates.)

According to Partridge and Peebles[375], and Weymann[376], the integrated emission from young galaxies would exceed the measured upper limits if galaxy formation occurred at a redshift less than 10.

16.5 NEAR ULTRAVIOLET BAND: $10^{15} \lesssim \nu \lesssim 3 \times 10^{15}$ Hz

Measurements in this band can be made from above the atmosphere. There are so far only upper limits[340,371,372] to the isotropic component (Figure 52). These limits are interesting because recent OAO results[377,378] show that galactic nuclei are often surprisingly intense emitters in the near ultraviolet. It is already possible to use the upper limits to set restrictions on the space densities of galaxies displaying this ultraviolet excess, and on permissible cosmological models.

These limits are also useful in connection with the density and temperature history of the intergalactic gas, especially redshifted Lyman α and helium line emission).

16.6 FAR ULTRAVIOLET BAND: $3\times10^{15} \lesssim \nu \lesssim 3\times10^{16}$ Hz

Between 912 Å (the Lyman limit) and ~ 100 Å the interstellar medium is so opaque that no extragalactic radiation penetrates to the solar vicinity, but an intergalactic radiation field in this waveband may have important indirect effects. In particular, as was first pointed out by Sunyaev[379], inward-spreading ionization fronts would tend to eat away hydrogen in the outlying regions of galaxies, and in the HI bridges (observed by the 21 cm line) between some pairs of galaxies. It does not seem possible at the present time to use the existence of these low-density HI regions to place firm limits on the far ultraviolet background, but the detection of a sharp edge to the HI profile of external galaxies could be adduced as evidence for an ionization front due to extragalactic ultraviolet radiation.

16.7 X-RAY BAND: $3\times10^{16} \lesssim \nu \lesssim 3\times10^{20}$ Hz

The x-ray band is the only one, apart from the microwave band, where a strong isotropic background dominates the contribution from our own Galaxy. This background was detected by the very first x-ray rocket experiment in 1962 (which also discovered the strong discrete source Sco X–1) and many observations have subsequently been made in various spectral regions (mainly using space vehicles, though at photon energies above ~ 30 keV (10^{19} Hz) the upper atmosphere is transparent enough to allow balloon observations, see Figure 6). An up-to-date summary of the available data is given in Figure 54.

Above 1 keV the spectrum is approximately fitted by a power law $I(\varepsilon) \propto \varepsilon^{-\alpha}$. The best estimate of α, yielding a good fit throughout the range 1–40 keV, is 0.7, though values as low as 0.3 have been quoted†. There appears to be a "break" in the spectrum at ~ 40 keV, because a straight extrapolation of the slope determined at lower energies lies significantly above the high energy x-ray measurements. The precise spectrum above

† These numbers refer to the *energy* spectrum. Sometimes the *photon number* spectrum is quoted, and this is of course one power steeper.

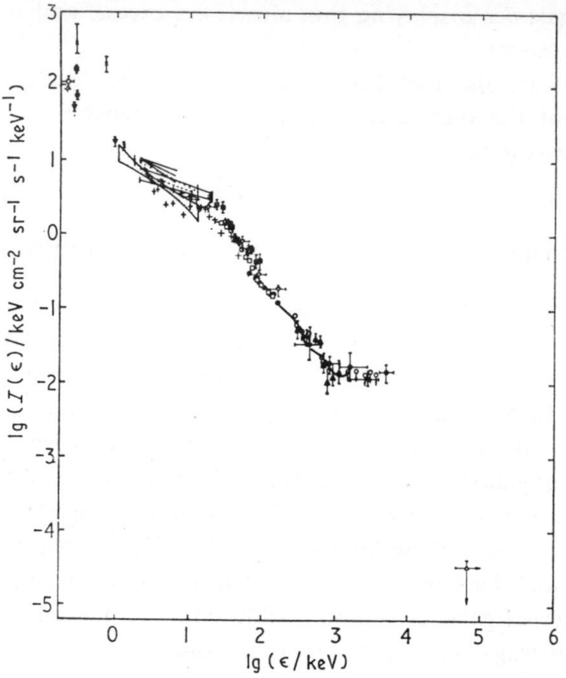

Figure 54 The x-ray background, reproduced from the review by Silk[380], to which the reader is referred for detailed references to the various data points plotted.

40 keV, however, is poorly determined. There is no reason at the moment to believe that the "break" is at all sharp, but it is a feature which any acceptable theory must explain.

The general power-law form of the spectrum over the 1 keV–1 MeV band indicates that a *non*-thermal process is operative. In contrast to the case of the microwave background, where it has not proved feasible to devise more than one plausible theory, it is all too easy for a competent astrophysicist to construct several models that are compatible with the existing x-ray data. The energy in the x-ray background is only $\sim 10^{-4}$ eV cm^{-3}—1 percent of intergalactic starlight, $\sim 10^{-2}$ percent of the microwave background, (see Table XXVII) and a mere 1 part in $10^8\,\Omega$ of the available rest-mass-energy. The problem is thus not an energetic one, but simply one of understanding how $\lesssim 1$ percent of the radiative output from sources† can be channelled into hard photons.

† Since any radiation present in the early dense phases of the universe would have been thermalized, there is no possibility of the x-ray background being primeval.

It could well be argued that theoretical discussions of the origin of the x-ray background is premature until we know more about extragalactic discrete sources. Until recently, the only such source definitely identified was M87[381] (with a power output $\sim 10^{43}$ erg sec^{-1}), but recently the UHURU satellite has detected[382,383] x-rays from a number of galaxies of different types, including NGC 5128 and the Seyfert galaxies NGC 4151 and NGC 1275 (the last two having x-ray powers of 10^{42} erg sec^{-1} and 10^{44} erg sec^{-1} respectively). There is evidence that all the nearby rich clusters of galaxies are extended sources. X-rays have also been confirmed from the QSO 3C 273. Until vastly more extensive data are available it will be impossible even to estimate the density and luminosity function of extragalactic sources; nor are the spectra of *any* extragalactic sources yet well determined. However it is quite conceivable that the bulk of the x-ray background is due to such sources. It is also conceivable that the x-rays are due to some still undiscovered class of discrete sources—for example, young supernovae in other galaxies. All discrete source models are, however constrained by the isotropy of the background on small angular scales. Schwartz[384] finds that the average fluctuations on an angular scale $\sim 15°$ amount to $\lesssim 3$ per cent, and Fabian and Sanford[385] find $\lesssim 5$ per cent fluctuations on a $5°$ angular scale. These results would be incompatible with a discrete source interpretation if the individual sources were too powerful, and too few in number. However an interpretation in terms of Seyfert galaxies ($\sim$ 1 per cent of all normal galaxies), or in terms of ordinary clusters of galaxies, is still tenable. These isotropy constraints, as well as the problems of producing the observed x-ray intensity, are eased if the sources evolve with epoch in such a manner that the main contribution to the flux comes large redshifts.†

Many papers have been written about theories of the KeV x-ray background which invoke emission mechanisms operating throughout intergalactic space, or at least in very extended diffuse regions. The most popular suggestion, due originally to Felten and Morrison[386], is that the x-ray photons result from "*inverse Compton scattering*" by relativistic electrons ("inverse" because in the laboratory frame, the electron *loses* energy to the scattered photon). The bulk of the radiation energy density in intergalactic space resides in the microwave background, so the main effect will be

† The isotropy of the x-ray background on large angular scales shows that not more than 5 per cent could come from a halo around our own galaxy.[384] These results also show that our peculiar velocity relative to the sources of the background is $\lesssim 800$ km sec^{-1}, and limit the anisotropy of the Hubble law in the same way as (though less accurately than) the microwave background isotropy measurements discussed in Chapter 17.

scattering of these very low energy ($\sim 10^{-3}$ eV) photons. When isotropic radiation is scattered by electrons of Lorentz factor γ, the photon energies are boosted by $\sim\frac{4}{3}\gamma^2$. Therefore, in order to obtain x-rays in the range 1 keV–1 MeV, electrons are needed with γ in the range 10^3–3×10^4. Such electrons are known to be present in extended radio sources, since this same range of γ is necessary in order to produce radio frequency synchroton radiation in magnetic fields of the strength 10^{-4}–10^{-6} G which is believed typical of radio galaxies. Thus there *must* be *some* contribution to the x-ray background from this effect, since there is independent evidence both for the soft photons and for electrons with the required energies.

But the pertinent questions are:

i) would the inverse Compton x-rays have the right *spectrum*?

and

ii) can there be enough electrons, either within radio sources or having escaped from defunct sources, to yield the observed x-ray *intensity*?

As regards the first question, the relativistic electrons in the radio sources are known (from the radio spectra) to have, in the mean, a power law spectrum $N(E)\,dE \propto E^{-2.5}\,dE$. The inverse Compton x-rays emitted concurrently by these electrons would have the same spectrum as the synchrotron radiation ($I(\nu) \propto \nu^{-0.75}$). However if the electrons escape from the sources with essentially their initial energy, and subsequently lose mose of their energy via the inverse Compton effect, the resulting x-ray spectrum would be $I(\nu) \propto \nu^{-1.25}$. (The inverse Compton lifetime of an electron moving through 3°K black body radiation is $\sim 10^{12}/\gamma$ years. Thus all the relevant electrons which have $\gamma \gtrsim 10^3$ would be *degraded* in much less than the Hubble time. This has the corollary that the inverse Compton effect is a highly efficient mechanism for producing x-rays—essentially 100 per cent of the energy of an electron moving in intergalactic space with $\gamma \gtrsim 1000$ emerges in this form). Though the inverse Compton model predicts a power law x-ray spectrum with roughly the right slope, there is no natural way of getting the apparent "break" at $\sim$ 40 keV, though some rather contrived models have been proposed which do succeed in reproducing this feature—either by assuming a corresponding "break" in the energy spectrum of the injected electrons, or by invoking some other loss mechanism (e.g. expansion) which is relatively more severe for the longer-lived electrons, and so tends so flatten the low energy x-ray spectrum.

Even if the *spectrum* can be fitted, the *intensity* of the observed x-ray background still poses something of a problem for the inverse Compton theory. Another related constraint is that the relativistic electrons must not produce too much synchroton radiation, or else the limits on the extragalactic radio background (see section 16.1) would be exceeded. This requires that $\gtrsim 1000$ times as much energy must go into x-rays as into radio emission (see Table XXVII), and leads to a limit $B \lesssim 2 \times 10^{-7}(1 + z)^2$ G on the mean magnetic field experienced by an electron, averaged over its inverse Compton lifetime. The factor $(1 + z)^2$, when z is the redshift at which the main x-ray production occurs, appears because the microwave background energy density is proportional to $(1 + z)^4$, and the permissible magnetic energy density can more or less be scaled in the same way. By appropriately choosing the parameters of radio sources (namely: weak magnetic fields $\lesssim 10^{-6}$ G and correspondingly large relativistic electron content; and lifetimes $\lesssim 10^6$ years) it is possible to construct models of the radio source population which can produce the observed ratio of energy densities in x-rays and in radio waves.[387–389] The cosmological evolution in the coordinate density of sources implied by the $N(S)$ relation (section 12.2), indicating that relativistic electrons are generated in greater profusion at earlier epochs, yields an advantageous factor, because the higher background temperature means that the inverse Compton process was more efficient relative to other loss mechanisms (e.g. synchrotron radiation) at large z.

Another diffuse process which has been proposed for the keV x-ray background involves bremsstrahlung by non-thermal electrons or protons moving through an intergalactic medium. In contrast to the inverse Compton process, this mechanism is highly *in*efficient because $\gtrsim 99$ per cent of the fast particles' energy goes directly into heating of the intergalactic gas. The energy requirements of this model seem exorbitant, even if the process is relegated to an early epoch.

There have recently been several observations of diffuse x-rays at energies $\lesssim 1$ keV. The interpretation of these results is difficult, because interstellar absorption is appreciable even at high galactic latitudes (and also because of the technical problem of obtaining a sufficiently transparent window in the detector). The reader is referred to the reviews by Setti and Rees,[390] or Silk[380] for a fuller discussion of these points. These observations are, however, of especial interest because there is some evidence of an *extra* contribution to the x-ray background below 1 keV (despite the uncertainties and inaccuracies, it seems that the $\frac{1}{4}$ keV flux, after correction for galactic attenuation, lies significantly above an extrapolation of the power-law spectrum obeyed at $\gtrsim 1$ keV (see Figure 54)).

Even if this extra contribution exists, its extragalactic origin and isotropy are not yet established beyond doubt, because the interstellar absorption introduces a large and uncertain direction-dependence which cannot be subtracted out precisely. However it has commonly been proposed that this "excess" soft x-ray emission is due to a diffuse intergalactic gas with density $\sim 10^{-5}$ particles cm^{-3} at 10^6 °K. If correct, this would be in the first (and, so far, the only) positive evidence for intergalactic matter (see section 15.3). There are, however, many other interpretations of the "excess" soft x-ray background. Thermal bremsstrahlung of the inferred intensity requires plasma whose emission measure $\int n_e^2 \, dl$ amounts to 0.3 cm^{-6} pc in a typical direction (with appropriate redshift factors if the main contribution comes from cosmological distances). This emission measure could be furnished by gas of density $\sim 10^{-4}$ particles cm^{-3} within clusters of galaxies (contributing only ~ 0.1 to Ω). Another possibility would be emission from the "halo" of our Galaxy (with density $\sim 10^{-2}$ cm^{-3} and dimensions $\sim 10^4$ pc) or possibly even smaller, denser and more localised regions of gas with $T \simeq 10^6$ °K, such as those formed behind shock fronts associated with the high velocity clouds. (See Rees *et al*[391] for further discussion of these possibilities.)

We confidently expect satellite observations during the next few years to clarify the present confusing situation below 1 keV. Such observations should also yield a more precise spectrum at harder energies, and pin down the position and sharpness of the apparent "break" at 40 keV. A genuinely sharp break would be exceedingly hard for any theory to explain, because all contributions to the background would need to have the same intrinsic spectrum, and to come from a narrow range of redshifts. It is also important to know the degree of isotropy, and the likely contribution from various types of extragalactic discrete source.

A variant of the inverse Compton model which can in principle account both for the keV background and for the "excess" soft x-ray background, has been described by Longair and Sunyaev.[392] In this picture, all the x-rays come from inverse Compton scattering by relativistic electrons generated in those galactic nuclei which are also strong infrared sources. The most energetic electrons lose a large fraction of their energy by scattering infrared photons during their escape from the nucleus, and produce a hard x-ray spectrum with a "break". The electrons then escape into intergalactic space, where they lose their remaining energy—an amount comparable to that already emitted—by scattering the microwave background photons. This energy will appear as soft x-rays. It is interesting that among the first extragalactic sources to be detected were Seyfert galaxies, which are intense infrared emitters *and* sources of relativistic electrons. These objects have a sufficiently high space density that fluctuations in the x-ray background would not be expected to have been observed in the experiments undertaken so far.

16.8 γ-RAY BAND: $\nu \gtrsim 3 \times 10^{20}$ Hz

The distinction between hard x-rays and γ-rays is a somewhat artificial one, but for the purposes of this exposition we draw the dividing line at 1 MeV.

Although celestial γ-rays have been detected in the 1–6 MeV band, and around 100 MeV, there is *no evidence* for any isotropic γ-ray background,† though the existing upper limits are already low enough to be interesting.

Vette *et al*[393] have reported cosmic 1–6 MeV γ-rays but their detectors were non-directional, so it is uncertain whether this flux is extragalactic or not. If so, it indicates a flatter spectrum rising above an extrapolation of the x-ray background spectrum.

Observations from the OSO III[394] and Cosmos-208[395] satellites allow us to set several upper limits to the isotropic background at various energies $\gtrsim$ 100 MeV. None of these limits fall below a straight (power law) extrapolation of the x-ray background spectrum.

It has long been realised that γ-ray astronomy is important in many spheres of high energy astrophysics, and even the present negative results are significant because a number of mechanisms may be envisaged which *could* yield a strong γ-ray background. Among these are the following:

i) The inverse Compton effect (or whatever other process—e.g., non-thermal bremsstrahlung—is responsible for the isotropic background in the 1 keV–1 MeV band) can equally well operate in the γ-ray band. In fact the 100 MeV upper limit due to Clark *et al*[394] lies on the extrapolation of a $\nu^{-1.2}$ spectrum from the hard x-ray region.

ii) Various γ-ray lines associated with nucleosynthetic processes may be detectable. In particular, if nickel is synthesized during silicon-burning, and transforms into iron via the chain

$$Ni^{56} \rightarrow Co^{56} \rightarrow Fe^{56}$$

one might observe the γ-rays associated with the radioactive decays (which have half-lives of 6.1 days and 77 days respectively) provided that they occur in an environment (e.g. a supernova envelope, as described in Chapter 6), which is sufficiently transparent.[48] Clayton and Silk[396] point out that, if all the iron in the Universe had been synthesized in this way, the associated γ-rays would constitute a background comparable in strength with that reported by Vette *et al.* The *spectrum* of the γ-ray background could then, in principle, tell us the epoch-dependence of the nucleosynthesis rate. How-

---

† Clark *et al.*[394] reported an isotropic flux in their detector but cannot exclude the possibility that this is caused by cosmic rays.

ever, since the hardest γ-ray that can arise in this manner is at 3.47 MeV, some other explanation would have to be sought for the excess observed in the 3.5–6 MeV channels.

iii) Levich and Sunyaev[397] show that a stochastic electromagnetic acceleration process, equivalent in quantum language to the "induced Compton effect", can heat a dilute gas in the vicinity of an intense radio or infrared source to relativistic temperatures. Moreover, Bisnovati-Kogan *et al*[398] show that a transparent relativistic plasma cannot attain a temperature exceeding $\sim$ 20 MeV, because pair production causes a catastrophic rise in the cooling rate as this temperature is approached. Thus relativistic thermal bremsstrahlung is a possible source of γ-rays. The expected characteristic cut-off at $\lesssim$ 20 MeV suggests this as another possible explanation of Vette *et al*'s results, if they are confirmed.

iv) At energies $\gtrsim$ 50 MeV, γ-rays resulting from π_0 decay may contribute to the background. The pions themselves could be generated in $p - p$ collisions between cosmic rays and thermal matter. The 100 MeV upper limits suffice to rule out a universal cosmic ray flux unless the intergalactic gas density is $\lesssim 10^{-6}$ cm^{-3}: if the cosmic rays had the same density throughout the Universe as they have in the galactic disc, and intergalactic gas contributed all the "missing mass", the γ-ray flux would exceed the observational limit by an order of magnitude. Any more definite statement requires specific cosmological assumptions.

(The possibility that cosmic rays are universal has recently received renewed consideration from several authors. There is no clinching argument against this view, at least for the case of the protons. Because of their short lifetime against inverse Compton losses, the *electrons* cannot be universal. Also, since the electron energy would all have been transformed into photons, observations of the background radiation set a limit to the mean energy, per cc, of relativistic electrons produced over a Hubble time. If the inverse Compton photons are all in the x-ray band, this limit is 10^{-4} eV cm^{-3}. Therefore it is impossible to produce the "universal" density of relativistic protons ($\sim$ 1 ev cm^{-3}) unless the acceleration mechanism favours protons over electrons by a factor $\gtrsim 10^4$).

Matter-antimatter annihilation is another source of π_0 γ-rays. About a third of the annihilation energy of a proton-antiproton pair goes into γ-rays with a mean energy of 180 MeV and a spectrum peaking at 70 MeV. The observed limits then tell us that only $\sim$ 1 part in $10^7\,\Omega$ of the matter in the Universe could have annihilated since the epoch ($z \simeq 100$) when the Universe became transparent to these photons (see section 16.9). (Stecker[399] has actually speculated that the 1–6 MeV x-ray background flux results from

π_0 decay γ-rays redshifted from $z \simeq 100$). No evidence has been found for the 0.511 MeV γ-rays (redshifted or unredshifted) from $e^+ - e^-$ annihilation, but this sets *less* sensitive limits to the amount of annihilation than the π_0 γ-rays. It is thcsc γ-ray limits which most severely constrain cosmological theories in which the Universe possesses overall matter-antimatter symmetry: they imply that matter and antimatter must, at least for $z \lesssim 100$, be separated on the scale of galaxies or even clusters. (See Steigman[400] for a fuller discussion of other consequences of cosmic antimatter).

v) Among many further processes (occurring either in discrete sources or in intergalactic space) which may contribute to the γ-ray background, we recall only the interesting possibility, suggested by Colgate[47] and already mentioned in Chapter 4, that a prompt "flash" of γ-rays may be emitted by a supernova explosion. The shock wave initiated by the collapse of the stellar core would become relativistic as it propagates out through the atmosphere. By the time it becomes visible, it would be moving with a Lorentz factor $\gamma \simeq 10^3$–10^4. The photons emitted by the heated material behind the shock would be blueshifted to energies $\gtrsim 10^3$ MeV, (and the "flash", being time-contracted by the Doppler effect, would last only a few microseconds). The integrated photon flux from all such phenomenon throughout the Universe could make an important contribution to the very high energy ($\sim$ 1 BeV) γ-ray background.

16.9 OPACITY EFFECTS ON BACKGROUND RADIATION

Before closing this Chapter on background radiation, we summarize the possible effects of intergalactic absorption in the various spectral bands. Note that *coherent* scattering (e.g., Thomson scattering of a photon with $h\nu \ll m_e c^2$), which would reduce the apparent strength of a discrete source, has *no effect* on the isotropic background.

Radio Free-free absorption ($\propto n_e^2 T_e^{-3/2}$) may be significant at frequencies up to 10 MHz if $n_e \simeq 10^{-5}$ cm^{-3} and $T_e \simeq 10^4$ °K.† There would be no significant absorption of radiation from discrete sources at frequencies $\gtrsim$ 10 MHz.

Microwave and millimeter No effect for $z \lesssim 100$ (distortions occurring at earlier epochs are mentioned in Chapter 17).

---

† See Section 15.3 for discussion of the likely temperature of intergalactic gas.

Infrared *Optical* *Near ultraviolet*	Unless there is an enormous amount of intergalactic dust (as Zwicky,[313] and Karachentsev and Lipovetski[401] have claimed), the only effect would be absorption of ultraviolet and visible light by dust within galaxies, which might intercept 10 per cent of the radiation from $z \lesssim 2$.[402]
Far ultraviolet *x-rays*	Possibly substantial photoelectric absorption by neutral hydrogen, neutral and singly ionized helium, and perhaps heavy ions, in intergalactic space.

γ-rays When $h\nu \sim m_e c^2$, a photon loses a substantial fraction of its energy when it undergoes Compton scattering by a thermal electron. This effect would reduce the intensity of the γ-ray (and hard x-ray) background if it originates at substantial redshifts.[403,404] If $\Omega_{IGG} \simeq 1$, unit optical depth to Thomson scattering [Eq. (15.2)] occurs at $z_{\tau=1} \simeq 7$. The maximum effect on the background due to the Compton recoil would then occur at photon energies $\sim m_e c^2(1 + z_{\tau=1})^{-1}$: at lower energies the recoil is insignificant, and at higher energies the Compton cross-section falls progressively below the Thomson limit.

High energy γ-rays can interact with soft background photons to produce pairs (each photon must have an energy $> m_e c^2$ in the centre-of-momentum frame). The consequent opacity[405] (shown in Figure 55) is more severe for

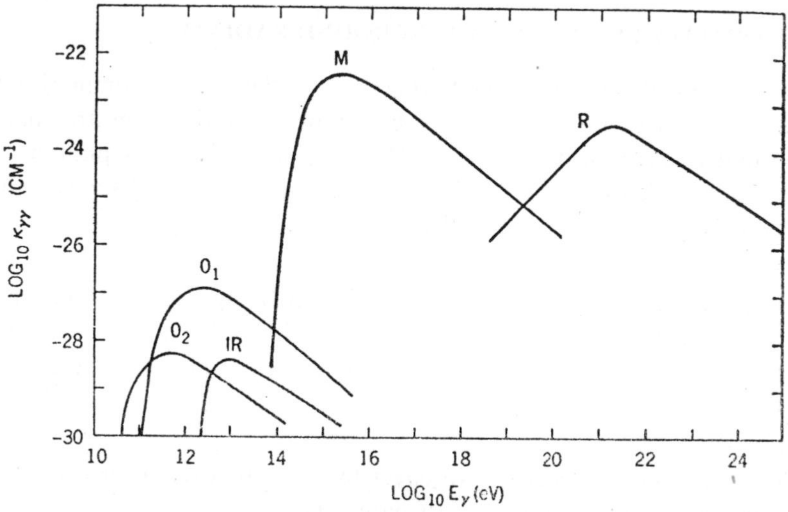

Figure 55 The opacity of the Universe to γ-rays due to pair production on background photons.[405] The various curves indicate the contributions to the opacity from different components of the background spectrum.

the higher-energy γ-rays, which can interact with lower-energy (and more numerous) background photons. Photons of $\gtrsim 10^{11}$ eV, which can produce pairs by interacting with optical background photons, cannot reach us from cosmological distances; and for photons of $\sim 10^{15}$ eV, which can interact with the microwave background, the mean free path is no larger than our Galaxy. (If the γ-ray background comes from a redshift z, the higher energy cut-off arising from interactions with the microwave background would be at an energy $\sim 10^{15}(1 + z)^{-2}$ eV).

CHAPTER 17

"Non-canonical" Models and the Very Early Universe

THE MAIN features of the "canonical" primordial fireball cosmology are:

i) *Isotropy*—implying that the expansion is characterised by a single parameter $a(t)$, and that all local aspects of the physics can be simulated by the material within a box whose sides all expand in proportion to this factor.

ii) *Approximate large-scale homogeneity*—in other words, the inhomogeneities which give rise to galaxies, etc. can always be treated merely as perturbations of the overall Friedmann model. This condition is fulfilled if the density irregularities on all scales always satisfy (14.1).

iii) *Expansion rate*—We have assumed, in particular in discussing the helium production and the distortions of the microwave background, that $a/\dot{a}$ is related to the mass-energy density by Einstein's equations.

iv) *Constant entropy*—The Universe has been assumed to evolve along a constant adiabat right from the "lepton era" and the epoch of helium formation until 'recent' times.

We now consider the evidence for these assumptions, and then outline some observational consequences of more general or unorthodox cosmologies.

17.1 EVIDENCE FOR THE ISOTROPY AND LARGE SCALE HOMOGENEITY OF THE UNIVERSE

The most important evidence on the isotropy of the Universe comes from the microwave background radiation. The temperature is uniform to $\lesssim 0.1$ per cent, over the areas of sky which have been surveyed, on all scales from ~ 3 arc minutes to 360°. These observations have already been mentioned in connection with discrete source models of the microwave background (Chapter 13). They set limits to 12 hour and 24 hour anisotropies, as well as to angular fluctuations on scales of a few arc minutes.

To interpret these limits, it is helpful to introduce the concept of the "last scattering surface". In a strictly isotropic universe, this can be defined as consisting of the material at a redshift z_{scat} such that $\tau_{es}(z_{\text{scat}}) = 1$. This surface, at a temperature $T(1 + z_{\text{scat}})$, can be regarded as the effective source of the microwave photons received at the Earth. The value of z_{scat} depends on the epoch at which the gas was reheated: in general, if $z_{\text{reheat}} \lesssim 8\,\Omega^{-1/3}$, then, from equation 15.2, $\tau_{es}(z_{\text{reheat}}) \lesssim 1$ and so we can see back to the recombination epoch at $z \simeq 1000$; but if $z_{\text{reheat}} \gtrsim 8\,\Omega^{-1/3}$, then $z_{\text{scat}} \simeq 8\,\Omega^{-1/3}$.

a. Small scale fluctuations

The absence of measured temperature fluctuations on small angular scales sets limits on the peculiar velocity, and on the gravitational potential irregularities, at $z \simeq z_{\text{scat}}$. However, the interpretation is complicated by the fact that the "last scattering surface" is not completely sharp. If the recombination proceeds gradually according to (13.8), then the "last scattering surface" has an effective thickness corresponding to the diameter of a region with mass $\sim 10^{15}\,\Omega^{-1/2} M_\odot$. ($\sim$ 10 per cent of the horizon size at that epoch). Thus, even if $z_{\text{scat}} \simeq 1000$, we expect the fluctuations associated with smaller scale inhomogeneities to be smeared out, because the photons we receive along a single line of sight would have undergone their last scattering in several uncorrelated regions. If $z_{\text{scat}} \ll 1000$, these effects are much *more* serious because the "last scattering surface" is even less sharply defined.

Taking account of the above "smearing" effect, Sunyaev and Zeldovich[407] have shown that, if $z_{\text{scat}} \simeq 1000$, the expected net doppler shifts at $z \simeq z_{\text{scat}}$ are

$$\frac{\Delta T}{T} \simeq 10^{-2} \left(\frac{\Delta \varrho}{\varrho}\right)_{z=z_{\text{scat}}} \left(\frac{M\Omega^{1/2}}{10^{15} M_\odot}\right)^{5/6} \tag{17.1}$$

for fluctuations associated with masses $M < 10^{15}\,\Omega^{-1/2} M_\odot$. For larger masses the smearing effects on the last scattering surface are less serious and the values of $\Delta T/T$ consequently larger. We can then distinguish two contributions to the temperature fluctuations, arising from (a) Doppler effects and (b) the adiabatic effects associated with the density fluctuations. These are respectively

$$\frac{\Delta T}{T} = 2 \times 10^{-2} \left(\frac{\Delta \varrho}{\varrho}\right)_{z=z_{\text{scat}}} \left(\frac{M\Omega^{1/2}}{10^{15} M_\odot}\right)^{1/3} \tag{17.2}$$

and

$$\frac{\Delta T}{T} = 10^{-3} \left(\frac{\Delta \varrho}{\varrho}\right)_{z=z_{\text{scat}}} (1 + 27\Omega) \left(\frac{M\Omega^{-1}}{10^{15} M_\odot}\right) \tag{17.3}$$

The fluctuations in the gravitational potential on the last scattering surface yield an additional contribution[408]

$$\frac{\Delta T}{T} = 2 \times 10^{-6} \left(\frac{\Delta \varrho}{\varrho}\right)_{z=z_{\rm scat}} \left(\frac{M}{10^{15} M_\odot}\right)^{2/3} \tag{17.4}$$

which is in fact the dominant one on scales exceeding the horizon size at $z_{\rm scat}$. The angular size associated with a fluctuation of mass M is

$$\theta = 10 \left(\frac{M\Omega^2}{10^{15} M_\odot}\right)^{1/3} \text{ arc min} \tag{17.5}$$

(θ is almost independent of $z_{\rm scat}$ provided that—as is true in all circumstances—$z_{\rm scat} \gtrsim \Omega^{-1}$.)

During the post-recombination epoch, density perturbations grow no faster than $\Delta\varrho/\varrho \propto (1 + z)^{-1}$, (see section 14.1), so in order to have condensed out by the present epoch the amplitudes at $z \simeq 1000$ must have been $\gtrsim 10^{-3}$. However, one finds that the predicted fluctuation amplitudes associated with embryonic galaxies and clusters would be below the present threshold of detectability, and may even be swamped by the background fluctuations due to discrete radio sources or to irregular reheating of the inter-galactic gas. The limits to $\Delta T/T$ *do* however set constraints on very large-scale ($\gg 10^{15} M_\odot$) irregularities.

If $z_{\rm scat} \ll 1000$, the predicted fluctuations on all scales $\lesssim 10^{18} M_\odot$ would be far below (17.1)–(17.4), and the inferences from the present limits are correspondingly less interesting (except on very massive scales indeed).

The above effects refer to processes occurring on the "last scattering surface". There is an additional effect, whereby *transparent* regions of enhanced density (with $z < z_{\rm scat}$) can perturb the temperature.[409] This is quite negligible except for scales $\gg 10^{15} M_\odot$, and arises from two competing effects: (a) the *increased travel time* of photons induced by the gravitational potential well, which implies that the photons we receive must have undergone their last scattering closer to us, and thus be *blue*-shifted relative to radiation arriving from other directions; and (b) the fact that, if $\Delta\varrho/\varrho$ is increasing with time, the potential well is *deepening* as the photons pass through it, which causes a relative *red*-shift. The net $\Delta T/T$ can be of either sign.

b. Implications of 12 hour and 24 hour isotropy

If the Hubble expression were anisotropic, then the factor by which the Universe had expanded since the epoch of last scattering would be different in different directions. Thus the observed background temperature would be *higher* in the directions where the expansion was *slower*.

The quoted limit of

$$\frac{\Delta T}{T} = (0.55 \pm 0.69) \times 10^{-3}$$

to the 12-hour anisotropy[410] severely constrains the types of anisotropic cosmological models which could apply to the actual universe.[411] In the simplest of the anisotropic models, the anisotropy in the expansion rate varies with z like $(1 + z)^{3/2}$. The observations would then imply a present fractional anisotropy in the Hubble constant of $\lesssim 10^{-3}(1 + z_{\text{scat}})^{-3/2}$. The z-dependence of the anisotropy is actually somewhat more complicated because the neutrinos, which would have decoupled when $T \gtrsim 10^{10}$ °K, exert an anisotropic pressure. Taking this into account, Rasband[411] finds that if $z_{\text{scat}} \simeq 1000$ the measured microwave isotropy implies at the 80 per cent confidence level that the present anisotropy is either below 5×10^{-1} or else close to 3.8×10^{-6}. (The anisotropy at epochs $z \gg z_{\text{scat}}$ may still, however be sufficient to affect processes such as helium formation).

The limit on the 24-hour anisotropy set by the Princeton Group,[410] of

$$\frac{\Delta T}{T} = (0.40 \pm 0.56) \times 10^{-3},$$

restricts the possible rotation of the universe as a whole, which is no doubt encouraging to advocates of "Mach's Principle". It also places a limit on the peculiar velocity of the earth—i.e. its velocity relative to the distant matter which constitutes the "last scattering surface". An observer moving with peculiar velocity v through a black body radiation field at temperature T still sees a black body spectrum in any direction, but the angular distribution of temperature is

$$T(\theta) = T\left(1 - \frac{v^2}{c^2}\right)^{1/2}\left(1 - \frac{v}{c}\cos\theta\right)^{-1} \qquad (17.6)$$

where θ, measured in the moving frame, is the angle between v and the direction of observation. The 24-hour isotropy data therefore imply $v \lesssim 300$ km sec^{-1}. We expect some contribution to v from the sun's motion around the galactic centre at 200–250 km sec^{-1}. It is likely that the galactic centre is itself moving relative to the centre-of-mass of the Local Groups of galaxies (of which our own Galaxy and Andromeda are the dominant members). This velocity is estimated to be ~ 80 km sec^{-1}, and the vector sum of our various velocity components relative to the Local Group is about 300 km sec^{-1} towards a direction whose galactic coordinates are $l_{\text{II}} \simeq 95°$, $b_{\text{II}} \simeq -8°$. It is of course possible that the Local Group as a whole moves with respect to distant matter, and there have been some attempts[412,413] to estimate the contribution arising from membership of the so-called "local supercluster"—a disk-like (and allegedly rotating) agglomeration of groups of galaxies centered on the Virgo Cluster.

Observing from a balloon at $\sim$ 3 cm wavelength, Henry[237] finds a velocity 320 ± 100 km sec^{-1} towards 10.5 h, $-20°$. Conklin,[238] working at 3.7 cm, also finds a positive effect with a component (210 ± 100) km sec^{-1} in the earth's equatorial plane and a direction consistent with Henry's result. The claimed velocity agrees with that predicted by de Vaucouleurs and Peters' analysis[413] of the dynamics of the "local supercluster". However, because of the uncertain correction for the emission from the galactic disc, it is prudent to regard these results merely as upper limits.

The remarkable message of these observations is, of course, the fact that v is so small.† This implies that the universe displays a remarkable degree of large scale uniformity, and that there are no substantial deviations from the Hubble law such as would be induced by very large-scale density inhomogeneities.

If we accept that the microwave background is primordial, then the 12 and 24 hour isotropy measurements tell us—with vastly higher precision than any other data—that the overall Hubble expansion has been exceedingly isotropic, at least since the epoch corresponding to $z \lesssim z_{scat}$. The small-scale measurements imply that, again for $z \lesssim z_{scat}$, the departures from homogeneity are "weak" in the the sense that they do not violate the inequality (14.1) (and the 24 hour data also show that we are not ourselves located within a "strong" inhomogeneity). These results give us much greater confidence in the relevance of models based on the Robertson-Walker metric than would have been warranted before the discovery of the microwave background.

The conventional view (as recounted in Chapter 14) supposes that, in the unobservable early phases, the universe was even smoother then the part we observe, the protogalaxies and protoclusters then being merely regions of infinitesimally enhanced density. However, it is possible, in principle, to take a contrary view, and suppose that the universe deviated drastically from homogeneity and isotropy when $z \gg z_{scat}$ but became progressively smoother and more isotropic as it expanded. We now consider some indirect manifestations of departures from the canonical expansion parameters, or of a "chaotic" initial state—in particular, its effects on the "fossil" helium abundance and on the microwave background spectrum.

---

† Note that this conclusion requires only that the microwave background originates at cosmological distances—it need not necessarily be primordial radiation. In fact, observations of the x-ray background (which is certainly in no sense primordial) yield a limit $v < 800$ km sec^{-1}.[414]

17.2 HELIUM PRODUCTION IN NON-CANONICAL MODELS

Our discussion of primordial nucleosynthesis (section 13.3) showed that the *expansion rate* plays a key role in determining the helium abundance. One characteristic property of anisotropic cosmologies is that the expansion rate at a given temperature is greatly speeded up as compared with the rate (13.6) which applies in the canonical model. (An expansion rate exceeding (13.6) also occurs in some Jordan-Brans-Dicke cosmologies where the effective value of G was higher in the past; and the presence of extra particles or fields, or of other species of leptons, all adding to the energy density, would speed up the expansion even in ordinary isotropic relativistic models).

Any speed-up in the expansion tends to *increase* the "frozen in" fraction of neutrons, because thermal equilibrium breaks down at a higher temperature than in the canonical model. Provided that there is still time for the neutrons to be processed into helium, this *raises* the primordial helium abundance. However, if the time-scale is speeded up by a factor $\gtrsim 10^6$, there is no time for these further reactions to proceed: there would then be *no* primordial helium (the neutrons instead decaying freely at $t \simeq 1000$ sec). Peebles[261] has given a useful set of graphs (Figure 56) for the helium and deuterium abundance as a function of the speed-up factor $S = \tau_{\text{exp}}/(\tau_{\text{exp}})_0$, where $(\tau_{\text{exp}})_0$ is the expansion time-scale given by (13.6). Thus, a *modest*

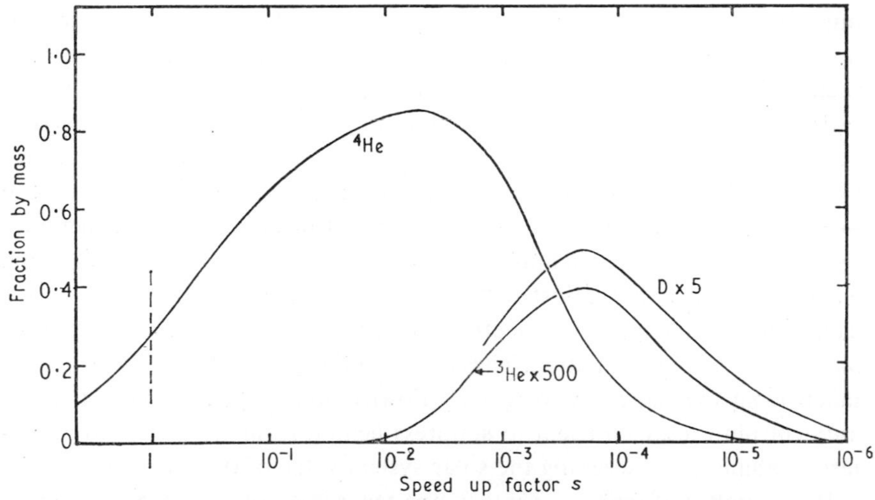

Figure 56 The predicted abundance of ^{4}He, ^{3}He and D as a function of the speed-up factor S for a universe with $\Omega \simeq 1$ and present temperature 3°K (from Peebles[261]). In order to include them on the figure, the predicted abundances of D and ^{3}He have been multiplied by factors of 5 and 500 respectively.

speed-up can increase the primeval helium abundance to $\sim$ 80 per cent, but a *drastic* speed-up reduces it to zero. One needs $S \lesssim 10^{-6}$ in order to both suppress the helium and bring the deuterium abundance below the solar system value of $\sim$ 0.02 per cent.

Thorne[415] has shown that it is *just* possible to devise anisotropic models in which the expansion is speeded up sufficiently to suppress the helium and the deuterium, but the anisotropy has nevertheless decayed by z_{scat} to a magnitude compatible with the microwave background isotropy. Zero helium abundance is also possible in some Brans-Dicke models. (It is clear from Figure 56 that a modest *slow-down* in the expansion rate could also reduce the helium abundance, but it is harder to think of ways of doing this.)

Figure 56 also shows that a moderate increase in the expansion rate, with S in the range 0.3–10^{-3}, would definitely produce *too much* helium, with more than 40–50 per cent helium (by mass) emerging from the fireball. There being no known way whereby the extra helium could be destroyed during the lifetime of the Galaxy, this would be incompatable with present observations. Thus one can set limits to the energy density in the fireball in the form of new species of non-interacting particles, gravitons, etc.[416] However, the limits that can be set on primordial *neutrinos* are rather more complicated.

If, as an "initial condition", the universe possessed a non-zero lepton number leading to an excess of ν_e or $\bar{\nu}_e$, there would be an associated speed-up in the expansion. But an even more severe effect on the helium abundance then arises from microphysical processes which change the neutron/proton ratio. A preponderence of ν_e would lead to rapid reactions of the type $\nu_e + n \to p + e^-$, converting the neutrons into protons; similarly, a preponderence of $\bar{\nu}_e$ would rapidly deplete the protons via the reaction $\bar{\nu}_e + p \to n + e^+$. A *large* excess ($\gg 10^8$ per baryon) of either ν_e or $\bar{\nu}_e$ therefore reduces the primordial helium abundance, even though it speeds up the expansion. A suitably adjusted "moderate" ($\sim 10^8$ per baryon) $\bar{\nu}_e$ excess can, however, maintain the neutron/proton ratio close to unity and thus *increase* the helium formation. These possibilities are illustrated in Figure 57, taken from the paper by Wagoner, Fowler and Hoyle.[262] Note that an antineutrino excess sufficient to suppress helium formation leads to a deuterium abundance exceeding the solar system value of 0.02 per cent.

It is a matter of opinion whether one regards a non-zero lepton number as a palatable feature of the initial conditions imposed on the "big bang". Fowler[417] has expressed the view that a neutrino or antineutrino excess of the same order as the photon density is perfectly acceptable. But one can equally well argue, contrarywise, that the lepton number and photon number are by no means on the same footing, because the former is strictly conserved in all but black hole processes, whereas the latter can be augmented by dissipative effects.

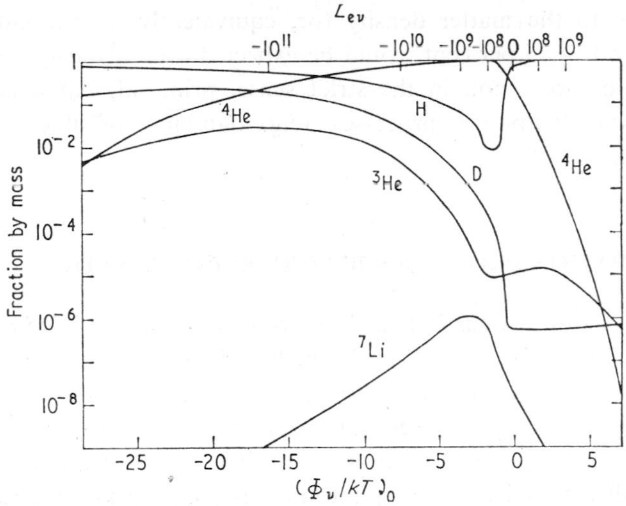

Figure 57 Element production in a universe with $\Omega = 1$ and present temperature 3°K in the case of neutrino ($\Phi_\nu > 0$) or antineutrino ($\Phi_\nu < 0$) degeneracy. Φ_ν is the Fermi level of the ν_e and $L_{e\nu}$ is the ratio of electron lepton number to baryon number.

Even if the universe had a "hot" origin, it is possible that some of the presently observed background radiation was generated by dissipative processes occurring after helium formation (i.e. at $\gtrsim$ 1000 sec) but still early enough to have been thermalized. The Universe would then have been evolving along a lower-entropy adiabat during the epoch of helium formation—in other words, there would be a higher density at a given temperature than in the canonical model. This tends to increase the helium abundance somewhat, but the effect is not drastic unless the entropy is so low that the expansion dynamics are matter-dominated at the helium formation epoch [so that the time scale at a given temperature is shorter than (13.6)], or the electrons are degenerate.

To summarize these results, the primeval helium production is highly sensitive to the expansion time scale. Only an expansion rate close to the canonical one leads to ~ 25 per cent helium by mass. Most "modified" cosmologies lead to *too much* helium, and can therefore be ruled out. There are some relativistic cosmologies, however, in which the expansion proceeds so rapidly (as a result of extreme anisotropy, for example) that no helium is formed. These cannot be ruled out, but one is then faced with the problem of explaining the helium content of our Galaxy in some other manner. The helium abundance would also change from $\sim$ 25 per cent if there were an 'excess' of degenerate neutrinos or antineutrinos. The helium abundance is insensitive to the matter density (or, equivalently, to the entropy) at $\sim 10^9$ °K, and $\sim$ 25 per cent would be expected even if most of the black body radiation were not, in the strict sense, primeval, but instead were generated via dissipative processes (e.g. damping of fluctuations) at $t \gtrsim 1000$ sec.

17.3 DISSIPATION PROCESSES BEFORE RECOMBINATION

As outlined in the discussion leading to Eqs. (15.7) and (15.8), energy injected into the universe at "late" epochs distorts the microwave background spectrum from a true black body, and this places limits on the amount of energy injected into the universe at $z \gtrsim 1000$. The reason is that the main energy loss is via repeated Compton scattering, which increases the radiation energy density but conserves photon numbers. Processes such as bremsstrahlung can generate some new photons, but are not efficient enough to produce a true black body spectrum at the temperature appropriate to the enhanced energy density. The photon production processes, being "two body" phenomena, will tend to be more efficient during the earlier denser phases of the expansion (even allowing for the shorter time available during these phases). Energy generated at sufficiently early epochs can therefore be more readily thermalized. The quantitative details of the thermalization have an important bearing on how much energy could have been dissipated (via, for example, shock waves or photon viscosity) at various stages, and thus on how irregular or turbulent the early universe could have been.

Sunyaev and Zeldovich[348,407] have considered this problem fully. They show that any energy injected at red-shifts

$$z \gtrsim 10^5 \, \Omega^{-4} \tag{17.7}$$

would (for all practical purposes) be thermalized, leading to a Planckian form for the background spectrum. However, at smaller redshifts the thermalization would be less efficient, and if the material in the universe is assumed homogeneous the absence of significant distortions on the Rayleigh-Jeans part of the microwave spectrum places an upper limit of

$$g = \int \frac{Q(t)\, dt}{aT_r^4(t)} \lesssim 2 \times 10^{-2} \, \Omega^{7/8} \tag{17.8}$$

to the energy production rate $Q(t)$ per unit volume within the redshift range $10^4\,\Omega^{-1/2} \lesssim z \lesssim 5.4 \times 10^4\,\Omega^{-6/5}$. This inequality sets a limit to the amplitude of primordial turbulence or (oscillatory) adiabatic perturbations which might have dissipated just prior to recombination. (Note, however, that the photon production rate depends on the *mean square* density, so (17.8) could be relaxed, and more dissipation permitted, if the plasma was already clumpy.)

The implications of (17.8) for various types of dissipative processes have been discussed by Sunyaev and Zeldovich in several papers.[286,348,407] At epochs early enough to satisfy (17.7), dissipation (even if it involves non-linear processes) is likely to be limited to small scales, because the horizon mass M_H is itself small at early times. When (17.7) holds, any energy dissipated can be thermalized—and we cannot rule out the possibility that *the*

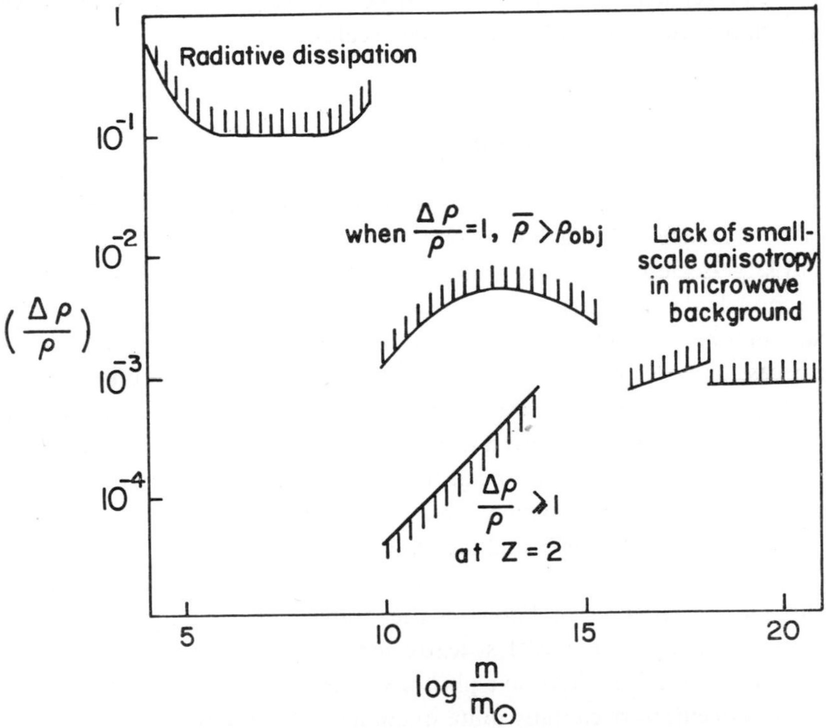

Figure 58 Limits on the amplitude of adiabatic fluctuations at the recombination epoch for a model with $\Omega = 1$.[286] The upper limits come from (a) the undistorted form of microwave background spectrum, (b) the low density of observed bound systems, and (c) the isotropy of the microwave background on smaller angular scales. The lower limit is set by the fact that the bound systems have already stopped expanding.

bulk of the thermal microwave background results from dissipation of small-scale, but large-amplitude, irregularities when $z \gtrsim 10^5 \Omega^{-4}$ (or $M_H \lesssim 10^{17} \Omega^{11} M_\odot$). This would of course modify the early thermal history shown in Figure 46. On the other hand, (17.8) does limit the initial amplitude of larger-scale adiabatic fluctuations which (in the manner described in Chapter 14) damp out at redshifts too small to satisfy (17.7). The consequent constraints on the spectrum of adiabatic fluctuations are plotted in Figure 58. Also plotted in this diagram are the *upper limits* on larger scales, inferred via (17.1)–(17.4) (assuming $z_{\text{scat}} \simeq 1000$) from the microwave background isotropy; as well as the *lower limits* inferred from the existence of bound systems, assumed to have formed from adiabatic fluctuations by gravitational instability according to (14.2). These results have the interesting consequence of ruling out any single smooth power-law spectrum of primordial perturbations. It appears that any such spectrum must be "truncated" at bigger masses ($> 10^{20} M_\odot$), to be consistent with the microwave background anisotropy on larger angular scales.

17.4 THE VERY EARLY UNIVERSE

The main conceptual problem in cosmology is not, perhaps, the presently observed *inhomogeneity* (galaxies, clusters, etc.) but, on the contrary, the overall large-scale *homogeneity* and *isotropy*. The difficulty is that in Friedmann models the mass M_H within a particle horizon shrinks to zero as one extrapolates back towards the initial instant. The "canonical" models therefore necessitate the unpalatable postulate that all parts of the universe are accurately synchronized, and start expanding at the same time and with the same entropy and curvature, even though there was then no communication or causal connection between neighbouring regions. In these Friedmann models, therefore, the observed overall uniformity is postulated, never explained; and is even unexplainable. Matter at one point, however willing to live and evolve in conformity with matter at another point in the Universe, has no means whatever of knowing the "conditions of life" beyond its own horizon. The parts of the "last scattering surface" in different directions, from which we receive microwave photons agreeing in temperature to better than 0.1 per cent, were causally quite unconnected at the time this radiation was emitted.

But there is no direct evidence that the Robertson-Walker metric is even remotely applicable arbitrarily close to $t = 0$, and the horizon may behave quite differently in more general models. Though it is not feasible to calculate the dynamics of a fully inhomogeneous universe, much attention has recently been devoted to homogeneous but *anisotropic* cosmologies, whose horizon structure is indeed found to differ qualitatively from the Friedmann case.

The anisotropy oscillations of the geometry of a closed model universe, small though they may be at any moment late in the history of expansion, grow, if they have a finite amplitude at all, larger and larger in importance as one follows the history of the system back in time, until the earliest days when they dominated every other form of excitation [from the a^{-6} dependence of (15.7)]. By comparison, the effective energy density of matter ($\propto a^{-3}$) and radiation ($\propto a^{-4}$) become negligible. The resulting idealised model universe has three degrees of freedom, corresponding to the radii of curvature along the three principal axes at any point. Three second-order ordinary differential equations govern the time rate of change of the three radii. The mechanical problem is reminiscent in many ways of the well-known problem of a rotating ellipsoid with three unequal moments of inertia.

Studies of this so-called "Mixmaster" universe (and of anisotropic models in general) were motivated by the hope that the anisotropy vibrations might enable photons—and so perhaps pressure waves as well—to make their way round the universe (travelling, during each phase of the oscillation, along the shortest available route) even at arbitrarily early times. This mixing could then homogenize the material content; and the anisotropy itself would decay automatically [a^{-6} dependence of (15.7)] during the expansion, being perhaps further attenuated by neutrino viscosity when $T \simeq 10^{10}$ °K. Thus, according to Misner and his associates, one needs to invoke drastic initial *anisotropy*, and perhaps inhomogeneity, to explain why the Universe is now, on large scales, so *isotropic*! Unfortunately the most detailed recent work suggests that only in rather special cases do the horizons in the mixmaster universe display the requisite behaviour for fully effective mixing, and this may indicate that some more complex model is called for. Also, some modes of anisotropy actually *amplify* during the expansion. Nevertheless the Mixmaster model raises many new questions, among which are the following:

i) How early in the history did the anisotropy oscillations fall to an amplitude where they no longer played a great part in cosmology?†

ii) How effective was it earlier in coupling neutrinos with matter and other radiation fields?[284,418]

---

† Note that when "anisotropy energy" dominates the dynamics then—in one sense at least—a simplification is introduced because the equation of state of matter and radiation, uncertain and controversial at high densities, is quite immaterial.

iii) Did the anisotropic curvatures have anything to do with the production of particles during a still earlier epoch?[419–422]

iv) How do initial small perturbations of density and velocity in the Mixmaster universe grow with time, and how would this affect our discussions in Chapter 14?

The character of the initial singularity in these more general and realistic models may be grossly different from the Friedmann situation—indeed the latter probably constitutes a "set of measure zero" in this respect, because sufficiently early stages are bound to be dominated by "anisotropy energy" unless, as in the Friedmann models, this is strictly zero. As shown by Hawking and Ellis,[423] however, an almost incontrovertable consequence of Einstein's theory is that, irrespective of the degree of uniformity in the early Universe there *must* have been a singularity in the past. This conclusion follows from the observed isotropy back to $z \simeq z_{scat}$—from which one can infer the presence of a "closed trapped surface" at $z \lesssim z_{scat}$—subject only to very general causality conditions, together with the restriction that the equation of state satisfies $p + \varrho \geqq 0$ and $3p + \varrho \geqq 0$.† This singularity theorem does not necessarily imply a "real" physical singularity, but it *does* imply conditions so extreme that Einstein's theory is no longer applicable (maybe because of quantum effects?). Nor can one even deduce that *all* the matter in the universe experienced these extreme conditions. It is conceivable that most of the matter previously underwent a contracting phase, after which a "bounce" ensued. However, the occurrence of singularities implies break-down of causality through the "bounce"—there is then no reason to believe that thermodynamic quantities, nor even the "fundamental" constants of microphysics, would have been conserved.

If we live in a closed Universe ($\Omega > 1$) the theory predicts that the eventual recontraction, during which clusters, galaxies and stars are successively destroyed and engulfed in a new "fireball", must terminate in another singularity (or singularities).

† If the material is not a perfect fluid, these conditions have the more general form: for every time-like vector V^a at each point, $R_{ab}V^aV^b \geqq 0$.

CHAPTER 18

The "Large Numbers": Coincidence or Consequence?

THE DYNAMICS of closed Friedmann universes are characterised by the parameter a_0, which measures the maximum radius attained before recontraction and the duration of the complete "cycle". The primordial fireball hypothesis introduces a second parameter, the initial entropy per particle $\mathcal{S} \simeq n\gamma/n_p$. (For our Universe this is $\sim 10^7$. The entropy generated by stars and other discrete objects up to the present epoch would not have increased this significantly, even though this small additional entropy—consequent to the fragmentation of debris from the fireball into gravitating bound systems—is a precondition for the existence of observers and other manifestations of thermodynamic non-equilibrium!)

These two parameters seem quite unrelated to the other basic constants of nature—which we have assumed throughout to be genuine 'constants' during the evolution of the Universe, allowing us to discuss quantitatively the local physics of the fireball at least back to $t \simeq 10^{-6}$ seconds. Certain well-known "coincidences" involving numbers of order 10^{40} (and simple powers thereof) have often been held to imply some intimate interrelation between microphysics and cosmology. The best known of these are the following.

The ratio of the electric to the gravitational forces between pairs of charged elementary particles is in the range 10^{36}–10^{42}: for two protons it is

$$N_1 = \frac{e^2}{Gm_p^2} = 10^{36} \simeq 10^{40} \qquad (18.1)$$

This ratio derives purely from laboratory physics. But a prima facie unrelated dimensionless ratio involving cosmology, the ratio of the Hubble radius $c\tau_H$ to the classical electron radius, is of the same order.

$$N_2 = \frac{c\tau_H}{(e^2/m_e c^2)} \simeq 10^{40} \simeq N_1, \qquad (18.2)$$

A fascinating line of argument, due principally to Dicke[424] and Carter,[425] suggests that, instead of calling for some time-dependence of the microphysical constants (as was argued by Dirac[426]), the number N_2 depends on epoch, and that the existence of observers would only be expected when it *is* comparable with the ratio N_1. The reasoning proceeds as follows.

The largest mass capable of supporting itself against collapse by degeneracy pressure—the "Landau mass" or "Chandrasekhar mass"—is (to within a factor of order unity depending on molecular weight, etc.)

$$M_L \simeq \left(\frac{e^2}{\hbar c}\right)^{-3/2} N_1^{3/2}\, m_p \tag{18.3}$$

The masses of all stars lie within one or two orders of magnitude of M_L. A body with mass $\lesssim 0.1 M_L$ does not get hot enough for hydrogen burning to occur at its center (the factor ~ 0.1 represents a product of microphysical constants which are all "of order unity", at least in the perspective of numbers like 10^{40}). Stars more massive than $M_{\max} \simeq 30 M_L$ (where the factor 30 is of purely arithmetical origin) would be supported by radiation pressure rather than gas pressure, and consequently prone to numerous instabilities. $M_{\max}$ is therefore an upper limit to the mass of a main sequence star. Jordan[427], who was the first to notice this coincidence, ascribed cosmological significance to it. However it is purely a deduction from stellar physics. The lifetime t_{star} of a main sequence star is roughly

$$\frac{(\text{nuclear energy available}) \times (\text{time for a photon to diffuse out of star})}{(\text{energy of radiation trapped within star})}$$

The stars with shortest lifetimes are those with $M \simeq M_{\max}$. For these, Thomson scattering provides the main opacity, and one finds

$$ct_{\text{star}} \simeq \eta \left(\frac{m_p}{m_e}\right) N_1 \left(\frac{e^2}{mc^2}\right). \tag{18.4}$$

η is the fraction of rest mass energy released in hydrogen burning (~ 0.007) so the factor $\eta(m_p/m_e)$ is of order 10.

If it were the case that

i) Galaxy formation occurs when the age of the Universe is $t \lesssim t_{\text{star}}$; and

ii) The evolution of "intelligent" observers occupies a time $\gtrsim t_{\text{star}}$,

then (18.4), together with the fact that a galaxy would "die", with most of its material trapped in degenerate cold stars or black holes, after a few generations of stars have had time to evolve, automatically implies that t must be comparable with t_{star} when observers become cognizant of the Universe. Since t and τ_H are comparable, except when a is very close to a_0, the "coincidence" (18.2) would then be fulfilled. The other "coincidences"—

like, for instance, the fact that the number of particles within the particle horizon is of order N_1^2—are then straightforward consquences of the field equations for a closed Friedmann universe.

It then follows that there would not be observers in the Universe unless

$$a_0 \gtrsim ct_{\text{star}}. \tag{18.5}$$

Can we place any limits on $\mathscr{S}$ from similar considerations? Our discussion in Chapter 14 indicates that, for galaxies to form via gravitational instability

a) The plasma must be neutral and decoupled from the radiation—i.e. $T_{\text{rad}} \lesssim 3000°\text{K}$; and

b) The dynamics of the universe must be matter-dominated—i.e. $\varrho_m c^2 \gtrsim aT^4$.

Relation (a) is automatically satisfied when t is as large as t_{star}. For (b) to be true at the present epoch the maximum allowable radiation energy density corresponds to a black body temperature of $T \simeq 40°\text{K}$. Thus (b) would hold provided that

$$\mathscr{S} \lesssim 10^{11} \tag{18.6}$$

There does not seem any firm lower limit to $\mathscr{S}$. If $\mathscr{S} \lesssim 1$ the material emerging from the fireball would be 100 per cent helium, but it is not clear that this would preclude life.

Note that these arguments do not really "explain" why our Universe should satisfy (18.5) and (18.6). They merely indicate that these inequalities, and the "coincidence" (18.2), need occasion no surprise in a "cognizable" universe. Carter[425] has developed similar arguments which show that some of the *micro*physical dimensionless numbers are severely constrained, in any 'cognizable' universe, by the requirement that (for example) complex nuclei should be stable.

Further progress along these lines must await theoretical developments which allow us to place the concept of an "ensemble of universes" on a firm basis. One may also venture to hope—even more boldly—that eventually we may understand the very early universe well enough to see what determines a_0 and $\mathscr{S}$, and perhaps even to uncover some profound connection between the fundamental constants of microphysics and the geometry of the cosmos.

CHAPTER 19

Beyond the End of Time*

19.1 GRAVITATIONAL COLLAPSE AS THE GREATEST CRISIS IN PHYSICS OF ALL TIME

THE UNIVERSE starts with a big bang, expands to a maximum dimension, then recontracts and collapses: no more awe-inspiring prediction was ever made. It is preposterous. Einstein himself could not believe his own prediction. It took Hubble's observations to force him and the scientific community to abandon the concept of a universe that endures from everlasting to everlasting.

Later work[428,429] generalizes the conclusion. A model universe that is closed, that obeys Einstein's geometrodynamic law, and that contains a nowhere negative density of mass-energy, inevitably develops a singularity. Density of mass-energy rises without limit. A computing machine calculating ahead step by step the dynamical evolution of the geometry comes to the point where it can not go on. Smoke, figuratively speaking, starts to pour out of the computer. Yet physics surely continues to go on if for no other reason than this: Physics is by definition that which does go on its eternal way despite all the shadowy changes in the surface appearance of reality.

Some day a door will surely open and expose the glittering central mechanism of the world in its beauty and simplicity. Towards the arrival of that day no development holds out more hope than the paradox of gravitational collapse. Why paradox? Because Einstein's equation says "this is the end" and physics says "there is no end". Why hope? Because among all paradigms for probing a puzzle, physics proffers none with more promise than a paradox.

No previous period of physics brought a greater paradox than 1911 (Table XXVIII). Rutherford had just been forced to conclude that matter, is made up of localized positive and negative charges. Matter as so constituted should undergo electric collapse in $\sim 10^{-12}$ sec, according to theory. Observation equally clearly proclaimed that matter is stable. No one took the paradox more seriously than Bohr. No one worked around and around the central mystery with more energy wherever work was possible. No one brought to bear a more judicious combination of daring and conservative-

* Adapted from J. A. Wheeler, Marchon Lecture, University of Newcastle-upon-Tyne, May 18, 1971 and Nuffield Lecture, Cambridge University, July 19, 1971.

Table XXVIII Collapse of universe predicted by classical theory, compared and contrasted with classically predicted collapse of atom

System	Atom (1911)	Universe (1970's)
Dynamic entity	System of electrons	Geometry of space
Nature of classically predicted collapse	Electron headed towards point center of attraction is driven in a finite time to infinite energy	Not only matter but space itself arrives in a finite proper time at a condition of infinite compaction
One rejected "way out"	Give up Coulomb law of force	Give up Einstein's field equation
Another proposal for a "cheap way out" that has to be rejected	"Accelerated charge need not radiate"	"Matter cannot be compressed beyond a certain density by any pressure, however high"
How this proposal violates principle of causality	Coulomb field of point charge cannot readjust itself with infinite speed out to indefinitely great distances to sudden changes in velocity of charge	Speed of sound cannot exceed speed of light; pressure cannot exceed density of mass-energy
A major new consideration introduced by recognizing quantum principle as overarching organizing principle of physics	Uncertainty principle; binding too close to center of attraction makes zero-point kinetic energy outbalance potential energy; consequent existence of a lowest quantum state; can't radiate because no lower state available to drop to	Uncertainty principle; propagation of representative wave packet in superspace does not lead deterministically to a singular configuration for the geometry of space; expect rather a probability distribution of outcomes, each outcome describing a universe with a different size, a different set of particle masses, a different number of particles, and a different length of time required for its expansion and recontraction

ness, nor a deeper feel for the harmony of physics. The direct opposite of harmony, cacophony, is the impression that comes from sampling the literature of the 'teens on the structure of the atom. (1) Change the Coulomb law of force between electric charges? (2) Give up the principle that an accelerated charge must radiate? Little inhibition there was against this and still wilder abandon of the well established. In contrast, Bohr held fast to these two principles. At the same time he insisted on the importance of a third principle, firmly established by Planck in quite another domain of physics, the quantum principle. With that key he opened the door to the world of the atom.

Great as was the crisis of 1911, today gravitational collapse confronts physics with its greatest crisis ever. At issue is the fate, not of matter alone, but of the universe itself. The dynamics of collapse, or rather of its reverse, expansion, is evidenced not by theory alone, but also by observation; and not by one observation, but by observations many in number and carried out by astronomers of unsurpassed ability and integrity. Collapse, moreover, is not unique to the large scale dynamics of the universe. A white dwarf star or a neutron star of more than critical mass is predicted to undergo gravitational collapse to a black hole (Chapter 5). Sufficiently many stars falling sufficiently close together at the center of the nucleus of a galaxy are expected to collapse to a black hole many powers of ten more massive than the sun. An active search is under way for evidence for a black hole in this galaxy, via double stars[95], *x*-ray emission[45,98] or gravitational waves[430]. The process that makes a black hole is predicted to provide an experimental model for the gravitational collapse of the universe itself, with one difference. For collapse to a black hole the observer has his choice whether (1) to observe from a safe distance—in which case his observations will reveal nothing of what goes on inside the horizon; or (2) to follow the falling matter on in—in which case he sees the final stages of the collapse, not only of the matter itself, but of the geometry surrounding the matter, to indefinitely high compaction, but only at the cost of his own early demise. For the gravitational collapse of a closed model universe no such choice is available to the observer. His fate is sealed. So too is the fate of matter and elementary particles, driven up to indefinitely high densities. The stakes in the crisis of collapse are hard to match: the dynamics of the largest object, space, and the smallest object, an elementary particle—and how both began.

19.2 ASSESSMENT OF THE THEORY THAT PREDICTS COLLAPSE

No one reflecting on the paradox of collapse ("collapse ends physics"; "collapse cannot end physics") can fail to ask, what are the limits of validity of Einstein's geometric theory of gravity. A similar question posed itself in the crisis of 1911. The Coulomb law for the force acting between two charges had been tested at distances of meters and millimeters, but what warrant was there to believe that it holds down to the 10^{-8} cm of atomic dimensions? Of course in the end it proves to hold not only at the level of the atom, and at the 10^{-13} cm level of the nucleus, but even down to 5×10^{-15} cm [Barber, Gittelman, O'Neill and Richter[431] and the CERN group[432] as reviewed by Farley[433] and by Brodsky and Drell[434]], a striking example of what Wigner[435] calls the "unreasonable effectiveness of mathematics in the natural sciences".

No theory more resembles Maxwell's electrodynamics in its simplicity, beauty and scope than Einstein's geometrodynamics. Few principles in physics are more firmly established than those upon which it rests: the local validity of special relativity, the equivalence principle, the conservation of momentum and energy, and the prevalence of second order field equations throughout physics. Those principles lead to the conclusion that the geometry of spacetime must be Riemannian and the geometrodynamic law must be Einstein's.

To say that the geometry is Riemannian is to say that the interval between any two nearby events C and D, anywhere in spacetime, stated in terms of the interval AB between two nearby fiducial events, at quite another point in spacetime, has a value CD/AB independent of the route of intercomparison. By this hydraheaded prediction Einstein's theory exposes itself to destruction in a thousand ways.

Geometrodynamics lends itself to being disproven in other ways as well. The geometry has no option about the control it exerts on the dynamics of particles and fields. The theory makes predictions about the equilibrium configurations and pulsations of compact stars (chapter 2). It gives formulas (chapter 11) for the deceleration of the expansion of the universe, for the density of mass-energy, and for the magnifying power of the curvature of space, the tests of which are not far off. It predicts gravitational collapse and the existence of black holes and a wealth of physics associated with these objects (chapter 5). It predicts gravitational waves (chapter 7). In the appropriate approximation it encompasses all of the well tested predictions of the Newtonian theory of gravity for the dynamics of the solar system, and predicts testable post-Newtonian corrections besides, including several already verified effects.

No inconsistency of principle has ever been found in Einstein's geometric theory of gravity. No purported observational evidence against the theory has ever stood the test of time. No other acceptable account of physics of comparable simplicity and scope has ever been put forward.

Continue this assessment of general relativity a little further before returning to the central issue, the limits of validity of the theory and their bearing on the issue of gravitational collapse. What has Einstein's geometrodynamics contributed to the understanding of physics?

First, it has dethroned spacetime from a post of preordained prefection high above the battles of matter and energy and marked it as a new dynamic entity participating actively in this combat.

Second, by tying energy and momentum to the curvature of spacetime, Einstein's theory has recognized the law of conservation of momentum and energy as an automatic consequence of the geometric identity that the boundary of a boundary is zero.

Third, it has recognized gravitation as a manifestation of the curvature of the geometry of spacetime rather than as something foreign and "physical" imbedded in spacetime.

Fourth, general relativity has reinforced the view that "physics is local"; that only that analysis of physics is simple which connects quantities at a given event with quantities at immediately adjacent events.

Fifth, obedient to the quantum principle, it recognize that spacetime and time itself are ideas valid only at the classical level of approximation; that the proper arena for the Einstein dynamics of geometry is not spacetime, but superspace; and that this dynamics is described in accordance with the quantum principle by the propagation of a probability amplitude through superspace. In consequence, the geometry of space is subject to quantum fluctuations in metric coefficients of the order

$$\delta g \sim \frac{(\text{Planck length, } L^* = (\hbar G/c^3)^{1/2} = 1.6 \times 10^{-33}\ \text{cm})}{(\text{linear extension of region under study})}$$

Sixth, standard Einstein geometrodynamics is partial as little to Euclidean topology as to Euclidean geometry. A multiply connected topology provides a natural description for electric charge as electric lines of force trapped in the topology of a multiply connected space (Figure 59). Any other description of electricity postulates a breakdown in Maxwell's equations at a site where charge is located, or postulates the existence of some foreign and "physical" electric jelly imbedded in space, or both. No one has ever found a way to describe electricity free of these unhappy features except to say that the quantum fluctuations in the geometry of space are so great at small distances that even the topology fluctuates. These fluctuations have to be viewed, not as tied to particles, and endowed with the scale of distances associated with particle physics ($\sim 10^{-13}$ cm) but as pervading all space ("foam-like structure of geometry") and characterized by the Planck distance ($\sim 10^{-33}$ cm). Thus a third type of gravitational collapse forces itself on one's attention, a collapse continually being done and being undone everywhere in space, and surely a guide to the outcome of collapse at the level of a star and at the level of the universe.

19.3 VACUUM FLUCTUATIONS: THEIR PREVALENCE AND FINAL DOMINANCE

If Einstein's theory thus throws light on the rest of physics, the rest of physics also throws light on geometrodynamics. No point is more central than this, that empty space is not empty. It is the seat of the most violent physics. The electromagnetic field fluctuates. Virtual pairs of positive and

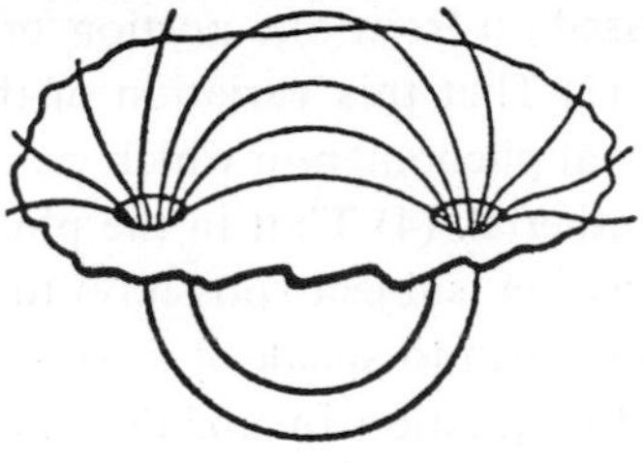

Figure 59 Electric charge viewed as electric lines of force trapped in the topology of a multiply connected space. [For the history of this concept see Wheeler,[438] reference 36.] The wormhole or handle is envisaged as connecting two very different regions in the same space. One of the wormhole mouths, viewed by an observer with poor resolving power, appears to be an electric charge. Out of this region of 3-space he finds lines of force emerging over the whole 4π solid angle. He may construct a boundary around this charge, determine the flux through this boundary, incorrectly apply the theorem of Gauss and "prove" that there is a charge "inside the boundary". It isn't a boundary. Someone caught within it—to speak figuratively—can go into that mouth of the wormhole, through the throat, out the other mouth and return by way of the surrounding space to look at his "prison" from the outside. Lines of force nowhere end. Maxwell's equations nowhere fail. Nowhere can one place his finger and say, "Here there is some charge". This classical type of electric charge has no direct relation whatsoever to quantized electric charge. There is a freedom of choice about the flux through the wormhole, and a specificity about the connection between one charge and another, which is quite foreign to the charges of elementary particle physics. For ease of visualization the number of space dimensions in the above diagram has been reduced from three to two. The third dimension, measured off the surface, has no physical meaning—it only provides an extra dimension in which to imbed the surface for more convenient diagrammatic representation. [For more detail see Misner and Wheeler; reprinted in Wheeler.[437]]

negative electrons in effect are continually being created and annihilated, and likewise pairs of mu mesons, pairs of baryons and pairs of other particles. All of these fluctuations coexist with the quantum fluctuations in the geometry and topology of space. Are they additional to those geometrodynamical zero-point disturbances or are they in some sense not now well understood mere manifestations of them?

Put the question in other words. Recall Clifford,[436] inspired by Riemann, speaking to the Cambridge Philosophical Society on 21 February 1870 "On the Space Theory of Matter", and saying, "I hold in fact (1) That small portions of space *are* in fact of a nature analogous to little hills on a surface which is on the average flat; namely, that the ordinary laws of geometry are not valid in them. (2) That this property of being curved or distorted is continually being passed on from one portion of space to another after the manner of a wave. (3) That this variation of the curvature of space is what really happens in that phenomenon which we call the *motion of matter*, whether ponderable or etherial. (4) That in the physical world nothing else takes place but this variation, subject (possibly) to the law of continuity." Ask if there is a sense in which to speak of a particle as constructed out of geometry. Or rephrase the question in updated language: "Is a particle a geometrodynamic exciton?" What else is there out of which to build a particle except geometry itself? And what else is there to give discreteness to such an object except the quantum principle?

The Clifford-Einstein space theory of matter has not been forgotten in recent years.

"In conclusion," one of the authors wrote a decade ago [Wheeler[437]], "the vision of Riemann, Clifford and Einstein, of a purely geometrical basis for physics, today has come to a higher state of development, and offers richer prospects—and presents deeper problems—than ever before. The quantum of action adds to this geometrodynamics new features, of which the most striking is the presence of fluctuations of the wormhole type throughout all space. If there is any correspondence at all between this virtual foam-like structure and the physical vacuum as it has come to be known through quantum electrodynamics, then there seems to be no escape from identifying these wormholes with "undressed electrons." Completely different from these "undressed electrons", according to all available evidence, are the electrons and other particles of experimental physics. For these particles the geometrodynamical picture suggests the model of collective disturbances in a virtual foam-like vacuum, analogous to different kinds of phonons or excitons in a solid.

"The enormous factor from nuclear densities $\sim 10^{14}$ g/cm^3 to the density of field fluctuation energy in the vacuum, $\sim 10^{94}$ g/cm^3, argues that elementary particles represent a percentage-wise almost completely negligible change in the locally violent conditions that characterize the vacuum. ['A particle (10^{14} g/cm^3) means as little to the physics of the vacuum (10^{94} g/cm^3) as a cloud (10^{-6} g/cm^3) means to the physics of the sky (10^{-3} g/cm^3)'.] In other words elementary particles do not form a really basic starting point for the description of nature. Instead, they represent a first order correction to vacuum physics. That vacuum, that zero order state of affairs, with its enormous densities of virtual photons and virtual positive-negative pairs and virtual wormholes, has to be described properly before one has a fundamental starting point for a proper perturbation theoretic analysis.

"These conclusions about the energy density of the vacuum, its complicated topological character, and the richness of the physics which goes on in the vacuum, stand in no evident contradiction with what quantum electrodynamics has to say about the vacuum. Instead the conclusions from the 'small distance' analysis (10^{-33} cm)—sketchy as it is—and from 'larger distance analysis' (10^{-11} cm) would seem to have the possibility to reinforce each other in a most natural way.

"The most evident shortcoming of the geometrodynamical model as it stands is this, that *it fails to supply any completely natural place* for spin $\frac{1}{2}$ in general and *for the neutrino* in particular."

Attempts to find a natural place for spin $\frac{1}{2}$ in Einstein's standard geometrodynamics founder because there is no natural way for a *change* in connectivity to take place within the context of classical differential geometry.

A uranium nucleus undergoing fission starts with one topology and nevertheless ends up with another topology. It makes this transition in a perfectly continuous way, classical differential geometry notwithstanding.

There are reputed to be two kinds of lawyers. One tells the client what not to do. The other listens to what the client has to do and tells him how to do it. From the first lawyer, classical differential geometry, the client goes away disappointed, still searching for a natural way to describe quantum fluctuations in the connectivity of space. Only in this way can he hope to describe electric charge as lines of electric force trapped in the topology of space. Only in this way does he expect to be able to understand and analyze the final stages of gravitational collapse. Pondering his problems, he comes to the office of a second lawyer, with the name "Pregeometry" on the door. Full of hope, he knocks and enters. What is pregeometry to be and say? Born of a combination of hope and need, of philosophy and physics and mathematics and logic, pregeometry will tell a story unfinished as of this writing, but full of incidents of evolution so far as it goes.

19.4 NOT GEOMETRY BUT PREGEOMETRY AS THE MAGIC BUILDING MATERIAL

An early survey† asked whether geometry can be constructed with the help of the quantum principle out of more basic elements which do not themselves have any specific dimensionality.

---

† *A Bucket of Dust*—an Early Attempt to Formulate the Concept of Pregeometry [Wheeler[438]]:

"... what line of thought could ever be imagined as *leading* to 4-dimensions—or any dimensionality at all—out of more primitive considerations? In the case of atoms one derives the yellow color of the sodium D-lines by analyzing the quantum dynamics of a system, no part of which is ever endowed with anything remotely resembling the attribute of color. Likewise any derivation of the 4-dimensionality of spacetime can hardly *start* with the idea of dimensionality."

"... recall the notion of a Borel set. Loosely speaking, a Borel set is a collection of points ('bucket of dust') which have not yet been assembled into a manifold of any particular dimensionality ... Recalling the universal sway of the quantum principle, one can imagine probability amplitudes for the points in a Borel set to be assembled into points with this, that and the other dimensionality. ... more conditions have to be imposed on a given number of points—as to which has which for a nearest neighbor—when the points one put together in a 5-dimensional array than when these same points are arranged in a 2-dimensional pattern. Thus one can think of each dimensionality as having a much

The focus of attention in this 1964 discussion was "dimensionality without dimensionality". However, the prime pressures to ponder pregeometry were and remain two features of nature, spin $\frac{1}{2}$ and charge, that speak out powerfully from every part of elementary particle physics.

A fresh perspective on pregeometry comes from a fresh assessment of general relativity. "Geometrodynamics is neither as important or as simple as it looks. Do not make it the point of departure in searching for underlying simplicity. Look deeper, at elementary particle physics." This is the tenor of interesting new considerations put forward by Sakharov[439], most easily summarized in the phrase, "Gravitation as the metric elasticity of space". In brief, as elasticity is to atomic physics, so—in Sakharov's view—gravitation is to elementary particle physics. The energy of an elastic deformation is nothing but energy put into the bonds between atom and atom by the deformation. The energy that it takes to curve space is nothing but perturbation in the vacuum energy of fields plus particles brought about by that curvature, according to Sakharov. The energy required for the deformation is governed in the one case by two elastic constants and in the other case by one elastic constant (the Newtonian constant of gravity) but in both cases, he reasons, the constants arise by combination of a multitude of complicated individual effects, not by a brave clean stroke on an empty slate.

---

higher statistical weight than the next *higher* dimensionality. On the other hand, for manifolds with 1, 2 and 3 dimensions the geometry is too rudimentary—one can suppose—to give anything interesting. Thus Einstein's field equations, applied to a manifold of dimensionality so low, demand flat space; only when the dimensionality is as high as 4 do really interesting possibilities arise. Can four, therefore, be considered to be the unique dimensionality which is at the same time high enough to give any real physics and yet low enough to have great statistical weight?

"It is too much to imagine that one has yet made enough mistakes in this domain of thought to explore such ideas with any degree of good judgment."

Consider a handle on the geometry. Let it thin halfway along its length to a point. In other words, let the handle dissolve into two bent prongs that touch at a point. Let these prongs separate and shorten. In this process two points part company that were once immediate neighbors. "However sudden the change is in classical theory, in quantum theory there is a probability amplitude function which falls off in the classically forbidden domain. In other words, there is some residual connection between points which are ostensibly very far away (travel from one 'tip' down one prong, then through the larger space to which these prongs are attached, and then up the other prong to the other tip). But there is nothing distinctive in principle about the two points that have happened to come into discussion. Thus it would seem that there must be a connection ... between *every* point and every other point. Under these conditions the concept of nearest neighbor would appear no longer to make sense. Thus the tool disappears with the help of which one might otherwise try to speak [un-] ambiguously about dimensionality."

One gives all the more favorable reception to Sakharov's view of gravity because one knows today, as one did not in 1915, how opulent in physics the vacuum is. In Einstein's day one had come in a single decade from the ideal God-given Lorentz perfection of flat spacetime to curved spacetime. It took courage to assign even one physical constant to that world of geometry that had always stood so far above physics. The vacuum looked for long as innocent of structure as a sheet of glass emerging from a rolling mill. With the discovery of the positive electron [Anderson[440]] one came to recognize a little of the life that heat can unfreeze in "empty" space. Each new particle and radiation that was discovered brought a new accretion to the recognized richness of the vacuum. Macadam looks smooth, but a bulldozer has only to cut a single furrow through the roadway to disclose all the complications beneath the surface.

Think of a particle as built out of the geometry of space; think of a particle as a "geometrodynamical exciton"? No model—it would seem to follow from Sakharov's assessment—could be less in harmony with nature, except to think of an atom as built out of elasticity! Elasticity did not explain atoms. Atoms explained elasticity. If likewise particles fix the constant in Einstein's geometrodynamic law (Sakharov), must it not be unreasonable to think of the geometrodynamic law as explaining particles?

Carry the comparison between geometry and elasticity one stage deeper (Figure 60). In a mixed solid there are hundreds of distinct bonds and all of them contribute to the elastic constants; some of them arise from Van der Waal's forces, some from ionic coupling, some from homopolar linkage; they have the greatest variety of strengths; but all have their origin in something so fantastically simple as a system of positively and negatively charged masses moving in accordance with the laws of quantum mechanics. In no way was it required or right to meet each complication of the chemistry and physics of a myriad of bonds with a corresponding complication of principle. By going to a level of analysis deeper than bond strengths one had emerged into a world of light, where nothing but simplicity and unity was to be seen.

Compare with geometry. The vacuum is animated with the zero point activity of distinct fields and scores of distinct particles and all of them, according to Sakharov, contribute to the Newtonian G, the "elastic constant of the metric". Some interact via weak forces, some by way of electromagnetic forces, and some through strong forces. These interactions have the greatest variety of strengths. But must not all of these particles and interactions have their origin in something fantastically simple? And must not this something, this "pregeometry", be as far removed from geometry as the quantum mechanics of electrons is far removed from elasticity?

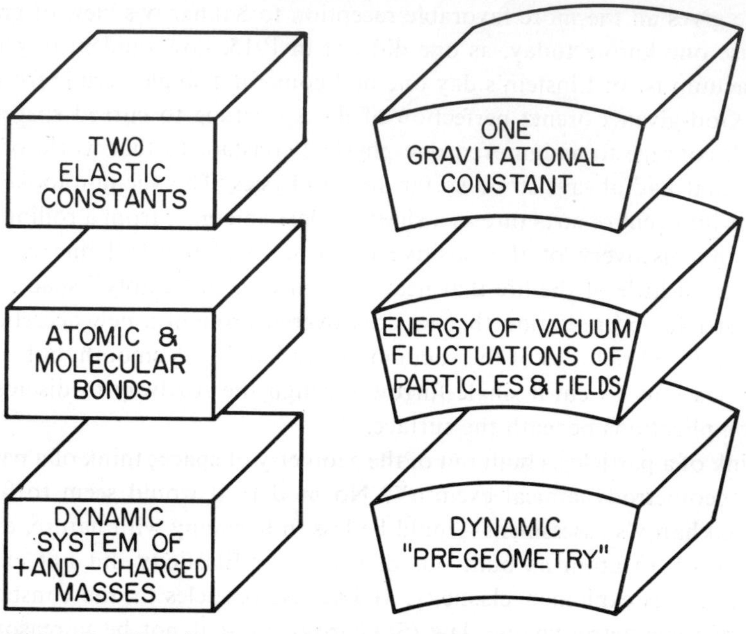

Figure 60 Elasticity and geometrodynamics, as viewed at three levels of analysis.

If one once thought of general relativity as a guide to the discovery of pregeometry, nothing might seem more dismaying than this comparison with an older realm of physics. No one would dream of studying the laws of elasticity to uncover the principles of quantum mechanics. Neither would anyone investigate the work hardening of a metal to learn about atomic physics. The order of understanding ran not

Work hardening (1 cm) → Dislocations (10^{-4} cm) → Atoms (10^{-8} cm)

but the direct opposite,

Atoms (10^{-8} cm) → Dislocations (10^{-4} cm) → Work hardening (1 cm)

One had to know about atoms to conceive of dislocations, and had to know about dislocations to understand work hardening. Is it not likewise hopeless to go from the "elasticity of geometry" to an understanding of particle physics, and from particle physics to the uncovering of pregeometry? Must not the order of progress again be the direct opposite? And is not the source of any dismay the apparent loss of guidance that one experiences in giving up geometrodynamics—and not only geometrodynamics, but geometry itself—as crutch to lean on as one hobbles forward? Yet there is so much chance that this view of nature is right that one has to take it seriously and explore its consequences. Never more than today does one have the incentive to explore pregeometry.

19.5 PREGEOMETRY AS THE CALCULUS OF PROPOSITIONS

"What really interests me is whether God had any choice in the creation of the world"—Einstein

Paper in white the floor of the room and rule it off in one foot squares. Down on one's hands and knees, write in each square a different set of equations conceived as able to govern the physics of the universe. At the end of one's labors one has worked oneself out into the door way. Stand up, look back on all those equations, some perhaps more hopeful than others, raise one's finger commandingly, and give the order "fly". Not one of those sets of equations will put on wings, take off or fly. Yet the universe "flies".

Some principle uniquely right and uniquely simple must, when one knows it, be also so obvious that it is clear that the universe is built, and must be built, in such and such a way and that it could not possibly be otherwise. But how discover that principle? If it was hopeless to learn atomic physics by studying work hardening and dislocations, it may be equally hopeless to learn the basic operating principle of the universe, call it pregeometry or call it what one will, by any amount of work in general relativity and particle physics.

Thomas Mann[441] in his essay on Freud utters what Niels Bohr would surely have called a great truth ("A great truth is a truth whose opposite is also a great truth") when he says, "Science never makes an advance until philosophy authorizes and encourages it to do so". If the equivalence principle and Mach's principle were the philosophical godfathers of general relativity, it is also true that what those principles do mean, and ought to mean, only became established after Einstein's theory had been in the world for some time. Therefore it would seem reasonable to expect the primary guidance in the search for pregeometry to come from a principle both philosophical and powerful, but one also perhaps not destined to be wholly clear in its contents or its implications until some later day.

Among all the principles that one can name out of the world of science it is difficult to think of one more compelling than *simplicity*; and among all the simplicities of dynamics and life and movement none is starker (Werner[442]) than the *binary choice* yes-no or true-false. It in no way proves that this choice for starting principle is correct, but it at least gives one some comfort in the choice, that Pauli's "non-classical two-valuedness" or "spin" so dominates the world of particle physics.

It is one thing to have a start, a tentative construction of pregeometry; but how goes one go on? How not to go on is illustrated by Figure 61. The

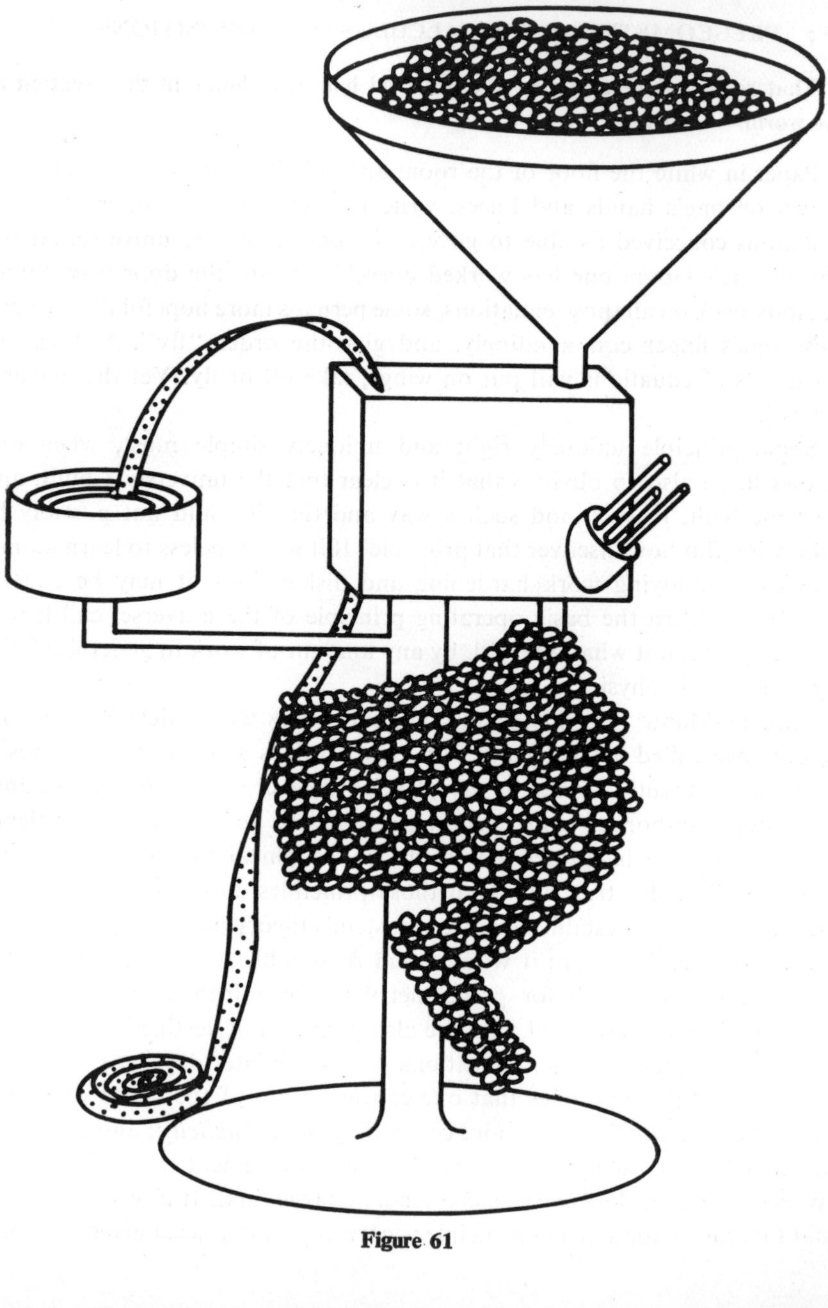

Figure 61

Figure 61 "Ten thousand rings"; or an example of a way to think of the connection between pregeometry and geometry, wrong because it is too literal minded and for other reasons spelled out in the text. The Vizier [story by Wheeler as reported by Kilmister[443]] speaks:

Take $N = 10{,}000$ brass rings. Take an automatic fastening device that will cut open a ring, loop it through another ring, and resolder the joint. Pour the brass rings into the hopper that feeds this machine. Take a strip of instruction paper that is long enough to contain $N(N-1)/2$ binary digits. Look at the instruction in the (jk)-th location on this instruction tape $(j, k = 1, 2, \ldots, N; j < k)$. When the binary digit at that location is 0, it is a signal leave to the j-th ring disconnected from the k-th ring. When it is 1, it is an instruction to connect that particular pair of rings. Thread the tape into the machine and press the start button. The clatter begins. Out comes a chain of rings 10,000 links long. It falls on the table and the machine stops. Pour in another 10,000 rings, feed in a new instruction tape, and push the button again. This time it is not a 1-dimensional structure that emerges, but a 2-dimensional one: a Crusader's coat of mail complete with neck opening and sleeves. Take still another tape from the library of tapes and repeat. Onto the table thuds a smaller version of the suit of mail, this time filled out internally with a solid network of rings, a 3-dimensional structure. Now forego the library and make one's own instruction tape, a random string of 0's and 1's. Guided by it, the fastener produces a "Christmas tree ornament", a collection of segments of 1-dimensional chain, 2-dimensional surfaces, and 3-, 4-, 5- and higher dimensional entities, some joined together, some free-floating. Now turn from a structure deterministically fixed by a tape to a probability amplitude, a complex number,

$$\psi\,(\text{tape}) = \psi(n_{12}, n_{13}, n_{14}, \ldots, n_{N-1,N}) \qquad (n_{ij} = 0, 1) \qquad (1)$$

defined over the entire range of possibilities for structures built of 10,000 rings. Let not these probability amplitudes be assigned randomly. Instead couple together amplitudes for structures that differ from each other by the breaking of a single ring by linear formulas that treat all rings on the same footing. The separate ψ's, no longer entirely independent, will still give non-zero probability amplitudes for "Christmas tree ornaments". Of greater immediate interest than these "unruly" parts of the structures are the following questions about the smoother parts: (1) In what kinds of structures is the bulk of the probability concentrated? (2) What is the dominant dimensionality of these structures in an appropriate correspondence principle limit? (3) In this semiclassical limit what is the form taken by the dynamical law of evolution of the geometry?" No principle more clearly rules out this model for pregeometry than the principle of simplicity (see text).

"sewing machine" builds objects of one or another definite dimensionality, or of mixed dimensionalities, according to the instructions that it receives on the input tape in yes-no binary code. Some of the difficulties of building up structure upon the binary element according to this model, or any one of a dozen other models, stand at out once. (1) Why $N = 10{,}000$ building units? Why not a different N? And if one feeds in one such arbitrary number at the start why not fix more features "by hand"? No natural stopping point is evident, nor any principle that would fix such a stopping point. Such arbitrariness contradicts the principle of simplicity and rules out the model. (2) Quantum mechanics is added from outside, not generated from inside (from the model itself). On this point too the principle of simplicity speaks against the model. (3) The passage from pregeometry to geometry is made in a too literal minded way, with no appreciation of the need for particles and fields to appear along the way. The model, in the words used by Bohr on another occassion, is "crazy, but not crazy enough to be right".

Noting these difficulties, and fruitlessly trying model after model of pregeometry to see if it might be free of them, one suddenly realizes that a machinery for the combination of yes–no or true-false elements does not have to be invented. It already exists. What else can pregeoemetry be, one asks himself, than the calculus of propositions? (Table XXIX).

Table XXIX "Pregeometry as the calculus of propositions"

A sample proposition taken out of a standard text on logic selected almost at random reads (Kneebone[444])

$$[X \to ((X \to X) \to Y)] \& (\bar{X} \to Z) \mathrm{eq} (\bar{X} \vee Y \vee Z) \& (\bar{X} \vee Y \vee \bar{Z}) \& (X \vee Y \vee Z) \& (X \vee \bar{Y} \vee Z)$$

The symbols have the following meaning:

$\bar{A}$	Not A
$A \vee B$	A or B or both ("A *vel* B")
$A \& B$	A and B
$A \to B$	A implies B ("if A, then B")
$A \leftrightarrow B$	B is equivalent to A ("B if and only if A")

Propositional formula $\mathscr{A}$ is said to be equivalent ("eq") to propositional formula $\mathscr{B}$ if and only if $\mathscr{A} \leftrightarrow \mathscr{B}$ is a tautology.

The letters A, B, etc. serve as connectors to "wire together" one proposition with another. Proceeding in this way one can construct propositions of indefinitely great length.

A switching circuit [see for example Shannon[445] or Hohn[446]] is isomorphic to a proposition.

Compare a short proposition or an elementary switching circuit to a molecular collision. No idea seemed more preposterous than thatof Daniel Bernouilli (1733) that heat is a manifestation of molecular collisions. Moreover, a three-body encounter is difficult to treat, a four-body collision is more difficult, and a five- or more molecule system is essentially intractable. Nevertheless, mechanics acquires new elements of simplicity in the limit in which the number of molecules is very great and in which one can use the concept of density in phase space. Out of statistical mechanics in this limit come such concepts as temperature and entropy. When the temperature is well-defined, the

Table XXIX (*continued*)

energy of the system is not a well-defined idea; and when the energy is well defined, the temperature is not. This complementarity is built inescapably into the principles of the subject. Thrust the finger into the flame of a match and experience a sensation like nothing else on heaven or earth; yet what happens is all a consequence of molecular collisions, early critics notwithstanding.

Any individual proposition is difficult for the mind to apprehend when it is long; and still more difficult to grasp is the content of a cluster of propositions. Nevertheless, make a statistical analysis of the calculus of propositions in the limit where the number of propositions is great and most of them are long. Ask if parameters force themselves on one's attention in this analysis (1) analogous in some small measure to the temperature and entropy of statistical mechanics but (2) so much more numerous, and everyday dynamical in character, as to reproduce the continuum of everyday physics.

Nothing could seem so preposterous at first glance as the thought that nature is built on a foundation as ethereal as the calculus of propositions. Yet, beyond the push to look in this direction provided by the principle of simplicity there are two pulls. First, bare bones quantum mechanics lends itself in a marvelously natural way to formulation in the language of the calculus of propositions, as witnesses not least the book of Jauch.[447] If the quantum principle were not in this way already automatically contained in one's proposed model for pregeometry, and if in contrast it had to be introduced from outside, by that very token one would conclude that the model violated the principle of simplicity, and would have to reject it. Second, the pursuit of reality seems always to take one away from reality. Who would have imagined describing something so much a part of the here and now as gravitation in terms of curvature of the geometry of spacetime? And when later this geometry came to be recognized as dynamic, who would have dreamed that geometrodynamics unfolds in an arena so ethereal as superspace? Little astonishment there should be, therefore, if the description of nature carries one in the end to logic, ethereal eyrie at the center of mathematics.

"An issue of logic having nothing to do with physics" was the assessment by many of a controversy of old about the axiom, "parallel lines never meet." Does it follow from the other axioms of Euclidean geometry or is it independent? "Independent," Bolyai and Lobachevsky proved. With this and the work of Gauss as a start, Riemann went on to create Riemannian geometry. Study nature, not Euclid, to find out about geometry, he advised; and Einstein went on to take that advice and to make geometry a part of physics.

"An issue of logic having nothing to do with physics" is one's natural first assessment of the startling limitations on logic discovered by Gödel,[448] Cohen[449] and others [for a review see for example Kacs and Ulam[450]]. The exact opposite must be one's assessment if the real pregeometry of the real physical world indeed turns out to be identical with the calculus of propositions.

"Physics as manifestation of logic" or "pregeometry as the calculus of propositions" is as yet [Wheeler[451]] not an idea, but an idea for an idea. It is put forward here only to make it a little clearer what it means to suggest that the order of progress will not be

$$\text{physics} \rightarrow \text{pregeometry}$$

but

$$\text{pregeometry} \rightarrow \text{physics}$$

19.6 THE BLACK BOX MODEL OF COLLAPSE: THE REPROCESSING OF THE UNIVERSE

No amount of searching has ever disclosed a "cheap way" out of gravitational collapse any more than earlier it revealed a cheap way out of the collapse of the atom. Physicists in that earlier crisis found themselves in the end confronted with a revolutionary pistol, "Understand nothing—or accept the quantum principle". Today's crisis can hardly force a lesser revolution. One sees no alternative except to say that geometry fails and pregeometry has to take its place to ferry physics through the final stages of gravitational collapse and on into what happens next. No guide is evident on this unchartered way except the principle of simplicity, applied to drastic lengths.

Whether the whole universe is squeezed down to the Planck dimension, or more, or less, before re-expansion can begin and dynamics can return to normal may be irrelevant for some of the questions one wants to consider. Physics has long used the "black box" to symbolize situations where one wishes to concentrate on what goes in and what goes out, disregarding what takes place in between.

At the beginning of the crisis of atomic collapse one conceived of the electron as headed on a deterministic path towards a point center of attraction, and unhappily destined to arrive at a condition of infinite kinetic energy in a finite time. After the advent of quantum mechanics one learned to summarize the interaction between center of attraction and electron in a "black box": fire in a wave train of electrons traveling one in direction, and get electrons coming out in this, that, and the other direction with this that, and the other well determined probability amplitude (Figure 62). Moreover, to predict these probability amplitudes quantitatively and correctly it was enough to translate the Hamiltonian of classical theory into the language of wave mechanics and solve the resulting equation, key to the "black box".

A similar "black box" view of gravitational collapse leads one to expect a "probability distribution of outcomes." Here, however, one outcome is distinguished from another, one has to anticipate, not by a single parameter, such as the angle of scattering of the electron, but by many. They govern, one foresees, such quantities as the size of the system at its maximum of expansion, the time from the start of this new cycle to the moment it ends in collapse, the number of particles present, and a thousand other features. The "probabilities" of these outcomes will be governed by a dynamic law, analogous to (1) the Schroedinger wave equation for the electron or, to cite another black box problem, (2) the Maxwell equations that couple together at a wave guide junction electro-magnetic waves running in other-

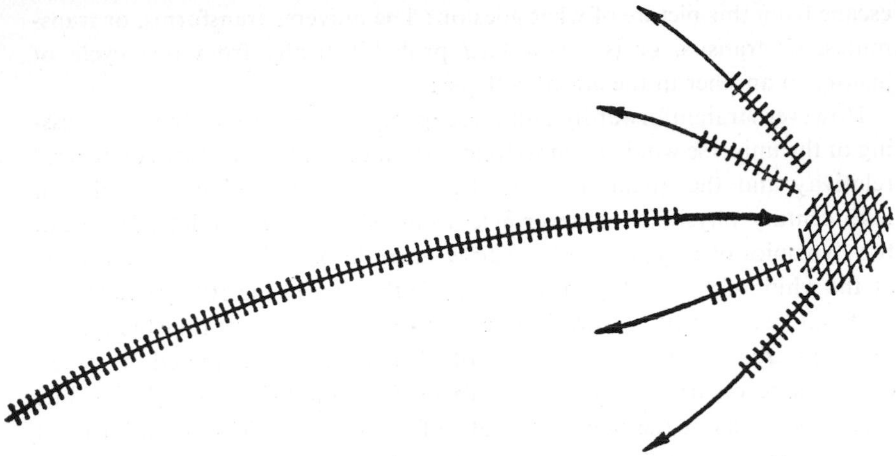

Figure 62 The "black box model" applied (1) to the scattering of an electron by a center of attraction and (2) to the collapse of the universe itself. The deterministic electron world line of classical theory is replaced in quantum theory by a probability amplitude, the wave crests of which are illustrated schematically in the diagram. The catastrophe of classical theory is replaced in quantum theory by a probability distribution of outputs. The same diagram illustrates the "black box account" of gravitational collapse mentioned in the text. The arena of the diagram is no-longer spacetime, but superspace. The incident arrow marks no longer a classical world line of an electron through spacetime but a classical "leaf of history of geometry" slicing through superspace. The wave crests symbolize no longer the elctron wave function propagating through spacetime, but the geometrodynamic wave function propagating through superspace. The cross hatched region is no longer the region where the one-body potential goes to infinity, but the region of gravitation collapse where the curvature of space goes to infinity. The outgoing waves describe no longer alternative directions for the new course of the scattered electron, but the beginnings of alternative new histories for the universe itself after collapse and "re-processing" end the present cycle.

wise separate wave guides. However, it is hardly reasonable to expect the necessary dynamic law to spring forth upon translating the Hamilton-Jacobi equation of general relativity into a Schroedinger equation. The reason is simple. Geometrodynamics in both its classical and its quantum version is built on standard differential geometry. That standard geometry leaves no room for any of those quantum fluctuations in connectivity that seem inescapable at small distance and therefore also inescapable in the final stages of gravitational collapse. Not geometry but pregeometry must fill the black box of gravitational collapse.

Little as one knows the internal machinery of the black box, one sees no escape from this picture of what goes on: The universe transforms, or transmutes, or transits, or is *reprocessed* probabilistically from one cycle of history to another in the era of collapse.

However straightforwardly and inescapably this picture of the reprocessing of the universe would seem to follow from the leading features of general relativity and the quantum principle, the two overarching principles of 20th century physics, it is nevertheless fantastic to contemplate. How can the dynamics of a system so incredibly gigantic be switched, and switched at the whim of probability, from one cycle that has lasted 10^{11} years to one that will last only 10^6 years? At first only the circumstance that the system gets squeezed down in the course of this dynamics to incredibly small distances reconciles one to a transformation otherwise so unbelievable. Then one looks at the upended strata of a mountain slope, or a bird not seen before, and marvels that the whole universe is incredible:

- mutation of a species
- metamorphosis of a rock
- chemical transformation
- spontaneous transformation of a nucleus
- radioactive decay of a particle
- reprocessing of the universe itself.

If it cast a new light on geology to know that rocks can be raised and lowered thousands of meters and hundreds of degrees, what does it mean for physics to think of the universe as from time to time "squeezed through a knot hole," drastically "reprocessed," and started out on a fresh dynamic cycle? Three considerations above all press themselves upon one's attention, prefigured in these compressed phrases,

- destruction of all constants of motion in collapse
- particles, and the physical "constants" themselves, as the "frozen-in part of the meteorology of collapse"
- "the biological selection of physical constants".

The gravitational collapse of a star, or a collection of stars, to a black hole extinguishes all details of the system (see Chapter 5) except mass and charge and angular momentum. Whether made of matter or antimatter or radiation, whether endowed with much entropy or little entropy, whether in smooth motion or chaotic turbulence, the collapsing system ends up as seen from outside, according to all indications, in the same standard state. The laws of conservation of baryon number and lepton number are transcended (see Chapter 5; also Wheeler[452]). No known means whatsoever will distinguish between black holes of the most different provenance if only they have the same mass, charge and angular momentum. But for a closed universe even these constants vanish from the scene. Total charge is automatically zero because lines of force have nowhere to end except upon charge. Total mass and total angular momentum have absolutely no definable meaning whatsoever for a closed universe. This conclusion follows not least because there is no asymptotically flat space outside where one can put a test particle into Keplerian orbit to determine period and precession.

Of all principles of physics the laws of conservation of charge, lepton number, baryon number, mass, and angular momentum are among the most firmly established. Yet with gravitational collapse the content of these conservation laws also collapses. The established is disestablished. No determinant of motion does one have left that could continue unchanged in value from cycle to cycle of the universe. Moreover, if particles are dynamic in construction, and if the spectrum of particle masses is therefore dynamic in origin, no option would seem left except to conclude that the mass spectrum is itself reprocessed at the time when "the universe is squeezed through a knot hole." A molecule in this piece of paper is a "fossil" from photochemical synthesis in a tree a few years ago. A nucleus of the oxygen in this air is a fossil from thermonuclear combustion at a much higher temperature in a star a few 10^9 years ago. What else can a particle be but a fossil from the most violent event of all, gravitational collapse?

That one geological stratum has one many-mile long slope, with marvelous linearity of structure, and another stratum has another slope, is either an everyday triteness taken as for granted by every passer-by, or a miracle, until one understands the mechanism. That an electron here has the same mass as an electron there is also a triviality or a miracle. It is a triviality in quantum electrodynamics because assumed rather than derived. However, it is a miracle on any view that regards the universe as from time to time "reprocessed." How can electrons at different times and places in the present cycle of the universe have the same mass if the spectrum of particle masses differs between one cycle of the universe and another?

Inspect the interior of a particle of one type, and magnify it up enormously, and in that interior see one view of the whole universe [compare the concept of monad of Leibniz[453], "The monads have no window through which anything can enter or depart"]; and do likewise for another particle of the same type. Are particles of the same pattern identical in any one cycle of the universe because they give identically patterned views of the same universe? No acceptable explanation for the miraculous identity of particles of the same type has ever been put forward. That identity must be regarded, not as a triviality, but as a central mystery of physics.

Not the spectrum of particle masses alone, but the physical "constants" themselves would seem most reasonably regarded as reprocessed from one cycle to another. Reprocessed relative to what? Relative, for example, to the Planck system of units,

$$L^* = (\hbar G/c^3)^{1/2} = 1.6 \times 10^{-33} \text{ cm}$$

$$T^* = (\hbar G/c^5)^{1/2} = 5.3 \times 10^{-44} \text{ sec}$$

$$M^* = (\hbar c/G)^{1/2} = 2.2 \times 10^{-5} \text{ g},$$

the only system of units, Planck[454] pointed out, free, like black body radiation itself, of all complications of solid-state physics, molecular binding, atomic constitution and elementary particle structure, and drawing for its background only on the simplest and most universal principles of physics, the laws of gravitation and black body radiation. Relative to the Planck units every constant in every other part of physics is expressed as a pure number.

No pure numbers in physics are more impressive than $\hbar c/e^2 = 137.0388$ and the so-called "big numbers" [Eddington[455]; Dirac[426 456]; Jordan[457]; Dicke[424 458]; Hayakawa[459]; Carter[425]]

$$\sim 10^{80} \text{ particles in the universe}$$

$$\sim 10^{40} \sim \frac{10^{28} \text{ cm}}{10^{-12} \text{ cm}} \sim \frac{\text{(radius of universe at maximum expansion}}{\text{("size,, of an elementary particle)}}$$

$$\sim 10^{40} \sim \frac{e^2}{GmM} \sim \frac{\text{(electric forces)}}{\text{(gravitational forces)}}$$

$$\sim 10^{20} \sim \frac{e^2/mc^2}{(\hbar G/c^3)^{1/2}} \sim \frac{\text{("size'' of an elementary particle)}}{\text{(Planck length)}}$$

$$\sim 10^{10} \sim \frac{\text{(number of photons in universe)}}{\text{(number of baryons in universe)}}$$

Some understanding of the relationships between these numbers has been won (Carter[425]). Never has any explanation appeared for their enormous magnitude. Nor will there ever, if the view is correct that reprocessing the universe reprocesses also the physical constants. These constants on that view are not part of the laws of physics. They are part of the initial value data. Such numbers are freshly given for each fresh cycle of expansion of the universe. To look for a physical explanation for the "big numbers" would thus seem to be looking for the right answer to the wrong question.

In the week between one storm and the next most features of the weather are ever changing, but some special patterns of the wind last the week. If the term "frozen features of the meteorology" is appropriate for them, much more so would it seem appropriate for the big numbers, the physical constants and the spectrum of particle masses in the cycle between one reprocessing of the universe and another.

A per cent or so change one way in one of the "constants," $\hbar c/e^2$, will cause all stars to be red stars; and a comparable change the other way will make all stars be blue stars, according to Carter[425]. In neither case will any star like the sun be possible. He raises the question whether life could have developed if the determinants of the physical constants had differed substantially from those that characterize this cycle of the universe.

Dicke[424,458] has pointed out that the right order of ideas may not be, here is the universe, so what must man be; but here is man, so what must the universe be. In other words: (1) What good is a universe without awareness of that universe? But: (2) Awareness demands life. (3) Life demands the presence of elements heavier than hydrogen. (4) The production of heavy elements demands thermonuclear combustion. (5) Thermonuclear combustion normally requires several 10^9 years of cooking time in a star. (6) Several 10^9 years of time will not and cannot be available in a closed universe, according to general relativity, unless the radius-at-maximum-expansion of that universe is several 10^9 light years or more. So why on this view is the universe as big as it is? Because only so can man be here!

In brief, the considerations of Carter and Dicke would seem to raise the idea of the "biological selection of physical constants." However, to "select" is impossible unless there are options to select between. Exactly such options would seem for the first time to be held out by the only overall picture of the gravitational collapse of the universe that one sees how to put forward today, the *pregeometry black box model of the reprocessing of the universe.*

Rich prospects stand open for investigation in gravitation physics, from neutron stars to cosmology and from post-Newtonian celestial mechanics to gravitational waves. Einstein's geometrodynamics exposes itself to destruction on a dozen fronts and by a thousand predictions. No predictions subject to early test are more entrancing than those on the formation and properties of a black hole, "laboratory model" for some of what is predicted for the universe itself. No field is more pregnant with the future than gravitational collapse. No more revolutionary views of man and the universe has one ever been driven to consider seriously than those that come out of pondering the paradox of collapse, greatest crisis of physics of all time.

All of these endeavors are based on the belief that existence should have a completely harmonious structure. Today we have less ground than ever before for allowing ourselves to be forced away from this wonderful belief.

EINSTEIN

References

1. Maxwell, J. C. *The Scientific Papers*, p. 241, Hermann, Paris, 1927.
2. Einstein, A. *The Meaning of Relativity* (fifth ed.), Princeton University Press, Princeton, N. J., p. 108, 1955.
3. Riemann, B. Ueber die Hypothesen, welche der Geometrie zu Grunde liegen, in *Gesammelte Mathematische Werke*, (H. Weber, ed.), Leipzig, 1876; second ed. of 1892, Dover Publ., New York, 1953.
4. Taylor, E. F., and Wheeler, J. A. *Spacetime Physics*, Ch. 3, W. H. Freeman and Co., San Francisco, 1966.
5. Harrison, B. K., Thorne, K. S., Wakano, M., and Wheeler, J. A. *Gravitation Theory and Gravitational Collapse*, University of Chicago Press, Chicago, 1965.
6. Jordan, P., Brans, C., and Dicke, R. H. *Schwerkraft und Weltall*, Braunschweig, 1955; *Phys. Rev.*, **124**, p. 925 (1960).
7. Hartle, J. B., and Thorne, K. S. *Ap. J.* **153**, p. 807 (1968).
8. Salmona, A. *Phys. Rev.* **154**, p. 1218 (1967).
9. Cameron, A. G. W., Cohen, J. H., Langer, W. D., and Rosen, L. R. *Astrophys. Space Sci.*, **6**, No. 2, p. 228 (1970).
10. Brueckner, K. A. *Phys. Rev.*, **100**, p. 36 (1955).
11. Wang, C. J., Rose, W. K., and Schlenker, S. L. *Ap. J. Lett.*, **160**, L17 (1970).
12. Reid, R. V. *Ann. Phys.*, **50**, p. 411 (1968).
13. Ambartsumyan, V. A., and Saakyan, G. S. *Soviet Astr., A. J.*, **5**, p. 601 (1962); Wheeler, J. A. in *Gravitation and relativity*, (H.-Y. Chiu and W. Hoffman, eds.), W. A. Benjamin, New York, 1964.
14. Hagedorn, R. *Thermodynamics of Strong Interactions at High Energy and its Consequences for Astrophysics*, Ref. Th. 1027-CERN (1969).
15. Rhoades, C. E., Jr., and Ruffini, R. *Bull. Am. Phys. Soc.*, Washington, D. C. Meeting, 1970; for discussion of $p \sim \varrho/ln\varrho$ (low T limit of Hagedorn) see *Ap. J. Lett.*, **163**, L83 (1971).
16. Gerlach, U. *Phys. Rev.*, **173**, p. 1325 (1968).
17. Ruffini, R., and Bonazzola, S. *Phys. Rev.*, **187**, p. 1767 (1969).
18. Hewish, A., Bell, S. J., Pilkington, J. D. H., Scott, P. F., and Collins, R. A. *Nature*, **217**, p. 709 (1968).
19. Manchester, R. N., and Taylor, J. H. *Astrophys. Letters*, **10**, p. 67 (1972).
20. Radhakrishnan, V., Cooke, D. J., Komesaroff, M., and Morris, D. *Nature*, **221**, p. 443 (1969).
21. Minkowski, R. *Ap. J.*, **96**, p. 199 (1942).
22. Boynton, P. E., Groth, J. E., Hutchinson, D. P., Nanos, G. P., Partridge, R. B., and Wilkinson, D. T. *Ap. J.*, **175**, p. 217 (1972).
23. Richards, D. W., Rankin, J. M., and Counselman, C. C. I.A.U. Circular No. 2/64, 1969.
24. Radhakrishnan, V., and Manchester, R. N. *Nature*, **222**, p. 228 (1969).
25. Reichley, P. E., and Downes, G. S. *Ibid.*, p. 229 (1969).
26. Richards, D. W., Pettengill, G. H., Counselman, C. C., and Rankin, J. M., *Ap. J. Lett.*, **160**, L1 (1970).

27. Fritz, G., Henry, R. C., Meekins, J. F., Chubb, T. A., and Friedman, H. *Science*, **164**, p. 709 (1969).
28. Rosseland, S., *Pulsation Theory of Variable Stars*, Dover, New York, 1964; Thorne, K. S., and Ipser, J. R. *Ap. J. Lett.*, **152**, L71 (1968);
Faulkner, J., and Gribbin, J. *Nature*, **218**, p. 734 (1968).
29. Landau, L. D., and Lifschitz, E. M. *The Classical Theory of Fields*, pp. 366–367, Addison-Wesley, Reading, Mass., (1966).
30. For review and references see, for example, Wheeler, J. A. *A. Rev. Astr., Astrophys.*, **4**, pp. 393–432 (1966).
31. Deutsch, A. *Ann. Astrophys.*, **18**, p. 1 (1955).
32. Ferrari, A., and Ruffini, R. *Ap. J. Lett.*, **158**, L71 (1969).
33. Ruderman, M. A. *Nature*, **218**, p. 1128 (1968).
34. Kirzhnitz, D. A. *Sov. Phys. JETP*, **11**, p. 365 (1960).
For summary of literature and more complete discussion see
Ruderman, M. A. *Extreme Regimes of Temperature and Pressure in Astrophysics* in Proceedings of International Conference on Thermodynamics, p. 429 (P. T. Landsberg, ed.), Butterworth, London, 1970.
35. Dyson, F. J. *Ann. Phys.*, New York **63**, p. 1 (1971).
36. Ginzburg, V. L., and Kirzhnitz, D. A. *Sov. Phys. JETP*, **20**, p. 1346 (1965).
37. Wolf, R. A., and Bahcall, J. N. *Phys. Rev.*, **140**, B1445 and B1452 (1965).
38. Tkachenko, K. *Sov. Phys. JETP*, **23**, p. 1049 (1966).
39. Baym, G., Pethick, C., and Pines, D. *Nature*, **224**, p. 673 (1969).
40. Baym, G., Pethick, C., Pines, D., and Ruderman, M. A. *Ibid.*, p. 872 (1969).
41. Oke, J. B., Neugebauer, G., and Becklin, E. E. *Ap. J. Lett.*, **156**, L41 (1969).
42. Colgate, S. A., and White, R. H. *Ap. J.*, **143**, p. 626 (1966).
43. May, M. M., and White, R. H. *Relativity Theory and Astrophysics*, (J. Ehlers, ed.), Vol. III, Stellar Structure, American Mathematical Society, 1967. *Phys. Rev.*, **141**, p. 1232 (1966).
44. LeBlanc, J. M., and Wilson, J. R. *Ap. J.*, **161**, p. 541 (1970).
45. Zel'dovich, Ya. B., and Novikov, I. D. *Stars and Relativity* (Transl. Eli Erlock: K. S. Thorne and W. D. Arnett, eds.), Chicago University Press, 1971.
46. Shklovskii, I. S. *Supernovae*, Vol. 21 Interscience Monographs and Texts in Physics and Astronomy, J. Wiley, New York, 1968.
47. Colgate, S. A. *Can. J. Physics*, **46**, 10, Part 3, S476 (1968) (Proc. Int. Cosmic Rays Conf., Calgary, 1967).
48. Clayton, D. D., Colgate, S. A., and Fishman, G. J. *Ap. J.*, **155**, p. 75 (1969).
49. Katgert, P., and Oort, J. H. *Bull. Astr. Inst. Netherl.*, **19**, p. 239 (1967).
50. Oppenheimer, J. R., and Snyder, H. *Phys. Rev.*, **56**, p. 455 (1939).
51. Klein, O. in *Werner Heisenberg und die Physik unserer Zeit*, F. Vieweg und Sohn, Branschweig, 1961.
52. Thorne, K. S. in *Les Houches Lectures*, Vol. III, (C. DeWitt, E. Schatzman and P. Véron, eds.), Gordon and Breach, 1967.
53. Beckedorff, D. L., and Misner, C. W. 1962 (unpublished).
54. Podurets, M. A. *Soviet Astr. A. J.*, **8**, p. 868 (1965).
55. Kruskal, M. D., and Fronsdal, C. *Phys. Rev.*, **119**, p. 1743 (1960). *Phys. Rev.*, **116**, p. 778 (1959).
56. Zel'dovich, Ya. B., and Novikov, I. D. *Soviet Phys. Dokl.*, **9**, p. 246 (1964).
57. Penrose, R. *Phys. Rev. Lett.*, **14**, p. 57 (1965).

58. Wheeler, J. A. Superspace and the Nature of Quantum Geometrodynamics in *Battelles Rencontres*: 1967 *Lectures in Mathematics and Physics* (C. DeWitt and J. A. Wheeler, eds.), W. A. Benjamin, New York, 1968.
59. Israel, W. *Phys. Rev.*, **164**, p. 1776 (1967).
60. Doroshkevich, A. G., Zel'dovich, Ya. B., and Novikov, I. D. *Zu. Eksp. Teor. Fiz. USSR*, **49**, pp. 170–181 (1965), (Transl. *Soviet Phys. JETP*, **22**, p. 122 (1966)).
61. Regge, T., and Wheeler, J. A. *Phys. Rev.*, **108**, p. 1063 (1957).
62. Price, R. *Phys. Rev.* **D5** p. 2419 (1972).
63. Ginzburg, V. L., and Ozernoi, L. M. *Soviet Phys. JETP*, **20**, p. 689 (1965).
Anderson, J. L., and Cohen, J. M. Preprint, 1970.
64. DeWitt, B., and Brehme, R. W. *Ann. Phys.* **9**. p. 220 (1960).
65. Einstein, A., and Pauli, W. *Ann. of Math.*, **44**, pp. 131–137 (1943).
66. Zerilli, F. *Phys. Rev.*, **D2**, p. 2411 (1970); reproduced here as A.7.
67. Brill, D. in *Perspectives in Geometry and Relativity*: *Essays in Honour of V. Hlavaty* (B. Hoffmann, ed.), Indiana University Press, 1966.
68. Ruffini, R., and Wheeler, J. A. *Bull. Am. Phys. Soc.*, **15**, Ser. II, p. 76 (1970).
69. Reines, F., Cowen, L., Jr., and Goldhaber, M. *Phys. Rev.*, **96**, p. 1157 (1954).
Reines, F., and Giamati, C. C. *Phys. Rev.*, **126**, p. 2178 (1962).
Feinberg, G., and Goldhaber, M. *Proc. Nat. Acad. Sc. USA.*, **45**, p. 1301 (1959).
70. Backenstoss, G., Frauenfelder, H., Hyams, B. D., Koestler, L. J., and Marin, P. C. *Nuovo Cim.*, **16**, p. 749 (1960).
71. Feinberg, G., and Sucher, J. *Phys. Rev.*, **166**, p. 1638 (1968) (Feinberg gives references to the early papers on the lepton-lepton interaction; the history of the subject goes back at least as far as W. Heisenberg, Lectures at the Cavendish Laboratory, Cambridge, 1934, unpublished).
Hartle, J. D. *Phys. Rev.*, **D1**, p. 394 (1970).
72. Kerr, R. P. *Phys. Rev. Lett.*, **11**, p. 237 (1963).
73. Newman, E. T., Couch, E., Chinnapared, R., Exton, A., Prakash, A., and Torrence, R. *J. Math. Phys.*, **6**, p. 918 (1965).
74. Carter, B. *Phys. Rev.*, **141**, p. 1242 (1966); *Phys. Rev.*, **174**, p. 1559 (1968).
75. Boyer, R. H., and Lindquist, R. W. *J. Math. Phys.*, **8**, p. 265 (1967).
76. Vishveshwara, C. V. *J. Math. Phys.*, **9**, p. 1319 (1968).
77. Penrose, R. *Nuovo Cim.*, I, Special Issue, p. 252 (1969).
78. Doroshkevich, A. G. *Astrofizika*, **1**, p. 255 (1965).
79. Godfrey, B. B. *Phys. Rev.*, **D1**, p. 2721 (1970). (see also Ph. D. Thesis, Princeton University, 1970, unpublished).
80. Borogorodsky, A. F. Izdatel'stvo Kievskohog Universitet, 1962.
81. Bardeen, J. M. *Nature*, **226**, p. 64 (1970).
82. Christodoulou, D. *Bull. Am. Phys. Soc.*, **15**, Ser. II, p. 661 (1970).
83. This estimate comes as a personal communication from Mr X of T. Gold, ed., *The Nature of Time*, Cornell University Press, Ithaca, New York (1967).
84. Lin, C. C., Mestel, L. and Shu, F. H. *Ap. J.* **142**, p. 1431 (1965).
85. Thorne, K. S. Ph. D. Thesis, Princeton University, 1965 (unpublished), (also more recent work which is in press).
86. Edgerton, H. E., and Killian, J. R. *Flash!*, Cushman and Flint, Boston, 1939.
87. Strutt, J. W. *Third Baron Rayleigh*, *Collected Papers*, Cambridge University Press, 1920.
88. Bohr, N. *Proc. R. Soc. London*, **A84**, p. 395 (1910).
89. Kasner, E. *Trans. Am. Math. Soc.*, **27**, p. 101 (1925).

90. Khalatnikov, I. M., and Lifschitz, E. M. *Phys. Rev. Lett.*, **24**, p. 76 (1970).
91. Misner, C. W. *Phys. Rev.*, **186**, pp. 1319 and 1328 (1969).
92. Baade, W., and Zwicky, F. *Proc. Nat. Acad. Sci.* USA, **20** 254 (1934).
93. Zel'dovich, Ya. B., and Guseynov, O. Kh. *Dokl. Akad. Nauk. SSSR*, **162**, p. 791 (1965).
94. Shklovskii, I. S. *Ap. J. Lett.*, **148**, L1 (1967).
95. Trimble, V. L., and Thorne, K. S. *Ap. J.*, **156**, p. 1013 (1969).
96. Gaposchkin, V. F. *Handbuch der Physik, Bandl.* **L-225**, Springer-Verlag, 1958.
97. Struve, O. *Stellar Evolution*, Princeton University Press, 1950.
98. Pringle, J. E. and Rees, M. J. *Astron. Astrophys.*, **21**, p. 1 (1972).
99. Prendergast, K. H., and Burbidge, G. R. *Ap. J. Lett.*, **151**, L83 (1968).
100. Salpeter, E. E. *Ap. J.*, **140**, p. 796 (1964).
101. Giacconi, R., Gursky, H., Kellogg, E., Murray, S., Schreier, E., and Tananbaum, H. The UHURU Catalogue of x-ray sources, *Ap. J.* **178**, p. 281 (1972).
102. Lynden-Bell, D. *Nature*, **223**, p. 690 (1969).
Lynden-Bell, D., and Rees, M. J. *Mon. Not. R. Astr. Soc.*, **152**, p. 461 (1971).
103. Oke, J. B. *Ap. J.*, **147**, p. 901 (1967).
104. Bahcall, J. N., Gunn, J. E., and Schmidt, M. *Ap. J. Lett.*, **157**, L77 (1969).
105. Cohen, M. H. *Ann Rev. Astr. Astrophys.*, **7**, p. 619 (1969) for a review and bibliography.
106. Kellermann, K. I., and Pauliny-Toth, I. I. K. *A. Rev. Astr. Astrophys.*, **6**, p. 417 (1968) for a review.
107. Friedman, H., and Byram, E. T. *Science*, **158**, p. 257 (1967).
108. Schmidt, M. *Ann. Rev. Astr. Astrophys.*, **7**, p. 527 (1969).
109. Kleinmann, D. E., and Low, F. J. *Ap. J. Lett.*, **159**, L165 (1970).
110. Greenstein, J. L., and Schmidt, M. *Ap. J.*, **140**, p. 1 (1964).
111. Einstein, A. *Sitzber. Preuss. Akad. Wiss.*, p. 588, 1916; p. 154, 1918.
112. Weber, J. *Phys. Rev.*, **117**, p. 306 (1960) (see also Ref. 127). Graviational Radiation Experiment, *in Relativity, Groups and Topology* (C. DeWitt and B. DeWitt, eds.), Gordon and Breach Publ., Inc., New York, 1964.
Sinsky, J., and Weber, J. *Phys. Rev. Lett.*, **18**, p. 795 (1967).
Weber, J. *Phys. Rev. Lett.*, **17**, p. 1228 (1966); in *Physics of the Moon* (S. F. Singer, ed.), American Astronautical Soc., Hawthorne, California, 1967; *Phys. Rev. Lett.*, **20**, p. 1307 (1968); (see also Refs. 114 and 117).
113. Breit, G., Yost, F. L., and Wheeler, J. A. *Phys. Rev.*, **49**, p. 174 (1936).
114. Weber, J. *Phys. Rev. Lett.*, **21**, p. 395 (1968).
115. Weber, J. *Phys. Rev. Lett.*, **24**, p. 276 (1970).
116. Weber, J. *Phys. Rev. Lett.*, **22**, p. 1320 (1969) (Table I entry No. 5 for 30 Jan. 1969).
117. Weber, J. *Phys. Rev. Lett.*, **18**, p. 498 (1967).
Weber, J., and Larson, J. V. *J. Geophys. Res.*, **71**, p. 6005 (1966).
118. de la Cruz, V., Chase, J. E., and Israel, W. *Phys. Rev. Lett.*, **24**, p. 423 (1970).
119. Weber, J. *Phys. Rev.* Lett. **25**, p. 180 1970.
120. Eddington, A. S. *Proc. R. Soc.*, **A102**, p. 268 (1922).
121. Landau, L., and Lifschitz, E. M. *The Classical Theory of Fields* (*2nd ed.*) (translation by M. Hamermesh), p. 318, Addison-Wesley, Cambridge (now Reading), Massachusetts 1951.
122. *Op. cit.*, pp. 320–323.
123. Petrov, A. Z. *Einstein Spaces* (translated from Russian by R. F. Kelleher), Pergamon Press, London, 1969.

124. See Ref. 121, p. 326.
125. Breit, G., and Wigner, E. P. *Phys. Rev.*, **49**, p. 519 (1936).
126. Kuhn, W. *Zeits. Phys.*, **33**, p. 408 (1925).
Reiche, F., and Thomas, W. *Zeits. Phys.*, **34**, p. 510 (1925).
127. Weber, J. in *General Relativity and Gravitational Radiation*, p. 128, Interscience Publishers, Inc., New York, 1961.
128. For a review see for example
Bolt, A. *Physics and Chemistry of the Earth, Vol. V* (L. H. Ahrens, F. Press and S. K. Runcorn, eds.), Pergamon Press, London, 1964.
Press, F. Resonant Vibrations of the Earth, *Scient. Am.*, **213**, No. 5, p. 28 (1965). (see also cited work of C. Pekeris and others).
Backus, G. E., and Gilbert, F. The Rotational Splitting of the Free Oscillations of the Earth, *Proc. Nat. Acad. Sci. USA*, **47**, p. 362 (1961).
129. Dyson, F. J. *Ap. J.*, **156**, p. 529 (1969).
130. Poincaré, H. *Acta Math.*, **7**, pp. 259–380 (1885).
Appel, P. E. *Traité de mécanique rationnelle*, Vol. V–VI., Gauthier-Villars, Paris, 1937.
Chandrasekhar, S. *Ellipsoidal Figures of Equilibrium*, Yale University Press, 1969.
131. Jeans, J. H. *Astronomy and Cosmology*, Cambridge University Press, 1919.
132. Smoluchowski, R. *Phys. Rev. Lett.*, **24**, p. 923 (1970).
Smoluchowski, R., and Welch, D. O. *Ibid.*, p. 1191 (1970).
133. Rhoades, C. E. ,Jr., and Ruffini, R. *Ap. J. Lett.*, **163**, L83 (1971).
134. van de Kamp, P., and Gaposchkin, S. *Handbuch der Physik*, **50**, p. 187 (1958); *Ibid.*, p. 225 (1958).
135. Kraft, R. P., Mathews, J., and Greenstein, J. L. *Ap. J.*, **136**, p. 312 (1962).
136. Zee, A., and Wheeler, J. A. cited in *A. Rev. Astr. Astrophys.*, **4**, pp. 423–427 (1966).
137. Morganstern, R. E., and Chiu, H.-Y. *Phys. Rev.* **157**, p. 1228 (1967).
138. Thorne, K. S. *Ap. J.*, **158**, p. 1 (1969).
139. Meltzer, D. W., and Thorne, K. S. *Ap. J.*, **145**, p. 514 (1966).
140. Thorne, K. S. *Phys. Rev. Lett.*, **21**, p. 320 (1968).
141. Thorne K. S. and Campolattaro A. *Ap. J.* **149** p. 591 (1967); *Ap. J.* **152**, p. 673 (1968).
142. Price, R., and Thorne, K. S. *Ap. J.*, **155**, p. 163 (1969).
143. Bohr, A., and Mottelson, B. *Nuclear Structure*, Vol. I, W. A. Benjamin, Inc., New York, 1969.
144. Peters, P. C., and Mathews, J. *Phys. Rev.*, **131**, p. 435 (1963).
145. Halpern, L., and Laurent, B. *Nuovo Cim.*, **33**, p. 728 (1964).
146. Ivanenko, D., and Sokolov, A. *Vest. Mos. Gos. Univ.*, No. **8** (1947).
Ivanenko, D., and Brodski, A. *C. R.* (*Dokl.*) *Akad. Sci.*, *USSR*, **92**, p. 731 (1953).
Ivanenko, D., and Sokolov, A. *Quantum Theory of Fields Part II*, Section 5 (in Russian) Moscow, 1952.
147. Bertotti, B., Brill, D., and Krotkov, R. in *Gravitation, An Introduction to Current Research* (L. Witten, J. Wiley and Sons, New York, 1962).
148. Seielstad, G. A., Sramek, R. A., and Weiler, K. W. *Phys. Rev. Lett.*, **24**, p. 1373 (1970).
149. Muhlemann, D. O., Ekers, R. D., and Formalont, E. B. *Ibid.*, p. 1377 (1970).
150. Pound, R. W., and Snider, J. L. *Phys. Rev.*, **140B**, p. 788 (1965).
151. Dicke, R. H., and Goldenberg, H. M. *Phys. Rev. Lett.*, **18**, p. 313 (1967).
152. Ingersoll, A. P., and Spiegel, E. A. *Ap. J.*, **163**, p. 375 (1971).

153. Shapiro, I. I. Radio and Radar Test of General Relativity—Oral Report given at the Third Cambridge Conference on Relativity, New York, 8 June 1970.
154. Shapiro, I. I. *Phys. Rev. Lett.*, **13**, p. 789 (1964), *Scient. Am.*, **219**, No. 1, p. 28 (1968).
Muhlemann, D. O., and Reichley, P. E. Space Program Summary 37–29, 37, 31, Vol. IV (Jet Propulsion Laboratory), 1964.
Aoki, S. *Astr. J.*, **69**, p. 221 (1964).
Ross, D. K., and Schiff, L. *Phys. Rev.*, **141**, p. 1215 (1965).
Clemence, G. M. *Astr. J.*, **72**, p. 1324 (1967).
155. Shapiro, I. I., Pettengill, G. H., Ash, M. E., Stone, M. L., Smith, W. B., Ingalls, R. P., and Brockelman, R. A. *Phys. Rev. Lett.*, **20**, p. 1265 (1968).
156. Anderson, J. D., Pearl, G. E., Efron, L., and Taushworthe, R. C. *Science*, **158**, p. 1689 (1967).
157. Anderson, J. D., and Lorell, J. *AIAA Journal*, **1**, p. 1372 (1963).
158. de Sitter, W. On Einsteins' Theory of Gravitation and its Astronomical Consequences, *Mon. Not. R. Astr. Soc.*, **76**, pp. 669–728 (1916).
159. Chazy, J. *La Théorie de la Relativité et la Mécanique Céleste*, Gauthier-Villars, Paris, 1928.
160. Moyer, D. UCLA MS Thesis, 1965; Jet Propulsion Laboratory Technical Memo., 312–548.
161. Tausner, M. J. MIT Technical Report 425, 1966.
162. Thorne, K. S., and Will, C. M. *Comm. Astrophys. and Space Physics*, **11**, p. 35 (1970).
163. Smullin, L., and Fiocco, G. *Proc. I.R.E.*, **50**, p. 1703 (1962).
164. Baierlein, R. *Phys. Rev.*, **162**, p. 1275 (1967).
Krogh, C., and Baierlein, R. *Phys. Rev.*, **175**, p. 1576 (1968).
165. Nordtvedt, K. *Phys. Rev.*, **169**, p. 1014 (1969).
166. Everitt, L. W. F., Fairbank, W. M., and Schiff, L. I. in *The Significance of Space Research for Fundamental Physics*, A. F. Moore and V. Hardy, ed. (ESRO book No. SP 52 (1971)), p. 33.
167. Hubble, E. *Proc. Nat. Acad. Sci.*, **15**, p. 168 (1929); *The Realm of the Nebulae*, O.U.P. (1936).
168. Milne, E. A., and McCrea, W. H. *Q. J. Math.* **5**, p. 73 (1934).
169. Sandage, A. R. *Ap. J.*, **152**, L149 (1968).
170. Sandage, A. R. *Observatory*, **88**, p. 91 (1968).
171. de Vaucouleurs, G. *Ap. J.*, **159**, p. 435 (1970).
172. Sandage, A. R. *Pont. Acad. Sci Scripta Varia* "*Nuclei of Galaxies*" D. O'Connell, ed., p. 601 (1971).
173. Van den Bergh, S. *Nature*, **225**, p. 503 (1970).
174. Fowler, W. A. *Gamow Memorial Volume* (in press).
175. Bondi, H., and Gold, T. *Mon. Not. R. Astr. Soc.*, **108**, p. 252 (1948).
176. Hoyle, F. *Mon. Not. R. Astr. Soc.*, **108**, p. 372 (1948).
177. Scott, E. L. in *Problems of Extragalactic Research*, G. C. McVittie, ed., p. 269 (Macmillan, N. Y. 1961).
178. Spinrad, H. *Publ. Astr. Soc. Pacific*, **78**, p. 367 (1966).
179. Tinsley, B. M. *Astrophys. J.*, **151**, p. 547 (1968).
180. Dashevskii, V. M., and Zeldovich, Ya. B. *Sov. Astron.*, **8**, p. 854 (1964).
181. Longair, M. S. *Rep. Prog. Phys.*, **34**, p. 1125 (1971).
182. Pooley, G. C., and Ryle, M. *Mon. Not. R. Astr. Soc.*, **139**, p. 515 (1968).
183. Gower, J. F. R. *Mon. Not. R. Astr. Soc.*, **133**, p. 151 (1966).
184. Grueff, G., and Vigotti, M. *Astrophys. Letters*, **2**, p. 113 (1968).

185. Shimmins, A. J., Bolton, J. G., and Wall, J. V. *Nature*, **217**, p. 818 (1968).
186. Galt, J. A., and Kennedy, J. E. D. *Astron. J.*, **73**, p. 135 (1968).
187. Bolton, J. G., Gardner, F., and Mackay, M. B. *Aust. J. Phys.*, **17**, p. 340 (1964).
188. Macleod, J. M., Swenson, G. W., Yang, K. S., and Dickel, J. R. *Astron. J.*, **70**, p. 756 (1965).
189. Braccesi, A., Ceccarelli, M., Fanti, R., and Giovannini, C. *Nuovo Cim.*, **41B**, p. 92 (1966).
190. Ryle, M. *Ann. Rev. Astron. Astrophys.*, **6**, p. 249 (1968).
191. Scheuer, P. A. G. in *Stars and Stellar Systems*, Vol. IX, A. and M. Sandage ed. (Chicago University Press, in press).
192. Jauncey, D. L. *Nature*, **216**, p. 877 (1967).
193. Hoyle, F. *Proc. Roy. Soc.*, **A308**, p. 1 (1969).
194. Hubble, E. *Ap. J.*, **84**, p. 517 (1936).
195. Schmidt, M. *Ap. J.*, **162**, p. 371 (1970).
196. Sandage, A. R., and Luyten, W. J. *Ap. J.*, **155**, p. 913 (1969).
197. Braccesi, A., and Formiggini, L. *Astron. Astrophys.*, **3**, p. 364 (1969).
198. Burbidge, E. M., and Lynds, C. R. *Sci. American*, **223**, p. 22 (1970).
199. Longair, M. S., and Scheuer, P. A. G. *Nature*, **215**, p. 919 (1967).
200. Schmidt, M. *Ap. J.*, **151**, p. 393 (1968).
201. Rees, M. J., and Schmidt, M. *Mon. Not. R. Astr. Soc.* **154**, p. 1 (1971).
202. Longair, M. S., and Scheuer, P. A. G. *Mon. Not. R. Astr. Soc.*, **151**, p. 45 (1970).
203. Lynden-Bell, D. *Mon. Not. R. Astr. Soc.*, **155**, p. 95 (1971).
204. Rowan-Robinson, M. *Nature*, **229**, p. 388 (1971).
205. DeVeny, J. B., Osborn, W. H., and Janes, K. *Publ. Astr. Soc. Pacific*, **83**, p. 611 (1970).
206. Sandage, A. R. *Ap. J.*, **133**, p. 355 (1961).
207. Miley, G. K. *Mon. Not. R. Astr. Soc.*, **152**, p. 477 (1971).
208. Legg, T. H. *Nature*, **226**, p. 65 (1970).
209. Longair, M. S., and Pooley, G. C. *Mon. Not. R. Astr. Soc.*, **145**, p. 309 (1969).
210. Baum, W. A. *External Galaxies and Quasi-Stellar Objects* D. S. Evans, ed. p. 393 (Reidel, Dordrecht; 1971).
211. Penzias, A. A., and Wilson, R. W. *Astrophys. J.*, **142**, p. 419 (1965).
212. Dicke, R. H., Peebles, P. J. E., Roll, P. G., and Wilkinson, D. T. *Astrophys. J.*, **142**, p. 414 (1965).
213. Gamow, G. *Phys. Rev.*, **74**, p. 505 (1948).
214. Gamow, G. *Vistas in Astronomy*, **2**, p. 1726 (1956).
215. Roll, P. G., and Wilkinson, D. T. *Phys. Rev. Letters*, **16**, p. 405 (1966).
216. Boynton, P. E., Stokes, R. A., and Wilkinson, D. T. *Phys. Rev. Letters*, **21**, p. 462 (1968).
217. Millea, M. F., McColl, M., Pederson, R. J., and Vernon, F. L. *Phys. Rev. Letters*, **26**, p. 919 (1971).
218. McKellar, A. *Publ. Dom. Astrophys. Obs.*, **7**, p. 251 (1941).
219. Herzburg, G. *Spectra of Diatomic Molecules*, 2nd Edition. Van Nostrand, New York, p. 496 (1950).
220. Field, G. B., and Hitchcock, J. L. *Ap. J.*, **146**, p. 1 (1966).
221. Bortolet, V. J., Clauser, J. F., and Thaddeus, P. *Phys. Rev. Letters*, **22**, p. 307 (1969).
222. Howell, T. F., and Shakeshaft, J. R. *Nature*, **216**, p. 753 (1967).
223. Penzias, A. A., and Wilson, R. W. *Astron. J.*, **72**, p. 315 (1967).
224. Howell, T. F., and Shakeshaft, J. R. *Nature*, **210**, p. 1318 (1966).

225. Stokes, R. A., Partridge, R. B., and Wilkinson, D. T. *Phys. Rev. Letters*, **19**, p. 1199 (1967).
226. Welch, W. J., Keachie, S., Thornton, D. D., and Wrixon, G. *Phys. Rev. Letters*, **18**, p. 1068 (1967).
227. Ewing, M. S., Burke, B. F., and Staelin, D. H. *Phys. Rev. Letters*, **19**, p. 1251 (1967).
228. Wilkinson, D. T. *Phys. Rev. Letters*, **19**, p. 1251 (1967).
229. Puzhano, V. I., Salomonovich, A. E., and Stankevich, K. S. *Sov. Astron.*, **11**, p. 905 (1967).
230. Kislyakov, A. G., Chernuskev, V. I., Lebsky, Y. V., Maltsev, V. A. and Serov, N. V. *Astron. Zh.*, **48**, p. 39 (1971).
231. Shivananden, K., Houck, J. R., and Harwit, M. O. *Phys. Rev. Letters*, **21**, p. 1460 (1968).
232. Pipher, J. L., Houck, J. R., Jones, B. W., and Harwit, M. O. *Nature*, **231**, p. 375 (1971).
233. Muehlner, D., and Weiss, R. *Phys. Rec. Letters*, **24**, p. 742 (1970).
234. Mather, J. C., Werner, M. W., and Richards, P. L. *Ap. J.*, **159**, L 59 (1971).
235. Blair, A. G., Beery, J. G., Edeskuty, F., Hiebert, R. D., Shipley, J. P., and Williamson, K. D. *Phys. Rev. Letters*, **27**, p. 1154 (1971).
236. Conklin, E. K., and Bracewell, R. N. *Phys. Rev. Letters*, **18**, p. 614 (1967); *Nature*, **216**, p. 777 (1967).
237. Henry, P. S. *Nature*, **231**, p. 516 (1971).
238. Conklin, E. K. *Nature*, **222**, p. 971 (1969).
239. Pariiskii, Y., and Pyatunina, T. *B. Soc. Astron.*, **14**, p. 1068 (1970).
240. Boughn, S. P., Fram, D. M., and Partridge, R. B. *Ap. J.*, **165**, p. 439 (1971).
241. Penzias, A. A., Schraml, J., and Wilson, R. W. *Ap. J.*, **157**, L 49 (1969).
242. Partridge, R. B., and Wilkinson, D. T. *Phys. Rev. Letters*, **18**, p. 557 (1967).
243. Sciama, D. W. *Nature*, **211**, p. 277 (1966).
244. Hazard, C., and Salpeter, E. E. *Ap. J.*, **157**, L87 (1969).
245. Gold, T., and Pacini, F. *Ap. J.*, **152**, L115 (1968).
246. Pariiskii, Y. *Sov. Astron.*, **12**, p. 219 (1969).
247. Wolfe, A. M., and Burbidge, G. R. *Ap. J.*, **156**, p. 345 (1969).
248. Smith, M. G., and Partridge, R. B. *Ap. J.*, **159**, p. 737 (1970).
249. Narlikar, J. V., and Wickramasinghe, N. C. *Nature*, **216**, p. 43 (1967).
250. Hoyle, F., and Wickramasinghe, N. C. *Nature*, **214**, p. 969 (1967).
251. Field, G. B. *Mon. Not. R. Astr. Soc.*, **144**, p. 411 (1969).
252. Layzer, D. *Astrophys. Letters*, **1**, p. 99 (1968).
253. Setti, G., and Woltjer, L. *Nature*, **227**, p. 586 (1970).
254. Hagedorn, R. *Cargèse Lectures in Physics*, VI, E. Schatzman, ed. p. 643, Gordon and Breach (1973).
255. Kundt, W. *Springer Tracts in Physics*, No. 49 (1971).
256. Omnes, R. *Phys. Reports* **3C**, p. 1 (1972)
257. Zeldovich, Y. B., Kurt, D., and Sunyaev, R. A. *J.E.T.P.*, **28**, p. 146 (1968).
258. Peebles, P. J. E. *Ap. J.*, **153**, p. 1 (1968).
259. Sunyaev, R. A., and Zeldovich, Y. B. *Astrophys. and Sp. Phys.*, **7**, p. 3 (1970).
260. Peebles, P. J. E., and Yu, J. T. *Ap. J.*, **162**, p. 815 (1970).
261. Peebles, P. J. E. *Ap. J.*, **146**, p. 542 (1966).
262. Wagoner, R. V., Fowler, W. A., and Hoyle, F., *Ap. J.*, **148**, p. 3 (1967).
263. Tayler, R. J. *Rep. Prog. Phys.*, **29**, p. 489 (1966).

264. Danziger, I. *J. Am. Rev. Astron. and Astrophys.*, **8**, p. 161 (1970).
265. Davis, R. J., Roger, L. C., and Radena, V. presented at meeting of APS, Washington, April 1971.
266. Christiansen, C. J., Nielsen, A., Bahnsen, A., Brown, W. K., and Rustad, B. N. *Phys. Letters*, **26B**, p. 11 (1967).
267. Tayler, R. J. *Nature*, **217**, p. 433 (1968).
268. Wagoner, R. V. unpublished.
269. Wagoner, R. V. *Ap. J.*, **151**, L103 (1968).
270. Arnett, W. D., and Clayton, D. D. *Nature*, **227**, p. 780 (1970).
271. Ne'eman, Y., and Tauber, G. *Ap. J.*, **150**, p. 755 (1967).
272. Lifshitz, E. *J.E.T.P.*, **10**, p. 116 (1946).
273. Zeldovich, Ya. B. *Adv. in Astron and Astrophys.*, **3**, p. 241 (1965).
274. Harrison, E. R. *Cargèse Lectures in Physics* VI, E. Schatzman, ed. p. 581, Gordon and Breach (1973).
275. Field, G. B. in *Stars and Stellar Systems*, Vol. IX, ed. A and M. Sandage (Chicago University Press, in press).
276. Einstein, A., and Strauss, E. G. *Rev. Mod. Phys.*, **17**, p. 120 (1945).
277. Silk, J. I. *Ap. J.*, **151**, p. 459 (1968).
278. Peebles, P. J. E., *J. R. Astron. Soc. Canada*, **63**, p. 4 (1969).
279. Michie, R. W. unpublished.
280. Weinberg, S. *Ap. J.*, **168**, p. 175 (1971).
281. Harrison, E. R. *Phys. Rev. Letters*, **18**, p. 1011 (1967).
282. Ozernoi, L. M., and Chernin, A. D. *Sov. Astron.*, **11**, p. 907 (1968).
283. Misner, C. W. *Nature*, **214**, p. 40 (1967).
284. Misner, C. W. *Ap. J.*, **151**, p. 431 (1968).
285. Field, G. B. *Ap. J.*, **165**, p. 29 (1971).
286. Sunyaev, R. A., and Zeldovich, Ya. B. *Astrophys. and Sp. Sci.*, **9**, p. 368 (1970).
287. Peebles, P. J. E. *Phys. Rev.*, **D.1.**, p. 397 (1970).
288. Ozernoi, L. M., and Chernin, A. D. *Sov. Astron.*, **12**, p. 901 (1969).
289. Ozernoi, L. M., and Chibisov, G. V. *Sov. Astron.*, **14**, p. 615 (1971).
290. Peebles, P. J. E. *Astrophys. and Sp. Sci*, **11**, p. 443 (1971).
291. Oort, J. H. *Astron. and Astrophys.*, **7**, p. 381 (1970).
292. Nariai, H. *Sci. Rept. Tohoku Univ.*, **39**, p. 213 (1965).
293. Tomita, K., Nariai, H., Sato, H., Matsuda, T., and Takeda, H. *Proc. Theor. Phys.*, **43**, p. 1511 (1970).
294. Eggen, O. J., Lynden-Bell, D., and Sandage, A. R. *Ap. J.*, **136**, p. 748 (1962).
295. Peebles, P. J. E. *Ap. J.*, **157**, p. 1075 (1969).
296. Peebles, P. J. E., and Dicke, R. H. *Ap. J.*, **154**, p. 891 (1968).
297. Doroshkevich, A. G., Zeldovich, Y. B., and Novikov, I. D. *Sov. Astron*, **11**, p. 233 (1967).
298. Peebles, P. J. E. *Astron. and Astrophys.*, **11**, p. 377 (1971).
299. Arp, H. C. *Atlas of Peculiar Galaxies* (California Institute of Technology 1966).
300. Burbidge, G. R. *Ann. Rev. Astron. and Astrophys.*, **8**, p. 369 (1970).
301. Burbidge, E. M., and Burbidge, G. R. in *Stars and Stellar Systems*, Vol. IX, A. and M. Sandage ed. (Chicago University Press, in press).
302. Abell, G. O. in *Stars and Stellar Systems*, Vol. IX, A. and M. Sandage, ed (Chicago University Press, in press).
303. Oort, J. H., Solvay Conference on *Structure and Evolution of the Universe*, p. 163 (1958).

304. Shapiro, S. L. *Astron, J.*, **76**, p. 271 (1971).
305. Arp, H. C., and Bertola, F. *Astrophys. J.*, **163**, p. 195 (1971).
306. Arp, H. C. *Ap. J.*, **142**, p. 402 (1965).
307. Allen, R. J. *Nature*, **220**, p. 147 (1968).
308. Field, G. B. *Ann. Rev. Astron. and Astrophys.*, **10**, p. 227 (1972).
309. Woolf, N. J. *Ap. J.*, **148**, p. 287 (1967).
310. Turnrose, B. E., and Rood, H. J. *Ap. J.*, **159**, p. 773 (1970).
311. Meekins, J. F., Fritz, G., Chubb, T. A., Friedman, H., and Henry, R. C., *Nature*, **231**, p. 107 (1971).
312. Hill, J. M., and Longair, M. S. *Mon. Not. R. Astron. Soc.*, **154**, p. 125 (1971).
313. Zwicky, F. *Morphological Astronomy*, Springer-Verlag, Berlin (1957).
314. Peebles, P. J. E. *Ap. J.*, **154**, L121 (1968).
315. Sunyaev, R. A., and Zeldovich, Ya. B. *Comments on Astrophys. and Sp. Phys.*, **1**, p. 159 (1969).
316. Van den Bergh, S. *Nature*, **224**, p. 891 (1969).
317. Wolfe, A. M., and Burbidge, G. R. *Ap. J.*, **161**, p. 419 (1970).
318. Burbidge, E. M., and Sargent, W. L. W. *Pont. Acad. Sci. Scripta Varia*, **35**, *Nuclei of Galaxies*, D. O'Connell, ed. (1971).
319. Verontsov-Velyaminov, B. A. *Astron. Tsirk*, No. 513, p. 5 (1965).
320. Gunn, J. E. *Ap. J.*, **164**, L113 (1971).
321. Arp. H. C. *Astrophys. Letters*, **7**, p. 221 (1971).
322. Penzias, A. A., and Scott, E. H. *Ap. J.*, **153**, L7 (1968).
323. Allen, R. J. *Astron. and Astrophys.*, **3**, p. 382 (1969).
324. Gunn, J. E., and Peterson, B. A. *Ap. J.*, **142**, p. 1633 (1965).
325. Scheuer, P. A. G. *Nature*, **207**, p. 963 (1965).
326. Field, G. B., Solomon, P. M., and Wampler, E. J. *Ap. J.*, **145**, p. 351 (1966).
327. Shklovski, I. S. *Astron. Tsirk*, No. 303, p. 3 (1964).
328. Bahcall, J. N., and Salpeter, E. E. *Ap. J.*, **142**, p. 1677 (1965).
329. Rees, M. J., and Sciama, D. W. *Ap. J.*, **147**, p. 353 (1967).
330. Bahcall, J. N., *Comments on Astrophys. and Sp. Phys.*, **2**, p. 221 (1970).
331. Bahcall, J. N., and May, R. M. *Ap. J.*, **152**, p. 37 (1968).
332. Bridle, A. H. *Mon. Not. R. Astr. Soc.*, **136**, p. 219.
333. Clark, T. A., Brown, L. W., and Alexander, J. K. *Nature*, **228**, p. 847 (1970).
334. Weinberg, S. *Gravitation and Cosmology* (Wiley, N. Y., 1972).
335. Haddock, F. T., and Sciama, D. W. *Phys. Rev. Letters*, **14**, p. 1007 (1965).
336. Weymann, R. J. *Ap. J.*, **147**, p. 887 (1967).
337. Bergeron, J. E. *Astron. and Astrophys.*, **3**, p. 42 (1969).
338. Rees, M. J., and Setti, G. *Astron. and Astrophys.*, **8**, p. 410 (1970).
340. Kurt, V. G., and Sunyaev, R. A., IAU symposium No. 36 *Ultraviolet Stellar Spectra*, Reidel and Co., Dordrecht, Holland (1970).
341. Peebles, P. J. E. *Ap. J.*, **151**, p. 45 (1969).
342. Sofue, J., Fujimoto, M., and Kawabata, K. *Publ. Astron. Soc. Japan*, **20**, p. 388 (1968).
343. Kawabata, K., Fujimoto, M., Sofue, Y., and Fuikui, M. *Publ. Astron. Soc. Japan*, **21**, p. 293 (1969).
344. Reinhardt, M., and Thiel, M. *Astrophys. Letters*, **7**, p. 101 (1970).
345. Harrison, E. R. *Mon. Not. R. Astr. Soc.*, **147**, p. 279 (1970).
346. Ginzburg, V. L. *Comments on Astrophys. and Sp. Phys.*, **2**, p. 1 (1970).
347. Weymann, R. J. *Ap. J.*, **145**, p. 560 (1966).
348. Zeldovich, Y. B., and Sunyaev, R. A. *Astrophys. and Sp. Sci.*, **4**, p, 301 (1969).

349. Doroshkevich, A. G., and Sunyaev, R. A. *Soviet Astron.*, **13**, p. 15 (1969).
350. Longair, M. S., and Sunyaev, R. A. *Astrophys. Letters*, **4**, p. 65 (1969).
351. Cowsik, R., Pal, Y., and Tandon, S. N. *Phys. Letters*, **13**, p. 265 (1964).
352. Gerstein, S. S., and Zeldovich, Ya. B. *J.E.T.P. Letters*, **4**, p. 20 (1966).
353. Zeldovich, Y. B. *Comments on Astrophys. and Sp. Phys.*, **2**, p. 12 (1970).
354. Zeldovich, Y. B., and Novikov, I. D. *Sov. Astron.*, **13**, p. 754 (1970).
355. Rees. M. J. *Mon. Not. R. Astr. Soc.*, **154**, p. 187 (1971).
356. Misner, C. W. *Phys. Rev. Letters*, **22**, p. 1071 (1969).
357. Zeldovich, Y. B., and Novikov, I. D. *Sov. Astron.*, **10**, p. 602 (1967).
358. Refsdal, S. *Ap. J.*, **159**, p. 357 (1970).
359. Reber, G. *J. Franklin Inst.*, **285**, p. 1 (1968).
360. Houck, J. R., and Harwit, M. O. *Science*, **164**, p. 1271 (1969).
361. McNutt, D. P., and Feldman, P. D. *J. Geophys. Res.*, **74**, p. 4791 (1969).
362. Walker, R. G., and Price, S. G. private communication to P. J. E. Peebles (ref. 364).
363. Harwit, M. D., McNutt, D. P., Shivanandan, K., and Zajar, B. J. *Astron. J.*, **71**, p. 1026 (1966).
364. Peebles, P. J. E. *Comments on Astrophys. and Sp. Phys.*, **3**, p. 20 (1971).
365. Peebles, P. J. E., and Partridge, R. B. *Ap. J.*, **148**, p. 713 (1967).
366. Low, F. J., and Tucker, W. H. *Phys. Rev. Letters*, **21**, p. 1538 (1968).
367. Houck, J. R., and Harwit, M. O. *Ap. J.*, **157**, L45.
368. Roach, F. E., and Smith, L. L. *Geophys. J.*, **15**, p. 227 (1968).
369. Lillie, C. F. *Bull. Amer. Astron. Soc.*, **1**, p. 132 (1969).
370. Sudbury, G. C., and Ingham, M. F. *Nature*, **226**, p. 526 (1970).
371. Hayakawa, S., Yamashito, K., Yoshioka, S. *Astrophys. and Sp. Sci.*, **5**, p. 493 (1969).
372. Kurt, V. G., and Sunyaev, R. A. *Sov. Astron.*, **11**, p. 928 (1967).
373. Low, F. J. *Ap. J.*, **159**, L171 (1970).
374. Petrosian, V., Bahcall, J. N., and Salpeter, E. E. *Ap. J.*, **155**, 157 (1969).
375. Partridge, R. B., and Peebles, P. J. E. *Ap. J.*, **148**, p. 377 (1967).
376. Weymann, R. J. Steward Observatory Preprint No. 6 (1966).
377. Code, A. D. *Publ. Astro. Soc. Pacific*, **81**, p. 475 (1969).
378. Code, A. D., and Savage, B. D. *Science*, **177**, p. 213 (1972).
379. Sunyaev, R. A. *Astrophys. Letters*, **3**, p. 33 (1969).
380. Silk, J. I. *Space. Sci. Rev.*, **11**, p. 671 (1970).
381. Byram, E. T., Chubb, T. A., and Friedman, H. *Science*, **152**, p. 66 (1966).
382. Gursky, H., Kellogg, E. M., Leong, C., Tananbaum, H., and Giacconi, R. *Ap. J.*, **165**, L43 (1971).
383. Kellogg, E. M., Gursky, H., Leong, C., Schreier, E., Tananbaum, H., and Giacconi, R. *Ap. J.*, **165**, L 49 (1971).
384. Schwartz, D. A. *Ap. J.*, **162**, p. 439 (1970).
385. Fabian, A. C., and Sanford, P. W. *Nature Phys. Sci.*, **231**, p. 52 and **234**, p. 20 (1971).
386. Felten, J. E., and Morrison, P. *Ap. J.*, **146**, p. 686 (1966).
387. Bergamini, R., Londrillo, P., and Setti, G. *Nuovo Cimento*, **52B**, p. 495 (1967).
388. Rees, M. J., and Setti, G. *Nature*, **219**, p. 127 (1968).
389. Felten, J. E., and Rees, M. J. *Nature*, **221**, p. 926 (1969).
390. Setti, G., and Rees, M. J. IAU Symposium No. 37 on *Non-Solar X-ray and X-ray Astronomy*, L. Gratton, ed. Reidel, Dordrecht, Holland (1970).
391. Rees, M. J., Sciama, D. W., and Setti, G. *Nature*, **217**, p. 326 (1968).
392. Longair, M. S., and Sunyaev, R. A. *Astrophys. Letters*, **4**, p. 65 (1969).

393. Vette, J. I., Gruber, D., Matteson, J. L., and Peterson, L. E. *Ap. J.*, **160**, L161 (1970).
394. Clark, G. W., Garmire, G. P., and Kraushaar, W. L. IAU Symposium No. 37, *Non-Solar X-ray and X-ray Astronomy*, L. Gratton. ed., p. 267, D. Reidel, Dordrecht, Holland (1970).
395. Bratolyubova-Tsulukidze, L., Grigorov, N. L., Kalinkin, L. F., Melioranskii, A. S., Priakhin, E. A., Savenko, I. A., and Yafarkin, V. Y., *13th General Assembly COS-PAR, Leningrad* (1970).
396. Clayton, D. D., and Silk, J. I. *Ap. J.*, **158**, L43 (1969).
397. Levich, E. V., and Sunyaev, R. A. *Astrophys. Letters*, **7**, p. 69 (1970).
398. Bisnovati-Kogan, G. S., Zeldovich, Y. B., and Sunyaev, R. A. *J.E.T.P. Letter*, **12**, p. 45 (1970).
399. Stecker, F. W. *Nature*, **224**, p. 870 (1969).
400. Steigman, G., *Cargese Lectures in Physics* VI, E. Schatzman, ed. p. 505 Gordon and Breach (1973).
401. Karachentsev, I. D., and Lipovetski, V. A. *Soviet. Astron.*, **12**, p. 909 (1969).
402. Wagoner, R. V. *Ap. J.*, **149**, p. 465 (1967).
403. Rees, M. J. *Astrophys. Letter*, **4**, p. 113 (1969).
404. Arons, J., and McCray, R. *Ap. J.*, **158**, L91 (1969).
405. Gould, R. J., and Schreder, G. P. *Phys. Rev.*, **155**, p. 1408 (1967).
406. Stecker, F. W. *Cosmic γ-Rays* (Mono 1971).
407. Sunyaev, R. A., and Zeldovich, Y. B. *Astrophys. and Sp. Sci.*, **7**, p. 3 (1970).
408. Sachs, R. K., and Wolfe, A. M. *Ap. J.*, **147**, p. 73 (1967).
409. Rees, M. J., and Sciama, D. W. *Nature*, **217**, p. 511 (1968).
410. Partridge, R. B. *Amer. Scientist*, **57**, p. 37 (1969).
411. Rasband, S. M. *Ap. J.*, **170**, p. 1 (1971).
412. Stewart, J. M., and Sciama, D. W. *Nature*, **216**, p. 748 (1967).
413. de Vaucouleurs, G., and Peters, W. L. *Nature*, **220**, p. 868 (1968).
414. Wolfe, A. M. *Ap. J.*, **159**, L61 (1970).
415. Thorne, K. S. *Ap. J.*, **148**, p. 51 (1967).
416. Doroshkevich, A. G., Novikov, I. D., Sunyaev, R. A., and Zeldovich, Y. B. *Highlights of Astronomy*, C. Jager ed., D. Reidel, Dordrecht, Holland (1971).
417. Fowler, W. A. *Comments on Astrophys. and Sp. Phys.*, **2**, p. 134 (1970).
418. Zeldovich, Y. B., and Novikov, I. D. *Sov. Phys. J.E.T.P.*, **26**, p. 408 (1968).
419. Wheeler, J. A. *Particles and Geometry in Relativity*, M. Carmeli, S. I. Fickler, and L. Witten eds. Plenum Press, New York (1970).
420. Parker, L. *Phys. Rev. Letter*, **21**, p. 562 (1968).
421. Sexl, R. U., and Urbantke, H. K. *Phys. Rev.*, **179**, p. 1247 (1969).
422. Zeldovich, Y. B. *J.E.T.P. Letter*, **12**, p. 307 (1970).
423. Hawking, S. W., and Ellis, G. F. R. *Ap. J.*, **152**, p. 25 (1968).
424. Dicke, R. H. *Nature*, **192**, p. 440.
425. Carter, B. unpublished preprint (1968).
426. Dirac, P. A. M. *Proc. Roy. Soc.*, **A165**, p. 199 (1938).
427. Jordan, P. *Die Herkunft der Sterne*, Stuttgart (1947).
428. Tolman, R. C. *Relativity, thermodynamics and cosmology*, Clarendon Press, Oxford, 1934, pp. 429–431.
Papapetrou, A., and Treder, H. *Ann. der Phys.*, **3**, p. 360 (1959).
Avez, A. *Acad. des sciences, Paris, Comptes rend.*, **250**, p. 3585 (1960).
Papapetrou, A. in *Les theories relativistes de la gravitation*, Editions du centre national de la recherche scientifique, Paris (1962), pp. 193–198.

429. Geroch, R. "Singularities in the spacetime of general relativity: their definition, existence, and local characterization". Ph. D. thesis in physics, Princeton University, 1967, unpublished, available in photocopy from University Microfilms, Ann Arbor, Michigan.
Hawking, S. W., and Penrose, R. *Roy. Soc. London, Proc.*, **A314**, p. 529 (1970).

430. Misner, C. W. *Phys. Rev. Letters*, **28**, p. 994 (1972).

431. Barber, W. C., Gittelman, B., O'Neill, G. K., and Richter, D. *Phys. Rev.*, **16**, p. 1127 (1968) and Proc. 14th Int. Conf. High Energy Physics, Vienna, 1968.

432. Bailey, J., Bartl, W., von Bochmann, G., Brown, R. C. A., Farley, F. J. M., Jostlein, H., Picasso, E., and Williams, R. W. *Phys. Lett.*, **28B**, p. 287 (1968).

433. Farley, F. J. M. *Rivista del Nuovo Cim.*, **1**, pp. 59–86 (1969).

434. Brodsky, S. J., and Drell, S. D. *The present status of quantum electrodynamics*, a chapter (pp. 147–194) in *Ann. Rev. Nuc. Sci.*, **20** (1970).

435. Wigner, E. P. *Comm. in Pure Appl. Math.*, **13**, p. 1 (1960).

436. Clifford, W. K. *Mathematical papers*, R. Tucker, ed., London, p. 21 (1882).
Clifford, W. K. *Lectures and essays*, L. Stephen and F. Pollock, eds., London, Vol. 1, pp. 244 and 322 (1879).

437. Wheeler, J. A. *Geometrodynamics*, Academic Press, New York, pp. 88 and 129 (1962)

438. Wheeler, J. A. "Superspace and the nature of quantum geometrodynamics", a chapter in C. DeWitt and J. A. Wheeler, eds., *Battelle rencontres: 1967 lectures in mathematics and physics*, Benjamin, New York (1968).

439. Sakharov, A. *Doklady Akad. Nauk SSSR*, **177**, p. 70 (1967); English translation in *Soviet Physics, Doklady*, **12**, pp. 1040–1041 (1968).

440. Anderson, C. D. *Phys. Rev.*, **491** (1933).

441. Mann, T. *Freud, Goethe, Wagner*, Knopf, New York (1937).

442. Werner, F. G. private remark to J. A. Wheeler on June 3, 1969 in an interval at the Cincinnati, Ohio, *Relativity Conference in the Midwest.*

443. Kilmister, C. W. "More about the King and the Vizier", *Gen. Rel. and Grav.*, **2**, pp. 35–36 (1971). Wheeler's story about the Vizier, and what the Vizier had to say about superspace, to which Kilmister alludes, was never published; but it appears in part in the caption of Figure 61, in the section within quotes. (18 May 1970 report of J. A. Wheeler at the Gwatt Seminar on the Bearings of Topology upon General Relativity).

444. Kneebone, G. T. *Mathematical logic and the foundations of mathematics: an introductory survey*, Van Nostrand, Princeton, New Jersey (1963), p. 40.

445. Shannon, C. *Trans. Amer. Inst. Elect. Eng.*, **57**, pp. 713–723 (1938).

446. Hohn, F. E. *Applied Boolean algebra: an elementary introduction*, 2nd ed., Macmillan, New York (1966).

447. Jauch, J. *Foundations of quantum mechanics*, Addison-Wesley, Reading, Massachusetts (1968).

448. Gödel, K. „Über formal unentscheidbare Sätze der Principia Mathematica und verwandte Systeme I," *Msch. Math. Phys.*, **38**, pp. 173–198 (1931). English translation by B. Meltzer, *on formally undecidable propositions*, with an introduction by R. B. Braithwaite, Basic Books, New York (1962); outlined and discussed in E. Nagel and J. R. Newman, *Gödel's proof*, New York University Press, (1958).

449. Cohen, P. *Set theory and the continuum hypothesis*, W. A. Benjamin, New York, (1966).

450. Kac, M., and Ulam, S. *Mathematics and logic: retrospect and prospects*, Praeger,

New York; abridged paperback edition by Pelican Books, Harmondsworth, Middlesex, (1971).
451. Wheeler, J. A. notebook entry, "pregeometry and the calculus of variations", 9: 10 a.m., April 10, 1971; seminar, Department of Mathematics, Kings College, London, May 10, 1971; letter to L. Thomas, "Pregeometry and propositions", June 11, 1971, unpublished.
452. Wheeler, J. A. *Accademia Nazionale Lincei-Quaderno*, **157**, p. 165 (1971).
453. Leibniz, G. W. *La monadologie* (1714); English translation available in several books; parts included in P. P. Wiener, ed., *Leibniz selections*, Scribners, New York (1951) (paperback).
454. Planck, M. *Preuss. Akad. Wiss. Berlin, Sitzungsber.*, p. 440 (1899).
455. Eddington, A. S. *Cambridge Phil. Soc., Proc.*, **27**, p. 15 (1931); *Relativity theory of protons and electrons*, Cambridge University Press (1936); *Fundamental theory*, Cambridge University Press (1946).
456. Dirac, P. A. M. *Nature*, **139**, p. 323 (1937).
457. Jordan, P. *Schwerkraft und Weltall*, Vieweg und Sohn, Braunschweig (1955); *Zeits. f. Physik*, **157**, p. 112 (1959).
458. Dicke, R. H. *Science*, **129**, p. 3349 (1959); *The theoretical significance of experimental relativity*, Gordon and Breach, New York (1964), p. 72.
459. Hayakawa, S. *Prog. Theor. Phys.*, **33** p. 538 (1965); *Prog. Theor. Phys. Suppl.* (1965), p. 532.

[illegible] Pictures and Fiction, in [illegible] Books, [illegible] (1971).

[illegible] Jr., A., [illegible] April 16, 1971 [illegible] Department of Mathematics, [illegible] May 10, 1971 [illegible] Theory of Taxonomy and [illegible] 11, 1956, unpublished.

[illegible]

[illegible] (1956) [illegible]

[illegible] Press, [illegible]

[illegible]

[illegible]

[illegible]

[illegible]

Index

Index

APPENDIX

Selected Reprints on Black Holes and Gravitational Waves

A-1

Reversible and Irreversible Transformations in Black-Hole Physics*

Demetrios Christodoulou
Joseph Henry Laboratories, Princeton University, Princeton, New Jersey 08540
(Received 17 September 1970)

The concepts of irreducible mass and of reversible and irreversible transformations in black holes are introduced, leading to the formula $E^2 = m_{ir}^2 + (L^2/4m_{ir}^2) + p^2$ for a black hole of linear momentum p and angular momentum L.

This note reports five conclusions: (1) The mass energy of a black hole of angular momentum L can be expressed in the form

$$m^2 = m_{ir}^2 + L^2/4m_{ir}^2, \qquad (1)$$

where m_{ir} is the irreducible mass [geometrical units: $L(\text{cm}) = (G/c^3)L_{\text{conv}}(\text{g cm}^2/\text{sec})$; $m(\text{cm}) = (G/c^2)M_{\text{conv}}(\text{g})$; $G/c^2 = 0.742 \times 10^{-28}$ cm/g] of the black hole. (2) Insofar as one looks apart from the atomicity of matter one can approach arbitrarily closely to reversible transformations that augment or deplete the rotational contribution to the square of the mass. (3) The attainable range of reversible transformation extends[1,2] from $L = 0$, $m^2 = m_{ir}^2$ to $L = m^2$, $m^2 = 2m_{ir}^2$. (Contrast to the formula for mass energy as it depends upon translation, $E^2 = m^2 + p^2$, where p is unlimited; and with the formula for the squared mass energy of a meson!) (4) An irreversible transformation is characterized (Fig. 1) by an increase in the irreducible mass of the black hole. (5) There exists no process which will decrease the irreducible mass.

Roger Penrose has pointed out[3] a way to extract energy from a black hole endowed with angular momentum. It makes use of the "ergosphere" (Ruffini and Wheeler; cf. Fig. 2, reproduced from their paper[4]), the region between the horizon (surface of black hole; boundary of region from which no particle or radiation can ever escape) and the surface of infinite red shift (coincident with the horizon only for case of the angular-momentum-free Schwarzschild black hole). A particle of energy E_0 is sent from infinity into the ergosphere and decays there into (1) a particle which emerges to infinity with a rest-plus-kinetic energy E_2 greater than E_0, together with (2) a particle ("rocket ejecta") which has an energy E_1, that is negative as measured at infinity $(E_1 = E_0 - E_2)$, but positive in the local Lorentz frame, and which is ejected into such a direction that it is captured into the black hole, thereby diminishing its mass. We consider the case where all masses can be regarded as infinitesimal compared with the mass of a black hole.

The energy E, as measured at infinity, of a particle of angular momentum p_φ and rest mass μ, having a turning point at r, is given by the

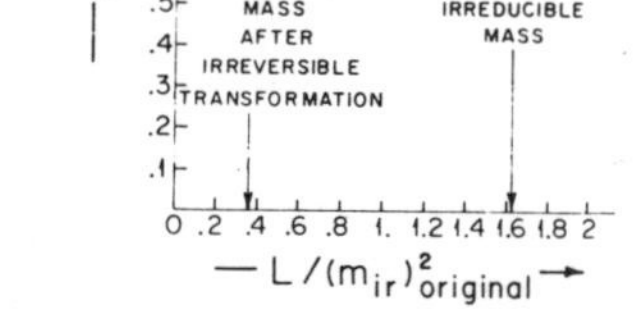

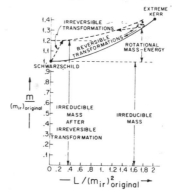

FIG. 1. Mass energy m versus angular momentum L for a black hole of specified irreducible mass m_{ir} illustrating the difference between reversible transformations and irreversible transformations (which increase the irreducible mass).

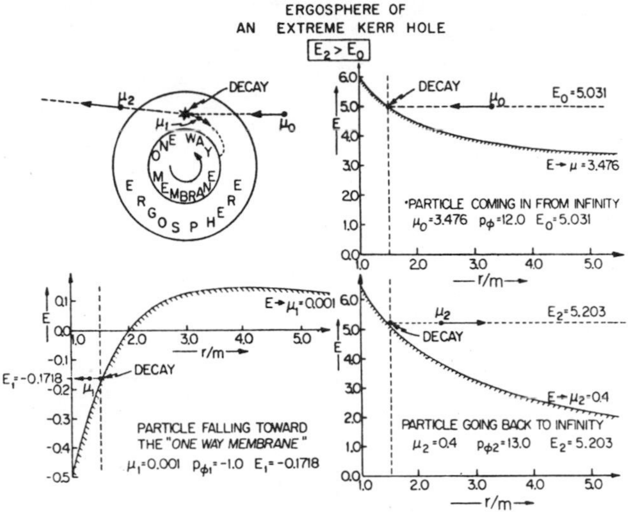

FIG. 2. (Reproduced from Ruffini and Wheeler, Ref. 4, with their kind permission.) Decay of a particle of rest-plus-kinetic energy E_0 into a particle which is captured into the black hole with positive energy as judged locally, but negative energy E_1 as judged from infinity, together with a particle of rest-plus-kinetic energy $E_2 > E_0$ which escapes to infinity. The cross-hatched curves give the effective potential (gravitational plus centrifugal) defined by the solution E of Eq. (2) for constant values of p_φ and μ.

equation[5] (where a is an abbreviation for L/m)

$$E^2[r^3+a^2(r+2m)]-4mEap_\varphi+(2m-r)p_\varphi^2-\mu^2r^2(r-2m)-a^2\mu^2r=\text{multiple of (radial momentum)}=0. \quad (2)$$

The Penrose process is most efficient when the reduction of mass is greatest for a given reduction in angular momentum. To meet this requirement the energy E_1 must be as negative as possible. This happens at the surface of the black hole itself,

$$r=r_+=m+(m^2-a^2)^{1/2}, \quad (3)$$

where the separation of "positive-" and negative-energy states goes to zero [vanishing of discriminant of Eq. (2) for E]. At this point the relation between energy and angular momentum reduces to

$$E_1=[a/(r_+^2+a^2)](p_\varphi)_1. \quad (4)$$

Applying the laws of conservation of energy and angular momentum to the assimilation of particle 1 by the black hole, we arrive at the relation

$$dm=\frac{(L/m)\,dL}{[m+(m^2-L^2/m^2)^{1/2}]^2+L^2/m^2} \quad (5)$$

Integration leads to the relation

$$(1-a^2/m_2)^{1/2}=(2m_{ir}^2/m^2)^{-1}$$

which, if condition (3) is fulfilled, is equivalent to expression (1).

I would like to thank Professor J. A. Wheeler and Dr. R. Ruffini for very helpful discussions and suggestions.

*Publication assisted by National Science Foundation and U. S. Air Force Office of Scientific Research Grants to Princeton University.

[1]J. Bardeen, "The Weight and Fate of a Relativistic Rotating Disk," relativity seminar, Princeton University, 5 May 1970 (unpublished).

[2]D. Christodoulou, Bull. Amer. Phys. Soc., 15, 661 (1970).

[3]R. Penrose, Riv. Nuovo Cimento 1, 252 (1969).

[4]R. Ruffini and J. A. Wheeler, in "The Significance of Space Research for Fundamental Physics" (European Space Research Organization, Paris, to be published).

[5]B. Carter, Phys. Rev. 174, 1559 (1968).

A-2

Gravitational Radiation from Colliding Black Holes

S. W. Hawking
Institute of Theoretical Astronomy, University of Cambridge, Cambridge, England
(Received 11 March 1971)

It is shown that there is an upper bound to the energy of the gravitational radiation emitted when one collapsed object captures another. In the case of two objects with equal masses m and zero intrinsic angular momenta, this upper bound is $(2-\sqrt{2})m$.

Weber[1-3] has recently reported coinciding measurements of short bursts of gravitational radiation at a frequency of 1660 Hz. These occur at a rate of about one per day and the bursts appear to be coming from the center of the galaxy. It seems likely[3,4] that the probability of a burst causing a coincidence between Weber's detectors is less than $\frac{1}{10}$. If one allows for this and assumes that the radiation is broadband, one finds that the energy flux in gravitational radiation must be at least 10^{10} erg/cm^2 day.[4] This would imply a mass loss from the center of the galaxy of about $20\,000 M_\odot$/yr. It is therefore possible that the mass of the galaxy might have been considerably higher in the past than it is now.[5] This makes it important to estimate the efficiency with which rest-mass energy can be converted into gravitational radiation. Clearly nuclear reactions are insufficient since they release only about 1% of the rest mass. The efficiency might be higher in either the nonspherical gravitational collapse of a star or the collision and coalescence of two collapsed objects. Up to now no limits on the efficiency of the processes have been known. The object of this Letter is to show that there is a limit for the second process. For the case of two colliding collapsed objects, each of mass m and zero angular momentum, the amount of energy that can be carried away by gravitational or any other form of radiation is less than $(2-\sqrt{2})m$.

I assume the validity of the Carter-Israel conjucture[6,7] that the metric outside a collapsed object settles down to that of one of the Kerr family of solutions[8] with positive mass m and angular momentum a per unit mass less than or equal to m. (I am using units in which $G=c=1$.) Each of these solutions contains a nonsingular *event horizon*, two-dimensional sections of which are topographically spheres with area[9]

$$8\pi m[m+(m^2-a^2)^{1/2}]. \qquad (1)$$

The event horizon is the boundary of the region of space-time from which particles or photons can escape to infinity. I shall consider only space-times which are asymptotically flat in the sense of being weakly asymptotically simple.[10] The metric on such a space M can be extended conformally to a manifold with boundary $\bar{M}$ which consists of M and two null hypersurfaces $\mathscr{I}^+$ and $\mathscr{I}^-$. These represent future and past null infinity, respectively. The event horizon can then be defined as $\dot{J}^-(\mathscr{I}^+)$, where a dot indicates the boundary, and for any set $\mathcal{S}$, $J^-(\mathcal{S})$ is the causal past of $\mathcal{S}$, i.e., the set of all points that can be reached from $\mathcal{S}$ by past-directed nonspacelike curves.

The situation that I consider is one in which there are initially two collapsed objects or "black holes" a considerable distance apart in asymptotically flat space. The black holes are assumed to have formed at some earlier time as a result of either gravitational collapses or the amalgamation of smaller black holes. This situation can be described in terms of a spacelike hypersurface Σ_i which is a Cauchy surface for the asymptotically flat region of space-time. On Σ_i there will be two separate regions, B_1 and B_2, which contain closed, trapped surfaces.[11] In fact the surface Σ_i need not exist within the regions B_1 and B_2 since what happens there cannot affect what happens outside. Just outside B_1 and B_2 will be two two-spheres which are the intersection of $\dot{J}^-(\mathscr{I}^+)$ with Σ_i. By the Carter-Israel conjecture the areas of these two spheres will be to a good approximation given by formula (1) for the values m_1, a_1 and m_2, a_2 of the masses and specific angular momenta of the two black holes. Suppose now that the two black holes capture each other to form a single black hole which settles down to a Kerr solution with parameters m_3 and a_3. There will then be a later spacelike hypersurface Σ_f on which there will be a single region B_3 containing closed trapped surfaces. Just outside B_3 will be a two-sphere $\Sigma_f \cap \dot{J}^-(\mathscr{I}^+)$ whose area is given by formula (1) with m_3 and a_3. Now $\dot{J}^-(\mathscr{I}^+)$ is generated by null geodesic segments which may have past end points but which can have no future end points.[10,12] I shall take it that there are no singularities in the exterior region $J^-(\mathscr{I}^+) \cap J^+(\Sigma_i)$. Whether such "naked" singularities can occur or whether the singularities are always hidden behind the event horizon is an unresolved question.[13] One would not expect to obtain a limit on the energy that could be carried to infinity by gravitational radiation if there were a naked singularity since what happens at a singularity is not governed by any known laws. I shall therefore consider only those situations where what happens in $J^-(\mathscr{I}^+) \cap J^+(\Sigma_i)$ is determined by data on Σ_i, i.e., where Σ_i is a Cauchy surface for $J^-(\mathscr{I}^+) \cap J^+(\Sigma_i)$. One can choose Σ_f so that it is a Cauchy surface for $J^-(\mathscr{I}^+) \cap J^+(\Sigma_f)$ and show that every null geodesic generating segment of $\dot{J}^-(\mathscr{I}^+)$ in $J^+(\Sigma_i) \cap J^-(\Sigma_f)$ must intersect Σ_f. By the Carter-Israel conjecture the solution in $J^-(\mathscr{I}^+) \cap J^+(\Sigma_f)$ resembles the exterior Kerr solution and so contains no singularities. Thus each null geodesic generating segment of $\dot{J}^-(\mathscr{I}^+)$ in $J^+(\Sigma_f)$ will be geodesically complete in the future direction. I assume also that the weak energy condition holds[11]: $T_{ab}K^aK^b \geq 0$ for every null vector K^a. It then follows that the convergence ρ of the null geodesic generators of $\dot{J}^-(\mathscr{I}^+)$ cannot be positive anywhere since if it were there would be a caustic and hence a future end point of some of the generators.[10] This shows that the area of $\Sigma_f \cap \dot{J}^-(\mathscr{I}^+)$ must be greater than or equal to the area of $\Sigma_i \cap \dot{J}^-(\mathscr{I}^+)$ (in fact, it must be strictly greater). Therefore,

$$m_3[m_3+(m_3^2-a_3^2)^{1/2}] > m_1[m_1^2+(m_1^2-a_1^2)^{1/2}] + m_2[m_2+(m_2^2-a_2^2)^{1/2}]. \quad (2)$$

By the conservation law for asymptotically flat space,[14,15] the energy emitted in gravitational or other forms of radiation is $m_1+m_2-m_3$. The efficiency $\epsilon = (m_1+m_2)^{-1}(m_1+m_2-m_3)$ is limited by Eq. (2). The highest limit on ϵ is $\frac{1}{2}$ which occurs if $m_1=m_2=a_1=a_2$, $a_3=0$. If $a_1=a_2=0$, then $\epsilon < (1-2^{-1/2})$. It should be stressed that these are upper limits. The actual efficiency might be much less.

Imagine a situation in which there were initially a large number of black holes. First these could combine in pairs, and then the resulting holes could combine, and so on. On dimensional grounds one could expect the efficiency to be the same at each stage. Thus one would extract a very large fraction of the original mass (this argument was suggested by C. W. Misner). At each stage the emitted bursts would be longer and stronger. This might explain Weber's observations.

One can also apply this result that the area on the event horizon increases to the case of a single black hole on which various particles and fields impinge. One then has that the final mass m_2 and specific angular momentum a_2 must satisfy the inequality

$$m_2[m_2+(m_2^2-a_2^2)^{1/2}] > m_1[m_1+(m_1^2-a_1^2)^{1/2}],$$

where m_1 and a_1 are the initial values. In particular, if a_2 is equal to zero, then $2m_2^2 > m_1[m_1 + (m_1^2 - a_1^2)^{1/2}]$. Using an idea of Penrose,[13] Christodolou[16] has shown that one can get arbitrarily near this limit. One can interpret this as extracting the rotational energy of the black hole.

[1]J. Weber, Phys. Rev. Lett. 22, 1320 (1969).

[2]J. Weber, Phys. Rev. Lett. 24, 276 (1970).

[3]J. Weber, Phys. Rev. Lett. 25, 180 (1970).

[4]G. W. Gibbons and S. W. Hawking, "The Detection of Short Bursts of Gravitational Radiation" (to be published).

[5]G. B. Field, M. J. Rees, and D. W. Sciama, Comments Astrophys. Space Phys. 1, 187 (1969).

[6]B. Carter, Phys. Rev. Lett. 26, 331 (1971).

[7]W. Israel, Phys. Rev. 164, 1776 (1967).

[8]R. H. Boyer and R. W. Lindquist, J. Math. Phys. 8, 265 (1967).

[9]R. Penrose, private communication.

[10]R. Penrose, in *Battelle Rencontres 1967*, edited by C. M. de Witt and J. A. Wheeler (Benjamin, New York, 1968).

[11]R. Penrose and S. W. Hawking, Proc. Roy Soc. Ser. A 314, 529 (1970).

[12]S. W. Hawking, Proc. Roy. Soc., Ser. A 300, 187 (1967).

[13]R. Penrose, Riv. Nuovo Cimento 1, Num. spec., 252 (1969).

[14]H. Bondi, M. G. J. van der Burg, and A. W. K. Metzner, Proc. Roy. Soc., Ser. A 269, 21 (1962).

[15]R. Penrose, Phys. Rev. Lett. 10, 66 (1963).

[16]D. Christodolou, Phys. Rev. Lett. 25, 1596 (1970).

A-3

PHYSICAL REVIEW D VOLUME 4, NUMBER 12 15 DECEMBER 1971

Reversible Transformations of a Charged Black Hole*

Demetrios Christodoulou
Joseph Henry Laboratories, Princeton University, Princeton, New Jersey 08540

and

Remo Ruffini
Joseph Henry Laboratories, Princeton University, Princeton, New Jersey 08540, and Institute for Advanced Study, Princeton, New Jersey 08540

(Received 1 March 1971; revised manuscript received 26 July 1971)

A formula is derived for the mass of a black hole as a function of its "irreducible mass," its angular momentum, and its charge. It is shown that 50% of the mass of an extreme charged black hole can be converted into energy as contrasted with 29% for an extreme rotating black hole.

The mass m of a rotating black hole can be increased and (Penrose[1]) decreased by the addition of a particle and so can its angular momentum L; but (Christodoulou[2]) there is no way whatsoever to decrease the irreducible mass m_{ir} in the equation

$$E^2 - p^2 = m^2 = m_{ir}^2 + L^2/4m_{ir}^2 \qquad (1)$$

for the mass of a black hole. The concept of reversible (m_{ir} unchanged) and irreversible transformations (m_{ir} increases), which was introduced and exploited by one of us to obtain this result, is extended here to the case where the object also has charge, to yield the following four conclusions:

(1) The rest mass of a black hole is given in

terms of its irreducible mass and its angular momentum L and charge e by the formula[3]

$$m^2=(m_{ir}+e^2/4m_{ir})^2+L^2/4m_{ir}^2\,. \tag{2}$$

(2) Reversibility implies and demands zero separation between the "negative-root states" and "positive-root states" of the particle defined by a quadratic equation of the form

$$\alpha E^2-2\beta E+\gamma=0\,, \tag{3}$$

a requirement which is met and can only be met at the horizon itself.

(3) There exists a one-to-one connection between (a) the irreducible mass (as defined here and previously exclusively through the theory of reversible and irreversible transformations), and (b) the proper surface area S of the horizon (shown by Hawking[4] never to decrease),

$$S=16\pi m_{ir}^2\,. \tag{4}$$

(4) The innermost stable circular orbit is the simplest place for a black hole to hold a particle bound and ready for capture. This orbit lies just outside the horizon only when the black hole is an extreme Kerr-Newman[5] black hole in the sense

$$(L^2/m^2)+e^2\ (=a^2+e^2 \text{ in the notation of Kerr})=m^2 \tag{5}$$

or parametrically

$$L=2m_{ir}^2\cos\chi\,,\quad e=\pm 2m_{ir}(\sin\chi)^{1/2}\,,$$
$$m=2^{1/2}m_{ir}(1+\sin\chi)^{1/2} \tag{6}$$

(χ has any value from 0 to $\frac{1}{2}\pi$). The binding energies of a particle in this most-bound stable orbit are given by the formula

$$\frac{E}{\mu}=\frac{(2a^2-m^2)\lambda+a(\lambda^2m^2+4a^2-m^2)^{1/2}}{4a^2-m^2}\,, \tag{7}$$

where $\lambda=\epsilon e/\mu m$ (cf. Fig. 1). The transformation on the black hole affected by capture of a particle from such an orbit becomes reversible only when the charge-to-mass ratios of the particle (ϵ/μ) and the black hole (e/m) attain the limits $|\epsilon/\mu|\to\infty$ and parameter $\chi\to 0$ ($e/m\to 0$, $L\to m^2$) such that $\epsilon e/m\mu\to-\infty$. The binding of the particle in this orbit is 100% of its rest mass.

In black-hole physics one has reversibility without reversibility. As compared to such frictional processes as a brick sliding on a pavement, or an accelerated charged particle radiating, or a freely falling deformed droplet of molasses reverting to sphericity, it is difficult to name an act more impressively lacking in reversibility than capture of a particle by a black hole. Lost beyond recall is not a part of the mass-energy of the moving system but all of it; and not only mass-energy but also identity. The resulting black hole, like the original black hole, is characterized by three "independent determinants," mass, charge, and angular momentum, and by nothing else; all particularities (anomalies in higher multipole moments; also baryon number, lepton number, and strangeness) are erased, according to all available indications.[6-11] To reverse the change in a black hole brought about by the addition of a particle with a given rest mass, charge, and angular momentum (μ, ϵ, p_φ) one does not and cannot cause the black hole to reexpel the particle. Nor is there any such thing as a particle with a negative rest mass that one can add to cancel the first addition. Add instead (B in Fig. 1) a particle of the original rest mass μ but of charge $-\epsilon$ and angular momentum $-p_\varphi$. This addition restores the determinants m,

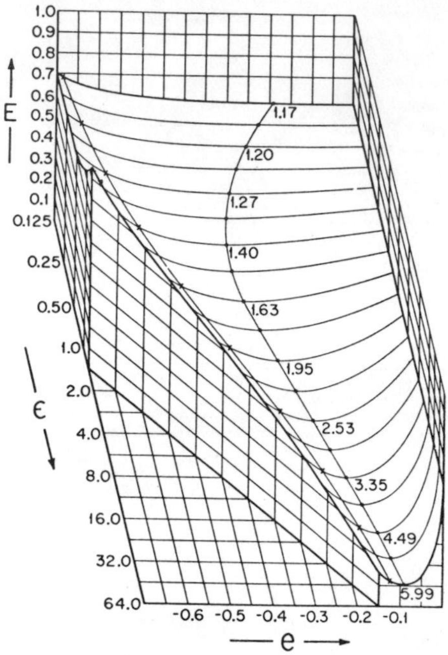

FIG. 1. Energy of a test particle of specified charge ϵ, corresponding to the circular orbit touching the one-way membrane of a black hole of specified charge e of the limiting configurations $a^2+e^2=m^2$, versus ϵ and e. Such orbits will be stable and will be the orbits of lowest energy if $E\le e(\mu/m)$ for $e\epsilon<0$. The crossed curve is the boundary between stable and unstable orbits. The numbers on the energy minima correspond to the values of the angular momentum for the most-bound orbit.

e, L of the black hole to their original value when and only when positive- and negative-root states have zero separation, a condition that is fulfilled only at the horizon itself, $r = r_{\text{horizon}} = r_+ = m + (m^2 - a^2 - e^2)^{1/2}$. At the horizon the positive- and negative-root surfaces $E = E_\pm(p_\varphi, \epsilon)$ meet at a "knife edge" (cf. Fig. 18 in Ruffini and Wheeler[12]) and the two "hyperbolas" of Fig. 2 degenerate to the "straight line" (acquires one more dimension and becomes a plane when the charge as well as the angular momentum of the test particle is taken into account),

$$E = \frac{ap_\varphi + e\epsilon r_+}{a^2 + r_+^2}. \tag{8}$$

Details follow.

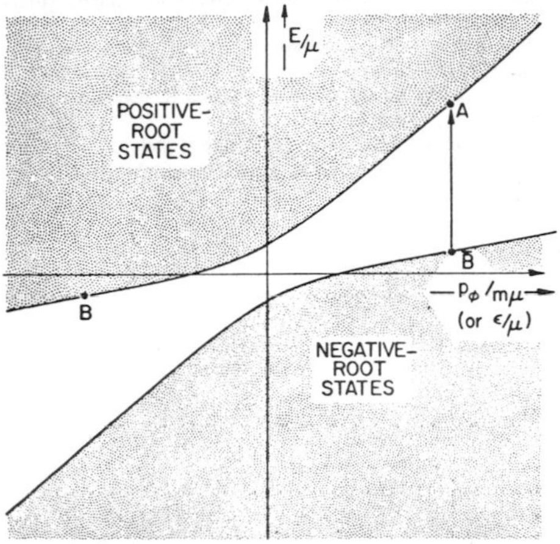

FIG. 2. Reversing the effect of having added to the black hole one particle (A) by the Penrose process of adding another particle (B) of the same rest mass but of opposite angular momentum and charge in a "positive-root negative-energy state." The diagram shows schematically the energy E of the particle (measured in the Lorentz frame tangent at $r \to \infty$) as a function of the charge ϵ of the particle and its angular momentum p_φ about the axis of the black hole (for simplicity one of these two dimensions has been suppressed in the frame). The particle under examination is in the equatorial plane of the black hole ($\theta = \frac{1}{2}\pi$) at a specified value of r. The energy lies on the indicated "positive-root" curve when p_r and p_θ are zero, otherwise in the dotted region above the curve. Addition of B is equivalent to subtraction of $\bar{B}$. Thus the combined effect of the capture of particles A and B is an increase in the mass of the black hole given by the vector $\bar{B}A$. This vector vanishes and reversibility is achieved when and only when the separation between positive-root states and negative-root states is zero [in this case the hyperbolas coalesce to the straight line given by Eq. (8)].

The Hamilton-Jacobi equation

$$g^{\alpha\beta}\left(\frac{\partial S}{\partial x^\alpha} + \epsilon A_\alpha\right)\left(\frac{\partial S}{\partial x^\beta} + \epsilon A_\beta\right) + \mu^2 = 0 \tag{9}$$

for the motion of the particle, separated and solved by Carter,[13] leads to the quadratic equation (3) for the energy with

$$\begin{aligned}
\alpha &= r^4 + a^2(r^2 + 2mr - e^2),\\
\beta &= (2mr - e^2)ap_\varphi + \epsilon er(r^2 + a^2),\\
\gamma &= -(r^2 - 2mr + e^2)p_\varphi^2 + 2\epsilon earp_\varphi + \epsilon^2 e^2 r^2\\
&\quad -(r^2 - 2mr + a^2 + e^2)(\mu^2 r^2 + Q)\\
&\quad -[(r^2 - 2mr + a^2 + e^2)p_r]^2,
\end{aligned} \tag{10}$$

see Fig. 1. Here Q is a constant of the motion (generalization of the usual expression for the square of the angular momentum) related to the polar momentum p_θ at any angle θ by the equation

$$Q = \cos^2\theta\,[a^2(\mu^2 - E^2) + p_\varphi^2(\sin\theta)^{-2}] + p_\theta^2. \tag{11}$$

The positive- and negative-root solutions of (a) coincide only when the discriminant of this equation vanishes. This condition is satisfied only at the horizon where the energy is given by Eq. (8). The derivation of formula (1) assumes and demands that the in-falling particles make an effectively infinitesimal change in the properties of the black hole. Applying the laws of conservation of the three determinants and writing $E = dm$, $p_\varphi = dL$, $\epsilon = de$, we obtain the partial differential equation

$$dm(L, e) = \frac{(L/m)dL + r_+ e\,de}{r_+^2 + L^2/m^2}. \tag{12}$$

Integration gives Eq. (1), provided that the following condition is satisfied:

$$\frac{L^2}{4m_{\text{ir}}^2} + \frac{e^4}{16 m_{\text{ir}}^4} \leqslant 1. \tag{13}$$

For it to be possible to reverse the transformation, both the original particle (positive energy) and the added particle (negative energy) must arrive at the horizon with zero radial momentum. Otherwise there is a nonzero kinetic energy that is irretrievably lost.

When one turns from reversibility as a criterion for an interesting transformation to merely the ability to extract energy, it becomes important to specify under what conditions a positive-root state has negative energy. From Eq. (10) it follows that the region (outside the one-way membrane) where positive-root states of negative energy are available to a particle of specified rest mass μ, charge ϵ, and angular momentum p_φ extends to the surface

$$(r^2-2mr+e^2+a^2)[p_\varphi{}^2+\sin^2\theta(r^2+a^2\cos^2\theta)\mu^2] = \sin^2\theta(\epsilon e r + a p_\varphi)^2. \tag{14}$$

When the transformation is reversible, the energy-extraction process has its maximum possible efficiency. Repetition of an energy-extraction process with maximum possible efficiency results in conversion into energy of 50% of the mass of an extreme charged black hole and 29% of that of an extreme rotating black hole. Thus, black holes appear to be the "largest storehouse of energy in the universe."[14]

ACKNOWLEDGMENT

We are indebted to John A. Wheeler for stimulating discussions and also for advice on the wording of this paper.

---

*Work supported by the National Science Foundation under Grant No. GP-30799X.

[1]R. Penrose, Nuovo Cimento, Rivista, Vol. 1, special issue, 1969.

[2]D. Christodoulou, Phys. Rev. Letters 25, 1596 (1970).

[3]D. Christodoulou and R. Ruffini, Bull. Am. Phys. Soc. 16, 34 (1971).

[4]S. Hawking, Phys. Rev. Letters 26, 1344 (1971).

[5]E. T. Newman, E. Couch, R. Chinnapared, A. Exton, A. Prakash, and R. Torrence, J. Math. Phys. 6, 918 (1965).

[6]J. Bekenstein, Bull. Am. Phys. Soc. 16, 34 (1971).

[7]B. Carter, Phys. Rev. Letters 26, 331 (1971).

[8]J. B. Hartle, Phys. Rev. D 1, 394 (1970).

[9]W. Israel, Phys. Rev. 164, 1776 (1967).

[10]C. Teitelboim, Bull. Am. Phys. Soc. 16, 35 (1971).

[11]J. A. Wheeler, in a chapter in the Mendeleev anniversary volume (Accademia delle Scienze di Torino, Torino, 1971).

[12]R. Ruffini and J. A. Wheeler, in *The Significance of Space Research for Fundamental Physics*, edited by A. F. Moore and V. Hardy (European Space Research Organization, Paris, 1970).

[13]B. Carter, Phys. Rev. 174, 1559 (1968).

[14]D. Christodoulou and R. Ruffini (unpublished).

PHYSICAL REVIEW D VOLUME 5, NUMBER 4 15 FEBRUARY 1972

Bound Geodesics in the Kerr Metric*

Daniel C. Wilkins
Institute of Theoretical Physics, Department of Physics, Stanford University, Stanford, California 94305
and Joseph Henry Laboratories, Princeton University, Princeton, New Jersey 08540
(Received 25 June 1971)

The bound geodesics (orbits) of a particle in the Kerr metric are examined. (By "bound" we signify that the particle ranges over a finite interval of radius, neither being captured by the black hole nor escaping to infinity.) All orbits either remain in the equatorial plane or cross it repeatedly. A point where a nonequatorial orbit intersects the equatorial plane is called a node. The nodes of a spherical (i.e., constant radius) orbit are dragged in the sense of the spin of the black hole. A spherical orbit near the one-way membrane traces out a helix-like path lying on a sphere enclosing the black hole.

I. INTRODUCTION

Do effects of general relativity proper play a central role in such celestial phenomena as the pulsars and quasars? This question merits serious consideration if only for the reason that there is no (macroscopic) device as effective as the gravitational field of a collapsed star for the release of energy.[1]

At present, two sorts of energy-releasing processes are known. In one, a particle falling into a black hole emits electromagnetic or gravitational waves; the hole does not participate actively. In the other, a possible mechanism for which has been proposed by Penrose, the black hole itself provides a source of energy. A particle near the one-way membrane breaks up into two particles. One can arrange that one of the fragments will escape to infinity with an energy larger than that possessed by the original particle provided the other fragment is captured by the black hole. It may be that this mechanism, more contrived than the radiation process, will have less direct importance for astrophysics.

The Penrose process does not violate conservation of energy. As shown by Christodoulou,[2] the extra energy of the escaping particle is taken from the rotational energy of the black hole. By repeating the Penrose process many times one may deplete a black hole of its entire rotational energy, which can amount for a charged black hole to as much as 50% of its rest mass.[3]

The energy release possible through radiation processes is equally remarkable. Christodoulou and Ruffini[3] have shown that a charged particle falling into a charged black hole can, under appropriate conditions, emit its entire rest-mass energy.

Preliminary to investigating the problem of radiation, one must understand the kinematics of test particles for which the effect of radiation is neglected. Darwin[4,5] has already considered the geodesics of a particle in the spherically symmetric Schwarzschild field. The Kerr-Newman field, which describes a body having charge, mass, and spin angular momentum, is richer in structure than the Schwarzschild field. In this paper, we study the bound geodesics or "orbits" in the charge-free Kerr metric.[6] We have restricted ourselves to the charge-free case both because this case can be a very relevant one for astrophysics, and because we would like to distinguish the purely relativistic effects of the gravitational field from the electrodynamic ones.

In Sec. II, we deduce a necessary condition for binding to occur. Subsequently, we specialize to spherical orbits, that is, orbits of constant radius. In Newtonian mechanics, one can understand much about the general orbit by studying the circular-orbit case. Here, too, one believes that most features of interest are present in the case of spherical orbits.

In the limit of large radius, a spherical orbit goes asymptotically to a Keplerian circle. Considering a sequence of orbits of ever smaller radius, we find that as the radius decreases, the line of the (ascending or descending) node is increasingly "dragged" in the sense of the spin of the black hole. (The line of the ascending node is a line of constant azimuth in the equatorial plane passing through that point at which the orbit, going from negative to positive latitudes, intersects the equatorial plane.)

At a certain radius, a horizon occurs. This horizon, known as the one-way membrane, has the property that anything penetrating the surface from outside will be unable to escape again. The most rapidly rotating Kerr particle for which causality does not break down[7] has an angular momentum equal to its mass squared (in geometrical units). Stable orbits are possible down to its one-way membrane. As the orbital radius approaches that of the horizon, the dragging of the nodes increases without limit. During the time that a particle makes one oscillation in latitude, it will be swept through many complete azimuthal revolutions. Consequently, an orbit near the horizon will have a helix-like shape with axis parallel to the spin of the black hole.

II. CONDITIONS FOR BINDING

Carter[7] has given the first integrals of the equations of motion of a particle in the Kerr-Newman field. For the sake of a clearer physical interpretation, we use his Eq. (9) to transform to Boyer-Lindquist coordinates; in these coordinates, the metric is symmetric under simultaneous inversion of the axial and stationary Killing vectors. The metric is

$$ds^2=\rho^2\Delta^{-1}dr^2+\rho^2d^2\theta$$
$$+\rho^{-2}\sin^2\theta[a\,dt-(r^2+a^2)d\phi]^2$$
$$-\rho^{-2}\Delta(dt-a\sin^2\theta d\phi)^2,$$

where

$$\rho^2=r^2+a^2\cos^2\theta,\qquad \Delta=r^2-2mr+a^2+e^2, \tag{1}$$

a, m, and e are, respectively, the specific angular momentum, mass, and charge of the black hole. Specializing to the charge-free case, the equations of motion take the form[8]

$$\rho^2\dot r=\pm\sqrt{R}, \tag{2a}$$
$$\rho^2\dot\theta=\pm\sqrt{\Theta}, \tag{2b}$$
$$\rho^2\dot\phi=(\Phi\sin^{-2}\theta-aE)+a\Delta^{-1}P, \tag{2c}$$
$$\rho^2\dot t=a(\Phi-aE\sin^2\theta)+(r^2+a^2)\Delta^{-1}P, \tag{2d}$$

with

$$\Theta=Q-\cos^2\theta[a^2(\mu^2-E^2)+\Phi^2\sin^{-2}\theta],$$
$$P=E(r^2+a^2)-\Phi a, \tag{3}$$
$$R=P^2-\Delta[\mu^2r^2+Q+(\Phi-aE)^2].$$

The dot denotes differentiation with respect to a parameter λ, defined in terms of the proper time by

$$\tau=\mu\lambda. \tag{4}$$

The signs in (2a) and (2b) can be chosen independently. The constant a denotes the angular momentum per unit mass of the central body. E, Φ, and Q are three constants of the particle's motion. E and Φ refer, respectively, to the energy and to the z component of angular momentum; Q is related to the θ velocity, $\dot\theta$.

We rewrite $R(r)$ in a form independent of the masses. Writing $R(r)$ out in detail,

$$R(r)=(E^2-\mu^2)r^4+2\mu^2mr^3+[a^2(E^2-\mu^2)-\Phi^2-Q]r^2$$
$$+2m[(aE-\Phi)^2+Q]r-a^2Q.$$

Dividing through by μ^2m^4,

$$R(r)/\mu^2m^4=(\hat E^2-1)\hat r^4+2\hat r^3+[\hat a^2(\hat E^2-1)-\hat\Phi^2-\hat Q]\hat r^2$$
$$+2[(\hat a\hat E-\hat\Phi)^2+\hat Q]\hat r-\hat a^2\hat Q, \tag{5}$$

where

$$\hat E=E/\mu,$$
$$\hat\Phi=\Phi/m\mu,\quad Q=Q/m^2\mu^2, \tag{6}$$
$$\hat r=r/m,\quad \hat a=a/m.$$

We shall henceforth take

$$\mu=m=1,$$

which is equivalent to using the caret variables. Since $\mu^2=1$ is greater than zero, the geodesics described by (1), (2), and (3) are timelike.

One can determine the values of E, Φ, and Q for which binding occurs by using the effective potentials.

We first discuss the effective radial potential $V(r)$. $V(r)$ is defined as that value of E such that $\dot r=0$ at radius r; by Eq. (2a) it follows that $R(r)=0$. Rewriting $R(r)$ in a more convenient form,

$$R(r)=[r^4+a^2(r^2+2r)]E^2-4a\Phi rE$$
$$+a^2\Phi^2-(r^2+Q+\Phi^2)(r^2-2r+a^2), \tag{7}$$

a quadratic in E. $R(r)$ has not one but two roots:

$$V_\pm(r,\Phi,Q)=\frac{4a\Phi r\pm\sqrt{D}}{2[r^4+a^2(r^2+2r)]}. \tag{8}$$

The discriminant D depends on Φ only through Φ^2. In general, the reality of the radial velocity leads by (2a) to the requirement that $R(r)$ be non-negative. If D is negative, this requirement is fulfilled

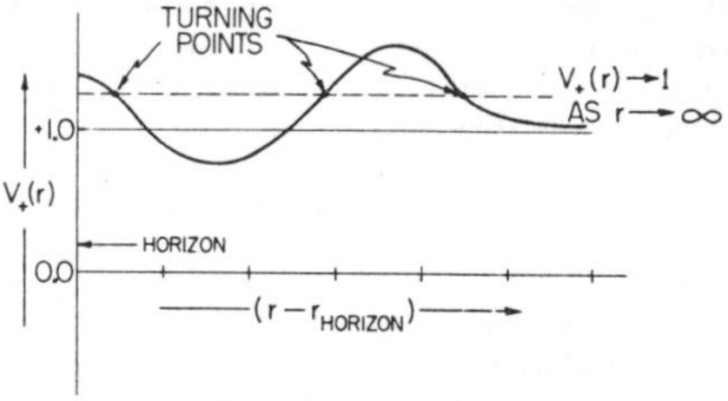

FIG. 1. Sketch of an effective radial potential which would bind a particle of energy greater than unity (dashed line). There must be at least three turning points. We show that such a potential is impossible by proving that there may be at most two turning points for $E>1$.

for all values of the energy. If, however, D is non-negative, the energy must satisfy either of the following conditions:

$$E \geq V_+(r, \Phi, Q) \tag{9a}$$

or

$$E \leq V_-(r, \Phi, Q). \tag{9b}$$

Whether or not the effective potential is real, we may always make the one-to-one correspondence,

$$E, \Phi, Q \leftrightarrow -E, -\Phi, Q. \tag{10}$$

Inspection of the equations of motion, (2) and (3), shows that with proper choice of sign, the 4-velocities of two such corresponding motions will differ only in sign.

We now prove that if $E^2 \geq 1$, and if the specific angular momentum of the black hole lies in the causality-preserving range $0 \leq a \leq 1$, then the motion is unbound.

The proof depends on setting an upper limit on the number of turning points. Consider a conceivable $V_+(r)$ sketched in Fig. 1. Notice that by (7) and (8), $V_+(r)$ goes asymptotically to unity at large radii. As indicated by the figure a bound state with $E \geq 1$ is only possible if there is a range of energies for which three or more turning points occur. We will prove in fact, that there are at most two. $E \leq -1$ also satisfies $E^2 \geq 1$. By the correspondence (10), however, if our theorem is proven for $E \geq 1$, it will necessarily also hold for $E \leq -1$.

Let us reexpress $R(r)$ in terms of the new coordinate x, defined by

$$r = x + 1.$$

Substituting this into (5) yields

$$R(x) = (E^2 - 1)x^4 + (4E^2 - 2)x^3 + [(6 + a^2)E^2 - a^2 - \Phi^2 - Q]x^2 + [(4 + 2a^2)(E^2 - 1) + 6 + 2a^2E^2 - 4a\Phi E]x + [(2aE - \Phi)^2 + (E^2 + 1 + Q)(1 - a^2)]. \tag{11}$$

In the language of classical algebra, whenever a term in the polynomial is followed by one of the opposite sign, that is described as a "variation of sign."[9] A polynomial (with real coefficients) cannot have more positive roots than there are variations of sign.[10] If $E \geq 1$, the first two terms are non-negative. Write the last three terms as

$$c_2x^2, \quad c_1x, \quad c_0.$$

As many as three variations of sign can occur if

$$c_2, c_0 < 0, \qquad c_1 > 0. \tag{12}$$

From (12),

$$c_1 - (c_0 + c_2) > 0,$$

or by (11),

$$Qa^2 > 3E^2 - 1 > 0. \tag{13}$$

If $a = 0$ (Schwarzschild), (13) is impossible. If $0 < a \leq 1$, (13) implies a positive Q, whence c_0 is non-negative; but c_0 non-negative contradicts (12). Since (12) cannot be satisfied, there may be at most two variations of sign and hence two positive roots. We have shown this for $r \geq 1$. It must be true *a fortiori* outside the horizon, since the latter occurs at

$$r = 1 + (1 - a^2)^{1/2} \geq 1.$$

When $E^2 < 1$, the first term is negative. Four variations would be possible only if the succeeding terms had the signs +, −, +, −. But if the second term is positive, then

$$E^2 > \tfrac{1}{2}.$$

Equation (13) continues to hold, showing that the last three terms cannot have the supposed signs. Hence, not more than three variations of sign are possible. One sees then that for given Q, Φ, and $|E| < 1$, there may be at most one region of binding.

By considering the motion in latitude, and using the preceding theorem, we will prove that Q must be ≥ 0 for binding.

By analogy with $V(r)$ the effective θ potential, $V^2(\theta)$, is defined as that value of E^2 which makes $\dot{\theta} = 0$ when the polar angle $= \theta$. By (2b) and (3),

$$0 = \Theta(\theta) = Q - \cos^2\theta\{a^2[1 - V^2(\theta)] + \Phi^2 \sin^{-2}\theta\}$$

or

$$V^2(\theta) = 1 + a^{-2}(\Phi^2 \sin^{-2}\theta - Q\cos^{-2}\theta). \tag{14}$$

The significance of $V^2(\theta)$ can be seen by writing

$$\Theta(\theta) = a^2 \cos^2\theta[E^2 - V^2(\theta)]. \tag{15}$$

A particle at polar angle θ can only satisfy

$$E^2 < V^2(\theta) \tag{16}$$

if it is confined to the equatorial plane and has $Q = 0$. That Q is not negative follows from (14) according to which $Q > 0$ implies

$$E^2 > V^2(\pi/2) = -\infty.$$

For all cases other than $Q = 0$,

$$E^2 \geq V^2(\theta). \tag{17}$$

But (14) shows that if $Q<0$, then

$$V^2(\theta)>1, \tag{18}$$

By our earlier theorem, (17) and (18) together imply that the particle is unbound. Thus, for binding we require

$$Q\geq 0. \tag{19}$$

It follows from (2b), (3), and the theorem that an orbit has $Q=0$ if and only if it is confined to the equatorial plane.

Typical examples of $V^2(\theta)$ with $Q>0$ are shown in Fig. 2.[11] A particle can only reach the axis ($\theta=0, \pi$) if $\Phi=0$. With the help of the figure we draw the conclusion: Every orbit either remains in the equatorial plane ($Q=0$) or crosses it repeatedly ($Q>0$).

III. SPHERICAL ORBITS

From here on we will treat the most interesting case, the extreme Kerr case:

$$a=1.$$

In addition, we restrict ourselves to spherical orbits. The particle's radial coordinate will be stable at some value r_0 if $R(r)$ vanishes at $r=r_0$

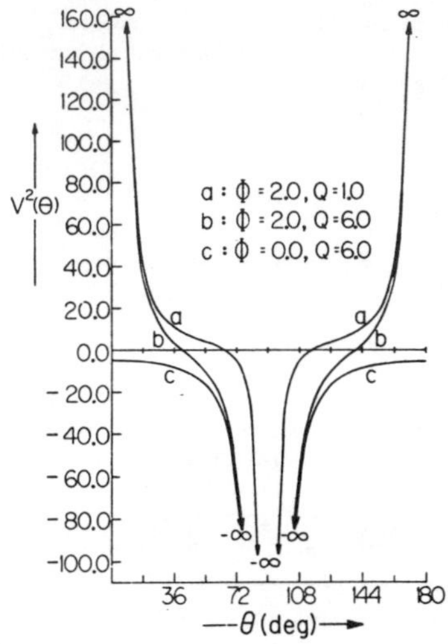

FIG. 2. Examples of the effective θ potential with $Q>0$. Only when $\Phi=0$ can particles of finite energy reach the axis ($\theta=0°$, $180°$). In this and subsequent figures, the spin parameter, a, equals unity.

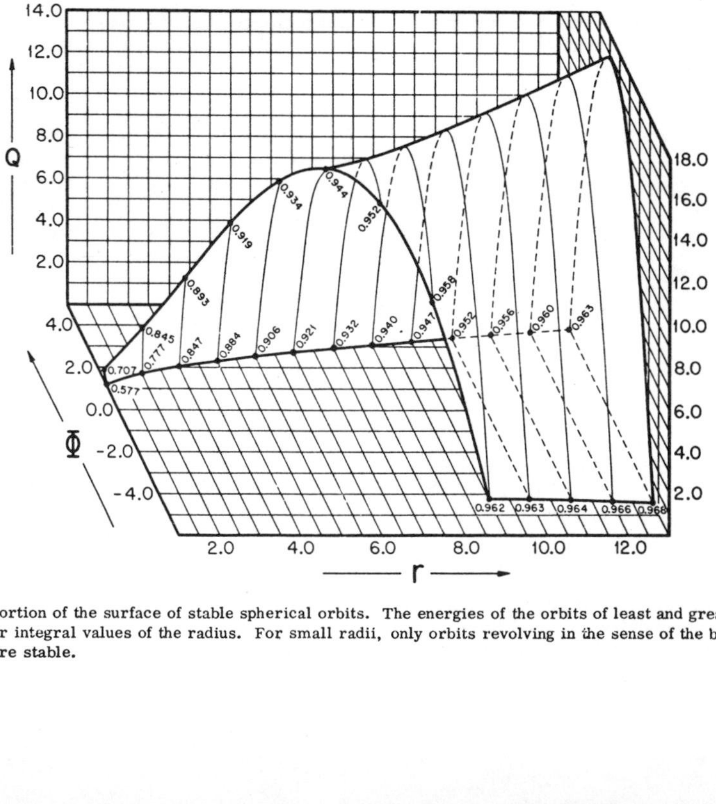

FIG. 3. Portion of the surface of stable spherical orbits. The energies of the orbits of least and greatest binding are given for integral values of the radius. For small radii, only orbits revolving in the sense of the black hole (those with $\Phi>0$) are stable.

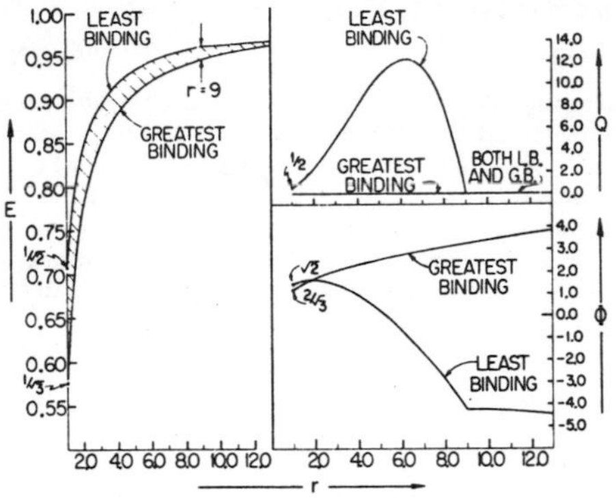

FIG. 4. Constants of motion of spherical orbits of least and greatest binding. As seen in the plot of Φ versus r and in Fig. 3, all orbits with radius $\lesssim 5.3$ are co-revolving ($\Phi > 0$). The discontinuity of slope at $r = 9$ for the orbits of least binding is not puzzling in view of the same discontinuity evident in Fig. 3.

and goes negative nearby. This will be the case if

$$R(r_0) = 0, \tag{20a}$$

$$\left.\frac{\partial R}{\partial r}\right|_{r=r_0} = 0, \tag{20b}$$

$$\left.\frac{\partial^2 R}{\partial r^2}\right|_{r=r_0} < 0. \tag{20c}$$

Replacing (20c) by

$$\left.\frac{\partial^2 R}{\partial r^2}\right|_{r=r_0} > 0 \tag{20c$'$}$$

yields instead the conditions for an unstable spherical orbit.

In the following we shall only be interested in the plus-root solutions, that is, those satisfying (9a). The behavior of the minus-root solutions is easily determined with the help of (10).

Solving Eqs. (20a) and (20b) simultaneously eliminates two of the four unknowns, E, Q, Φ, and r. Imposing the stability requirement (20c) and the boundedness requirement (19) determines a subset of this two-parameter set of trajectories.[12] In Fig. 3, the set of stable spherical orbits of small radius is represented as a two-dimensional surface in (r, Q, Φ) space. The intersections of the surface with the $Q=0$ plane are the equatorial orbits. For $r \geq 9$, there are two such lines, standing for the co-revolving[13] ($\Phi > 0$) and counterrevolving ($\Phi < 0$) equatorial orbits.

For a fixed radius the energy varies monotonically along the surface. The co-revolving equatorial orbits have the largest binding energy, $1-E$. For a given radius ≥ 9, the counterrevolving equatorial orbit has the smallest binding energy. For a radius < 9, such an orbit is unstable (and thus not shown in Fig. 3). Instead, the orbit situated at an inflection point of the effective radial potential, that is, with

$$\frac{\partial^2 R}{\partial r^2} = 0,$$

is the least tightly bound (for a given radius < 9); the set of these orbits constitutes the curved edge of the opening in the surface. (See Fig. 4.)

For large radii, the surface goes to the Schwarzschild limit:

$$Q(r) = \frac{r^2}{r-3} - \Phi^2 \geq 0,$$
$$E^2(r) = 1 - \frac{r-4}{r(r-3)}. \tag{21}$$

Stable orbits occur all the way to the horizon, at $r = 1$. The one-parameter family of horizon-skimming orbits is

$$2/\sqrt{3} \leq \Phi \leq \sqrt{2}, \tag{22a}$$

$$Q = \tfrac{3}{4}\Phi^2 - 1, \tag{22b}$$

$$E = \tfrac{1}{2}\Phi. \tag{22c}$$

The lower limit on Φ results from the restriction to $Q \geq 0$. To understand the upper limit one must consider an orbit just outside the horizon. Setting $r = 1 + \lambda$, with $0 < \lambda \ll 1$, one can show that

$$\left.\frac{\partial^2 R}{\partial r^2}\right|_{r=1+\lambda} = 2(\Phi^2 - 2)\lambda + O(\lambda^2). \tag{23}$$

Applying (20c) yields the upper limit, $\Phi = \sqrt{2}$.

Equations (22b) and (23) do not of themselves rule out the alternative range

$$-\sqrt{2} \leq \Phi \leq -2/\sqrt{3}. \tag{22a$'$}$$

Consideration of $V_+(r)$, however, shows that (22a) describes a solution of (9a), and (22a$'$) a solution of (9b).

From (21), observe that in the Schwarzschild case, the squared angular momentum, $\Phi^2 + Q$, depends only on the radius. By contrast, it follows from Eqs. (22) that, for orbits at the horizon, $\Phi^2 + Q$ varies almost by a factor of 2 in the extreme Kerr case.

IV. DRAGGING OF THE NODES

Lense and Thirring[14] have shown that in the weak-field limit, the nodes of a circular orbit are dragged in the sense of the spin by an angle

$$\Delta\Omega \cong 2(a/m)(m/r)^{3/2} \tag{24}$$

per revolution. An exact expression for $\Delta\Omega$, correct for all distances, can be given in the Kerr case.

Assume, say, that θ is decreasing. Dividing (2c) by (2b) then gives

$$\frac{d\phi}{d\theta} = -\frac{\Phi \sin^{-2}\theta - E + P\Delta^{-1}}{[Q - \cos^2\theta(1 - E^2 + \Phi^2 \sin^{-2}\theta)]^{1/2}}. \tag{25}$$

Setting

$$z = \cos^2\theta \tag{26}$$

and assuming $\theta \leq \frac{1}{2}\pi$, we integrate to obtain

$$\phi = \tfrac{1}{2}\Phi \int \frac{dz}{(1-z)Y(z)} + \frac{P\Delta^{-1} - E}{2} \int \frac{dz}{Y(z)}, \tag{27}$$

where

$$Y(z) = [\beta z^3 - (\alpha + \beta)z^2 + Qz]^{1/2}, \tag{28}$$

with $\alpha = \Phi^2 + Q$ and $\beta = 1 - E^2$. Turning points in θ occur when the denominator of (25) vanishes or, equivalently, when

$$\beta z^2 - (\alpha + \beta)z + Q = 0. \tag{29}$$

Since α, β, and Q are all non-negative and since $\alpha \geq Q$, the roots, $z_\pm$, of (29) are real and non-negative. The range of z for which the motion takes place includes the equatorial value, $z = 0$:

$$0 \leq z \leq z_-. \tag{30}$$

Equation (30) corresponds to one-quarter of a complete oscillation in latitude. From (2c) it is clear that the azimuth changes by the same amount in each quarter oscillation; that is, one gets the same change whatever the signs of $\dot{\theta}$ and $\theta - \frac{1}{2}\pi$.

Using a standard table of integrals,[15] the change of azimuth during one-quarter oscillation of latitude is cast into a more intelligible form:

$$\Delta\phi = \frac{1}{(\beta z_+)^{1/2}}[\Phi\Pi(-z_-, k) + (P\Delta^{-1} - E)K(k)], \tag{31}$$

where

$$k^2 = z_-/z_+$$

and

$$K(k) = \int_0^{\pi/2} \frac{dx}{(1 - k^2 \sin^2 x)^{1/2}},$$

$$\Pi(n, k) = \int_0^{\pi/2} \frac{dx}{(1 + n \sin^2 x)(1 - k^2 \sin^2 x)^{1/2}}$$

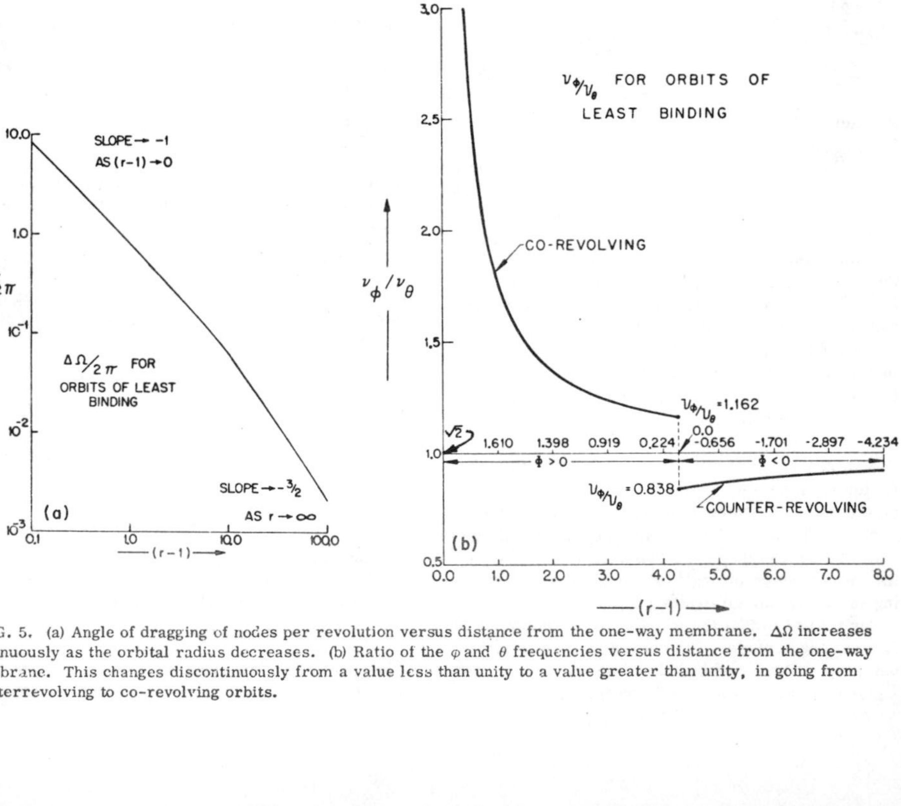

FIG. 5. (a) Angle of dragging of nodes per revolution versus distance from the one-way membrane. $\Delta\Omega$ increases continuously as the orbital radius decreases. (b) Ratio of the φ and θ frequencies versus distance from the one-way membrane. This changes discontinuously from a value less than unity to a value greater than unity, in going from counterrevolving to co-revolving orbits.

are elliptic integrals of the first and third kinds, respectively.

An orbit is called co-revolving if $\Delta\phi$ is positive. The first term on the right-hand side of (31) has the sign of Φ; one can show that the second term is always positive for a spherical orbit satisfying (9a). When Φ is negative the first term dominates the second; even when Φ approaches zero, the integral by which it is multiplied blows up so that the first term remains dominant. It follows that the sign of Φ determines that of $\Delta\phi$ and hence whether an orbit is co-revolving.

If the θ and ϕ frequencies were equal, $\Delta\phi$ would equal $\frac{1}{2}\pi$. Thus the ratio of the frequencies is given in general by

$$\nu_\phi/\nu_\theta = |\Delta\phi|/\tfrac{1}{2}\pi, \tag{32}$$

Substituting values of E, Φ, and Q for various orbits, one finds that

$$\begin{aligned}\nu_\phi/\nu_\theta &< 1 \quad \text{for} \quad \Phi < 0 \\ &> 1 \quad \text{for} \quad \Phi > 0.\end{aligned} \tag{33}$$

Equation (33) signifies that the nodes are always dragged in the sense of the spin.

The angle of advance of the nodes per nodal period is

$$\Delta\Omega = 2\pi|\nu_\phi/\nu_\theta - 1|. \tag{34}$$

One finds, as one would expect, that $\Delta\Omega$ always varies continuously when the constants of motion change continuously. ν_ϕ/ν_θ, however, undergoes a finite discontinuity when Φ passes through zero. Figure 5 displays the contrasting behavior of $\Delta\Omega$ and ν_ϕ/ν_θ for the least tightly bound orbits. The plot of ν_ϕ/ν_θ for the most tightly bound orbits does not show the discontinuity since such orbits are all co-revolving.

For large radii, one may replace E, Φ, and Q by their Schwarzschild values. From (21), (31), (32), and (34), the leading term of $\Delta\Omega$ is just the Lense-Thirring result, (24), with $a = m = 1$.

Contrast (24) with the effect in Newtonian theory of a mass quadrupole moment:

$$\Delta\Omega \propto \frac{\cos i}{r^2}, \tag{35}$$

where i is the angle of inclination. The $\Delta\Omega$ of (35) differs from that of (24) in these respects: (a) higher-order dependence on the radius; (b) dependence on the inclination of the orbit, e.g., the nodes do not regress for a polar orbit; and (c) the sense of rotation of the nodes depends on the motion of the orbiting particle – the nodes move contrary to the azimuthal velocity.

For small radii, only the term Δ^{-1} in (31) diverges. Using Eq. (8), the effective potential at $r = 1 + \lambda$ (expanded about $r = 1$) is

$$V(r) \simeq \tfrac{1}{2}\Phi + \frac{\partial V}{\partial r}\lambda + \frac{1}{2}\frac{\partial^2 V}{\partial r^2}\lambda^2.$$

Since the slope of $V(r)$ vanishes at $r = 1 + \lambda$,

$$\lambda \simeq -\frac{\partial V}{\partial r}\Big/\frac{\partial^2 V}{\partial r^2}.$$

Thus

$$E = V(r) \simeq \tfrac{1}{2}\Phi - \frac{1}{2}\frac{\partial^2 V}{\partial r^2}\lambda^2 = \tfrac{1}{2}\Phi + O(\lambda^2).$$

Using this to evaluate P to order λ, one finds

$$P\Delta^{-1} = \Phi\lambda^{-1} + O(1). \tag{36}$$

There results the asymptotic formula

$$\frac{\nu_\phi}{\nu_0} \simeq \frac{1}{r-1}\,\frac{2K(k)}{\pi(\beta z_+)^{1/2}} \tag{37}$$

Substitution of the values (22) for orbits at the horizon reveals that the coefficient of $(r-1)^{-1}$ varies from about 0.817 for $\Phi = 2/\sqrt{3}$ to about 0.835 for $\Phi = \sqrt{2}$.

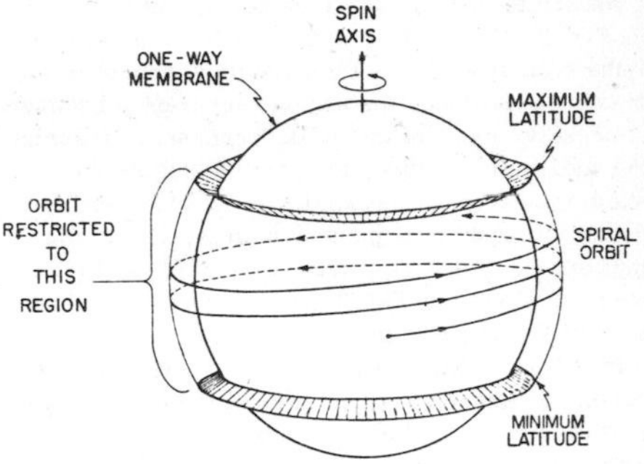

FIG. 6. Sketch of path of spherical orbit near one-way membrane (not drawn to scale.)

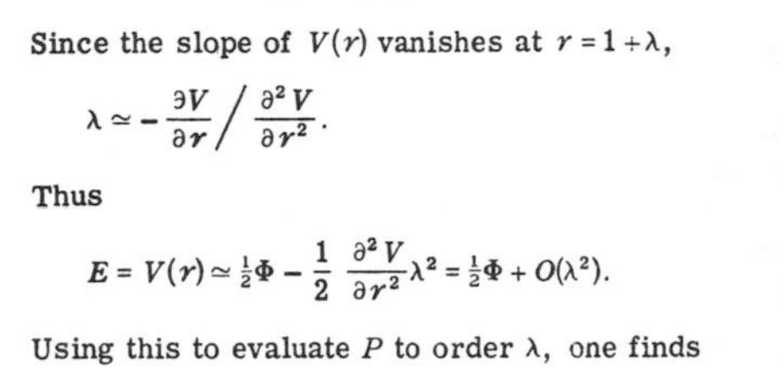

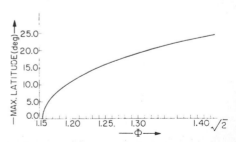

FIG. 7. Maximum latitude (= – minimum latitude) of orbits at the one-way membrane.

An orbital path close to the horizon is sketched in Fig. 6. The particle traces out a kind of helix lying on a sphere. As the particle approaches the maximum latitude, the angular separation between successive loops of the helix decreases. Reaching the maximum latitude, the particle begins the winding descent to the minimum latitude, located below the equatorial plane symmetrically to the maximum. Using Eqs. (22), (26), (29), and (30), we have plotted in Fig. 7 the maximum latitudes of orbits at the horizon.

It is known that there is a close resemblance between the linearized gravitational theory and classical electromagnetism. This prompts one to ask whether there occur orbits analogous to the spirals of electrons in the earth's magnetic field. The best place to look for such exotic behavior is near the horizon. From Eqs. (3), (9a), and (8) it follows that $P\Delta^{-1}$ always diverges at the horizon. Divide Eq. (2c) by Eq. (2d) to obtain

$$\frac{d\phi}{dt} \approx \tfrac{1}{2} \text{ for } r \approx 1.$$

One can see from this that the retrograde motion required for any kind of looping is impossible: Dragging forces a particle near the horizon to revolve always in the same sense as the black hole.

Unlike the weak-field region it is not true in general that $\Delta\Omega$ depends only on the radius. For example, for $r = 9$, $\Delta\Omega/2\pi$ decreases from 0.0814 for the counterrevolving equatorial orbit to 0.0607 for the co-revolving equatorial orbit.[16]

V. PERIODS

The proper θ period is obtained by integration of (2b). Squaring (2b), multiplying through by $\cos^2\theta \sin^2\theta$, and making the change of variables (26), one obtains

$$\dot{z}^2 = \frac{4}{r^2 + z^2} Y(z)$$

with $Y(z)$ as in (27). Hence, apart from a sign,

$$d\tau = \frac{(r^2 + z)dz}{2Y(z)} .$$

Integrating,

$$\tfrac{1}{4}\tau_{\theta,p} = \int_0^{z_-} \frac{(r^2 + z)dz}{2Y(z)}$$

or

$$\tau_{\theta,p} = \frac{4}{(\beta z_+)^{1/2}} (r^2 + z_+) K(k) - 4\left(\frac{z_+}{\beta}\right)^{1/2} E(k), \qquad (38)$$

where

$$E(k) = \int_0^{\pi/2} (1 - k^2 \sin^2 x)^{1/2} dx$$

is the complete elliptic integral of the second kind.

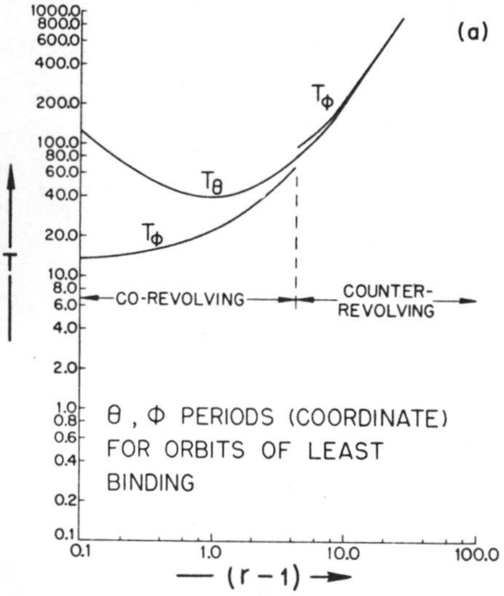

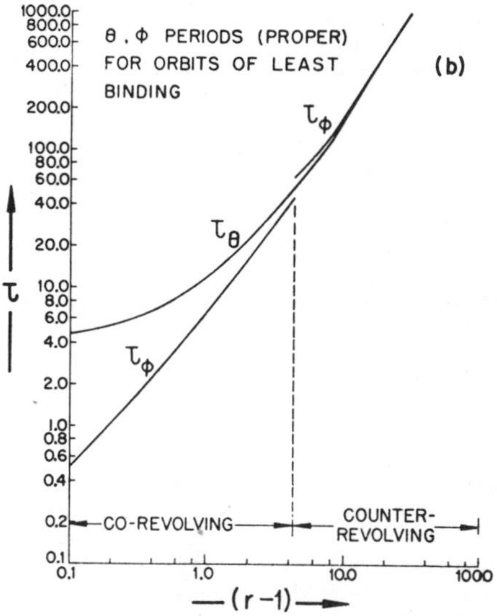

FIG. 8. The θ, ϕ periods (coordinate and proper) for orbits of least binding. The ϕ period is greater (less) than the θ period for counterrevolving (co-revolving) orbits. For the most tightly bound orbits (not shown), which are all co-revolving and equatorial, the ϕ period is everywhere less than the θ period.

To obtain the coordinate period, divide Eq. (2d) by Eq. (2b). Integration then yields

$$\tau_{\theta,c}=4\left[E\left(\frac{z_+}{\beta}\right)^{1/2}+\frac{1}{(\beta z_+)^{1/2}}[\Phi+P\Delta^{-1}(r^2+1)-E]\right]K(k)$$

$$-4E\left(\frac{z_+}{\beta}\right)^{1/2}E(k). \qquad (39)$$

To find the proper azimuthal period divide $\tau_{\theta,p}$ by

$$\nu_\phi/\nu_\theta=\tau_\theta/\tau_\phi,$$

which is given by (31) and (32); likewise for the coordinate period.

The periods determined above refer to $\mu=m=1$. The scaling law for times is obtained in similar manner to Eqs. (6). Equation (2b) is to be cast into the mass-independent form

$$\hat{\rho}^2\frac{d\hat{\theta}}{d\hat{\lambda}}=(\hat{\Theta})^{1/2}, \qquad (40)$$

where $\hat{\Theta}$ is the function Θ expressed in terms of caret variables. The polar angle is already scale free. From (1), (3), and (6) one sees that

$$\hat{\rho}^2=m^{-2}\rho^2,$$

$$\hat{\Theta}=(m\mu)^{-2}\Theta.$$

It follows that (2b) can be cast into the form (40) by setting

$$d\hat{\lambda}=\mu m^{-1}d\lambda=m^{-1}d\tau.$$

The second equality results from (4). Putting $m=1$,

$$d\hat{\lambda}=d\hat{\tau}$$

or

$$d\hat{\tau}=m^{-1}d\tau. \qquad (41)$$

Similarly, from (2d), (3), and (6),

$$d\hat{t}=m^{-1}dt. \qquad (42)$$

Equations (41) and (42) together with Eq. (6) constitute a general rule applicable to geodesics with $\mu^2>0$.

Figure 8 shows the periods for the least-bound orbits with radii ≤ 30. Here we see that the discontinuity in ν_ϕ/ν_θ mentioned earlier is due entirel to a discontinuity in ν_ϕ.

Note added in proof. One can show from (7) that

$$D=4r\Delta\{\Phi^2r^3+(r^2+Q)[r^3+a^2(r+2)]\}.$$

Equations (8) and (19) then imply that for an orbit the effective radial potential is everywhere real outside the horizon.

ACKNOWLEDGMENTS

Without the encouragement, guidance, and infectious enthusiasm of Professor R. Ruffini, this work would not have been accomplished. I am also indebted to Professor J. A. Wheeler and Dr. Demetrios Christodoulou for helpful discussions. To those at Stanford University and at Princeton who enabled me to spend some months in the stimulating atmosphere of Princeton's Department of Physics, I express my heartfelt gratitude.

---

*Preparation for publication assisted in part by NSF Grants No. GP14361 and No. GP7669.

[1]R. Ruffini and J. A. Wheeler, in *The Significance of Space Research for Fundamental Physics*, edited by A. F. Moore and V. Hardy (European Space Research Organization, Paris, 1970).

[2]D. Christodoulou, Phys. Rev. Letters 25, 1596 (1970).

[3]D. Christodoulou and R. Ruffini, Phys. Rev. D 4, 3552 (1971).

[4]C. Darwin, Proc. Roy. Soc. (London) A249, 180 (1958).

[5]C. Darwin, Proc. Roy. Soc. (London) A263, 39 (1961).

[6]By "bound" we mean that the particle remains within a finite radial distance of the black hole but is never captured.

[7]B. Carter, Phys. Rev. 174, 1559 (1968).

[8]We have corrected the signs of the terms containing P in Eqs. (2c) and (2d).

[9]The polynomial must be written in the order of decreasing (or increasing) powers. Terms with zero coefficients are not included.

[10]R. Descartes, *The Geometry of Rene Descartes*, translated by D. E. Smith and M. L. Latham (Open Court, Chicago, 1925). Descartes merely states the rule. For a proof see, e.g., H. E. Buchanan and L. C. Emmons, *A Brief Course in Advanced Algebra* (Houghton-Mifflin, Boston, 1937).

[11]The symbol $V^2(\theta)$ does not stand for the square of a real-valued function; rather it is meant to suggest a function having the dimensions of squared energy. That it can take on negative values is consequently not problematical.

[12]The artificiality of considering the θ and r motions separately here reveals itself in the curious circumstance that a trajectory may be "stable" and yet unbound.

[13]By co-revolving we mean that during one complete oscillation in latitude, the particle's azimuth changes in the sense of the spin of the black hole, that is to say, it increases. In Sec. IV we explain why an orbit is co-revolving if and only if $\Phi>0$.

[14]J. Lense and H. Thirring, Physik. Z. 19, 156 (1918).

[15]W. Grobner and N. Hofreiter, *Integraltafel* (Springer, Vienna, 1949), Vol. 1.

[16]To give a meaning to $\Delta\Omega$ for equatorial orbits, one imagines the motion to deviate infinitesimally from the equatorial plane.

Reprinted from:
PHYSICAL REVIEW D VOLUME 4, NUMBER 8 15 OCTOBER 1971

A-5

Theory of the Detection of Short Bursts of Gravitational Radiation

G. W. Gibbons
Department of Applied Mathematics and Theoretical Physics, University of Cambridge, England

and

S. W. Hawking
Institute of Theoretical Astronomy, University of Cambridge, England
(Received 30 November 1970)

It is argued that the short bursts of gravitational radiation which Weber reports most probably arise from the gravitational collapse of a body of stellar mass or the capture of one collapsed object by another. In both cases the bulk of the energy would be emitted in a burst lasting about a millisecond, during which the Riemann tensor would change sign from one to three times. The signal-to-noise problem for the detection of such bursts is discussed, and it is shown that by observing fluctuations in the phase or amplitude of the Brownian oscillations of a quadrupole antenna one can detect bursts which impart to the system an energy of a small fraction of kT. Applied to Weber's antenna, this method could improve the sensitivity for reliable detection by a factor of about 12. However, by using an antenna of the same physical dimensions but with a much tighter electromechanical coupling, one could obtain an improvement by a factor of up to 250. The tighter coupling would also enable one to determine the time of arrival of the bursts to within a millisecond. Such time resolution would make it possible to verify that the radiation was propagating with the velocity of light and to determine the direction of the source.

I. INTRODUCTION

In this paper we discuss the problem of detecting short bursts of gravitational radiation. This is rather different from the detection of continuous radiation, to which most attention has been given.[1] Such bursts have been reported by Weber,[2-4] who uses detectors at Maryland and Chicago in coincidence. Analysis of the time of arrival of these bursts suggests that they may be coming from the direction of the galactic center. We shall argue that the only events which are likely to produce bursts of waves strong enough for Weber to detect and at the rate ~1 per day (reported by Weber) are the collapse of bodies of stellar mass M or the capture of one collapsed object by another. In both these cases we would expect most of the energy of the gravitational radiation to be emitted in a time τ of the order of $10^{-5}M/M_{\odot}$ sec, during which time the sign of the Riemann tensor components of the radiation field would reverse only a few times. In other words, we expect something like a double or triple pulse rather than a long oscillating signal. We shall analyze the response of a gravitational-wave detector to such a burst.

We show that by observing fluctuations of the phase or amplitude of the Brownian motion of the antenna, one can detect bursts of gravitational waves that would probably not have been detected if one had merely observed the rms amplitude of the detector as Weber does. Applied to a detector like Weber's, which has a very low electromechanical coupling, this method would improve the sensitivity for reliable detection by a factor of about 12. However, by using a detector consisting of two metal bars connected by a piezoelectric transducer, one could possibly improve the sensitivity by as much as 250. This way of obtaining improved sensitivity would seem to be much easier and cheaper than cooling the detector to very low temperatures as has been proposed by a number of workers.

In Sec. II we consider possible sources for the bursts that Weber reports and discuss the nature of the signals they would produce. The response to such a burst of a simple quadrupole detector is considered in Sec. III. In Sec. IV we treat the signal/noise ratio of the detector by a method of fluctuations. An alternative treatment, which gives the same answers, is given in Sec. V in terms of an equivalent circuit for the detector.

II. THE EXPECTED SIGNALS

Gravitational radiation is produced whenever massive bodies accelerate under gravitational or nongravitational forces. However, because of the weakness of the gravitational constant, the rate of energy radiated is normally very small. For example, the earth revolving around the sun radiates about 1 kW at a frequency of 3 cycles per year. The weakness of the gravitational constant also means that gravitational-wave detectors are very inefficient: A flow of 2×10^4 erg/cm^2 sec for a duration of 10 sec in a bandwidth of 0.016 Hz at 1660 Hz is needed to excite Weber's detectors to an amplitude of the same order as that of the Brownian motion of the apparatus. Weber observes the amplitude of his detectors rising above threshold within 0.5 sec of one another. If the duration of the bursts were much longer than 0.5 sec, the amplitudes of the two detectors would rise slowly and would probably not cross the threshold within 0.5 sec of each other as the initial Brownian motion of the two detectors would be different. This would be inconsistent with Weber's report that shortening the coincidence time did not significantly affect the rate of coincidences. A burst of duration less than 0.5 sec must have a bandwidth of at least 0.3 Hz and so each burst must carry an energy of at least 4×10^6 erg/cm^2. If one allowed for an equal energy in the other polarization which Weber does not observe, the total energy would be at least 8×10^6 erg/cm^2. It is not reasonable to suppose that there is gravitational radiation only in a very narrow bandwidth centered on the frequency of Weber's detector. Therefore we presume that either each burst has a much larger bandwidth, say of the order of 1000 Hz, and so carries an energy of the order of 1.2×10^{10} erg/cm^2 per polarization, or that each burst has a narrow bandwidth but that there are something like a thousand bursts at other frequencies for every one that Weber observes. In either case the energy flux is at least 10^{10} erg/cm^2 per day.

A source emitting at 1660 Hz, the frequency at which Weber observes, must presumably be undergoing collective dynamical motions at that frequency. This would seem to imply that its size was less than 100 km, which is the distance light would travel in half a cycle. Possible sources of this size might be a neutron star, a star undergoing gravitational collapse, or the capture of one collapsed object by another. A neutron star would have an available nuclear and gravitational energy equal to about 1% of its rest mass, i.e., about 2×10^{52} erg. Therefore to produce a gravitational-wave burst of 1.2×10^{10} erg/cm^2 it would have to be within a distance of 3×10^{20} cm, i.e., 100 parsecs (pc). A distance of 100 pc includes about one part in 10^4 of the galactic disk. If, therefore, sources are distributed uniformly throughout the galactic disk, ten thousand of them are using up all their available energy every day, a number which seems unreasonably high. This number is not changed if one supposes that the neutron stars emit in a narrower bandwidth; for although the sources that Weber observes can be further away, there will be a corresponding number at other frequencies which he does not observe. The only reasonable way in which neutron stars might be responsible for the bursts that Weber observes is if we happen to be untypically near a neutron star that was undergoing violent eruptions: A neutron star at 1 pc would have enough energy to produce a burst of 4×10^6 erg/cm^2 once a day for about 150 000 years.

Gravitational collapse, or the capture of one collapsed object by another, would have the advantage over neutron stars as a source of gravitational radiation that a much greater fraction of the rest-mass energy might be released. This makes the energy problem easier but the amount is still embarrassingly high.[5-7] A source at the galactic center needs an energy of about 1.4×10^{56} erg ($70\,M_\odot$) to produce a burst of 1.2×10^{10} erg/cm^2 at the earth.

In a collapse or a capture most of the energy would not be released until matter has fallen to near the Schwarzchild radius or one collapsed object was near to the Schwarzchild radius of the other one. Thus one would expect most of the energy of the gravitational radiation to be in a period τ of the order of the dynamical time at this stage, i.e., $\tau\sim GMc^{-3}\sim10^{-5}MM_\odot^{-1}$ sec.

The energy flux in the gravitational wave at the observer is

$$F(t)\equiv c^7(4\pi G)^{-1}\left\{\left[\int^t R_{1010}(u)du\right]^2+\left[\int^t R_{1020}(u)du\right]^2\right\}\text{ erg/cm}^2\text{ sec}\,, \tag{1}$$

where the wave is taken to be traveling in the three directions and the Riemann tensor components R_{1010} and R_{1020} represent the two states of polarization of the wave. From this formula it can be seen that for a burst of finite energy the time integrals of the Riemann tensor components over the duration of the burst must be zero. This means that the sign of the Riemann tensor components must reverse during the burst.

In the linearized theory one has the relation

$$R_{\alpha 0\beta 0}=\frac{G}{3c^6}\frac{1}{r}\frac{d^4}{dt^4}D_{\alpha\beta}\,, \tag{2}$$

where $D_{\alpha\beta}$ is the quadrupole moment of the source and r is the distance from the source.

From this formula it follows that if the quadrupole is initially and finally time-independent (as one would expect for a gravitational collapse), then not only is the time integral of the components of the Riemann tensor zero but the second and third integrals as well, i.e.,[8]

$$\int_0^\tau dt\left[\int_0^t dt'\left(\int_0^{t'} R_{\alpha 0\beta 0}(t'')dt''\right)\right]=0\,. \tag{3}$$

This implies that the components of the Riemann tensor must change sign at least three times during the burst. For gravitational capture, on the other hand, the quadrupole moment will initially be a quadratic function of the time and so only the first time integral of the components of the Riemann tensor will be zero, and the components of the Riemann tensor need change sign only once during the burst.

In the nonlinear theory one does not have the simple relation (2) between the source and the radiation field. Thus Eq. (3) will not in general be satisfied for a gravitational collapse, but one could expect the transition from linear to nonlinear theory to preserve the *qualitative* feature that the sign of the components of the Riemann tensor change sign at least three times during the burst.

One might therefore be able to distinguish observationally between collapse or capture by examining the wave form of the burst. The simplest models for the bursts in the two cases are shown in Fig. 1.

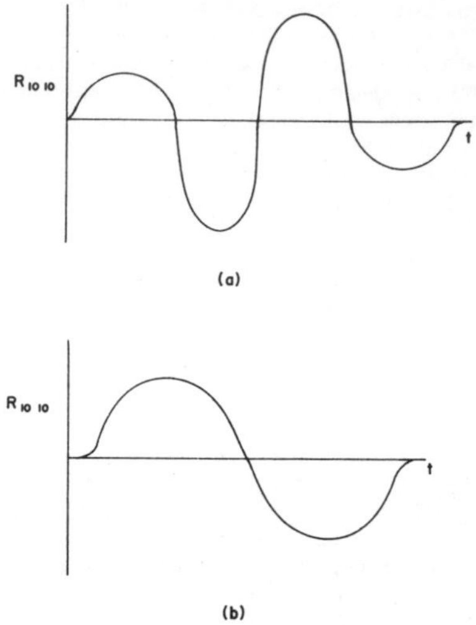

FIG. 1. Possible wave forms of the radiation field arising from (a) gravitational collapse and (b) gravitational capture.

III. THE RESPONSE OF THE DETECTOR

A simple quadrupole gravitational-wave detector consists basically of two masses m separated by a distance l and connected by a spring to give an oscillation angular frequency ω_0. The masses and the spring need not exist separately but can be combined in the form of a solid bar. The length of the bar will then be determined by the speed of sound in the material and the desired resonant frequency ω_0. The gravitational wave provides a relative force between the two masses which is proportional to their separation. Thus the equation of motion of the system is

$$\frac{d^2x}{dt^2}+\frac{\omega_0}{Q}\frac{dx}{dt}+\omega_0^2 x=-c^2 l R_{1010}\,, \tag{4}$$

where x is the change in separation of the masses, Q is the quality factor arising from the mechanical damping, and we take the line of center of the two masses to lie in the 1 direction.

The motion induced by a gravitational burst is

$$x=A\exp[-\omega_0(2Q)^{-1}+i\omega_0]\,t\,, \tag{5}$$

where

$$A=-c^2 l(i\omega_0)^{-1}\int_0^t e^{\omega_0 u(2Q)^{-1}}R_{1010}(u)e^{i\omega_0 u}du\,. \tag{6}$$

The duration τ of the burst we are considering will normally be much less than the damping time $Q\omega_0^{-1}$. This means that the motion of the detector immediately after the burst will be nearly independent of Q and will in fact be simply proportional to the Fourier components of the burst at the resonant frequency ω_0. From the discussion in Sec. II we expect the spectrum of the burst to have a maximum at a frequency of the order of $\omega_1=2\pi\tau^{-1}$ and to have a width $\Delta\omega$ of the same order as ω_1. The energy imparted to the detector by the burst, $4^{-1}m\omega_0^2|A|^2$, will be a maximum if the resonant frequency ω_0 is chosen to be of the same order as ω_1. Then

$$4^{-1}m\omega_0^2|A|^2\simeq 4\pi^3 G m l^2 \mathcal{E}(c^3\tau)^{-1}\,, \tag{7}$$

where $\mathcal{E}$ is the energy per unit area. Clearly the energy imparted to the detector will be greater the greater the length l, but l is limited by the speed of sound in the material, c_s, of the detector and the desired resonant frequency ω_0, $l\leqslant c_s 2\pi\omega_0^{-1}$; then,

$$\tfrac{1}{4}m\omega_0^2|A|^2\leqslant 4\pi^3 c^{-1}G m\mathcal{E}c_s^2 c^{-2}\tau\,. \tag{8}$$

We expect that the energy $\mathcal{E}$ and the duration of the burst τ will both be proportional to the mass of the emitting object; thus Eq. (8) suggests that it is more favorable to construct long detectors and to look for bursts of long duration arising from the collapse of objects of large mass. On the other hand, very massive collapses probably occur less frequently and it is more difficult to isolate the effects of extraneous mechanical and electrical disturbances if the frequency is low. We would therefore suggest that about 1000 Hz is probably the optimum frequency at which to look for gravitational bursts. This is in fact about the frequency at which Weber observes. At this frequency one would be looking for bursts emitted by systems of $100M_{\odot}$.

IV. THE SIGNAL/NOISE RATIO

Assuming that extraneous electrical and mechanical disturbances can be eliminated, the main sources of noise are the Brownian movements of the detector masses, the Johnson noise associated with the electrical losses in the transducer, and the noise produced in the first stage of the amplifier. We shall discuss first the Brownian movements and show that their effect can be minimized by using a detector with a fairly high Q and observing its motion with a good time resolution.

The two masses and the spring of the detector form a simple harmonic oscillator. By the law of equipartition of energy the oscillation will have an average energy of oscillation of kT provided that it is in equilibrium with its surroundings. The amplitude x_B of the thermal oscillations will be given by

$$m\omega_0^2 x_B^2 = 4kT . \tag{9}$$

For $m \sim 10^6$ g, $T \sim 300\,^0K$, and $\omega_0 \sim 10^4$, this gives $\sim 4\times 10^{-14}$ cm. One might think that one could detect only bursts that induced an amplitude greater than this or equivalently that imparted to the detector an energy larger than kT. However, it is in fact possible to detect much smaller bursts.

For although the average energy of the thermal oscillation is kT in equilibrium, the *noise power*, i.e., the rate at which energy enters and leaves the thermal oscillations of the detector, is very small if Q is large (we are grateful to P. Aplin for pointing this out). This can be seen as follows: One can regard the thermal oscillations as arising from a large number of small impulses applied to the detector with random phases and amplitudes. The oscillations induced by each small impulse will die away in the damping time $Q\omega_0^{-1}$. Thus at any time the energy of the detector arises effectively from the random impulses applied in the last $Q(2\pi)^{-1}$ cycles. Since the phases of the oscillations induced by the small impulses are uncorrelated, it follows that the mean value of the energy of the oscillation induced by the small impulses in one cycle is $kT2\pi Q^{-1}$. Thus by observing the change over n cycles of the phase or amplitude of the oscillations of the detector one could detect a gravitational burst that imparted to the detector in that time an energy greater than $2\pi nkTQ^{-1}$.[9] Obviously, using a smaller value of n gives a better time resolution.

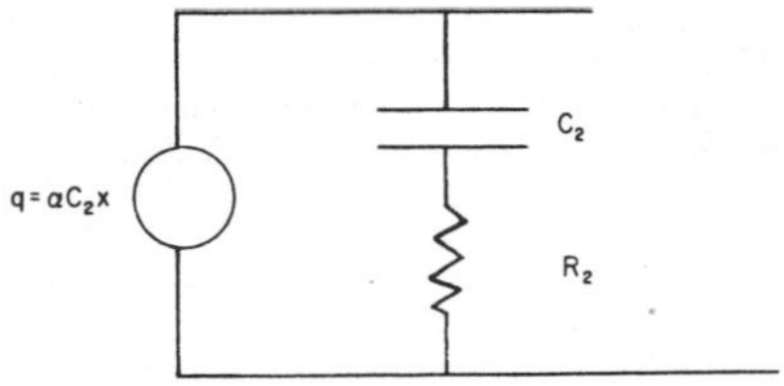

FIG. 2. The equivalent circuit of the transducer.

The most practical way to observe the motions of the detector seems to be to use a piezoelectric transducer[1] whose output is fed into a high-impedance amplifier. A piezoelectric transducer can be represented electrically as a charge generator in parallel with a capacitance and a resistance in series (Fig. 2). Here C_2 is the electrical capacitance of the transducer, α is its voltage output per unit change of separation x of the two masses of the detector, and $R_2 = (\omega_0 C_2)^{-1}\tan\delta$ represents the electrical losses in the transducer, where $\tan\delta$ is the "dissipation factor."

The Johnson noise of the resistor produces a mean-square voltage in a bandwidth $\Delta\omega$ of

$$V_T^2 = 2\pi^{-1}kTR_2\Delta\omega = 2\pi^{-1}kT\tan\delta\,\Delta\omega(\omega_0 C_2)^{-1} . \tag{10}$$

This may be compared to the mean-square voltage produced by the Brownian movements of the detector masses

$$V_B^2 = \alpha^2 x_B^2 , \tag{11}$$

where x_B is given by (9). Thus,

$$V_B^2 = 4\alpha^2 kT(m\omega_0^2)^{-1} \tag{12}$$

By observing fluctuations over n cycles, one can detect against the Brownian noise a burst that imparts to the detector an energy greater than $2\pi nkTQ^{-1}$. For such a burst the signal output voltage V_S satisfies $V_S^2 \geq 2\pi nV_B^2Q^{-1}$. In order to detect a signal against both the transducer noise and the Brownian noise, one needs

$$V_S^2 \geq V_T^2 + 2\pi nQ^{-1}V_B^2 .$$

The greatest sensitivity, therefore, is obtained by choosing the value of n for which

$$V_T^2 = 2\pi nV_B^2Q^{-1} .$$

This gives

$$n^2 = Q\tan\delta\,(8\pi^3\beta)^{-1} , \tag{13}$$

where $\beta = \alpha^2 C_2 (m\omega_0^2)^{-1}$ represents the proportion of elastic energy of the detector that can be extracted electrically from the transducer in one cycle. Thus one can detect gravitational waves which impart to the detector an energy of

$$(2\tan\delta)^{1/2}(\pi\beta Q)^{-1/2}kT .$$

For Weber's detector, $Q = 10^5$, $\beta = 5 \times 10^{-6}$, and $\tan\delta = 5 \times 10^{-3}$.[10] Thus the optimum value of n is 600. This would give a time resolution of about 0.4 sec (about what Weber uses) and would enable one to detect bursts which imparted to the detector an energy of about $\frac{1}{12}kT$. For a broad spectrum burst of the type we have considered this would correspond to an energy flow in one polarization $\mathcal{E}$ of 10^9 erg/cm^2. If, however, instead of observing fluctuations in the phase or amplitude of the oscillations of the detector one observes only whether the amplitudes of these oscillations increase suddenly (as Weber does), one could detect bursts of this intensity only if the phase of the oscillations induced by the burst coincided with the phase of the Brownian oscillations of the detector. This would mean that the probability of observing such a burst on a single detector would be about 1 in 4, and the probability of obtaining a coincidence of two detectors would be about 1 in 16. If, therefore, the bursts that Weber reports correspond to an energy of $\frac{1}{12}kT$, there must be 16 of them for each one that Weber observes, and the total energy flow must still be greater than 10^{10} erg/cm^2 per day.

In order to obtain greater sensitivity one would like to make β nearer unity. In fact, $\beta = \kappa^2\gamma$, where γ is the fraction of the elastic energy of the detector that is actually in the piezoelectric material and κ is a piezoelectric coupling constant which depends only on the material of the transducer. Values of κ range from 0.1 for quartz to about 0.6 for lead zirconate titanate. In order to make β larger, one wants to use a material like lead zirconate titanate and make γ higher. One way of doing this would be to use the transducer as a spring connecting two metal bars (Fig. 3). (This configuration was suggested by P. Aplin.) By suitably choosing the cross-sectional area of the transducer one can arrange that γ is 0.2 and so obtain a value of β of 0.07. Since the mechanical losses of the piezoelectric material are likely to be much greater than those in the metal bars, the mechanical Q of the detector would be the mechanical Q of the piezoelectric material divided by γ. One can obtain lead zirconate titanate with a Q of about 1000 and so the effective Q might be about 5000, although it would probably be rather lower because of mechanical losses in the cement joining the transducers to the bar, etc. The optimum value

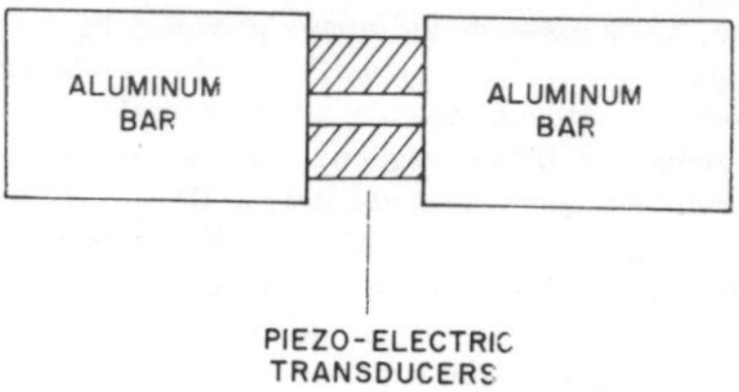

FIG. 3. A possible configuration for a detector with a high value of β, i.e., a tight electromechanical coupling.

of n for such a detector would be about unity. This would give a time resolution of about 10^{-3} sec and enable one to detect bursts which imparted in that time to the detector an energy greater than $\frac{1}{250}kT$. For $m \sim 10^6$ g and an effective value of l of about 100 cm,[11] this would correspond to an energy flow in one polarization of 4×10^7 erg/cm^2 in 1 msec at the earth, and to a total energy emitted in one polarization of at least $0.2 M_\odot$ if the source is at the galactic center. Of course, to be sure that the observed fluctuations of phase or amplitude of the oscillations of the detector are really due to gravitational bursts and not simply due to larger than average input of noise energy, one would have to set a threshold of several times the rms noise fluctuation in phase and amplitude. If

$$B \exp[-\omega_0(2Q)^{-1} + i\omega_0 t]$$

represents the oscillations of the detector induced by the noise energy entering in one cycle, then B will be distributed according to

$$P(z)dz = z_0^{-1} e^{-z/z_0} dz , \tag{14}$$

where $z = |B|^2$, $z_0 = |B_0|^2$, and B_0 is the rms amplitude of the oscillation induced in one cycle. A threshold for $|B|^2$ of $25|B_0|^2$, therefore, would be exceeded by chance less than once a year. Such a threshold would enable one to detect sources at the galactic center which emitted at least $5 M_\odot$ in gravitational radiation.

V. THE EQUIVALENT CIRCUIT

One can also discuss the signal/noise ratio in terms of an equivalent electrical circuit for the detector.[12] This may be derived as follows. The equivalent circuit of the transducer contains a charge generator $q = \alpha C_2 x$, where x obeys Eq. (4). This equation is the same as that for the displaced charge q in a series LCR circuit with a voltage generator V if $L_1 C_1 = \omega_0^{-2}$, $Q = R_1^{-1} L_1^{1/2} C_1^{-1/2}$, and $V = -c^2 l R_{1010} (\alpha C_2 L_1)^{-1}$. The remaining relation between L, C, and R is specified by requiring that the energy in the LCR circuit is equal to the energy of the motion of the detector. This gives

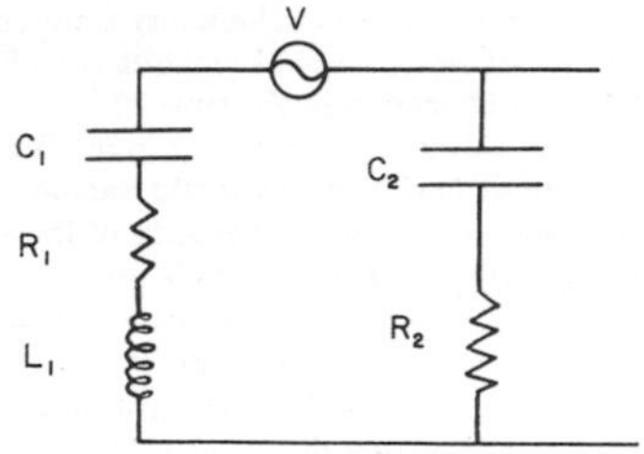

FIG. 4. The equivalent circuit of the detector and transducer.

$$L_1 = (\omega_0^2 \beta C_2)^{-1},$$
$$C_1 = \beta C_2,$$
$$R_1 = (Q\omega_0\beta C_2)^{-1},$$

and

$$V = -c^2 l R_{1010}(2\omega_0)^{-1} m^{1/2}(\beta C_2)^{-1/2}.$$

The equivalent circuit of the detector and transducer is therefore given by Fig. 4.[13]

We shall assume that the output of this equivalent circuit is fed into a high-impedance amplifier with gain A and that the amplifier output is divided in the ratio Z_3/Z_4 (Fig. 5). In this figure Z_1 represents the impedance of the series $L_1C_1R_1$ and Z_2 represents the impedance of the series C_2R_2. The impedances Z_3 and Z_4 are chosen as $Z_3 = D(Z_1 + Z_2)^{-1}$ and $Z_4 = HZ_2^{-1}$, where D and H are constant with $D >> H$. The Z_3, Z_4 circuit "undoes" the effect of the resonance of the detector and gives an output signal voltage

$$V_S = AHD^{-1} V$$
$$= -AHD^{-1}c^2 l R_{1010}(Z\omega_0)^{-1} m^{1/2}(\beta C_2)^{-1/2}.$$

It is not necessary to use such an inverse circuit but it is convenient for discussing the signal/noise ratio. The impedances Z_3 and Z_4 could be realized physically by a parallel LCR circuit and a parallel LR circuit respectively, though one might need to use a superconducting inductance in Z_3 to obtain a sufficiently high Q. It would probably be more convenient to simulate the Z_3, Z_4 circuit electronically.

Superimposed on the output signal voltage will be the Johnson noise produced by R_1 which represents the Brownian noise of the detector. This will produce at the output a flat noise spectrum with a mean-square voltage

$$V_B^2 = (AHD^{-1})^2 2kT\pi^{-1}(Q\omega_0\beta C_2)^{-1}$$

per unit bandwidth. The transducer noise produced by the resistance R_2 will give at the output a mean-square voltage of

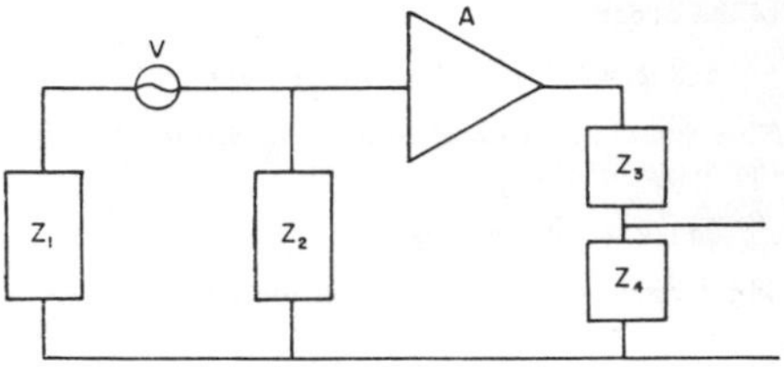

FIG. 5. The equivalent circuit for discussing the signal/noise ratio.

$$V_T^2 = (AHD^{-1})^2 2kT\pi^{-1}\tan\delta\,(\omega_0 C_2)^{-1}|Z_1|^2|Z_2|^{-2}$$
$$= (AHD^{-1})^2 2kT\pi^{-1}\tan\delta\,(\omega_0 C_2)^{-1}(\beta\omega_0)^{-2}$$
$$\times [\omega^2 Q^{-2} + \omega_0^{-2}(\omega^2 - \omega_0^2)^2]$$

per unit bandwidth. This has a sharp minimum at the resonant frequency ω_0. The noise produced by the amplifier will have a rather similar spectrum at the output. Using modern techniques, it seems possible to reduce the amplifier noise below the transducer noise and it will be neglected.

Suppose now that the output of the circuit in Fig. 5 is fed into a filter of bandwidth $\Delta\omega$. If the signal is of the form suggested in Sec. 2, i.e., a burst of one to three cycles, its Fourier transform will have a maximum at a frequency ω_1 of the order of $2\pi\tau^{-1}$ and a half-width of the same order. Therefore, if the filter pass band is centered at ω_1, the *amplitude* of the transmitted signal will be $V_S\Delta\omega\omega_1^{-1}$ and the *power* will be $V_S^2(\Delta\omega)^2\omega_1^{-2}$ for $\Delta\omega \leq \omega_1$. This behavior distinguishes short bursts from continuous incoherent radiation where the power is proportional to $\Delta\omega$. It is the reason why it is desirable to use a fairly large value of $\Delta\omega$, i.e., good time resolution.

If the resonant frequency ω_0 is chosen to be equal to ω_1,[14] the filter will transmit a Brownian noise power approximately equal to

$$(AHD^{-1})^2 2kT\pi^{-1}(Q\omega_0\beta C_2)^{-1}\Delta\omega$$

and a transducer noise power

$$(AHD^{-1})^2 kT(3\pi)^{-1}\tan\delta\,\beta^{-2}C_2^{-1}(\Delta\omega)^3\omega_0^{-3}.$$

The optimum value of $\Delta\omega$ will be the smaller of ω_0 and the value for which the transmitted noise power equals the transmitted transducer noise power. This gives

$$(\Delta\omega\omega_0^{-1})^2 = 6\beta(Q\tan\delta)^{-1}$$

which agrees almost exactly with Eq. (13). With this value of $\Delta\omega$ one could detect against the noise a short burst in which the amplitude of R_{1010} was of the order of

$$4\sqrt{2}\,\omega_0^2 kT(2\beta)^{1/2}(c^4 ml^2)^{-1}(3\pi^2\tan\delta)^{-1/2}.$$

This would correspond to an energy imparted to the detector of

$$2kT(2\tan\delta)^{1/2}(3\pi^2\beta Q)^{-1/2},$$

which agrees well with the estimate in Sec. IV.

VI. CONCLUSION

We have shown that by using an antenna with tight electromechanical coupling one could detect sources at the center of the Galaxy which emitted bursts carrying an energy of about $5M_\odot$. This energy still seems rather high, but it is considerably less than that needed to explain the events reported by Weber. One could increase the sensitivity by somewhat lowering the threshold or using several detectors in coincidence. The short time resolution of these detectors would make it possible to measure the difference in the time of arrival of a burst at two stations separated by more than 300 km. A chain of four stations would enable one to determine the direction and velocity of the signal. This would enable one to eliminate seismic disturbances which travel with the velocity of sound and would provide a test of general relativity, since there are other theories in which gravitational radiation travels with a different velocity from that of electromagnetic waves.[15]

ACKNOWLEDGMENT

We are greatly indebted to P. Aplin, who aroused our interest in the problem of the detection of short bursts of gravitational radiation and who suggested many of the ideas in this paper.

---

[1] J. Weber, *General Relativity and Gravitational Waves* (Interscience, New York, 1961).

[2] J. Weber, Phys. Rev. Letters 22, 1320 (1969).

[3] J. Weber, Phys. Rev. Letters 24, 276 (1970).

[4] J. Weber, Phys. Rev. Letters 25, 180 (1970).

[5] D. W. Sciama, G. B. Field, and M. J. Rees, Phys. Rev. Letters 23, 1514 (1969).

[6] D. W. Sciama, G. B. Field, and M. J. Rees, Comments Astrophys. Space Phys. 1, 187 (1969).

[7] D. W. Sciama, Nature 224, 1263 (1969).

[8] This equation also holds in the linearized Brans-Dicke theory.

[9] This point has also been made by V. Braginskiĭ and V. N. Rudenko, Usp. Fiz. Nauk 100, 395 (1970) [Soviet Phys. Usp. 13, 165 (1970)].

[10] These values can be deduced from the equivalent circuit given in J. Weber, Lett. Nuovo Cimento 4, 653 (1970).

[11] The effective value of l is about $2\pi^{-1}$ times the total length of the detector.

[12] J. Weber, Lett. Nuovo Cimento 4, 653 (1970).

[13] The resonant frequency of this circuit is ω_0 when the output of the transducer is short circuited. When it is not short circuited the resonant frequency is $(1+\beta)^{1/2}\omega_0$.

[14] Taking $\omega_0 < \omega_1$ improves the signal/(Brownian noise) ratio but makes the signal/(transducer noise) ratio worse. One cannot make $\omega_0 > \omega_1$ without reducing the length of the detector.

[15] S. Deser and B. E. Laurent, Ann. Phys. (N.Y.) 50, 76 (1968).

A-6

Reprinted from The Physical Review D, Vol. 2, No. 10, 2141–2160, 15 November 1970

Gravitational Field of a Particle Falling in a Schwarzschild Geometry Analyzed in Tensor Harmonics*†

Frank J. Zerilli‡
Joseph Henry Laboratories, Princeton, New Jersey 08540

(Received 5 January 1970; revised manuscript received 4 August 1970)

We are concerned with the pulse of gravitational radiation given off when a star falls into a "black hole" near the center of our galaxy. We look at the problem of a small particle falling in a Schwarzschild background ("black hole") and examine its spectrum in the high-frequency limit. In formulating the problem it is essential to pose the correct boundary condition: gravitational radiation not only escaping to infinity but also disappearing down the hole. We have examined the problem in the approximation of linear perturbations from a Schwarzschild background geometry, utilizing the decomposition into the tensor spherical harmonics given by Regge and Wheeler (1957) and by Mathews (1962). The falling particle contributes a δ-function source term (geodesic motion in the background Schwarzschild geometry) which is also decomposed into tensor harmonics, each of which "drives" the corresponding perturbation harmonic. The power spectrum radiated in infinity is given in the high-frequency approximation in terms of the traceless transverse tensor harmonics called "electric" and "magnetic" by Mathews.

I. INTRODUCTION

It was pointed out by Dyson[1] that a pulse of gravitational radiation will result from the capture of a star by a black hole (that is, a collapsed star). We consider the problem of a particle of mass m_0 falling along a geodesic of a Schwarzschild geometry produced by a larger mass m. The particle emits gravitational radiation as it falls until it is absorbed through the Schwarzschild surface at $2m$. The question of boundary conditions is interesting here. In a Euclidean topology we would require outgoing waves at infinity and regularity at the origin. In the Schwarzschild case there is no origin. However, the Schwarzschild surface at $2m$ has the property that future timelike or null trajectories pass through it only toward the interior region. Hence a natural boundary condition to replace regularity at the origin is to require that there are only ingoing waves at the Schwarzschild surface, that is, nothing coming out of the black hole.

Zel'dovich and Novikov[2] have considered the problem of the radiation of gravitational waves by bodies moving in the field of a collapsing star. They base their calculations on the formula, given by Landau and Lifshitz,[3] for the gravitational power radiated in terms of the third time derivative of the quadrupole moment of the system. Unfortunately, such considerations can only be valid for bodies which move at distances large compared to the Schwarzschild radius of the central body. But a substantial part of the radiation comes from the region r near $2m$. It is for this reason that we consider the field produced by the falling particle as a perturbation on the background Schwarzschild geometry so that we are not restricted to large distances from $2m$. The source term $T_{\mu\nu}$ is given by an integral over the world line of the particle, the integrand containing a four-dimensional invariant δ function. The source term is then guaranteed to be divergence-free if the world line is a geodesic in the background geometry.

Because of the spherical symmetry of the Schwarzschild field, the field equations for the perturbation $h_{\mu\nu}$ are in the form of a rotationally invariant differential operator on $h_{\mu\nu}$ set equal to the source term $T_{\mu\nu}$. We use this rotational invariance to separate the angular variables in the field equations. The usefulness of scalar harmonics Y_{LM} in separating, for example, Laplace's equation lies in the fact that they transform under a particular irreducible representation of the rotation group. Thus a rotationally invariant operator on Y_{LM} gives a quantity which transforms under the same irreducible representation and hence is a linear combination of Y_{LM} of the same order L. When dealing with tensor fields, we use tensor harmonics which transform under a particular representation of the rotation group.

In Appendix G we discuss the solutions of the $L=0$ and $L=1$ equations. There is no $L=0$ odd-parity-type (magnetic) harmonic. By suitable gauge transformations, it is possible to solve explicitly the partial differential equations in r and t for the $L=0$ and $L=1$ even-parity-type (electric) harmonics and the $L=1$ magnetic harmonics. The $L=0$ electric equations give the expected result. Let $R(t)$, $\Theta(t)$, and $\Phi(t)$ denote the Schwarzschild coordinates of the falling particle at Schwarzschild time t. Then inside the sphere $r<R(t)$ the perturbation from the background is zero, while outside it simply represents an augmentation of the Schwarzschild mass by $m_0\gamma_0$, the mass-energy of the falling particle. The $L=1$ magnetic equations give as a solution zero perturbation for $r<R(t)$, but give an $h_{0\phi}$ term outside of $R(t)$ which goes as $(\sin^2\theta/r)$ and which, according to the criterion given in Landau and Lifshitz,[3] represents a metric with angular momentum m_0a, where

* Based on the author's Ph.D. thesis, Princeton University, 1969 (unpublished).

† Work supported in part by NSF Grant No. GP 7769 and NSF Graduate Fellowship.

‡ Present address: Physics Department, University of North Carolina, Chapel Hill, N. C. 27514.

[1] F. Dyson (private communication).

[2] Ya. Zel'dovich and I. Novikov, Dokl. Akad. Nauk SSSR 155, 1033 (1964) [Soviet Phys. Doklady 9, 246 (1964)].

[3] L. Landau and E. Lifshitz, *The Classical Theory of Fields* (Addison-Wesley, Reading, Mass., 1962).

$m_0 a$ is the conserved angular momentum of the falling particle. The $L=1$ electric equations also give zero inside $R(t)$ and a nonzero $h_{\mu\nu}$ outside $R(t)$ which can be removed by a gauge transformation which is interpretable by a distant observer as a shift of the origin of the coordinate system. When the particle is far from $2m$, that is, $R(t)$ is large, the shift looks like a transformation to the center-of-momentum system where the particles orbit each other with distances from the center of momentum which are in inverse proportion to their relativistic masses.

For $L\geq 2$ we cannot solve the equations explicitly. We look at the Fourier transform in Schwarzschild time t of the equations which results in ordinary differential equations in r. In the high-frequency limit we obtain asymptotic expansions for the solutions of the homogeneous equations. The angular dependence is, of sourse, given by the tensor harmonics; by a gauge transformation to a gauge in which the leading terms of the expansion are traceless and divergenceless, the fields are expressed in terms of the transverse traceless electric and magnetic harmonics. We can write the expression for the radiated energy using the Landau-Lifshitz pseudotensor for r large compared to $2m$. Isaacson[4] has shown that this is a suitable expression everywhere, in the high-frequency limit, provided it is suitably averaged. Since we are dealing with Fourier transforms, this averaging amounts to taking field amplitudes times their complex conjugates. The stress tensor becomes singular at $r=2m$. If we transform to Kruskal coordinates, we observe that for outgoing waves the singularity is at $2m$ but $t=+\infty$, while for ingoing waves it is singular at $2m$ but at $t=-\infty$. This behavior agrees with Trautman's[5] result for the propagation of a discontinuity in the Riemann tensor in a Schwarzschild field. If we look at the leading terms of the correction to the Riemann tensor for our solution, we see that it has the same type of behavior as Trautman's discontinuity, and that the Riemann tensor is type N or radiation type in the Petrov-Pirani classification.

Using a Green's function formed from the high-frequency-limit solutions of the homogeneous equations, we obtain amplitudes for the ingoing (at $r=2m$) and outgoing (at $r=\infty$) radiation for a particle falling radially into the black hole. We use what is basically the saddle-point procedure to evaluate asymptotically the integral expressions for these amplitudes. We give in Appendix J a rough estimate for the power spectrum, which goes as a power of the frequency for low frequencies, reaches a peak at $\omega\sim 3/16\pi m$, and falls off exponentially for high frequencies. We estimate the total energy radiated to be $(1/625)$ (m_0^2/m) times a factor of order 1.

[4] R. Isaacson, Phys. Rev. 166, 1263 (1968); 166, 1272 (1968).

[5] A. Trautman, Lectures on General Relativity, lecture notes, Kings College, London, 1958 (unpublished).

II. WAVE EQUATIONS FOR RADIATION TREATED AS PERTURBATION

We write the metric in the form

$$g_{\alpha\beta}=(g_{\alpha\beta})_{\text{Schwarzschild}}+h_{\alpha\beta}\,,$$

where

$$(g_{\alpha\beta})_{\text{Schwarzschild}}dx^\alpha dx^\beta=-(1-2m/r)dt^2 +(1-2m/r)^{-1}dr^2+r^2(d\theta^2+\sin^2\theta\, d\varphi^2)\,.$$

The complete description in space and time of the gravitational wave generated by the infalling particle is provided by giving the perturbation in the metric, the symmetric tensor $\mathbf{h}=\{h_{\alpha\beta}\}$, as a function of r, θ, ϕ, and t. This tensor is expanded in tensor harmonic functions of angle $\mathbf{a}_{LM}^{(0)}(\theta,\phi)$, $\mathbf{a}_{LM}^{(1)}(\theta,\phi)$, $\mathbf{a}_{LM}$, $\mathbf{b}_{LM}^{(0)}$, $\mathbf{b}_{LM}$, $\mathbf{c}_{LM}^0$, $\mathbf{c}_{LM}$, $\mathbf{d}_{LM}$, $\mathbf{g}_{LM}$, and $\mathbf{f}_{LM}$. These ten harmonics are (a) complete over the space of symmetric tensor fields on a two-sphere and (b) orthonormal. They have been treated elsewhere[6–9] and are listed for convenience in Appendix A. Three of these harmonics are labeled "magnetic" and the other seven are labeled "electric." There is some ambiguity in the terminology in the literature. See Appendix B. The general first-order small perturbation in the geometry (Appendix C) is given by the following prescription: (a) Take each row in turn in Table I; (b) multiply each factor in the first column (tensor harmonic; function of θ and ϕ: details in Appendix A) by the factor in the second column ("coefficient in expansion in harmonics"; function of r and t); (c) sum all ten such products ("totalized part of metric perturbation of harmonic indes L, M"); and (d) sum over all L and M. This analysis presumes that one has a way to get the ten radial factors in this expansion directly as functions of r and t. For this purpose it proves simplest to express these factors as Fourier integrals, as, for example,

$$h_{0LM}(r,t)=(2\pi)^{-1/2}\int_{-\infty}^{+\infty} h_{0LM}(\omega,r)e^{-i\omega t}d\omega\,,$$

$$h_{0LM}(\omega,r)=(2\pi)^{-1/2}\int_{-\infty}^{+\infty} h_{0LM}(r,t)e^{i\omega t}dt\,,$$

with similar expressions for the other nine coefficients.

We have chosen the form of the coefficient functions to agree with the notation of Regge and Wheeler.[6] They go on to specialize the gauge (choice of four coordinates; see Appendix D for methods and options) so as to annul the four radial factors $h_{2LM}(r,t)$, $h_{0LM}^{(m)}(r,t)$, $h_{1LM}^{(m)}(r,t)$, and $G_{LM}(r,t)$ (reduction from 3 to 2 in number of radial factors in "magnetic" part of perturbation; and from 7 to 4 in number of radial factors in "electric" part). This choice of gauge simplifies the differ-

[6] T. Regge and J. A. Wheeler, Phys. Rev. 108, 1063 (1957).

[7] J. Mathews, J. Soc. Ind. Appl. Math. 10, 768 (1962).

[8] F. Zerilli, J. Math Phys. 11, 2203 (1970).

[9] J. Stachel, Nature 220, 779 (1968).

TABLE I. Components in expression of perturbation of metric in terms of tensor harmonics. The first three terms describe the "magnetic" part of the perturbation, and the last seven terms describe the "electric" part. The coefficients in this expansion ("radial factors") are listed as functions of r and t, but for the Fourier analysis (see text) the same formulas apply except that now the typical factor, for example, h_{0LM}, is to be understood as a function of ω and r. The third column shows the special values of these factors (whether functions of r, t, or of ω, r) in Regge-Wheeler (RW) gauge. The fourth column applies only in the (ω,r) representation; it gives the radial factors in radiation gauge, insofar as they have been expressed explicitly (details in Appendix I) in terms of the radial factors in RW gauge, as the latter may be evaluated from the differential equations given in the text. The outgoing radiation is governed exclusively by the third and by the next to the last row in the table. The tensor harmonics that come into these two terms are the transverse traceless harmonics $\mathbf{d}_{LM}(\theta,\varphi)$ and $\mathbf{f}_{LM}(\theta,\varphi)$, identical up to a sign with the harmonics $\mathbf{T}$ of Mathews.

Tensor harmonic	Coefficient	Specialization of coefficient to RW gauge	Value of coefficient for case of radiation gauge expressed of coefficients for RW gauge
$\mathbf{c}_{LM}^{(0)}(\theta,\varphi)$	$(-1/r)[2L(L+1)]^{1/2}h_{0LM}(r,t)$	same	zero
$\mathbf{c}_{LM}(\theta,\varphi)$	$(i/r)[2L(L+1)]^{1/2}h_{1LM}(r,t)$	same	$h_{1LM}^{(\mathrm{rad})}(\omega,r)=-(L-1)(L+2)(\omega r)^{-2}$ $\times(1-2m/r)h_{1LM}(\omega,r)$ (neglect at large ωr)
$-\mathbf{T}_{LM}^{(e)}=\mathbf{d}_{LM}(\theta,\varphi)$	$(\frac{1}{2}r^{-2})[2L(L+1)(L-1)(L+2)]^{1/2}h_{2LM}(r,t)$	zero	$h_{2LM}^{(\mathrm{rad})}(\omega,r)=(2i/\omega)h_{0LM}(\omega,r)$
$\mathbf{a}_{LM}^{(0)}(\theta,\varphi)$	$(1-2m/r)H_{0LM}(r,t)$	same	order of $1/\omega^2r^2$ at large ωr
$\mathbf{a}_{LM}^{(1)}(\theta,\varphi)$	$-2^{1/2}iH_{1LM}(r,t)$	same	order of $1/\omega^2r^2$ at large ωr
$\mathbf{a}_{LM}(\theta,\varphi)$	$(1-2m/r)^{-1}H_{2LM}(r,t)$	same	order of $1/\omega^2r^2$ at large ωr
$\mathbf{b}_{LM}^{(0)}(\theta,\varphi)$	$-(i/r)[2L(L+1)]^{1/2}h_{0LM}^{(m)}(r,t)$	zero	zero
$\mathbf{b}_{LM}(\theta,\varphi)$	$(1/r)[2L(L+1)]^{1/2}h_{1LM}^{(m)}(r,t)$	zero	order of $1/\omega^2r^2$ at large ωr
$\mathbf{T}_{LM}^{(m)}=\mathbf{f}_{LM}(\theta,\varphi)$	$[\frac{1}{2}L(L+1)(L-1)(L+2)]^{1/2}G_{LM}(r,t)$	zero	$[\frac{1}{2}L(L+1)(L-1)(L+2)]^{1/2}(1/i\omega r)K_{LM}(\omega,r)$ at large ωr
$\mathbf{g}_{LM}(\theta,\varphi)$	$\sqrt{2}K_{LM}(r,t)-2^{-1/2}L(L+1)G_{LM}(r,t)$	only K term	order of $1/\omega^3r^3$ at large ωr

ential equations. However, it has the feature that the perturbation in the metric increases with distance from the center of attraction in the "electric" part of $\mathbf{h}$; and in the "magnetic" part it keeps an unchanging order of magnitude. By contrast, for the calculation of the radiation one needs a gauge in which the magnitude of the perturbation falls off as $1/r$. The quantities needed in radiation gauge are expressed in Table I in terms of the radial factors in the Regge-Wheeler (RW) gauge because integrations seem easiest to do in the RW gauge. Only for the odd waves is the gauge transformation given explicitly; for the even waves the gauge transformation is spelled out only asymptotically in the limit of large ωr, the limit relevant for radiation (details of gauge transformations in Appendix D).

The perturbation in the geometry is "driven" by the source term in Einstein's field equations, 8π times the tensor of stress and energy, which tensor here—for a test particle moving on a goedesic—takes the form

$$T^{\mu\nu}=m_0\int_{-\infty}^{\infty}\delta^{(4)}(x-z(\tau))\frac{dz^\mu}{d\tau}\frac{dz^\nu}{d\tau}d\tau$$

$$=m_0\frac{dT}{d\tau}\frac{dz^\mu}{dt}\frac{dz^\nu}{dt}\frac{\delta(r-R(t))}{r^2}\delta^{(2)}(\Omega-\Omega(t)),\quad(1)$$

where the notation is as follows: $\delta^{(4)}$ is the invariant δ function, defined by

$$\iiiint\delta^{(4)}(x)(-g)^{1/2}d^4x=1,$$

τ is the proper time along the world line

$$z^\mu=z^\mu(\tau)=(T(\tau),R(\tau),\Theta(\tau),\Phi(\tau)),$$

Ω is an abbreviation for (θ,ϕ), and $\delta^{(2)}(\Omega)=\delta(\cos\theta)\delta(\phi)$. The stress-energy tensor is expressed in spherical harmonics in Appendix E and the procedure is given for evaluating its Fourier transform.

Appendix F gives the ordinary differential equations for the radial factors in the (ω,r) representation, RW gauge. In the odd case, three coupled equations for two radial factors are given; in the even case, where we have specialized to the problem of a mass falling straight in (and where in the absence of this specialization we would have had seven coupled equations for four radial factors), we have six interdependent equations for three radial factors.

An exact solution for the $L=0$ and $L=1$ terms is given in Appendix G. These terms describe the changes in mass, velocity, and angular momentum of the center of attraction produced by the arriving particle.

The harmonics of order $L=2, 3, \ldots$ describe the radiation. Asymptotically for large r in radiation gauge the perturbation $\mathbf{h}$ in the metric is the sum of two simple terms, each involving only one transverse traceless tensor harmonic:

$$\mathbf{h}^{(m)}(r,\theta,\varphi,t)=(2\pi)^{-1/2}\int_{-\infty}^{+\infty}\mathbf{h}^{(m)}(\omega,r,\theta,\varphi)e^{-i\omega t}d\omega$$

$$\sim(2\pi)^{-1/2}\left[\int_{-\infty}^{+\infty}\frac{2i}{\omega}h_{0LM}(\omega,r)e^{-i\omega t}d\omega\right]\mathbf{d}_{LM}(\theta,\varphi)$$

and

$$\mathbf{h}^{(e)}(r,\theta,\varphi,t)\sim(2\pi)^{-1/2}[\tfrac{1}{2}L(L+1)(L-1)(L+2)]^{1/2}$$

$$\times\left[\int_{-\infty}^{+\infty}\frac{1}{i\omega r}K_{LM}(\omega,r)e^{-i\omega t}d\omega\right]\mathbf{f}_{LM}(\theta,\varphi).$$

Asymptotically for large r, we find that radial factors have the form

$$h_{0LM}(\omega,r)\sim -rA_{LM}^{(m)}(\omega)\exp i\omega r^* \quad (2)$$

and

$$K_{LM}(\omega,r)\sim A_{LM}^{(e)}(\omega)\exp i\omega r^* \quad (3)$$

for magnetic and electric waves, respectively, where

$$r^*=r+2m\ln(r/2m-1). \quad (4)$$

The coefficients $A_{LM}(\omega)$ in these asymptotic expressions are found by integrating the radial wave equation. In terms of these coefficients, we can calculate intensity as a function of angle and as a function either of time or of frequency (case of pulse radiation; see Appendix H for case of multiply periodic motion). Thus we have

$$-\frac{d^2E}{d\omega d\Omega}=\frac{1}{32\pi}\sum_{LML'M'}[L(L+1)(L-1)(L+2)\times L'(L'+1)(L'-1)(L'+2)]^{1/2}\times\{A_{L'M'}^{(e)*}(\omega)A_{LM}^{(e)}(\omega)\mathbf{T}_{L'M'}^{(e)*}:\mathbf{T}_{LM}^{(e)}+A_{L'M'}^{(m)*}(\omega)A_{LM}^{(m)}(\omega)\mathbf{T}_{L'M'}^{(m)*}:\mathbf{T}_{LM}^{(m)}\} \quad (5)$$

and

$$-\frac{dE}{d\omega}=\frac{1}{32\pi}\sum_{LM}L(L+1)(L-1)(L+2)\times[|A_{LM}^{(e)}(\omega)|^2+|A_{LM}^{(m)}(\omega)|^2]. \quad (6)$$

To determine the distribution of the energy in time rather than in frequency, we must form the Fourier integrals for electric and magnetic waves, and construct the stress-energy tensor from these time-dependent fields. In a more extended account,[10] a treatment is also given for the amount of gravitational energy "going down the black hole."

The part of the radiation of more direct physical interest, that goes to great distances, is evidently determined completely by the coefficients $A_{LM}^{(m)}(\omega)$ and $A_{LM}^{(e)}(\omega)$ ("amplitudes of magnetic and electric waves"). We find these coefficients by solving the wave equations in the two cases, driven by the specified sources, and comparing the asymptotic behavior of the so obtained solutions with the asymptotic expressions (2) and (3).

Appendix I gives the analysis in question for the special case of a particle falling straight into the center of attraction. There is no magnetic term, and all the electric amplitudes also vanish, except for those with $M=0$, for which we find in the limit of high frequencies

$$A_L^{(\text{out})}\sim -4m_0(L+\tfrac{1}{2})^{1/2}e^{-4\pi m\omega}[\tfrac{1}{3}\sqrt{2}e^{5\pi i/8}\times\Gamma(\tfrac{3}{4})(m\omega)^{-3/4}+\tfrac{1}{12}\pi(m\omega)^{-1}+\cdots]. \quad (7)$$

From this result we find the radiation emitted in the Lth mode to be

$$-(dE/d\omega)_L\sim(1/32\pi)L(L+1)(L-1)(L+2)\times|A_L^{(\text{out})}(\omega)|^2 \quad (8)$$

at high frequencies. Appendix J discusses the qualitative behavior expected for this contribution at lower frequencies, the total output, and its comparison with the results of Zel'dovich and Novikov[2] based on the formulas cited by Landau and Lifshitz.[3]

For a fuller treatment it will be necessary to solve the radial equations for frequencies where an expansion asymptotic in ω is not appropriate. For this purpose it is possible to use directly the coupled systems of equations given in Appendix F. However, one gains insight into the structure of the equations by transforming from several dependent variables (coupled radial factors) to a single function. It satisfies a second-order wave equation with an effective potential $V_L(r)$ that lends itself to ready visualization. It has a peak ("barrier summit") at an r value of the order $r\sim 3m$ and goes to zero both at $r=\infty$ and as r approaches $2m$. "Large ω" in the asymptotic expansion previously employed means $\omega^2\gg V_{\text{peak}}$. When ω^2 is comparable to or less than V_{peak}, standard JWKB or numerical methods are appropriate. The details of the wave equation complete this paper.

For magnetic waves the new radial factor is $R_{LM}^{(m)}(\omega,r)$. In terms of it, the two old radial factors (RW gauge) are

$$h_{1LM}=r^2R_{LM}^{(m)}/(r-2m) \quad (9)$$

and

$$h_{0LM}=-\frac{i}{\omega}\frac{d}{dr^*}(rR_{LM}^{(m)})-\frac{8\pi r(r-2m)}{\omega[\frac{1}{2}L(L+1)(L-1)(L+2)]^{1/2}}D_{LM}(\omega,r), \quad (10)$$

where $D_{LM}(\omega,r)$ is the source term listed in Appendix E. The new radial factor obeys the wave equation [in terms of the new variable $r^*=r+2m\ln(r/2m-1)$]

$$\frac{d^2R_{LM}^{(m)}}{dr^{*2}}+[\omega^2-V_L^{(m)}(r)]R_{LM}^{(m)}=-\frac{8\pi i}{[\frac{1}{2}L(L+1)(L-1)(L+2)]^{1/2}}\frac{r-2m}{r^2}\times\left\{\frac{d}{dr}[r(r-2m)D_{LM}]+2(r-2m)D_{LM}+(r-2m)[(L-1)(L+2)]^{1/2}Q_{LM}\right\}, \quad (11)$$

where

$$V_L^{(m)}(r)=\left(1-\frac{2m}{r}\right)\left(\frac{L(L+1)}{r^2}-\frac{6m}{r^3}\right) \quad (12)$$

For electric waves the new radial factor is

[10] F. Zerilli, Ph.D. thesis, Princeton University, 1969 (unpublished).

$R_{LM}^{(e)}(\omega,r)$. In terms of it, the old radial factors are

$$K_{LM}=\frac{\lambda(\lambda+1)r^2+3\lambda mr+6m^2}{r^2(\lambda r+3m)}R_{LM}^{(e)}+\frac{r-2m}{r}\frac{dR_{LM}^{(e)}}{dr},\quad(13)$$

$$H_{1LM}=-i\omega\frac{\lambda r^2-3\lambda mr-3m}{(r-2m)(\lambda r+3m)}R_{LM}^{(e)}-i\omega r\frac{dR_{LM}^{(e)}}{dr},\quad(14)$$

$$H_{0LM}=\frac{\lambda r(r-2m)-\omega^2r^4+m(r-3m)}{(r-2m)(\lambda r+3m)}K_{LM}+\frac{m(\lambda+1)-\omega^2r^3}{i\omega r(\lambda r+3m)}H_{1LM}-\tilde{B}_{LM},\quad(15)$$

$$H_{2LM}=H_{0LM}+16\pi r^2\times[\tfrac{1}{2}L(L+1)(L-1)(L+2)]^{-1/2}F_{LM},\quad(16)$$

where

$$\lambda=\tfrac{1}{2}(L-1)(L+2),$$

and

$$\tilde{B}_{LM}=\frac{8\pi r^2(r-2m)}{\lambda r+3m}\{A_{LM}+[\tfrac{1}{2}L(L+1)]^{-1/2}B_{LM}\}-\frac{4\pi\sqrt{2}}{\lambda r+3m}\frac{mr}{\omega}A_{LM}^{(1)}.\quad(17)$$

The new radial factor obeys the wave equation

$$\frac{d^2R_{LM}^{(e)}}{dr^{*2}}+[\omega^2-V_L^{(e)}(r)]R_{LM}^{(e)}=S_{LM},\quad(18)$$

where the source term is

$$S_{LM}=-i\frac{r-2m}{r}\frac{d}{dr}\left[\frac{(r-2m)^2}{r(\lambda r+3m)}\left(\frac{ir^2}{r-2m}\tilde{C}_{1LM}+\tilde{C}_{2LM}\right)\right]+i\frac{(r-2m)^2}{r(\lambda r+3m)^2}\left[\frac{\lambda(\lambda+1)r^2+3\lambda mr+6m^2}{r^2}\tilde{C}_{2LM}+i\frac{\lambda r^2-3\lambda mr-3m^2}{r-2m}\tilde{C}_{1LM}\right],$$

where

$$V_L^{(e)}(r)=\left(1-\frac{2m}{r}\right)\times\frac{2\lambda^2(\lambda+1)r^3+6\lambda^2mr^2+18\lambda m^2r+18m^3}{r^3(\lambda r+3m)^2}.\quad(19)$$

The quantities $\tilde{C}_{1LM}$ and $\tilde{C}_{2LM}$ are combinations of source coefficients:

$$\tilde{C}_{1LM}=-\frac{8\pi}{\sqrt{2}\omega}A_{LM}^{(1)}-\frac{1}{r}\tilde{B}_{LM},\quad(20)$$

$$\tilde{C}_{2LM}=\frac{8\pi r^2}{i\omega}\frac{[\tfrac{1}{2}L(L+1)]^{-1/2}}{r-2m}B_{LM}^{(0)}-\frac{ir}{r-2m}\tilde{B}_{LM}.\quad(21)$$

ACKNOWLEDGMENT

I am grateful to Professor John A. Wheeler for proposing the problem and for many suggestions and much encouragement.

APPENDIX A: TENSOR HARMONICS

If $\mathbf{T}$ is a symmetric second-rank covariant tensor, it can be expanded in tensor harmonics[6–8] as follows:

$$\mathbf{T}=\sum_{LM}A_{LM}^{(0)}\mathbf{a}_{LM}^{(0)}+A_{LM}^{(1)}\mathbf{a}_{LM}^{(1)}+A_{LM}\mathbf{a}_{LM}+B_{LM}^{(0)}\mathbf{b}_{LM}^{(0)}+B_{LM}\mathbf{b}_{LM}+Q_{LM}^{(0)}\mathbf{c}_{LM}^{(0)}+Q_{LM}\mathbf{c}_{LM}+G_{LM}\mathbf{g}_{LM}+D_{LM}\mathbf{d}_{LM}+F_{LM}\mathbf{f}_{LM},\quad(\text{A1})$$

where

$$\mathbf{a}_{LM}^{(0)}=\begin{pmatrix}Y_{LM}&0&0&0\\0&0&0&0\\0&0&0&0\\0&0&0&0\end{pmatrix},\quad(\text{A2a})$$

$$\mathbf{a}_{LM}^{(1)}=(i/\sqrt{2})\begin{pmatrix}0&Y_{LM}&0&0\\Y_{LM}&0&0&0\\0&0&0&0\\0&0&0&0\end{pmatrix},\quad(\text{A2b})$$

$$\mathbf{a}_{LM}=\begin{pmatrix}0&0&0&0\\0&Y_{LM}&0&0\\0&0&0&0\\0&0&0&0\end{pmatrix},\quad(\text{A2c})$$

$$\mathbf{b}_{LM}^{(0)}=ir[2L(L+1)]^{-1/2}\times\begin{pmatrix}0&0&(\partial/\partial\theta)Y_{LM}&(\partial/\partial\phi)Y_{LM}\\0&0&0&0\\*&0&0&0\\*&0&0&0\end{pmatrix},\quad(\text{A2d})$$

$$\mathbf{b}_{LM}=r[2L(L+1)]^{-1/2}\times\begin{pmatrix}0&0&0&0\\0&0&(\partial/\partial\theta)Y_{LM}&(\partial/\partial\phi)Y_{LM}\\0&*&0&0\\0&*&0&0\end{pmatrix},\quad(\text{A2e})$$

$$\mathbf{c}_{LM}^{(0)}=r[2L(L+1)]^{-1/2}\times\begin{pmatrix}0&0&(1/\sin\theta)(\partial/\partial\phi)Y_{LM}&-\sin\theta(\partial/\partial\theta)Y_{LM}\\0&0&0&0\\*&0&0&0\\*&0&0&0\end{pmatrix},\quad(\text{A2f})$$

$$\mathbf{c}_{LM}=ir[2L(L+1)]^{-1/2}\times\begin{bmatrix}0&0&0&0\\0&0&(1/\sin\theta)(\partial/\partial\phi)Y_{LM}&-\sin\theta(\partial/\partial\theta)Y_{LM}\\0&*&0&0\\0&*&0&0\end{bmatrix},\tag{A2g}$$

$$\mathbf{d}_{LM}=-ir^2[2L(L+1)(L-1)(L+2)]^{-1/2}\times\begin{bmatrix}0&0&0&0\\0&0&0&0\\0&0&-(1/\sin\theta)X_{LM}&\sin\theta\, W_{LM}\\0&0&*&\sin\theta\, X_{LM}\end{bmatrix},\tag{A2h}$$

$$\mathbf{g}_{LM}=(r^2/\sqrt{2})\begin{bmatrix}0&0&0&0\\0&0&0&0\\0&0&1&0\\0&0&0&\sin^2\theta\end{bmatrix}Y_{LM},\tag{A2i}$$

$$\mathbf{f}_{LM}=r^2[2L(L+1)(L-1)(L+2)]^{-1/2}\times\begin{bmatrix}0&0&0&0\\0&0&0&0\\0&0&W_{LM}&X_{LM}\\0&0&*&\sin^2\theta\, W_{LM}\end{bmatrix},\tag{A2j}$$

$$X_{LM}=2\frac{\partial}{\partial\phi}\left(\frac{\partial}{\partial\theta}-\cot\theta\right)Y_{LM},$$

$$W_{LM}=\left(\frac{\partial^2}{\partial\theta^2}-\cot\theta\frac{\partial}{\partial\theta}-\frac{1}{\sin^2\theta}\frac{\partial^2}{\partial\phi^2}\right)Y_{LM}.$$

The symbol * denotes components derived from the symmetry of the tensors. The above set of tensors is orthonormal in the inner product

$$(T,S)\equiv\int\int T^*{:}S d\Omega,$$

where

$$T{:}S\equiv\eta^{\mu\lambda}\eta^{\nu\kappa}T_{\mu\nu}S_{\lambda\kappa}$$

and $\eta_{\mu\nu}$ is the Minkowski metric. Thus $A_{LM}^{(0)}=(\mathbf{a}_{LM}^{(0)},\mathbf{T})$, $B_{LM}^{(0)}=(\mathbf{b}_{LM}^{(0)},\mathbf{T})$, etc.

In place of $\mathbf{f}_{LM}$, Regge and Wheeler[6] use the harmonic (up to a normalization factor)

$$\mathbf{e}_{LM}=[2L(L+1)-2]^{-1/2}\{[(L-1)(L+2)]^{1/2}\mathbf{f}_{LM}-[L(L+1)]^{1/2}\mathbf{g}_{LM}\}.$$

Two of the harmonics in Eqs. (A2) are the transverse traceless electric and magnetic harmonics given by Mathews[7]:

$$\mathbf{d}_{LM}=-\mathbf{T}_{LM}^{(e)},\quad \mathbf{f}_{LM}=\mathbf{T}_{LM}^{(m)}.$$

Since our background metric is spherically symmetric, Eqs. (2.4) are in the form $Q[\mathbf{h}]=-16\pi\mathbf{T}$, where Q is a rotationally invariant operator. We have denoted the metric perturbation tensor by $\mathbf{h}$ and the source tensor by $\mathbf{T}$. Thus we can separate the angular variables θ, ϕ by expanding $\mathbf{h}$ and $\mathbf{T}$ in tensor harmonics.

APPENDIX B: EVEN-ODD CONVENTION

Table II is a correlation of the even-odd convention as used here and elsewhere.[11]

TABLE II. Correlation between terminology used here for the two types of harmonics and the terminology used elsewhere in the literature.

Listed in Table I	First three	Last seven
Parity	$(-1)^{L+1}$	$(-1)^L$
Number of such tensorial harmonics	3	7
Regge and Wheeler (Ref. 6)	Odd parity	Even parity
Mathews (Ref. 7); Zerilli (Ref. 8)	Electric	Magnetic
Thorne and Campolattaro (Ref. 11)	Odd or magnetic	Even or electric
Present paper	Magnetic	Electric
Emitted when mass falls straight in to Schwarzschild geometry?	No	Yes
Name employed for vector harmonic of same parity in case of electromagnetic radiation[a]	Magnetic	Electric
Emitted when charge falls straight towards copper sphere?	No	Yes

[a] Employing a definition of "magnetic" and "electric" such that "a vector potential A equal to one of these harmonics yields electromagnetic fields of precisely these types (magnetic or electric)."

APPENDIX C: PERTURBATIONS ON SCHWARZSCHILD METRIC

In this appendix we discuss the equations for linear perturbations from a background geometry that enables us to treat the gravitational interaction of two systems in an approximation which is good if the difference between the geometry of the base system (Schwarzschild geometry) and that of the interacting systems (mass falling in Schwarzschild geometry) is small. We will not discuss the question of what constitutes a "small" perturbation since this would be most adequately treated in the theory of "superspace," the set whose elements are three-dimensional Riemannian manifolds.[12] It will suffice to assume the norm in a particular coordinate system to be $\|g_{\mu\nu}\|=\sup\sum_{\mu\nu}|g_{\mu\nu}(x)|^2$, and we will consider a perturbation $\lambda h_{\mu\nu}$ small if there is a gauge transformation to an admissible coordinate system in which $|\|g_{\mu\nu}+\lambda h_{\mu\nu}\|-\|g_{\mu\nu}\||\rightarrow 0$ as $\lambda\rightarrow 0$.

Now let us consider linear perturbations[13] on a metric.

[11] K. Thorne and A. Campolattaro, Astrophys. J. 149, 591 (1967).

[12] J. A. Wheeler, "Superspace and the Nature of Quantum Geometrodynamics," in *Battelle Rencontres* (Benjamin, New York, 1968); A. E. Fischer, Ph.D. thesis, Princeton University, 1969 (unpublished).

[13] We use units in which $G=c=1$. We follow the convention proposed by C. W. Misner, K. S. Thorne, and J. A. Wheeler in "An Open Letter to Relativity Theorists," 1968 (unpublished): Latin indices run from 1 to 3 and indicate spacelike coordinates; Greek indices run from 0 to 2. The metric has signature +2 (spacelike convention). The connection coefficients and Riemann tensor are $\Gamma^\lambda{}_{\mu\nu}=\frac{1}{2}g^{\lambda\sigma}(g_{\sigma\mu,\nu}+g_{\sigma\nu,\mu}-g_{\mu\nu,\sigma})$ and $R^\alpha{}_{\lambda\mu\nu}=\Gamma^\alpha{}_{\lambda\nu,\mu}-\Gamma^\alpha{}_{\lambda\mu,\nu}+\Gamma^\alpha{}_{\sigma\mu}\Gamma^\sigma{}_{\lambda\nu}-\Gamma^\alpha{}_{\sigma\nu}\Gamma^\sigma{}_{\lambda\mu}$. The contracted Riemann tensor is $R_{\mu\nu}=R^\alpha{}_{\mu\alpha\nu}$ and the Einstein equations are $G_{\mu\nu}(g_{\alpha\beta})\equiv R_{\mu\nu}-\frac{1}{2}g_{\mu\nu}R=8\pi T_{\mu\nu}$.

Let $g_{\mu\nu}$ be a solution of the Einstein equations, and let $\tilde{g}_{\mu\nu}=g_{\mu\nu}+h_{\mu\nu}$. Then $G_{\mu\nu}(\tilde{g}_{\alpha\beta})=8\pi\tilde{T}_{\mu\nu}$, and $h_{\mu\nu}$ must satisfy

$$\Delta G_{\mu\nu}(g_{\alpha\beta},h_{\alpha\beta})\equiv G_{\mu\nu}(g_{\alpha\beta}+h_{\alpha\beta})-G_{\mu\nu}(g_{\alpha\beta})=8\pi(\tilde{T}_{\mu\nu}-T_{\mu\nu})\equiv 8\pi\Delta T_{\mu\nu}.$$

If $T_{\mu\nu}$ is "small," we can assume that $h_{\mu\nu}$ is "small." Now

$$\Delta G_{\mu\nu}(g_{\alpha\beta},h_{\alpha\beta})=\delta G_{\mu\nu}(g_{\alpha\beta})(h_{\rho\sigma})+O[h_{\rho\sigma}^2],$$

where $\delta G_{\mu\nu}(g_{\alpha\beta})$ is a linear operator on $h_{\rho\sigma}$ [$\delta G_{\mu\nu}(g_{\alpha\beta})$ is the derivative of the mapping which takes the tensor $g_{\mu\nu}$ into the tensor $G_{\mu\nu}$]. If $h_{\mu\nu}$ is small, then this linear operator is a good approximation to $\Delta G_{\mu\nu}(g_{\alpha\beta},h_{\alpha\beta})$. We thus consider solutions of

$$\delta G_{\mu\nu}(g_{\alpha\beta})(h_{\rho\sigma})=+8\pi\Delta T_{\mu\nu}. \qquad \text{(C1)}$$

The expression for $\delta G_{\mu\nu}(g_{\alpha\beta})(h_{\rho\sigma})$ is straightforward to calculate and has been given in several places.[14] Let the semicolon denote covariant differentiation in the base metric $g_{\mu\nu}$. Then

$$-\delta R_{\mu\nu}(g_{\alpha\beta})(h_{\rho\sigma})=\tfrac{1}{2}(h_{\mu\nu;\alpha}{}^{;\alpha}-h_{\mu\alpha;\nu}{}^{;\alpha}-h_{\nu\alpha;\mu}{}^{;\alpha}+h^{\alpha}{}_{\alpha;\mu;\nu}) \qquad \text{(C2)}$$

and Eq. (C1) becomes

$$\delta R_{\mu\nu}(g_{\alpha\beta})(h_{\rho\sigma})-\tfrac{1}{2}g_{\mu\nu}(h_{\alpha\lambda}{}^{;\alpha;\lambda}-h_{\alpha}{}^{\alpha}{}_{;\lambda}{}^{;\lambda})-\tfrac{1}{2}h_{\mu\nu}R+\tfrac{1}{2}g_{\mu\nu}h_{\alpha\beta}R^{\alpha\beta}=+8\pi\Delta T_{\mu\nu}.$$

Now by commuting covariant derivatives and denoting $h_{\mu\alpha}{}^{;\alpha}\equiv f_\mu$, we write Eq. (C1) as

$$-[h_{\mu\nu;\alpha}{}^{;\alpha}-(f_{\mu;\nu}+f_{\nu;\mu})+2R^{\rho}{}_{\mu}{}^{\alpha}{}_{\nu}h_{\rho\alpha}+h^{\alpha}{}_{\alpha;\mu;\nu}-R^{\rho}{}_{\nu}h_{\mu\rho}-R^{\rho}{}_{\mu}h_{\nu\rho}]-g_{\mu\nu}(f_\lambda{}^{;\lambda}-h^{\alpha}{}_{\alpha;\lambda}{}^{;\lambda})-h_{\mu\nu}R+g_{\mu\nu}h_{\alpha\beta}R^{\alpha\beta}=+16\pi\Delta T_{\mu\nu}. \qquad \text{(C3)}$$

The part in square brackets is $2\delta R_{\mu\nu}$. Finally, if the background is Ricci flat, $R_{\mu\nu}=0$, then

$$[h_{\mu\nu;\alpha}{}^{;\alpha}-(f_{\mu;\nu}+f_{\nu;\mu})+2R^{\rho}{}_{\mu}{}^{\alpha}{}_{\nu}h_{\rho\alpha}+h^{\alpha}{}_{\alpha;\mu;\nu}]+g_{\mu\nu}(f_\alpha{}^{;\alpha}-h^{\alpha}{}_{\alpha;\lambda}{}^{;\lambda})=-16\pi\Delta T_{\mu\nu}. \qquad \text{(C4)}$$

Consider now "gauge" transformations. If we make a coordinate transformation $x'^{\mu}=x^{\mu}+\xi^{\mu}$, then the transformed tensor field is $\tilde{g}'_{\mu\nu}=\tilde{g}_{\mu\nu}-\xi_{\mu;\nu}-\xi_{\nu;\mu}$ keeping terms to first order in ξ_μ. We can assume $\tilde{g}_{\mu\nu}'=g_{\mu\nu}+h'_{\mu\nu}$, then $h'_{\mu\nu}=h_{\mu\nu}-\xi_{\mu;\nu}-\xi_{\nu;\mu}$. We note[4] that if $R_{\mu\nu}(g_{\alpha\beta})=0$, it is consistent with (C4) to choose a gauge in which $h_{\mu\nu}{}^{;\nu}=0$ and $h_{\mu}{}^{\mu}=0$.

We will now restrict our attention to a background geometry which has spherical symmetry[15] and whose metric in a local coordinate system can be written

$$-e^{\nu}dt^2+e^{\lambda}dr^2+r^2(d\theta^2+\sin^2\theta\, d\phi^2). \qquad \text{(C5)}$$

In particular, for the Schwarzschild geometry, $e^{\nu}=1-2m/r$ and $e^{\lambda}=e^{-\nu}$. In this case the coordinate system is singular at $r=2m$. However, we know that the manifold is smooth at $2m$ since there are coordinate systems which cover the region near $r=2m$ which are nonsingular and in which the metric tensor is analytic. In fact, the maximal analytic extension of this manifold is covered by one coordinate patch, that of Kruskal.[16] The singularity at $r=0$, however, is real and is a three-dimensional manifold which is the boundary of the four-dimensional space-time.[17] All the physics dealt with in this paper takes place in the region from $r=2m$ to $r=\infty$.

Expanding the perturbation $h_{\mu\nu}$ and the source tensor $\Delta T_{\mu\nu}$ in tensor harmonics, the field equations (C4) become, in the case of magnetic-parity harmonics,

$$\frac{\partial^2 h_0}{\partial r^2}-\frac{\partial^2 h_1}{\partial r\partial t}-\frac{2}{r}\frac{\partial h_1}{\partial t}+\left[\frac{4m}{r^2}-\frac{L(L+1)}{r}\right]\frac{h_0}{r-2m}=-8\pi r[\tfrac{1}{2}L(L+1)]^{-1/2}Q_{LM}{}^{(0)}, \qquad \text{(C6a)}$$

$$\frac{\partial^2 h_1}{\partial t^2}-\frac{\partial^2 h_0}{\partial r\partial t}+\frac{2}{r}\frac{\partial h_0}{\partial t}+(L-1)(L+2)(r-2m)\frac{h_1}{r^3}=8\pi i(r-2m)[\tfrac{1}{2}L(L+1)]^{-1/2}Q_{LM}, \qquad \text{(C6b)}$$

$$\left(1-\frac{2m}{r}\right)\frac{\partial h_1}{\partial r}-\left(1-\frac{2m}{r}\right)^{-1}\frac{\partial h_0}{\partial t}+\frac{2m}{r^2}h_1=8\pi i r^2[\tfrac{1}{2}L(L+1)(L-1)(L+2)]^{-1/2}D_{LM}. \qquad \text{(C6c)}$$

For the electric harmonics, we have

$$\left(1-\frac{2m}{r}\right)^2\frac{\partial^2 K}{\partial r^2}+\left(1-\frac{2m}{r}\right)\left(3-\frac{5m}{r}\right)\frac{1}{r}\frac{\partial K}{\partial r}-\left(1-\frac{2m}{r}\right)^2\frac{1}{r}\frac{\partial H_2}{\partial r}-\left(1-\frac{2m}{r}\right)\frac{1}{r^2}(H_2-K)-\left(1-\frac{2m}{r}\right)\frac{1}{2r^2}L(L+1)(H_2+K)=8\pi A_{LM}{}^{(0)}, \qquad \text{(C7a)}$$

$$\frac{\partial}{\partial t}\left(\frac{\partial K}{\partial r}+\frac{1}{r}(K-H_2)-\frac{m}{r(r-2m)}K\right)-\frac{L(L+1)}{2r^2}H_1=\frac{8\pi i}{\sqrt{2}}A_{LM}{}^{(1)}, \qquad \text{(C7b)}$$

$$\left(1-\frac{2m}{r}\right)^{-2}\frac{\partial^2 K}{\partial t^2}-\frac{1-m/r}{r-2m}\frac{\partial K}{\partial r}-\frac{2}{r-2m}\frac{\partial H_1}{\partial t}+\frac{1}{r}\frac{\partial H_0}{\partial r}+\frac{1}{r(r-2m)}(H_2-K)+\frac{L(L+1)}{2r(r-2m)}(K-H_0)=8\pi A_{LM}, \qquad \text{(C7c)}$$

[14] C. Lanczos, Z. Physik 31, 112 (1925); E. Lifshitz, J. Phys. USSR 10, 116 (1946); P. C. Peters, Phys. Rev. 146, 938 (1966).

[15] For a discussion of spherically symmetric space-times, see J. P. Vajk, Ph.D. thesis, Princeton University, 1968 (unpublished).

[16] M. D. Kruskal, Phys. Rev. 119, 1743 (1960); for a discussion of the Kruskal picture, see R. Fuller and J. A. Wheeler, *ibid.* 128, 919 (1962).

[17] R. Geroch, J. Math. Phys. 9, 450 (1968).

$$\frac{\partial}{\partial r}\left[\left(1-\frac{2m}{r}\right)H_1\right]-\frac{\partial}{\partial t}(H_2+K)$$

$$=-8\pi ir[\tfrac{1}{2}L(L+1)]^{-1/2}B_{LM}{}^{(0)}, \quad \text{(C7d)}$$

$$-\frac{\partial H_1}{\partial t}+\left(1-\frac{2m}{r}\right)\frac{\partial}{\partial r}(H_0-K)$$

$$+\frac{2m}{r^2}H_0+\frac{1-m/r}{r}(H_2-H_0)$$

$$=-8\pi(r-2m)[\tfrac{1}{2}L(L+1)]^{-1/2}B_{LM}, \quad \text{(C7e)}$$

$$-\left(1-\frac{2m}{r}\right)^{-1}\frac{\partial^2 K}{\partial t^2}+\left(1-\frac{2m}{r}\right)\frac{\partial^2 K}{\partial r^2}+\left(1-\frac{m}{r}\right)\frac{2}{r}\frac{\partial K}{\partial r}$$

$$-\left(1-\frac{2m}{r}\right)^{-1}\frac{\partial^2 H_2}{\partial t^2}+\frac{2\partial^2 H_1}{\partial r\partial t}-\left(1-\frac{2m}{r}\right)\frac{\partial^2 H_0}{\partial r^2}$$

$$+\frac{2}{r-2m}\left(1-\frac{m}{r}\right)\frac{\partial H_1}{\partial t}-\frac{1}{r}\left(1-\frac{m}{r}\right)\frac{\partial H_2}{\partial r}$$

$$-\frac{1}{r}\left(1+\frac{m}{r}\right)\frac{\partial H_0}{\partial r}-\frac{1}{2r^2}L(L+1)(H_2-H_0)$$

$$=-8\pi\sqrt{2}G_{LM}, \quad \text{(C7f)}$$

$$\tfrac{1}{2}(H_0-H_2)$$

$$=-8\pi r^2[\tfrac{1}{2}L(L+1)(L-1)(L+2)]^{-1/2}F_{LM}. \quad \text{(C7g)}$$

These equations are written in the RW gauge. The terms on the right-hand side are the "radial factors" in the expansion of the source in tensor harmonics (see Appendix A). We obtain, for each (L,M), three magnetic-parity equations for two unknown functions h_0 and h_1, and seven electric-parity equations for four unknown functions H_0, H_1, H_2, and K. Consistency is assured for the vacuum equations since the Einstein equations satisfy the Bianchi identities; in the case where we have a nonzero source term, this implies that the divergence of the source stress tensor **T** must be zero. It may be verified that the divergence of the stress tensor vanishes if and only if the source particle follows a geodesic.

APPENDIX D: GAUGE TRANSFORMATION

The general perturbation is expressed by

$$\mathbf{h}=\sum_{LM}[\mathbf{h}_{LM}{}^{(e)}+\mathbf{h}_{LM}{}^{(m)}], \quad \text{(D1)}$$

where the magnetic $[(-1)^{L+1}$ parity$]$ terms are

$$\mathbf{h}_{LM}{}^{(m)}=(i/r)[2L(L+1)]^{1/2}[ih_{0LM}(r,t)\mathbf{c}_{LM}{}^{(0)}(\theta,\phi)+h_{1LM}(r,t)\mathbf{c}_{LM}(\theta,\phi)-(i/2r^2)\times[(L-1)(L+2)]^{1/2}h_{2LM}(r,t)\mathbf{d}_{LM}(\theta,\phi)] \quad \text{(D2a)}$$

and the electric $[(-1)^L$ parity$]$ terms are

$$\mathbf{h}_{LM}{}^{(e)}=(1-2m/r)H_{0LM}\mathbf{a}_{LM}{}^{(0)}-\sqrt{2}iH_{1LM}\mathbf{a}_{LM}{}^{(1)}+(1-2m/r)^{-1}H_{2LM}\mathbf{a}_{LM}-(1/r)[2L(L+1)]^{1/2}\times(ih_{0LM}{}^{(m)}\mathbf{b}_{LM}{}^{(0)}-h_{1LM}{}^{(m)}\mathbf{b}_{LM})+[\tfrac{1}{2}L(L+1)(L-1)(L+2)]^{1/2}G_{LM}\mathbf{f}_{LM}+\sqrt{2}[K_{LM}-\tfrac{1}{2}L(L+1)G_{LM}]\mathbf{g}_{LM}. \quad \text{(D2b)}$$

We have chosen the form of the coefficient functions to agree with the notation of Regge and Wheeler.[6]

We can simplify (D2a) and (D2b) by using the freedom to choose a particular gauge. Since we can choose four linearly independent vector fields ξ, we can eliminate four of the ten coefficient functions (see Table I) in (D2). Let ∇ denote the covariant derivative in the Schwarzschild geometry. Then the gauge-transformed perturbation is

$$\mathbf{h}'=\mathbf{h}-2[\nabla\xi]_s,$$

where the subscript s denotes symmetrization.

There is one vector harmonic of order (LM) and parity $(-1)^{L+1}$, viz., $(0,\mathbf{L}Y_{LM})$. Thus let

$$\xi_{LM}{}^{(m)}=(i/r)\Lambda_{LM}(r,t)(0,\mathbf{L}Y_{LM}(\theta,\phi)).$$

Then

$$2[\nabla\xi_{LM}{}^{(m)}]_s=(i/r)[2L(L+1)]^{1/2}\{i(\partial\Lambda_{LM}/\partial t)\mathbf{c}_{LM}{}^{(0)}+r^2(\partial/\partial r)(\Lambda_{LM}/r^2)\mathbf{c}_{LM}+(1/r)\times[(L-1)(L+2)]^{1/2}\Lambda_{LM}\mathbf{d}_{LM}\}. \quad \text{(D3)}$$

Regge and Wheeler's choice of gauge makes h_{2LM} zero. Thus

$$h_{LM}{}^{(m)}=(i/r)[2L(L+1)]^{1/2}(ih_{0LM}\mathbf{c}_{LM}{}^{(0)}+h_{1LM}\mathbf{c}_{LM}).$$

There are three vector harmonics of degree (LM) and parity $(-1)^L$; they are e_tY_{LM}, e_rY_{LM}, and $(0,\nabla Y_{LM})$ (e_t and e_r are unit vectors along t and r). Thus let

$$\xi_{LM}{}^{(e)}=M_0(r,t)Y_{LM}e_t+M_1(r,t)Y_{LM}e_r+M_2(r,t)(0,\nabla Y_{LM}).$$

Then verify that

$$2[\nabla\xi_{LM}{}^{(e)}]_s=2\left(\frac{\partial M_0}{\partial t}-\frac{m}{r^3}(r-2m)M_1\right)\mathbf{a}_{LM}{}^{(0)}-\sqrt{2}i\left[\frac{\partial M_1}{\partial t}+\frac{\partial M_0}{\partial r}-\frac{2m}{r(r-2m)}M_0\right]\mathbf{a}_{LM}{}^{(1)}$$

$$+2\left[\frac{\partial M_1}{\partial r}+\frac{m}{r(r-2m)}M_1\right]\mathbf{a}_{LM}-\frac{i}{r}[2L(L+1)]^{1/2}\left(\frac{\partial M_2}{\partial t}+M_0\right)\mathbf{b}_{LM}{}^{(0)}+\frac{1}{r}[2L(L+1)]^{1/2}\left(\frac{\partial M_2}{\partial r}-\frac{2}{r}M_2+M_1\right)\mathbf{b}_{LM}$$

$$+\frac{\sqrt{2}}{r^2}[2(r-2m)M_1-L(L+1)M_2]\mathbf{g}_{LM}+\frac{1}{r^2}[2L(L+1)(L-1)(L+2)]^{1/2}M_2\mathbf{f}_{LM}. \quad \text{(D4)}$$

We can choose the three functions M_{0LM}, M_{1LM}, and M_{2LM} to eliminate three of the coefficients in (D2b). Regge and Wheeler choose the gauge so that

$$\mathbf{h}_{LM}^{(e)}=(1-2m/r)H_{0LM}\mathbf{a}_{LM}^{(0)}-\sqrt{2}iH_{1LM}\mathbf{a}_{LM}^{(1)} +(1-2m/r)^{-1}H_{2LM}\mathbf{a}_{LM}+\sqrt{2}K_{LM}\mathbf{g}_{LM}.$$

The preceding considerations hold in general for $L\geq 2$. For $L=0$ and $L=1$, the situation is somewhat simpler since there are fewer independent harmonics. For $L=0$, it is clear that $\mathbf{h}_{00}^{(m)}\equiv 0$ since $\mathbf{c}_{LM}^{(0)}$, $\mathbf{c}_{LM}$, and $\mathbf{d}_{LM}$ are zero for $L=0$, while

$$\mathbf{h}_{00}^{(e)}=(1-2m/r)H_0\mathbf{a}_{00}^{(0)}-\sqrt{2}iH_1\mathbf{a}_{00}^{(1)} +(1-2m/r)^{-1}\mathbf{a}_{00}+\sqrt{2}K\mathbf{g}_{00}, \quad \text{(D5)}$$

since $\mathbf{b}_{LM}^{(0)}$, $\mathbf{b}_{LM}$, and $\mathbf{f}_{LM}$ are zero for $L=0$. For $L=1$, since $\mathbf{d}_{1M}\equiv 0$ and $\mathbf{f}_{1M}\equiv 0$, we have

$$\mathbf{h}_{1M}^{(m)}=(2i/r)(ih_0\mathbf{c}_{1M}^{(0)}+h_1\mathbf{c}_{1M}), \quad \text{(D6)}$$

while

$$\mathbf{h}_{1M}^{(e)}=(1-2m/r)H_0\mathbf{a}_{1M}^{(0)}-\sqrt{2}iH_1\mathbf{a}_{1M}^{(1)} +(1-2m/r)^{-1}\mathbf{a}_{1M}-(2i/r)h_0^{(m)}\mathbf{b}_{1M}^{(0)} +(2/r)h_1^{(m)}\mathbf{b}_{1M}+\sqrt{2}K\mathbf{g}_{1M}. \quad \text{(D7)}$$

Thus our gauge transformations allow us to eliminate one of the functions h_0 or h_1 for the magnetic harmonics and allow us to eliminate three of the six (instead of seven for $L\geq 2$) functions for the electric harmonics.

APPENDIX E: STRESS-ENERGY TENSOR EXPRESSED IN TENSOR HARMONICS

The following prescription gives the stress-energy tensor expressed in terms of tensor harmonics: (a) Take each row in turn in Table III; (b) multiply the factor in the first column ("coefficient in expansion in harmonics"; a function of r and t) by the factor in the second column (tensor harmonic; function of θ and ϕ); (c) sum all ten such products ("totalized part of stress energy tensor of index L, M"); and (d) sum over all L and M.

For the Fourier transforms of these radial factors in the expansion of the source, we multiply by $(2\pi)^{-1/2}$ and by $e^{i\omega t}dt$ and integrate from $-\infty$ to $+\infty$. The analysis is simplest when $r=R(t)$ is a monotonic function of time. Then we write $dt=dR/(dR/dt)$. The function $\delta(r-R(t))$ integrates out immediately. The net

TABLE III. Components in expression of stress-energy tensor of test particle in terms of tensor harmonics. The symbol γ is an abbreviation for the quantity $\gamma=dT(\tau)/d\tau$. The first five and the last two terms drive the electric part of the perturbation in the geometry; the remaining three terms drive the magnetic part. The arrangement of the terms is chosen to make more readily apparent the similarities in form between one coefficient and another. In the table Y_{LM} denotes the usual normalized spherical harmonic, X_{LM} and W_{LM} are functions derived from Y_{LM} as listed at the end of Appendix A, and $*$ denotes complex conjugate.

Description	Dependence of "driving term" on r and t	Tensor harmonic (Appendix A)
Electric	$A_{LM}(r,t)=m_0\gamma\left(\frac{dR}{dt}\right)^2(r-2m)^{-2}\delta(r-R(t))Y_{LM}^*(\Omega(t))$	$\mathbf{a}_{LM}(\theta,\phi)$
Electric	$A_{LM}^{(0)}=m_0\gamma\left(1-\frac{2m}{r}\right)^2 r^{-2}\delta(r-R(t))Y_{LM}^*(\Omega(t))$	$\mathbf{a}_{LM}^{(0)}(\theta,\phi)$
Electric	$A_{LM}^{(1)}=\sqrt{2}im_0\gamma\frac{dR}{dt}r^{-2}\delta(r-R(t))Y_{LM}^*(\Omega(t))$	$\mathbf{a}_{LM}^{(1)}(\theta,\phi)$
Electric	$B_{LM}^{(0)}=[\frac{1}{2}L(L+1)]^{-1/2}im_0\gamma\left(1-\frac{2m}{r}\right)r^{-1}\delta(r-R(t))dY_{LM}^*(\Omega(t))/dt$	$\mathbf{b}_{LM}^{(0)}(\theta,\phi)$
Electric	$B_{LM}=[\frac{1}{2}L(L+1)]^{-1/2}m_0\gamma(r-2m)^{-1}\frac{dR}{dt}\delta(r-R(t))dY_{LM}^*(\Omega(t))/dt$	$\mathbf{b}_{LM}(\theta,\phi)$
Magnetic	$Q_{LM}^{(0)}=[\frac{1}{2}L(L+1)]^{-1/2}m_0\gamma\left(1-\frac{2m}{r}\right)r^{-1}\delta(r-R(t)) \times\left[\frac{1}{\sin\Theta}\frac{\partial Y_{LM}^*}{\partial\Phi}\frac{d\Theta}{dt}-\sin\Theta\frac{\partial Y_{LM}^*}{\partial\Theta}\frac{d\Phi}{dt}\right]$	$\mathbf{c}_{LM}^{(0)}(\theta,\phi)$
Magnetic	$Q_{LM}=[\frac{1}{2}L(L+1)]^{-1/2}im_0\gamma(r-2m)^{-1}\delta(r-R(t))\frac{dR}{dt} \times\left[\frac{1}{\sin\Theta}\frac{\partial Y_{LM}^*}{\partial\Phi}\frac{d\Theta}{dt}-\sin\Theta\frac{\partial Y_{LM}^*}{\partial\Theta}\frac{d\Phi}{dt}\right]$	$\mathbf{c}_{LM}(\theta,\phi)$
Magnetic, transverse, traceless	$D_{LM}=-[\frac{1}{2}L(L+1)(L-1)(L+2)]^{-1/2}im_0\gamma\delta(r-R(t))\left\{\frac{1}{2}\left[\left(\frac{d\Theta}{dt}\right)^2-(\sin\Theta^2)\left(\frac{d\Phi}{dt}\right)^2\right] \times\frac{1}{\sin\Theta}X_{LM}^*[\Omega(t)]-\sin\Theta\frac{d\Phi}{dt}\frac{d\Theta}{dt}W_{LM}^*[\Omega(t)]\right\}$	$\mathbf{d}_{LM}(\theta,\phi)$
Electric, transverse, traceless	$F_{LM}=[\frac{1}{2}L(L+1)(L-1)(L+2)]^{-1/2}m_0\gamma\delta(r-R(t))\left\{\frac{d\Theta}{dt}\frac{d\Phi}{dt}X_{LM}^*[\Omega(t)] +\frac{1}{2}\left[\left(\frac{d\Theta}{dt}\right)^2-(\sin\Theta)^2\left(\frac{d\Phi}{dt}\right)^2\right]W_{LM}^*[\Omega(t)]\right\}$	$\mathbf{f}_{LM}(\theta,\phi)$
Electric	$G_{LM}=\frac{m_0\gamma}{\sqrt{2}}\delta(r-R(t))\left[\left(\frac{d\Theta}{dt}\right)^2+(\sin\Theta)^2\left(\frac{d\Phi}{dt}\right)^2\right]Y_{LM}^*[\Omega(t)]$	$\mathbf{g}_{LM}(\theta,\phi)$

result is to transform each radial factor in the table in the following three respects: (a) Multiply by $(2\pi)^{-1/2}$; (b) replace $e^{i\omega t}$ by $e^{i\omega T(r)}$, where $T(r)$ is the function inverse to $R(t)$; and (c) replace the δ function by $1/(dR/dt)$. For example, in the case when the particle falls from $r=\infty$ to $2m$ starting at $t=-\infty$ with zero velocity, we have

$$A_L^{(0)}(\omega,r)=(m_0/2\pi)[(L+\tfrac{1}{2})(r/2m)]^{1/2}(1/r^2)e^{i\omega T(r)},$$

$$A_L^{(1)}(\omega,r)=-i(m_0/2\pi)(2L+1)^{1/2}(1/r^2)\times(1-2m/r)^{-1}e^{i\omega T(r)}, \quad \text{(E1)}$$

$$A_L(\omega,r)=(m_0/2\pi)[(L+\tfrac{1}{2})(2m/r)]^{1/2}\times(r-2m)^{-2}e^{i\omega T(r)}.$$

APPENDIX F: FOURIER-TRANSFORMED FIELD EQUATIONS FOR RADIAL FACTORS IN METRIC PERTURBATION

We write the Fourier transform of the field equations listed in Appendix C. The magnetic equations are

$$\omega^2 h_{1LM}-\frac{i\omega dh_{0LM}}{dr}+\frac{2i\omega h_{0LM}}{r}-(r-2m)(L-1)(L+2)\frac{h_{1LM}}{r^3}=-8\pi i[\tfrac{1}{2}L(L+1)]^{-1/2}(r-2m)Q_{LM}(\omega,r), \quad \text{(F1a)}$$

$$(r-2m)\left(\frac{d^2h_{0LM}}{dr^2}+\frac{i\omega dh_{1LM}}{dr}+\frac{2i\omega h_{1LM}}{r}\right)+\left[\frac{4m}{r^2}-\frac{L(L+1)}{r}\right]h_{0LM}=-8\pi[\tfrac{1}{2}L(L+1)]^{-1/2}\times r^2Q_{LM}^{(0)}(\omega,r), \quad \text{(F1b)}$$

$$-(r-2m)\frac{dh_{1LM}}{dr}-\frac{i\omega r^2h_{0LM}}{r-2m}-\frac{2mh_{1LM}}{r}=8\pi i[\tfrac{1}{2}L(L+1)(L-1)(L+2)]^{-1/2}r^3D_{LM}(\omega,r). \quad \text{(F1c)}$$

The electric parity equations are

$$\left(1-\frac{2m}{r}\right)^2\frac{d^2K}{dr^2}+\frac{1}{r}\left(1-\frac{2m}{r}\right)\left(3-\frac{5m}{r}\right)\frac{dK}{dr}-\frac{1}{r}\left(1-\frac{2m}{r}\right)^2\frac{dH_2}{dr}-\frac{1}{r^2}\left(1-\frac{2m}{r}\right)(H_2-K)-L(L+1)\frac{1}{2r^2}\left(1-\frac{2m}{r}\right)(H_2+K)=8\pi A_{LM}^{(0)}(\omega,r), \quad \text{(F2a)}$$

$$-\omega^2\left(1-\frac{2m}{r}\right)^{-2}K+L(L+1)\frac{1}{2r^2}\left(1-\frac{2m}{r}\right)^{-1}(K-H_0)+\frac{1}{r(r-2m)}(H_2-K)+\frac{2i\omega}{r}\left(1-\frac{2m}{r}\right)^{-1}H_1-\frac{1}{r}\left(1-\frac{m}{r}\right)\left(1-\frac{2m}{r}\right)^{-1}\frac{dK}{dr}+\frac{1}{r}\frac{dH_0}{dr}=8\pi A_{LM}(\omega,r), \quad \text{(F2b)}$$

$$-i\omega\left[\frac{dK}{dr}+\frac{1}{r}(K-H_2)-\frac{m}{r^2}\left(1-\frac{2m}{r}\right)^{-1}K\right]-L(L+1)\left(\frac{1}{2r^2}\right)H_1=4\pi\sqrt{2}iA_{LM}^{(1)}(\omega,r), \quad \text{(F2c)}$$

$$\frac{d}{dr}\left[\left(1-\frac{2m}{r}\right)H_1\right]+i\omega(H_2+K)=-8\pi i[\tfrac{1}{2}L(L+1)]^{-1/2}rB_{LM}^{(0)}(\omega,r), \quad \text{(F2d)}$$

$$i\omega H_1+\frac{2m}{r^2}H_0+\left(1-\frac{2m}{r}\right)\frac{d}{dr}(H_0-K)+\frac{1}{r}\left(1-\frac{m}{r}\right)(H_2-H_0)=-8\pi[\tfrac{1}{2}L(L+1)]^{-1/2}\times(r-2m)B_{LM}(\omega,r), \quad \text{(F2e)}$$

$$r^2\omega^2\left(1-\frac{2m}{r}\right)^{-1}(H_2+K)+r(r-2m)\frac{d^2}{dr^2}(K-H_0)-\frac{2i\omega r^2dH_1}{dr}-2i\omega r\left(1-\frac{m}{r}\right)\left(1-\frac{2m}{r}\right)^{-1}H_1+2(r-m)\frac{dK}{dr}-r\left(1-\frac{m}{r}\right)\frac{dH_2}{dr}-r\left(1+\frac{m}{r}\right)\frac{dH_0}{dr}-\tfrac{1}{2}L(L+1)(H_2-H_0)=-\frac{16\pi r^2}{\sqrt{2}}G_{LM}(\omega,r), \quad \text{(F2f)}$$

$$\tfrac{1}{2}(H_{0LM}-H_{2LM})=-16\pi[2L(L+1)(L-1)(L+2)]^{-1/2}\times r^2F_{LM}(\omega,r). \quad \text{(F2g)}$$

These equations were first given by Regge and Wheeler for the case of vacuum perturbations. There were some minor errors in the magnetic-parity equations given by Regge and Wheeler. A corrected version has been given by Vishveshwara.[18] The equations given here differ from those of Regge-Wheeler-Vishveshwara only in that we consider variations in $G_{\mu\nu}$ rather than $R_{\mu\nu}$. The sources terms are zero in (F2d)–(F2i) in the case for motion without angular momentum along the z

[18] C. V. Vishveshwara, Ph.D. thesis, University of Maryland, 1968 (unpublished).

axis. Also in this case, the solution of the magnetic-parity equations which satisfies the boundary conditions is the zero solution. Thus we will only have to consider the electric equations for the case of the particle falling straight in.

For the magnetic-parity waves we have three equations. Equation (F1b) is a consequence of (F1a) and (F1c), provided that the source term satisfies the divergence condition. Thus we have a system of two first-order linear equations. The two first-order equations (F1a) and (F1c) can be expressed as a simple second-order Schrödinger-type equation (see text).[6]

For the electric-parity waves we have six equations and three unknown functions. Three of the first-order equations are sufficient to determine a solution provided the divergence conditions on the source term are satisfied. Further, as noted by Regge and Wheeler,[6] we obtain an algebraic equation relating the three unknown functions. Let us take (F2c)–(F2e) as our basic equations and solve them for the first derivatives of K, H_2, and H_1 as follows:

$$\frac{dK}{dr}+\frac{1}{r}\left(1-\frac{3m}{r}\right)\left(1-\frac{2m}{r}\right)^{-1}K-\frac{1}{r}H_2 +L(L+1)\frac{1}{2i\omega r^2}H_1=-\frac{4\pi\sqrt{2}}{\omega}A_{LM}{}^{(1)}, \qquad \text{(F3a)}$$

$$\frac{dH_2}{dr}+\frac{1}{r}\left(1-\frac{3m}{r}\right)\left(1-\frac{2m}{r}\right)^{-1}K -\frac{1}{r}\left(1-\frac{4m}{r}\right)\left(1-\frac{2m}{r}\right)^{-1}H_2 +\left[i\omega\left(1-\frac{2m}{r}\right)^{-1}+L(L+1)\frac{1}{2i\omega r^2}\right]H_1$$
$$=-\frac{4\pi\sqrt{2}}{\omega}A_{LM}{}^{(1)}-8\pi r[\tfrac{1}{2}L(L+1)]^{-1/2}B_{LM} +16\pi[\tfrac{1}{2}L(L+1)(L-1)(L+2)]^{-1/2} \times\left[\frac{d}{dr}(r^2F_{LM})-\frac{r-3m}{r(-r2m)}F_{LM}\right], \qquad \text{(F3b)}$$

$$\frac{dH_1}{dr}+i\omega\left(1-\frac{2m}{r}\right)^{-1}(K+H_2)+\frac{2m}{r^2}\left(1-\frac{2m}{r}\right)^{-1}H_1 =-\frac{8\pi ir^2}{r-2m}[\tfrac{1}{2}L(L+1)]^{-1/2}B_{LM}{}^{(0)}. \qquad \text{(F3c)}$$

If we substitute these in (F2a), we obtain a relation between the source terms which is the divergence condition. If we substitute (F3) into (F2b), we obtain

$$F(r)=16\pi r(r-2m)A_{LM}-\frac{8\pi\sqrt{2}m}{\omega}A_{LM}{}^{(1)} +\frac{16\pi r(r-2m)}{[\frac{1}{2}L(L+1)]^{1/2}}B_{LM}-16\pi[(L-1)(L+2)+6m/r] \times[\tfrac{1}{2}L(L+1)(L-1)(L+2)]^{-1/2}r^2F_{LM}, \qquad \text{(F4)}$$

where

$$F(r)=-\left[\frac{6m}{r}+(L-1)(L+2)\right]H_2+\left[(L-1)(L+2) -2\omega^2r^2\left(1-\frac{2m}{r}\right)^{-1}+\frac{2m}{r}\left(1-\frac{3m}{r}\right)\left(1-\frac{2m}{r}\right)^{-1}\right]K +\left[2i\omega r+L(L+1)\frac{m}{i\omega r^2}\right]H_1.$$

A straightforward way of showing the consistency of (F3) and (F4) is to eliminate one of the functions H, H_1, and K from two of the equations (F3) and use the divergence condition to show that the third equation is satisfied identically. Thus we reduce our original system to a system of two first-order equations for two unknown functions. We can proceed to write this as a single second-order equation,[19] given in the text [Eq. (18)].

APPENDIX G: SOLUTIONS OF $L=0$, 1 EQUATIONS

The fact that for $L=0$ and $L=1$ there are fewer independent tensor harmonics makes it possible for us to give exact solutions of the perturbation equations in these cases. We have noted previously that the $L=0$ magnetic-parity harmonic is identically zero. Thus we are left with the $L=1$ magnetic equations and the $L=0$ and $L=1$ electric equations. We will, for these three cases, give the solutions to the homogeneous equations and the solutions for the case where there is a source term which is produced by a point mass m_0 falling on a geodesic of the background.

A. Monopole (Mass) Perturbation

Let us consider, first, the $L=0$ electric-parity equations. The general form of the perturbation is given by Eq. (D5). Making a gauge transformation $\xi^{(e)}=M_0Y_{00}e_t+M_1Y_{00}e_r$, we can choose $M_0(r,t)$ and $M_1(r,t)$ so that $H_1(r,t)=K(r,t)=0$. Since $\mathbf{b}_{00}{}^{(0)}\equiv\mathbf{b}_{00}\equiv\mathbf{f}_{00}\equiv0$, the only trivial magnetic equations for $L=0$ are Eqs. (C7a)–(C7c) and (C7f). Equation (C7f) is satisfied identically by a solution of the other three provided that the source term satisfies the divergence condition. In the source-free case, Eqs. (C7a) and (C7b) give

$$H_2=2(4\pi)^{1/2}c/(r-2m),$$

[19] F. Zerilli, Phys. Rev. Letters 24, 737 (1970).

where c is a constant, while (C7c) gives

$$H_0=H_2+f(t),$$

where f is an arbitrary function of time. A gauge transformation of the form $\xi^{(e)}=M_0 e_t Y_{00}$ allows us to eliminate $f(t)$. From Eq. (D4) we see that if

$$M_0(r,t)=\tfrac{1}{2}(1-2m/r)\int f(t)dt,$$

then in the new gauge we have

$$H_0=H_2=2(4\pi)^{1/2}c/(r-2m), \tag{G1}$$

and so

$$\mathbf{h}_{00}^{(e)}=(16\pi)^{1/2}cr^{-1}\mathbf{a}_{00}^{(0)}+(1-2m/r)^{-2}+(16\pi)^{1/2}cr^{-1}\mathbf{a}_{00}, \tag{G2}$$

and up to the linear approximation in which we are dealing, this is simply

$$\begin{aligned}\bar{g}_{00}&=g_{00}+h_{00}=1-2m/r+2c/r,\\ \bar{g}_{11}&=g_{11}+h_{11}=(1-2m/r+2c/r)^{-1}.\end{aligned} \tag{G3}$$

Thus we obtain the result which we would expect: The $L=0$ perturbation, being spherically symmetric, represents only a change in the Schwarzschild mass, a result required by Birkhoff's theorem. Regge and Wheeler[6] stated this result by explicitly assuming the solutions to be time independent.

Now let us consider the case where there is a point mass m_0 falling along the geodesics of the Schwarzschild geometry. From Eq. (C7) we have

$$(\partial/\partial r)[(r-2m)H_2]=-(16\pi)^{1/2}m_0\gamma(1-2m/r)\times\delta(r-R(t)),$$

and integrating we obtain

$$\begin{aligned}(r-2m)H_2&=0, \quad r<R(t)\\ &=-(16\pi)^{1/2}m_0[1-2m/R(t)]dT/ds, \quad r>R(t).\end{aligned}$$

But $[1-2m/R(t)]dT/ds=\gamma_0$, and γ_0 is a constant of the motion which is an energy parameter; for example, $\gamma_0=1$ if the particle falls from infinity starting with zero velocity $dR/ds=0$ at $r=\infty$. Thus

$$\begin{aligned}H_2(r,t)&=0, \quad r<R(t)\\ &=-2(4\pi)^{1/2}m_0\gamma_0/(r-2m), \quad r>R(t).\end{aligned} \tag{G4}$$

From Eq. (C7c) we obtain

$$\begin{aligned}H_0(r,t)&=0, \quad r<R(t)\\ &=-2(4\pi)^{1/2}m_0\gamma_0/(r-2m)+f(t), \quad r>R(t)\end{aligned} \tag{G5}$$

where

$$f(t)=2(4\pi)^{1/2}m_0\gamma\frac{(dR/dt)^2}{[1-2m/R(t)]^2R(t)}.$$

As in the vacuum case, $f(t)$ is eliminated by a gauge transformation and, referring to Eqs. (G1)–(G3), we see that the solution is that of a Schwarzschild field of mass m inside the two-sphere of radius $R(t)$ and is a Schwarzschild field of mass $m+m_0\gamma_0$ outside the sphere of radius $R(t)$.

B. Magnetic Dipole (Angular Momentum) Perturbation

Now we discuss the $L=1$ magnetic-parity equations. Equation (D6) gives the general form of the perturbation. We perform a gauge transformation $\xi_{1M}^{(m)}=(i/r)\Lambda_{1M}(r,t)(0,\mathbf{L}Y_{1M})$, which makes $h_0(r,t)=0$. Then

$$\mathbf{h}_{1M}^{(m)}=(2i/r)h_1(r,t)\mathbf{c}_{1M}.$$

Since $\mathbf{d}_{1M}\equiv0$, the only nontrivial electric equations for $L=1$ are (C6a) and (C6b). Thus, in the source-free case, (C6a) gives $h_1(r,t)=f(r)t+g(r)$, where f and g are arbitrary functions of r. Then (C6b) gives $f'(r)+(2/r)f(r)=0$, whose solution is $f(r)=3c/r^2$ where c is a constant. The function $g(r)$ is entirely arbitrary. However, $g(r)$ can be eliminated by a guage transformation; at the same time we will transform to a gauge in which the perturbation is easily interpretable. We choose the gauge function

$$\Lambda(r,t)=-\frac{ct}{r}+r^2\int g(r)r^{-2}dr.$$

Then in the new guage, $h_1=0$ while $h_0=-\partial\Lambda/\partial t=c/r$. Thus

$$\mathbf{h}_{1M}^{(m)}=(2ic/r^2)\mathbf{c}_{1M}^{(0)}. \tag{G6}$$

This metric has only 0θ and 0ϕ components. We will see that this perturbation can be interpreted as adding angular momentum to the background metric. If we transform this perturbation to Kruskal coordinates, we find that it is singular at $2m$. However, Vishveshwara[18] has shown that a suitable gauge transformation brings it to a form which is regular everywhere in Kruskal coordinates.

Landau and Lifshitz[3] show that, for a weak field, the coefficients of the $1/r^2$ terms of the $dtdx^i$ components of the metric tensor are related to the angular momentum tensor (in a Cartesian frame at infinity):

$$h_{0i}=\sum_{j=1}^{3}\frac{2M_{ij}n_j}{r^2}, \quad i=1,2,3 \tag{G7}$$

where n_i are the components of e_r in Cartesian coordinates: $n_x=\sin\theta\ \sin\phi$, $n_y=\sin\theta\cos\phi$, $n_z=\cos\theta$. Transforming (G7) to spherical coordinates, we obtain

$$\begin{aligned}h_{0r}&=0,\\ h_{0\theta}&=(2/r)(l_x\sin\phi-l_y\cos\phi),\\ h_{0\phi}&=(2/r)[-l_z\sin^2\theta+(l_x\cos\phi+l_y\sin\phi)\sin\theta\cos\theta],\end{aligned} \tag{G8}$$

where $l_x=M_{yz}$, $l_y=M_{zx}$, and $l_z=M_{xy}$. We now note that (G8) can be written as a sum of the tensor harmonics $\mathbf{c}_{1M}^{(0)}$ as follows:

$$\mathbf{h}=(4i/r^2)(\tfrac{4}{3}\pi)^{1/2}(l_{+1}\mathbf{c}_{11}^{(0)}+l_z\mathbf{c}_{10}^{(0)}+l_{-1}\mathbf{c}_{-1,1}^{(0)}), \tag{G9}$$

where $l_{\pm 1}=\mp\frac{1}{2}\sqrt{2}(l_x\pm il_y)$. We see that the solution (G6) gives a perturbation which has angular momentum.

Let us now proceed to the case where there is a source term due to a particle orbiting in the Schwarzschild field. Since the motion lies in a plane, assume that $\Theta(t)=\frac{1}{2}\pi$ and then from (C6b) we obtain

$$\frac{1}{r^2}\frac{\partial^2}{\partial t\partial r}(r^2h_1)=8\pi m_0\left(\frac{3}{4\pi}\right)^{1/2}\delta(r-R(t))\frac{a}{R^2},$$

where $a=R^2d\Phi/ds$ is the z component of angular momentum per unit mass of the orbiting particle. Since $\Theta=\frac{1}{2}\pi$, the z component is the only nonzero component (and the only nonzero source terms Q_{1M} and $Q_{1M}^{(0)}$ have the index $M=0$). Integrating with respect to r, we obtain

$$\begin{aligned} h_1(r,t)&=0, \quad r<R(t)\\ &=8\pi(3/4\pi)^{1/2}m_0at/r^2, \quad r>R(t). \end{aligned} \qquad (G10)$$

As in the homogeneous case, if we perform a gauge transformation with

$$\begin{aligned} \Lambda(r,t)&=0, \quad r<R(t)\\ &=-2m_0a(3/4\pi)^{-1/2}t/r, \quad r>R(t) \end{aligned}$$

we obtain, in the new gauge,

$$\begin{aligned} \mathbf{h}^{(m)}&=0, \quad r<R(t)\\ &=(2i/r)(4\pi/3)^{1/2}(2m_0a/r)\mathbf{c}_{10}^{(0)}, \quad r>R(t). \end{aligned} \qquad (G11)$$

Referring to Eq. (G9), we see that this perturbation is zero inside the sphere $r<R(t)$ while it has, for $r>R(t)$, angular momentum

$$l_z=m_0a,$$

which is precisely the conserved angular momentum of the orbiting particle.

C. Electric Dipole (Coordinate) Perturbation

Proceeding now to the last of the nonradiative cases, we discuss the $L=1$ magnetic-parity equations. Equation (D7) gives the general form of the perturbation. Using our gauge freedom, we can make $K_{1M}=0$. Since $\mathbf{f}_{1M}\equiv 0$, there are six field equations: (C7a)–(C7f). Three of the equations determine a solution, and consistency is assured by the divergence condition on the coefficients $A_{1M}^{(0)}$, $A_{1M}^{(1)}$, etc. Looking at the homogeneous case, we see by substitution that any solution which satisfies (C7a) and (C7b) automatically satisfies (C7d). Also (C7e) and (C7d) imply (C7f). To show this, differentiate (C7e) with respect to r and (C7d) with respect to t. Finally, from (C7e) and (C7c) we obtain

$$3mH_0=mH_2-r^2\partial H_1/\partial t. \qquad (G12)$$

Thus the system is reduced to (C7a)–(C7c) and (G12). But it is easily verified that any solution of (C7a), (C7b), and (G12) identically satisfies (C7). In the homogeneous case, the solution of (C7a) is

$$H_2(r,t)=f(t)/(r-2m)^2,$$

where f is an arbitrary function of t. Then (C7b) and (G12) give, for H_0 and H_1,

$$\begin{aligned} H_1(r,t)&=-rf'(t)/(r-2m)^2,\\ H_0(r,t)&=f(t)/3(r-2m)^2+r^3f''(t)/3m(r-2m)^2. \end{aligned}$$

This will also be the form of the solution to the inhomogeneous equations, where the function $f(t)$ will be determined by the source term.

We will now see that this perturbation can be removed entirely by a guage transformation which we will attempt to interpret as a translation to the c.m. system of the two bodies. Thus we find a vector field $\xi_1^{(e)}=M_0e_tY_{1M}+M_1e_rY_{1M}+M_2\nabla Y_{1M}$ such that

$$2[\nabla\xi_{1M}^{(e)}]_s=\mathbf{h}_{1M}^{(e)}.$$

In fact, let

$$\begin{aligned} M_2&=r^2g(t)/(r-2m),\\ M_1&=r^2g(t)/(r-2m)^2,\\ M_0&=-r^2g'(t)/(r-2m), \end{aligned}$$

where $g(t)=-f(t)/6m$. Thus the $L=1$ magnetic perturbations are strictly removable by a gauge transformation.

Now we give an interpretation of this gauge transformation. In the limit of large r ($r\gg 2m$), we can write

$$\xi_{1M}^{(e)}\sim -g_M'(t)re_tY_{1M}+g_M(t)e_rY_{1M}+g_M(t)r\nabla Y_{1M}. \qquad (G13)$$

Just for the purposes of the following discussion, let ∇ denote the operator whose covariant components are $(-\partial/\partial t,\ \partial/\partial x,\ \partial/\partial y,\ \partial/\partial z)$ in a Cartesian coordinate system; and since we consider r large, we will raise and lower indices with the Minkowski metric $(+1, -1, -1, -1)$. Then (G13) may be written compactly

$$\xi_{1M}^{(e)}\sim\nabla[g_M(t)rY_{1M}(\theta,\phi)].$$

But $rY_{1M}=(3/4\pi)^{1/2}x_M$, where $x_0=z$, $x_{\pm 1}=\mp\frac{1}{2}\sqrt{2}\times(x\pm iy)$. For example, in contravariant components,

$$\xi_{10}^{(e)}\sim -(3/4\pi)^{1/2}(g_0'(t)z,0,0,g_0(t)).$$

Thus we identify this with a displacement along the z axis by $\rho_0(t)=(3/4\pi)^{1/2}g_0(t)$. That is, $x'^\mu=x^\mu+\xi^\mu$ or

$$\begin{aligned} t'&=t-\rho_0'(t)z, \quad x'=x,\\ y'&=y, \quad z'-z=\rho_0(t). \end{aligned}$$

This is a Lorentz transformation along the z axis; $\rho_0'(t)$ is the velocity v. The factor of $(1-v^2)^{1/2}$ does not appear here since we are dealing with linear perturbations and this factor is of quadratic rather than linear order. Similarly, $\xi_{11}^{(e)}+\xi_{1,-1}^{(e)}$ represents a translation in the (x,y) plane by $(\rho_x(t),\rho_y(t))$, where $\rho_x=-(3/8\pi)^{1/2}\times(g_{+1}-g_{-1})$ and $\rho_y=-i(3/8\pi)^{1/2}(g_{+1}+g_{-1})$. Thus we identify the gauge transformation as the analog, in the

Schwarzschild geometry, of a Lorentz transformation in flat space since this is what it looks like to the distant observer.

Now let us consider the inhomogeneous equations. The solution of Eq. (C7a) is

$$H_{2M}(r,t)=0,\quad r<R(t)$$
$$=f_M(t)/(r-2m)^2,\quad r>R(t)$$

where

$$f_M(t)=-8m\pi_0\gamma(t)Y_{1M}^*(\Omega(t))[R(t)-2m]^2/R(t).$$

Now from (C7b) we obtain

$$H_{1M}(r,t)=0,\quad t<R(r)$$
$$=-rf_M'(t)/(r-2m)^2,\quad r>R(t).$$

Note that the source terms contribute only a δ function on the surface $r=R(t)$ to H_{1M}. The same is true for H_0, and using (G12), we obtain

$$H_{0M}(r,t)=0,\quad r<R(t)$$
$$=[f_M(t)+(r^3/m)f_M''(t)]/3(r-2m)^2,\quad r>R(t).$$

Thus the solution is of the same form as that for the homogeneous case, and $f_M(t)$ is determined by the source term. Now define $\rho(t)$ by

$$\rho(t)=(m_0/m)\gamma(t)[R(t)-2m]^2/R(t).$$

Then $f_M(t)=-8\pi m\rho(t)Y_{1M}^*(\Omega(t))$, and the gauge transformation which eliminates this perturbation is (covariant components in spherical coordinates)

$$\begin{aligned}\xi_0&=-\frac{\partial}{\partial t}[r^2(r-2m)^{-1}\rho(t)\sigma(t,\theta,\phi)],\\ \xi_1&=\left[\left(1-\frac{2m}{r}\right)^{-2}\rho(t)\sigma(t,\theta,\phi)\right]\\ \xi_2&=\frac{\partial}{\partial\theta}\left[r\left(1-\frac{2m}{r}\right)^{-1}\rho(t)\sigma(t,\theta,\phi)\right],\\ \xi_3&=\frac{\partial}{\partial\phi}\left[r\left(1-\frac{2m}{r}\right)^{-1}\rho(t)\sigma(t,\theta,\phi)\right],\end{aligned}\tag{G14}$$

where $\sigma(t,\theta,\phi)=\frac{4}{3}\pi\sum_M Y_{1M}^*(\Omega(t))Y_{1M}(\theta,\phi)$. But $r\rho(t)\times\sigma(t,\theta,\phi)=\mathbf{x}\cdot\xi(t)$, where $\xi(t)=(\rho(t),\Theta(t),\Phi(t))$. Thus for r large compared with $2m$, we can write

$$\xi\sim\nabla[\xi(t)\cdot\mathbf{x}],$$

where ∇ is defined as above in the discussion after Eq. (G13). We see that this represents the Lorentz transformation

$$t'=t-\mathbf{x}\cdot\xi'(t),$$
$$\mathbf{x}'=\mathbf{x}-\xi(t),$$

since, in contravariant components,

$$\nabla[\xi(t)\cdot\mathbf{x}]=-[\xi'(t)\cdot\mathbf{x},\xi(t)].$$

Thus we identify the gauge transformation as an "analog" Lorentz transformation to a moving frame given by the displacement

$$\xi(t)=\{(m_0/m)\gamma(t)[R(t)-2m]^2/R(t),\ \Theta(t),\ \Phi(t)\}.$$

When the falling particle m_0 is far from the central mass m, that is, $R(t)\gg 2m$, then

$$\xi(t)\sim[m_0(/m)\gamma(t)R(t),\Theta(t),\Phi(t)],$$

which is the usual classical mechanical transformation to the center-of-momentum frame of the two bodies. Thus we are tempted to interpret the gauge transformation (G14) as the analog of a "transformation to the center-of-momentum system."

APPENDIX H: RADIATION IN MULTIPLY PERIODIC MOTION

The present investigation is concerned with the pulse of gravitational radiation given out when a small mass m_0 plunges into the black hole associated with a much larger mass m. In this connection we analyze the mechanical quantities descriptive of the motion of m_0, and analyze the gravitational radiation itself into Fourier integrals. The results of such an analysis for radiation emitted in a hyperbolic orbit are already known not to be at all complicated when the departure of the space from flatness is so slight that the geometry can be idealized as nearly Lorentzian. In that limit, the familiar textbook formula for the rate of radiation,

$$-\frac{dE}{dt}=\frac{1}{45}\frac{d^3Q^{pq}(t)}{dt^3}\frac{d^3Q^{pq}(t)}{dt^3},$$

in terms of the quadrupole moment

$$Q^{pq}(t)=\sum_{\text{all masses}}m(3x^px^q-\delta^{pq}x^sx^s),$$

lends itself readily to Fourier analysis:

$$Q^{pq}(t)=(2\pi)^{-1/2}\int_{-\infty}^{+\infty}Q^{pq}(\omega)e^{-i\omega t}d\omega,$$

$$Q^{pq}(\omega)=(2\pi)^{-1/2}\int_{-\infty}^{+\infty}Q^{pq}(t)e^{i\omega t}dt.$$

The amount of energy emitted per unit interval of frequency ν $(=\omega/2\pi)$, integrated over the entire time of the pulse, is

$$-dE/d\nu=(4\pi/45)Q^{pq*}(\omega)Q^{pq}(\omega).$$

This expression reduces at low frequencies (ω small compared to the reciprocal of the time required for the orbit to undergo the major part of its change in direction) to[20]

$$-\frac{dE}{d\nu}=\frac{2}{45}\sum_{p,q}\Big[\sum_{\text{masses}} m(3\dot{x}^p\dot{x}^q-\delta^{pq}\dot{x}^s\dot{x}^s)\Big|_{\text{before}}^{\text{after}}\Big]^2,$$

where the dot means differentiation with respect to t. We expect the formulas given in the present text to reduce to these simple expressions in the appropriate limit. However, the present analysis should by no means be considered to be confined to the case of aperiodic motion and pulse radiation. On the contrary, we have in atomic physics (see, for example, Born's *Atommechanik*) an example of the analysis of multiply periodic motion, and of the radiation given out in such motion, which can be taken over practically without change to the present problem. The energy of the system is expressed as a function of three action variables,

$$E=E(J_r,J_\theta,J_\varphi),$$

of which the last two measure total angular momentum and angular momentum about the z axis. The circular frequencies associated with the three modes of motion are

$$\omega_r=\partial E/\partial J_r,\quad \omega_\theta=\partial E/\partial J_\theta,\quad \omega_\varphi=\partial E/\partial J_\varphi.$$

In the limit of nearly flat space, where Cartesian coordinates are appropriate, one has

$$x=\sum_{n_r,n_\theta,n_\varphi} X_{n_rn_\theta n_\varphi}\exp[-i(n_r\omega_r+n_\theta\omega_\theta+n_\varphi\omega_\varphi)t],$$

and similar Fourier series for the other two coordinates and for the third time rate of change of the quadrupole moment as well. The rate of emission of quadrupole gravitational radiation at the circular frequency $\omega=n_r\omega_r+n_\theta\omega_\theta+n_\varphi\omega_\varphi$ is

$$\left(-\frac{dE}{dt}\right)_{\omega=n_\mu\omega_\mu}=\frac{2}{45}\left(\frac{d^3Q^{pq*}}{dt^3}\right)_{n_r,n_\theta,n_\varphi}\left(\frac{d^3Q^{pq}}{dt^3}\right)_{n_r,n_\theta,n_\varphi}$$

The generalization of this formula to the Schwarzschild geometry allows itself in principle to be read out of the present work, once the determination of the appropriate Fourier coefficients in the multiply periodic motion has been carried through.

APPENDIX I: AMPLITUDES FOR SPECIAL CASE OF PARTICLE FALLING STRAIGHT IN

We do a standard asymptotic expansion in the parameter ω for large ω. This simultaneously gives us an asymptotic expansion for large r. First consider the magnetic equations, in particular, the homogeneous case.

[20] R. Ruffini and J. A. Wheeler, "Cosmology from Space Platforms," European Space Research Organization report (unpublished).

Let $\epsilon=+1$ for outgoing waves and $\epsilon=-1$ for incoming waves; then we may write

$$h_1\sim r(1-2m/r)^{-1}[1-(\epsilon/i\omega r)(\lambda+1-3m/2r)+O(1/\omega^2)]\exp(\epsilon i\omega r^*),\quad \text{(I1)}$$

$$h_0\sim-\epsilon r[1-(\epsilon/i\omega r)(\lambda+m/r)+O(1/\omega^2)]\exp(\epsilon i\omega r^*).$$

Now we notice that for large r these functions go as $O(r)$. If we transform the perturbation to a Cartesian coordinate system, this implies that the perturbation goes as $O(1)$. We can easily see this by looking at Eq. (D2a).

Now in order to apply the usual pseudotensor criteria for energy and momentum radiated, we must have a perturbation that goes as $O(1/r)$ in order that the space be asymptotically flat. We remedy the difficulty by a gauge transformation. This gauge transformation does several things. It makes the new $h_0^{(N)}=0$ and the new $h_1^{(N)}=O(1/\omega r)$ and thus these functions do not contribute in the radiation field since $h_1=O(1/\omega r)$ implies that the perturbation is $O(1/\omega r^2)$. At the same time, it introduces a nonzero $h_2^{(N)}$ in the new gauge which is $O(r)$ and hence, as can be seen by looking at Eq. (D2a), the perturbation is then $O(1/r)$, which enables us to use the pseudotensor to calculate the energy or momentum radiated. Note that the angular dependence of the radiation term is given by Mathews's electric harmonic and that in this gauge the perturbation is divergenceless and traceless to $O(1/r)$. The form of the gauge transformation which does all this is easily found. From Eq. (D3) we see that setting $\Lambda_{LM}=-(1/i\omega)h_{0LM}$ makes $h_0^{(N)}$ zero. Also, using Eq. (F1a), we see that

$$h_{1LM}^{(N)}=-(2\lambda/\omega^2r^2)(1-2m/r)h_{1ML},$$

which is $O(1/r)$. Finally,

$$h_{2LM}^{(N)}=2\Lambda_{LM}=-(2/i\omega)h_{0LM}.$$

We will then denote the canonical solution of the homogeneous magnetic equations by $\mathbf{h}_{LM}^{(m)}(\omega,r,\phi)$ and

$$\mathbf{h}_{LM}^{(m)(\epsilon)}\sim(\epsilon/2i\omega r)\exp(\epsilon i\omega r^*)\times[2L(L+1)(L-1)(L+2)]^{1/2}\times\mathbf{T}_{LM}^{(e)}(\theta,\phi)+O(1/\omega^2r^2).\quad \text{(I2)}$$

Gauge transformations of a similar nature have been discussed by Edelstein[21] and by Price and Thorne.[22] By looking at (I1), we see that the second term of the asymptotic expansion is less than the first term when $2m\omega>\lambda+1$. Thus we can expect the asymptotic expansion to be a good approximation for $4m\omega>L(L+1)$. This means that the approximation is good if ω^2 is above the peak of the effective potential $V(r)$ given by Regge and Wheeler.

Now let us turn to the magnetic-parity equations. Again we look at the homogeneous equations.

[21] L. Edelstein, Ph.D. thesis, University of Maryland, 1969 (unpublished).

[22] R. Price and K. Thorne, Astrophys. J. **155**, 163 (1969).

Letting $\epsilon=+1$ denote outgoing waves and $\epsilon=-1$ denote ingoing waves, we have the result

$$K^{(\epsilon)}\sim\left\{1+\epsilon\left(\frac{3m}{2i\omega r^2}\right)+\frac{1}{\omega^2}\left[\frac{\lambda(\lambda+1)}{2r^2}+\frac{m(2\lambda-1)}{2r^3}+\frac{15m^2}{8r^4}\right]+O(1/\omega^3)\right\}\exp(\epsilon i\omega r^*),$$

$$H^{(\epsilon)}\sim-\epsilon H_1^{(\epsilon)}+O(1/\omega^3),\tag{I3}$$

$$H_1^{(\epsilon)}\sim-i\omega r\left(1-\frac{2m}{r}\right)^{-1}\left\{1-\frac{\epsilon}{i\omega}\left(\frac{\lambda}{r}+\frac{3m}{2r^2}\right)-\frac{1}{\omega^2}\left[\frac{\lambda(\lambda+1)}{2r^2}+\frac{m(\lambda+1)}{2r^3}-\frac{3m^2}{8r^4}\right]+O(1/\omega^3)\right\}\times\exp(\epsilon i\omega r^*).$$

The perturbation given by (I3) is $O(r)$ as $r\to\infty$, as can be seen by looking at Eq. (D2b). Again we must find a suitable gauge transformation so that the perturbation is $O(1/r)$ as $r\to\infty$. This means that in the new gauge all the coefficient functions in (D2b) must be $O(1/r)$ except for $h_0^{(\epsilon)}$ and $h_1^{(\epsilon)}$, which must be $O(1)$ in order for $\mathbf{h}_{LM}^{(\epsilon)}\sim O(1/r)$.

We will be able to find a gauge transformation which makes $h_0^{(\epsilon)}=0$ and which makes the perturbation divergenceless, traceless, and proportional to Mathews's transverse traceless magnetic harmonic, at least up to the order of the radiation terms. Most important of all, it gives the proper asymptotic dependence for the perturbation as $r\to\infty$. From (D2b) and (D4) we obtain seven equations relating M_0, M_1, M_2, H, H_1, and K to the transformed coefficient functions [denoted with a superscript (N)]. Since K is $O(1)$ and we want $K^{(N)}$ to be $O(1/r)$, we must require

$$M_1=-\tfrac{1}{2}r(1-2m/r)^{-1}[1+\mu_1(r)/\omega+\mu_2(r)/\omega^2]\times\exp(\epsilon i\omega r^*),$$

where $\mu_1(r)=O(1/r)$, $\mu_2(r)=O(1/r^2)$ are still free to be chosen. The requirement $H_0^{(N)}=H_2^{(N)}$ fixes M_0 in terms of M_1, and it follows also that $H_1^{(N)}=-\epsilon H^{(N)}+O(1/\omega^3)$. The condition $h_0^{(\epsilon)(N)}\equiv0$ fixes $M_2=(1/i\omega)M_1$. We can then choose $\mu_1(r)$ and $\mu_2(r)$ so that $H^{(N)}$ and $H_1^{(N)}$ are $O(1/\omega^2)$ [and hence also $O(1/r^2)$]. Then $G^{(N)}$ and $h_1^{(\epsilon)(N)}$ are fixed by the remaining relations: $K^{(N)}-\tfrac{1}{2}L(L+1)G^{(N)}=O(1/\omega^3r^3)$, $(1/r)h_1^{(\epsilon)(N)}=O(1/\omega^2r^2)$, and finally

$$K^{(N)(\epsilon)}=\frac{\epsilon}{2i\omega r}L(L+1)\left[1-\frac{\epsilon(2\lambda r+m)}{2i\omega r^2}+O(1/\omega^2r^2)\right]\times\exp(\epsilon i\omega r^*).\tag{I4}$$

Thus

$$\mathbf{h}_{LM}^{(\epsilon)}=\frac{\epsilon}{2i\omega r}\exp(\epsilon i\omega r^*)\times[2L(L+1)(L-1)(L+2)]^{1/2}\mathbf{T}_{LM}^{(m)}+O(1/\omega^2r^2).\tag{I5}$$

The trace of $\mathbf{h}_{LM}^{(\epsilon)}$ is $O(1/\omega^3r^3)$ and $\operatorname{div}(\mathbf{h}_{LM}^{(\epsilon)})=O(1/\omega^2r^2)$.

Using these results, we construct a high-frequency-limit Green's function, and, applying the boundary condition of outgoing waves at $r=\infty$ and ingoing waves at $r=2m$, we obtain amplitudes for the ingoing and outgoing radiation fields as integrals. One of these integrals is evaluated in an asymptotic expansion in ω by the method of saddle points in a beautiful procedure given by van der Waerden.[23] The other integral is evaluated in an asymptotic expansion using a theorem given by Copson.[24] First we discuss the stress tensor for radiation and some pecularities of the wave solutions of the homogeneous equations.

Isaacson[4] shows that in the high-frequency limit (wavelength of the radiation short compared with the curvature of the background space) for small perturbations a suitable stress tensor for gravitational radiation is

$$T_{\mu\nu}=(32\pi)^{-1}\{h_{\rho\sigma;\mu}h^{\rho\sigma}{}_{;\nu}\}_{\rm av}+O(h^3),\tag{I6}$$

where $\{\ \}_{\rm av}$ denotes an average over several wavelengths of the radiation. The expression (I6) is valid in the gauge where $\operatorname{tr}(\mathbf{h})$ and $\operatorname{div}(\mathbf{h})$ are zero. The result (I6) is also equivalent to using the Landau-Lifshitz[3] pseudotensor in the limit of large r. Since we are dealing with the Fourier transforms of the fields, averaging corresponds to taking the field amplitudes times their complex conjugates; that is,

$$T_{\mu\nu}=(32\pi)^{-1}h^*{}_{\rho\sigma;\mu}h^{\rho\sigma}{}_{;\nu}.$$

We are interested in the energy density and flux given by the components of $T_{\mu\nu}$ with μ and ν equal to zero or one. If we keep only the leading terms in ω, we can replace the covariant derivative with respect to t by $-i\omega$ and the covariant derivative with respect to r by $\epsilon i\omega(1-2m/r)^{-1}$. Now let

$$h(\omega,r,\theta,\phi)=\sum_{LM}\{A_{LM}^{(e)}(\omega)\mathbf{h}_{LM}^{(e)}(\omega,r,\theta,\phi)+A_{LM}^{(m)}(\omega)\mathbf{h}_{LM}^{(m)}(\omega,r,\theta,\phi)\},$$

where $\mathbf{h}_{LM}^{(e)}$ and $\mathbf{h}_{LM}^{(m)}$ are solutions which, in the absence of sources, are given asymptotically by (I5) and (I2). Let us consider the energy flux for large r. Then the power per unit solid angle per unit frequency

[23] B. L. van der Waerden, Appl. Sci. Res. **B2**, 33 (1960); also discussed by H. A. Lauwerier, *Asymptotic Expansions* (Mathematisch Centrum, Amsterdam, 1966).

[24] E. T. Copson, *Asymptotic Expansions* (Cambridge U. P., New York, 1965), p. 21.

for a particular electric or magnetic LM multipole is

$$\frac{dS_{LM}(\omega,\Omega)}{d\Omega}=\epsilon(32\pi)^{-1}L(L+1)(L-1)(L+2)\times|A_{LM}(\omega)|^2\mathbf{T}_{LM}^*:\mathbf{T}_{LM}, \quad (I7)$$

while the total power per unit frequency is

$$\lim_{r\to\infty} 2\,\mathrm{Re}\left\{r^2\int\int T_{10}d\Omega\right\}$$

or

$$S(\omega)=\epsilon(32\pi)^{-1}\sum_{LM} L(L+1)(L-1)(L+2)\times\{|A_{LM}^{(e)}(\omega)|^2+A_{LM}^{(m)}(\omega)|^2\}. \quad (I8)$$

By appealing to conservation of energy,[25] we can say that this expression also gives the energy flux through the Schwarzschild surface $r=2m$; that is, to find the power radiated into the $2m$ surface by some source, we take the ingoing wave solution of the homogeneous equations which has the same amplitude as that of the solution with the source term. Then we can calculate the power flowing inward by looking at the amplitude of the homogeneous solutions at large r where energy has a well-defined meaning and where a well-defined method of calculating the energy exists. Price and Thorne[22] discuss the polarization of waves described by tensor harmonics and also discuss the linear and angular momentum carried by these waves.

Let us look at the 00, 01, and 11 components of the stress tensor $T_{\mu\nu}$ for a particular harmonic. Let

$$U_{LM}(\theta,\phi)=(64\pi)^{-1}L(L+1)(L-1)(L+2)\times\mathbf{T}_{LM}^{(m)*}:\mathbf{T}_{LM}^{(m)}.$$

Then

$$T_{00}=r^{-2}U_{LM}(\theta,\phi),$$
$$T_{01}=-\epsilon r^{-2}(1-2m/r)^{-1}U_{LM}(\theta,\phi),$$
$$T_{11}=(r-2m)^{-2}U_{LM}(\theta,\phi).$$

If we transform this tensor to Kruskal coordinates, we obtain

$$T^K{}_{00}=16m^2r^{-2}(u-\epsilon v)^{-2}U_{LM}(\theta,\phi),$$
$$T^K{}_{01}=-\epsilon T^K{}_{00},\quad T^K{}_{11}=T^K{}_{00}.$$

Thus we see that the stress tensor is singular along $u=v$ (that is, $r=2m$, $t=\infty$) for outgoing waves ($\epsilon=+1$) while the stress tensor is singular along $u=-v$ (that is, $r=2m$, $t=-\infty$) for ingoing waves ($\epsilon=-1$). This singularity is a manifestation of the fact that the perturbation $h_{\mu\nu}$ is singular in just this manner at $r=2m$. This shows up in the higher-order terms which we ignored in our asymptotic expansions, that is, in the functions $H^{(N)}$ and $H_1^{(N)}$. Doroshkevich, Zel'dovich, and Novikov[26] arrive at a similar result and argue, therefore, that, in gravitational collapse, the higher moments of the gravitational field must be attenuated with collapse.

Vishveshwara[18] has shown that the solutions of the time-independent perturbation equations for $L\geq 2$ cannot be nonsingular at $r=2m$. He points out that these singularities pose no problem: It is possible to build wave packets which stay bounded away from $u=-v$ for ingoing waves or $u=+v$ for outgoing waves, and hence no singularity appears. For example, one way of producing a packet bounded in the manner described is to hold a particle at constant distance r_2 and then release it at a certain time, let it fall for a while (during this time it radiates), and then keep it at constant distance r_1 less than r_2. It has also been pointed out[27] that because of the sinusoidal behavior of the perturbations, energy pours out (say, for the outgoing waves) forever toward $r=\infty$ at a uniform rate, and thus in the Kruskal picture there must be an infinite amount of radiation in the region $0<u-v<\epsilon m$ for any $\epsilon>0$. In any case we will show that the singular behavior of the perturbations is not unexpected. Trautman[5] has examined the propagation of a discontinuity in the Riemann tensor for a Schwarzschild geometry. The result, if expressed in an orthonormal tetrad basis along the t, r, θ, ϕ directions, is that

$$\Delta R_{(\mu)(\lambda)(\alpha)(\beta)}(r)=\Delta R_{(\mu)(\lambda)(\alpha)(\beta)}(r_0)(r_0-2m)/(r-2m).$$

This is radiationlike, that is, $O(1/r)$, for large r, but is singular at $r=2m$. Now compare the leading term of the Riemann tensor for our asymptotic solutions with this result. The first-order correction to the Riemann tensor is

$$R_{\alpha\beta\gamma\delta}{}^{(1)}=-\tfrac{1}{2}\{h_{\alpha\gamma,\beta\delta}+h_{\beta\delta,\alpha\gamma}-h_{\beta\gamma,\alpha\delta}-h_{\alpha\delta,\beta\gamma}\}.$$

If we keep only the leading terms in ω [let us consider the electric solution (I5)] and if we transform $R_{\mu\nu\alpha\beta}$ to the orthonormal tetrad components and use the correspondence

$$\begin{Bmatrix}A\\ \alpha\beta\end{Bmatrix}=\begin{Bmatrix}1 & 2 & 3 & 4 & 5 & 6\\ 23 & 31 & 12 & 01 & 02 & 03\end{Bmatrix},$$

then we can write $R_{(\alpha)(\beta)(\gamma)(\delta)}$ as the 6×6 matrix:

$$\|R_{AB}\|\sim-\frac{1}{2}\frac{\epsilon i\omega}{r-2m}\exp(\epsilon i\omega r^*)\times\begin{bmatrix}0&0&0&0&0&0\\0&-\sigma&-\tau&0&\epsilon\tau&-\epsilon\sigma\\0&-\tau&\sigma&0&-\epsilon\sigma&-\epsilon\tau\\0&0&0&0&0&0\\0&\epsilon\tau&-\epsilon\sigma&0&\sigma&\tau\\0&-\epsilon\sigma&-\epsilon\tau&0&\tau&-\sigma\end{bmatrix} \quad (I9)$$

[25] C. W. Misner (private communication); this argument is due to L. Edelstein.

[26] A. Doroshkevich, Ya. Zel'dovich, and I. Novikov, Zh. Eksperim. i Teor. Fiz. 49, 170 (1965) [Soviet Phys. JETP 22, 122 (1966)].

[27] K. Thorne (private communication).

We recognize precisely the same $1/(r-2m)$ behavior that occurred in the shock-front propagation, and we also see that this tensor is type N in the Petrov-Pirani classification[28] and corresponds to a gravitational wave propagating in the (ϵ) direction.

Denote the outgoing and ingoing wave solutions of the homogeneous electric equation by ψ_{out} and ψ_{in}, where ψ is the column vector

$$\psi=\begin{pmatrix} R_{LM}^{(e)} \\ dR_{LM}^{(e)}/dr^* \end{pmatrix}.$$

Then let $\Psi(r)=(\psi_{\text{out}}|\psi_{\text{in}})$, where the notation means the matrix whose columns are ψ_{out} and ψ_{in}, respectively. Then the required solution to the inhomogeneous system is

$$\psi(r)=\int_{2m}^{r}\Psi(r)\Psi^{-1}(\rho)s(\rho)d\rho+\psi_{\text{hom}}(r),$$

where

$$s(\rho)=\begin{pmatrix} 0 \\ S_{LM} \end{pmatrix}$$

and where $\psi_{\text{hom}}(r)$ is a solution of the homogeneous system chosen so that $\psi(r)$ satisfies the ingoing wave condition at $2m$ and the outgoing wave condition at ∞. Now denote

$$\Psi^{-1}(\rho)s(\rho)=\begin{pmatrix} c_1(\rho) \\ c_2(\rho) \end{pmatrix},$$

where

$$s(\rho)=\begin{pmatrix} 0 \\ S_{LM}(\rho) \end{pmatrix},$$

S_{LM} being the source term given in Eq. (18). Then

$$\psi(r)=\left\{\int_{2m}^{r}c_1(\rho)d\rho\right\}\psi_{\text{out}}(r)+\left\{\int_{2m}^{r}c_2(\rho)d\rho\right\}\psi_{\text{in}}(r)+\psi_{\text{hom}}(r).$$

Thus to ensure outgoing waves for large r and ingoing waves for r near $2m$ we choose

$$\psi_{\text{hom}}=-\left\{\int_{2m}^{\infty}c_2(\rho)d\rho\right\}\psi_{\text{in}}(r)$$

and then, up to terms of $O(1/\omega)$, we have the result

$$\psi(r)=A_L^{(\text{out})}(\omega,r)\psi_{\text{out}}(r)+A_L^{(\text{in})}(\omega,r)\psi_{\text{in}}(r),\quad\text{(I10)}$$

where

$$A_L^{(\text{out})}(\omega,r)=-2m_0(L+\tfrac{1}{2})^{1/2}\int_{2m}^{r}\exp\{i\omega[T(\rho)-\rho^*]\}\times\tilde{c}_1(\rho)(\lambda\rho+3m)^{-1}(1-2m/\rho)^{-1}d\rho,$$

$$A_L^{(\text{in})}(\omega,r)=2m_0(L+\tfrac{1}{2})^{1/2}\int_{r}^{\infty}\exp\{i\omega[T(\rho)+\rho^*]\}\times\tilde{c}_2(\rho)(\lambda\rho+3m)^{-1}d\rho,$$

and

$$\tilde{c}_1(\rho)=[1+(2m/\rho)^{1/2}]\{1+(1/i\omega\rho)[m/2\rho+(\lambda\rho+\sqrt{2}m^{1/2}\rho^{1/2}+m)(\rho+3m/\lambda)^{-1}]\},$$

$$\tilde{c}_2(\rho)=[1+(2m/\rho)^{1/2}]^{-1}\{-1+(1/i\omega\rho)[m/2\rho+(\lambda\rho-\sqrt{2}m^{1/2}\rho^{1/2}+m)(\rho+3m/\lambda)^{-1}]\},$$

$$T(\rho)=-4m(\rho/2m)^{1/2}-\tfrac{4}{3}m(\rho/2m)^{3/2}-\ln[(\rho/2m)^{1/2}-1]+\ln[(\rho/2m)^{1/2}+1].$$

These are, in the high-frequency limit, the amplitudes for the outgoing and ingoing waves. We have

$$A_L^{(\text{out})}(\omega)=\lim_{r\to\infty}A_L^{(\text{out})}(\omega,r)\quad\text{(I11)}$$

and

$$A_L^{(\text{in})}(\omega)=\lim_{r\to 2m}A_L^{(\text{in})}(\omega,r).\quad\text{(I12)}$$

Equations (I7) and (I8) then give us the power radiated in a unit frequency interval. In principle one can calculate the field $\psi(r)$ to any order in $1/\omega$ and thus obtain integrals for the amplitudes $A_L^{(\text{out})}$ and $A_L^{(\text{in})}$. These integrals can also be expanded in an asymptotic expansion. Let us first consider $A_L^{(\text{out})}$. We note that the integral is not absolutely convergent, the integrand going like $1/\rho$ times an oscillatory factor for $\rho\to+\infty$ while going to a constant times an oscillatory factor for $\rho^*\to-\infty$ (that is, $\rho\to 2m$). This behavior is indicative of a δ function in ω for $\omega=0$. Since our approximation is valid for large ω, we will ignore this contribution, and this will be done in a natural manner in the procedure to be discussed. Let us make the transformation $x^2=\rho/(2m)$ followed by $e^y+1=x$ and let $k=2m\omega$. Then (I11) becomes

$$A_L^{(\text{out})}(\omega)=-2m_0(L+\tfrac{1}{2})^{1/2}I(k,\lambda),$$

where

$$I(k,\lambda)=\int_{-\infty}^{\infty}e^{ikf(y)}g(y,k,\lambda)dy,$$

$$f(y)=-\tfrac{2}{3}x^3-x^2-2x-2y,$$

$$g(y,k,\lambda)=4x^2(2\lambda x^2+3)^{-1}\{1+(1/ikx^2)\times[\tfrac{1}{4}x^{-2}+\lambda(2\lambda x^2+3)^{-1}(2\lambda x^2+2x+1)]+O(1/k^2)\}.$$

Thus we have an integral which is in a suitable form for asymptotic expansion by the saddle-point method. Now $f'(y)=0$ implies $e^{y_0}=-1$, which implies $y_{0n}=(2n+1)i\pi$ for any integer n. Thus the saddle points are at $(2n+1)i\pi$, where n is an integer. We see that a singularity of $g(y,k,\lambda)$ coincides with the saddle points. However, van der Waerden[23] has shown that we can still use the saddle-point method. Using van der Waerden's method, we make the transformation $w=-if(y)$

---

[28] F. Pirani, Phys. Rev. 105, 1089 (1957).

[assume $\omega>0$; for $\omega<0$, $A(\omega)=A^*(-\omega)$]. Then

$$I(k,\lambda)=\int_{-i\infty}^{+i\infty} e^{-kw}g(y(w),k,\lambda)(dy/dw)dw.$$

The saddle points in the y plane become branch points in the w plane. These branch points are at $w_n=2(2n+1)\pi$ for n an integer. Now make branch cuts along the real axis between the branch points, and deform the contour $C'=(-i\infty,i\infty)$ into the contour C by pushing it to the right as far as possible without passing any branch points. The contour C comes in from $+\infty$ below the positive real axis, goes around $w_0=2\pi$, and goes back out to $+\infty$ above the positive real axis. Clearly C is reached from C' by going downhill on the real part of e^{-kw}. This at most eliminates contributions to the integral from the infinite parts of the contour which, as we noted above, we expect to give a δ-function type of behavior near $\omega=0$. In an asymptotic expansion of the integral the most important contribution comes from the left-most branch point $w_0=2\pi$, the other points giving exponentially smaller terms (that is, asymptotic to zero compared with the contributions from $w_0=2\pi$). We obtain for the outgoing wave amplitude

$$A_L^{(\mathrm{out})}\sim -4m_0(L+\tfrac{1}{2})^{1/2}e^{-4\pi m\omega}\{\tfrac{1}{3}\sqrt{2}e^{5\pi i/8}\Gamma(\tfrac{3}{4})(m\omega)^{-3/4}+\tfrac{1}{12}\pi(m\omega)^{-1}+\cdots\}. \quad (I13)$$

The dominant feature of this amplitude is that it decreases exponentially with increasing frequency.

Now let us turn to the asymptotic evaluation of $A_L^{(\mathrm{in})}(\omega)$ given in Eq. (I12). Here the integrand is well behaved near $2m$ (that is, as $\rho^*\to-\infty$) but goes as $1/\rho$ for $\rho\to\infty$ with an oscillatory factor whose exponent is

$$T(\rho)+\rho^*=-4m(\rho/2m)^{1/2}-\tfrac{4}{3}m(\rho/2m)^{3/2}+\rho+4m\ln[1+(\rho/2m)^{1/2}].$$

The method of evaluation is again an example of the method of steepest descents, but in this case it is the end point of the contour which is most important. To evaluate the integral, we use a theorem of Copson[24] on asymptotic expansions which is just the analog, for integrals along the imaginary axis, of the result that the asymptotic behavior of the Laplace transform of a function depends on the behavior of the function near the origin. Let us go back to (I12) and make the substitution $x^2=\rho/(2m)$; then

$$A_L^{(\mathrm{in})}(\omega)=2m_0(L+\tfrac{1}{2})^{1/2}\int_1^\infty e^{ikf(x)}g(x,k,\lambda)dx,$$

where

$$f(x)=-2x-\tfrac{2}{3}x^3+x^2+2\ln(1+x),$$

$$g(x)=4x^2(2\lambda x^2+3)^{-1}(1+x)^{-1}\{-1+(1/ikx^2)\times[\tfrac{1}{4}x^{-2}+\lambda(2\lambda x^2+3)^{-1}(2\lambda x^2-2x+1)]+O(1/k^2)\}.$$

We evaluate the terms to $O(1/k^2)$ and obtain

$$A_L^{(\mathrm{in})}(\omega)=-2m_0(L+\tfrac{1}{2})^{1/2}ie^{2mi\beta\omega}(2\lambda+3)^{-1}(m\omega)^{-1}\times[1+\tfrac{1}{2}i(\lambda+5/4)(m\omega)^{-1}+O(1/\omega^2)], \quad (I14)$$

where $\beta=5/3-2\ln 2$.

APPENDIX J: SOME QUALITATIVE CONSIDERATIONS

Although the low-frequency part of the spectrum has not been adequately described in the preceding calculations, we will give here some estimates of a very qualitative nature which have been suggested in a conversation with John Wheeler.

We have seen [Eq. (I13)] that the amplitude $A_L^{(\mathrm{out})}(\omega)$ for outgoing waves at ∞ goes like $2m_0^{1/2}\times(m\omega)^{-3/4}e^{-4\pi m\omega}$. From Eq. (I8), the power per unit frequency is then

$$S_L(\omega)\sim 0.01kL^4|A_L(\omega)|^2\sim k0.04m_0^2L^5\times(m\omega)^{-3/2}e^{-8\pi m\omega}, \quad (J1)$$

where k is a numerical factor of order 1. We have also seen that the asymptotic approximation is good if $8m\omega\gtrsim L^2$. Thus, for a fixed frequency ω, we expect that the power as a function of the degree L first goes up as L^5, reaches a peak, and starts falling off rapidly with increasing L at the "barrier" $8m\omega\sim L_B^2$. Thus the power per unit ω, summed over all L's, is approximately

$$S(\omega)\sim\int_0^{L_B} S_L(\omega)dL$$

or

$$S(\omega)\sim 4m_0^2(m\omega)^{3/2}e^{-8\pi m\omega}k. \quad (J2)$$

As a function of ω, this looks like a power law for small ω, reaches a peak at $\omega=3/16\pi m$. and decreases exponentially thereafter. Further, we may integrate this power spectrum over ω and obtain an estimate for the total energy radiated. Thus we obtain

$$E=\int_0^\infty S(\omega)d\omega\sim 4(8\pi)^{5/2}\Gamma(\tfrac{5}{2})(m_0^2/m)k\sim k0.0016m_0^2/m. \quad (J3)$$

Compare this with a calculation using the linearized theory of Landau and Lifshitz[3]; they give

$$-\frac{dE}{dt}=\frac{1}{45}\sum_{i,j=1}^{3}\left(\frac{d^3D_{ij}}{dt^3}\right)^2,$$

where

$$D_{ij}=\iiint\rho(x)(3x_ix_j-\delta_{ij}r^2)d^3x.$$

For the case of radial motion starting at $r=\infty$ with zero velocity, we have

$$t^{1/3}\simeq-\tfrac{2}{3}(2mr)^{1/2}$$

for $r \gg 2m$. Thus

$$dE/dt = -(1/30)(m_0/m)^2(2m/r)^5 = -(1/30) \times (m_0/m)^2(2m/3t)^{10/3}.$$

Integrating this expression, we obtain the energy radiated in falling from ∞ to r:

$$E(r) = (1/70)(m_0/m)^2(2m/3)^{10/3}t^{-7/3} = (1/105)(m_0^2/m)(2m/r)^{7/2}. \quad \text{(J4)}$$

Thus the energy radiated in falling to ten Schwarzschild radii (it is reasonable to expect that the linearized approximation is fairly good up to this ponit) is

$$\Delta E(20m) \cong (1/330000)(m_0^2/m),$$

which is less than 0.3% of the total radiation given by (J3). If we use (J4) to calculate the radiation up to four Schwarzschild radii, we obtain

$$\Delta E(8m) \cong (1/14000)(m_0^2/m),$$

which is less than 6% of the total given by (J3). Thus all indications are that a substaintal portion of the radiation comes from the part of the particle's trajectory which is between one and four Schwarzschild radii.

APPENDIX K: TIME-INDEPENDENT PERTURBATIONS FOR $L \geq 2$

The case where the perturbation is assumed time independent, $\partial h/\partial t = 0$, has been discussed by Regge and Wheeler[6] and by Vishveshwara.[18] We present here the solutions of the electric time-independent equations which can be given in terms of hypergeometric functions (for the homogeneous equations).

Setting the terms in (C7) which contain derivatives with respect to time equal to zero, we obtain the following equations:

$$H_1 = 0, \quad \text{(K1)}$$

$$\frac{d}{dr}(H-K) + 2mr^{-2}\left(1-\frac{2m}{r}\right)^{-1} H = 0, \quad \text{(K2)}$$

$$\left(1-\frac{m}{r}\right)\left(1-\frac{2m}{r}\right)^{-1}\frac{dK}{dr} - \frac{dH}{dr} - \tfrac{1}{2}(L-1)(L+2)(r-2m)^{-1}(K-H) = 0. \quad \text{(K3)}$$

From these equations we obtain

$$\frac{d^2H}{dr^2} + \frac{d}{dr}\left[\frac{2}{r}\left(1-\frac{m}{r}\right)\left(1-\frac{2m}{r}\right)^{-1} H\right] - (L-1)(L+2)r^{-1}(r-2m)^{-1}H = 0, \quad \text{(K4)}$$

or, letting $M = r(r-2m)H$ and $x = r/2m$, we have

$$x(1-x)\frac{d^2M}{dx^2} + (2x-1)\frac{dM}{dx} + (L-1)(L+2)M = 0. \quad \text{(K5)}$$

This is a form of the hypergoemetric equation, and a particular solution is[29]

$$M(x) + x^2F(L+1, -L; 3; x). \quad \text{(K6)}$$

This is a polynomial in r of degree $L+2$ and goes to ∞ as $r \to \infty$. The other solution of the equation is

$$M = r^{-L+1}F(L+1, L-1; 2L+2; 2m/r). \quad \text{(K7)}$$

This solution, however, goes as $(r-2m)^{-L}$ as $r \to 2m$. Vishveshwara interprets these results as showing that there cannot exist any time-independent perturbations (for $L \geq 2$) on the Schwarzschild metric. Similar results hold for the electric-parity equations. We give the expressions for the hypergeometric functions in the above solutions:

$$F(-L, L+1; 3; z) = \sum_{n=0}^{L} \frac{\Gamma(L+n)\Gamma(L+n+1)\Gamma(3)}{\Gamma(L)\Gamma(L+1)\Gamma(n+3)} \frac{z^n}{n!}. \quad \text{(K8)}$$

The polynomial (K8) is the Jacobi polynomial

$$[\tfrac{1}{2}(L+1)(L+2)]^{-1}P_L^{(2,-2)}(1-2z). \quad \text{(K9)}$$

Also,[30]

$$F(L+1, L-1; 2L+2; 1/z) = -\frac{(2L+1)!}{(L!)^2(L+2)!(L-2)!} \times \frac{d^{L-2}}{dt^{L-2}}\left\{(1-t)^L \frac{d^{L+2}}{dt^{L+2}}\left[\frac{\ln(1-t)}{t}\right]\right\}_{t=1/z} \quad \text{(K10)}$$

[29] *Handbook of Mathematical Functions*, edited by M. Abramowitz and I. Stegun (Dover, New York, 1965), Chap. 15, p. 562.

[30] *Higher Trancendental Functions*, edited by A. Erdélyi *et al.* (McGraw-Hill, New York, 1953), Vol. I, Chap. 2.

A-7

Errata - F.J. Zerilli, Phys. Rev. 2, 2141 (1970)

P. 2144 Eq. (10) should read

$$h_{OLM} = \frac{i}{\omega} \frac{d}{dr^*} (rR_{LM}^{(m)}) + \frac{8\pi r(r-2m)}{\omega \left[L(L+1)\ (L-1)\ (L+2)/2\right]^{1/2}} D_{LM}(\omega,r)$$

Eq. (11) should read

$$\frac{d^2R_{LM}^{(m)}}{dr^{*2}} + \left[\omega^2 - V_L^{(m)}(r)\right] R_{LM}^{(m)}$$

$$= \frac{8\pi i}{\left[L(L+1)(L-1)\ (L+2)/2\right]^{\frac{1}{2}}} \frac{r-2m}{r^2} \left\{ r^2 \frac{d}{dr} \left(1 - \frac{2m}{r}\right) D_{LM} - (r-2m) \left[(L-1)(L+2)\right]^{1/2} Q_{LM} \right\}$$

P. 2145 Eq. (14) $\lambda r^2 - 3\lambda mr - 3m$ should read
$\lambda r^2 - 3\lambda mr - 3m^2$

P. 2145 Eq. (20) should read

$$\tilde{C}_{1LM} = -\frac{8\pi}{\sqrt{2}\omega} A_{LM}^{(1)} - \frac{1}{r} \tilde{B}_{LM} + 16\pi r \left[L(L+1)(L-1)(L+2)/2\right]^{-\frac{1}{2}} F_{LM}$$

Eq. (21) should read

$$C_{2LM} = \frac{8\pi r^2}{j\omega} \frac{\left[L(L+1)/2\right]^{-\frac{1}{2}}}{r-2m} B_{LM}^{(0)} + \frac{ir}{r-2m} \tilde{B}_{LM}$$

$$-16\pi i \frac{r^3}{r-2m} \left[L(L+1)(L-1)(L+2)/2\right]^{-\frac{1}{2}} F_{LM}$$

P. 2146 Eq. (A2j), $\sin^2\theta\, W_{LM}$ should read $-\sin^2\theta\, W_{LM}$

P. 2147 Eq. (C6a), right hand side should read $-8\pi \frac{r^2}{r-2m} \left[L(L+1)/2\right]^{-\frac{1}{2}} Q_{LM}$

P. 2148 Second line of Eq. D2a should read $+h_{1LM}(r,t)\, C_{LM}(\theta,\phi) - (1/2r)$

P. 2149 Table III, sign of D_{LM} should be +, not -

P. 2150 Eq. (F1c) left hand side, sign should be changed:

$$(r-2m)^{dh}\ 1LM/dr + \frac{i\omega r^2 h_{OLM}}{r-2m} + \frac{2m\, h_{1LM}}{r} = \ldots\ldots$$

P. 2151 Eq. (F3b), right hand side, $\frac{r-3m}{r(-r2m)}$ should read $\frac{r-3m}{r(r-2m)}$

P. 2160 Eq. (K6) should read:

$$M(X) = X^2F(L+1, -L;3;X)$$

VOLUME 27, NUMBER 21 PHYSICAL REVIEW LETTERS 22 NOVEMBER 1971

Gravitational Radiation from a Particle Falling Radially into a Schwarzschild Black Hole*

Marc Davis and Remo Ruffini
Joseph Henry Laboratories, Princeton University, Princeton, New Jersey 08540

and

William H. Press† and Richard H. Price‡
Kellogg Radiation Laboratory, California Institute of Technology, Pasadena, California 91109

(Received 24 September 1971)

We have computed the spectrum and energy of gravitational radiation from a "point test particle" of mass m falling radially into a Schwarzschild black hole of mass $M \gg m$. The total energy radiated is about $0.0104mc^2(m/M)$, 4 to 6 times larger than previous estimates; the energy is distributed among multipoles according to the empirical law $E_{2^l\text{-pole}} \approx (0.44m^2c^2/M)e^{-2l}$; and the total spectrum peaks at an angular frequency $\omega = 0.32c^3/GM$.

In view of the possibility that Weber may have detected gravitational radiation,[1] detailed calculations of the gravitational radiation emitted by fully relativistic sources are of considerable interest. Three such calculations have been published in the past: waves from pulsating neutron stars, by Thorne[2]; waves from rotating neutron stars, by Ipser[3]; and waves from a physically unrealistic collapse problem (important for the points of principle treated), by de la Cruz, Chase, and Israel.[4] To these, this paper adds a fourth calculation: the waves emitted by a body falling radially into a nonrotating black hole. This calculation is particularly important for two reasons: (i) It is the first accurate calculation of the spectrum and energy radiated by any realistic black-hole process (though upper limits on the energy output have been derived by Hawking[5]); (ii) Weber's events involve such high fluxes that black holes are more attractive as sources than are neutron stars.

A first analysis of the radial-fall problem was done by Ruffini and Wheeler[6] with a simple idealization: The particle's motion is derived from the Schwarzschild metric, but its radiation is calculated using the flat-space linearized theory of gravity. This scheme yielded a total energy radiated of $0.00246mc^2(m/M)$ and a spectrum peaked at an angular frequency $0.15c^3/GM$. Zerilli,[7] using the formalism of Regge and Wheeler,[8] gave the mathematical foundations for a fully relativistic treatment of the problem. Unfortunately, Zerilli's equations are sufficiently complicated as to make a calculation of the energy release inaccessible to analytic means.

We have used Zerilli's equations (corrected for errors in the published form), and by numerical techniques we have (i) computed the wave form of gravitational radiation, (ii) evaluated the amplitude of this wave asymptotically at great distances, and (iii) used this amplitude to compute the outgoing wave intensity in units of energy per unit frequency per unit of solid angle.

Zerilli describes the 2^l-pole component of gravitational waves by a radial function $R_l(r)$ which is a combination of the Fourier transform of metric perturbations in the Regge-Wheeler formalism. The function $R_l(r)$ satisfies the remarkably simple Zerilli wave equation ($G = c = 1$)

$$d^2R_l/dr^{*2} + [\omega^2 - V_l(r)]R_l = S_l, \tag{1}$$

with

$$r^* = r + 2M\ln(r/2M - 1). \tag{2}$$

$V_l(r)$ is an "effective potential" defined by

$$V_l(r) = (1 - 2M/r)[2\lambda^2(\lambda+1)r^3 + 6\lambda^2Mr^2 + 18M^2r + 18M^3]/r^3(\lambda r + 3M). \tag{3}$$

Here, $\lambda = \frac{1}{2}(l-1)(l+2)$ and $S_l(r)$ is the 2^l-pole component of the source of the wave. We are interested in the particular case of a particle initially at infinity ($t = +\infty, r = +\infty$) and falling radially into a Schwarzschild black hole ($t = +\infty, r = 2M$). For this simple case the source may be written as

$$S_l(r) = \frac{4M}{\lambda r + 3M}(l+\tfrac{1}{2})^{1/2}\left(1 - \frac{2M}{r}\right)\left[\left(\frac{r}{2M}\right)^{1/2} - \frac{i2\lambda}{\omega(\lambda r + 3M)}\right]e^{i\omega T(r)} \tag{4}$$

Here $t = T(r)$ describes the particle's radial trajectory giving the time as a function of radius along the

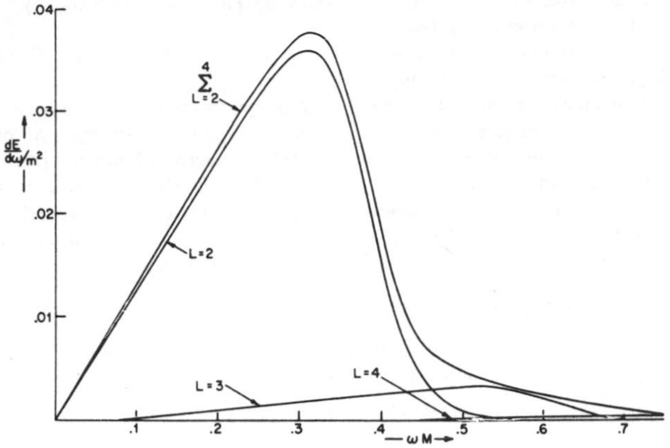

FIG. 1. Spectrum of gravitational radiation emitted by a test particle of mass m falling radially into a black hole of mass M (in geometrical units $c = G = 1$).

geodesic

$$T(r) = -\frac{4}{3}\left(\frac{r}{2M}\right)^{3/2} - 4\left(\frac{r}{2M}\right)^{1/2} + 2\ln\left\{\left[\left(\frac{r}{2M}\right)^{1/2} + 1\right]\left[\left(\frac{r}{2M}\right)^{1/2} - 1\right]^{-1}\right\}. \quad (5)$$

The effect of gravitational radiation reaction on the particle's motion is therefore ignored. This is justified by the final result: The total energy radiated, of order m^2c^2/M, is negligible compared to the particle's final kinetic energy, of order mc^2. Equation (1) is solved with boundary conditions of purely outgoing waves at infinity and purely ingoing waves at the Schwarzschild radius:

$$R_l \to \begin{cases} A_l^{\text{out}}(\omega)\exp(i\omega r^*) \text{ as } r^* \to +\infty, \\ A_l^{\text{in}}(\omega)\exp(-i\omega r^*) \text{ as } r^* \to -\infty. \end{cases} \quad (6)$$

The energy spectrum is determined by Zerilli's formula,

$$\left(\frac{dE}{d\omega}\right)_{2^l\text{-pole}} = \frac{1}{32\pi}\frac{(l+2)!}{(l-2)!}\omega^2|A_l^{\text{out}}(\omega)|^2.$$

Two distinct methods were used to calculate $A_l^{\text{out}}(\omega)$: (i) direct integration of Eq. (1) with a numerical search technique to determine both the phase and the amplitude of the outgoing wave at infinity that would give a purely ingoing wave at the black-hole surface [details of this analysis done by two of us (M.D. and R.R.) will be published elsewhere]; (ii) integration by a Green's-function technique (see Zerilli[7]). This method allows the coefficient A_l^{out} to be computed directly as an integral involving the source term Eq. (4) and certain homogeneous solutions to Eq. (1).

All these calculations gave results in agreement within a few percent. The results are summarized in Figs. 1–3. The total energy radiated away in gravitational waves is

$$E_{\text{total}} \approx 0.0104mc^2(m/M). \quad (8)$$

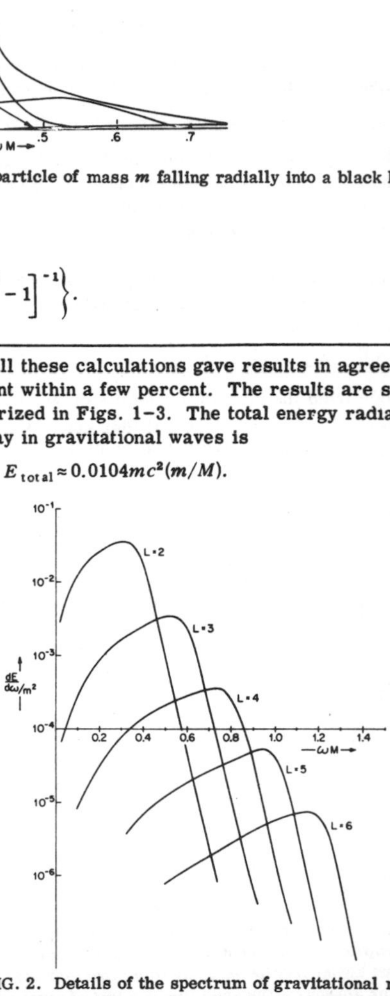

FIG. 2. Details of the spectrum of gravitational radiation integrated over all angles for the lowest five values of the multipoles.

This is about 6 times larger than Zerilli's estimate of the energy and 4 times larger than the estimate of Ruffini and Wheeler based on a purely linearized theory. The spectrum of the outgoing radiation is the superposition of a series of overlapping peaks, each peak corresponding to a certain multipole order l. Roughly 90% of the total energy is in quadrupole ($l=2$) radiation and 9% is in octupole ($l=3$). The total energy contributed by each multipole falls off quickly with l obeying the empirical relation (Fig. 3)

$$E_{2^l\text{-pole}} \approx (0.44m^2c^2/M)e^{-2l}. \tag{9}$$

The spectrum shown in Fig. 1 is for the energy integrated over all angles. An observer at a particular angle θ from the path of the particle's fall will see a slightly different spectrum because of the different angular dependence of the various 2^l-poles. For example, a pure 2^l-pole has the angular dependence

$$(dE/d\Omega)_{2^l\text{-pole}} = E_{2^l\text{-pole}}[(l-2)!/(l+2)!]\{2\partial_\theta{}^2Y_0{}^l(\theta,\varphi)+l(l+1)Y_0{}^l(\theta,\varphi)]^2. \tag{10}$$

As shown in Fig. 2, the energy contribution of progressively higher multipoles peaks at progressively higher angular frequencies, with the approximate relation

$$\omega(E_{2^l\text{-pole}},\text{peak}) \approx \{c^2[V_l(r)]_{\max}\}^{1/2} \approx lc^3(27)^{-1/2}/GM \text{ for large } l. \tag{11}$$

Each energy peak may be interpreted as due to a train of gravitational waves produced by 2^l-pole normal-mode vibrations of the black hole which the in-falling body excites (see Press[9]). Averaging over angular factors and summing the various l's, one finds that the total spectrum is peaked at $\omega=0.32c^3/GM$, and falls off at higher ω according to the empirical law

$$dE_{\text{total}}/d\omega \sim \exp(-9.9GM\omega/c^3) \tag{12}$$

Aside from the interesting details of our numerical results, the very fact that they are well behaved is important. Extrapolation of the flat-space linearized theory indicates that only a small fraction of a test body's rest mass [$\sim(m/M)mc^2$] should be converted to wave energy during "fast" parts of its orbit (parts with durations $\sim GM/c^3$). It has been an open question whether this estimate holds in the region of strong fields very near the black hole. If the estimates were wrong, our results would have been divergent, with either increasing l or increasing ω. In fact, our results are strongly convergent.

The other side of the coin is equally important: Although our computation verifies the linearized theory's dimensional estimate, it shows that a completely relativistic treatment can give quantitative amounts of gravitational radiation substantially larger than the linearized theory would predict.[10]

This research was performed independently and simultaneously at Caltech and Princeton, using different integration techniques but arriving at identical results. We thank Kip S. Thorne and Jayme Tiomno for helpful suggestions.

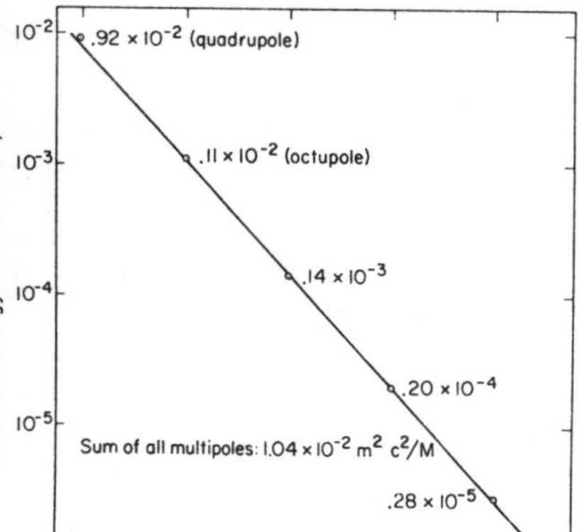

FIG. 3. Total energy radiated by each multipole. Quadrupole radiation contributes 90% of the total energy, and higher multipoles contribute progressively smaller amounts. The solid line is a plot of conste^{-2l}, an empirical fit to the data.

*Work supported in part by the National Science Foundation under Grants No. GP-19887, No. GP-28027, No. GP-27304, and No. GP-30799X.

†Fannie and John Hertz Foundation Fellow.

‡Present address: Department of Physics, University of Utah, Salt Lake City, Utah 84112.

[1]J. Weber, Phys. Rev. Lett. 22, 1320 (1969), and 24, 276 (1970), and 25, 180 (1970).

[2]K. S. Thorne, Atrophys. J. 158, 1 (1969).

[3]J. R. Ipser, Astrophys. J. 166, 175 (1970).

[4]V. de la Cruz, J. E. Chase, and W. Israel, Phys. Rev. Lett. 24, 423 (1970).

[5]S. Hawking, to be published.

[6]R. Ruffini and J. A. Wheeler, in *Proceedings of the Cortona Symposium on Weak Interactions*, edited by L. Radicati (Accademia Nazionale Dei Lincei, Rome, 1971).

[7]F. J. Zerilli, Phys. Rev. D 2, 2141 (1970).

[8]T. Regge and J. A. Wheeler, Phys. Rev. 108, 1063 (1957).

[9]W. H. Press, to be published.

[10]M. Davis, R. Ruffini, and J. Tiomno, to be published, will give further details on the intensity and pattern of radiation for this problem and more general particle orbits.

THE ASTROPHYSICAL JOURNAL, 170:L105–L108, 1971 December 15

A-9

LONG WAVE TRAINS OF GRAVITATIONAL WAVES FROM A VIBRATING BLACK HOLE

WILLIAM H. PRESS
California Institute of Technology
Received 1971 October 12

ABSTRACT

The vibrations of a black hole of mass M, perturbed from spherical symmetry, have been studied numerically. Initial perturbations of high spherical-harmonic index ($l \gg 1$) which contain Fourier components of long wavelength ($2\pi M \gtrsim \lambda \gg 2\pi M/l$) produce long-lasting vibrations. The vibrational energy is radiated away gradually as a long, nearly sinusoidal wave train of gravitational radiation with angular frequency $\omega \approx (27)^{-1/2}\, l/M$.

A Schwarzschild black hole, perturbed from spherical symmetry, will radiate gravitational waves to restore sphericity. This fact follows from the recent work of Price (1971), which applied generally to perturbations of any integer-spin, zero-rest-mass field, including gravity. The exact dynamics of this process, for gravitational perturbations, is governed by equations due to Zerilli (1970*a*, *b*) (even-parity case) and to Regge and Wheeler (1957) (odd-parity).

A priori, one might expect the black hole to divest itself of the unwanted perturbations in a single large belch, a burst of radiation of duration $\sim M$, the hole's mass or gravitational radius (units with $G = c = 1$). This Letter reports numerical computations which exhibit a totally different behavior: Initial perturbations of multipolarity $l \gg 1$ which contain Fourier components of wavelength $2\pi M/l \ll \lambda \lesssim 2\pi M$ are radiated only gradually, yielding a long and nearly sinusoidal wave train of gravitational radiation. The characteristic angular frequency ω of the wave train depends on the mass of the black hole and on the multipolarity of the perturbation, but is otherwise independent of the form of the initial perturbation: $\omega \approx (27)^{-1/2}\, l/M$. Loosely speaking, the black hole vibrates around spherical symmetry in a quasi-normal mode, and the mode is slowly damped by gravitational radiation.

The Zerilli and Regge-Wheeler equations governing black-hole perturbations have the form

$$\varphi^l{}_{,tt} - \varphi^l{}_{,r^*r^*} + V^l(r^*)\varphi^l = 0\,. \tag{1}$$

Here φ^l is a scalar quantity which describes the l-pole components of the gravitational radiation. (The components of the metric tensor are obtained by applying particular differential operators to φ^l; see Price 1971 or Thorne 1971.) The radial coordinate r^* is defined in terms of the Schwarzschild coordinate r by

$$r^* = r + 2M \ln\left(\frac{r}{2M} - 1\right). \tag{2}$$

Thus $r = 2M$ corresponds to $r^* = -\infty$, and $r = +\infty$ to $r^* = +\infty$. $V^l(r^*) \equiv \hat{V}^l(r)$ is the so-called curvature potential,

$$\hat{V}^l(r) = \begin{cases} \left(1 - \dfrac{2M}{r}\right) \dfrac{(2\Lambda^2(\Lambda + 1)r^3 + 6\Lambda^2 M r^2 + 18\Lambda M^2 r + 18M^3)}{r^3(\Lambda r + 3M)^2} & \text{even-parity (Zerilli)} \\ \left(1 - \dfrac{2M}{r}\right)\left(\dfrac{l(l+1)}{r^2} - \dfrac{6M}{r^3}\right) & \text{odd-parity (Regge-Wheeler)}\,, \end{cases} \tag{3}$$

where $\Lambda = \frac{1}{2}(l - 1)(l + 2)$. Asymptotically for large l, the even- and odd-parity potentials become identical.

To study black-hole vibrations, we choose a set of initial conditions at time $t = 0$: $\varphi^l(r^*, t = 0)$ and $\varphi^l{}_{,t}(r^*, t = 0)$ for $-\infty < r^* < +\infty$. We then solve equation (1) numerically to determine the subsequent evolution. Solutions have been computed from a variety of initial conditions, and for various values of l. It is immediately clear that initial conditions containing predominantly short-wavelength Fourier components (e.g., a narrow peak or a high-frequency sine wave) are uninteresting: the potential term in equation (1) has only a slight dispersive influence, so the perturbation is radiated outward and inward with essentially its original profile (i.e., this case *does* yield a single belch). This expected behavior has been verified numerically.

Initial perturbations of greater interest are broad, "thick" ones which contain long-wavelength Fourier components; these cannot propagate as free waves in the region of the potential. Two such initial conditions are shown in Figure 1. The curvature potential $V^l(r^*)$ is indicated by the crosshatched curve. In general $V^l(r^*)$ is peaked at about $r^* = 2M$ and drops off exponentially in the inward direction ($r^* \to -\infty$), and as r^{*-2} in the outward direction ($r^* \to \infty$). In these examples, the initial time derivative of the perturbations is chosen zero.

The subsequent evolution of the perturbations, as computed numerically, can be described and understood as follows. At any given r^*, the perturbation initially oscillates (no propagation leftward or rightward!) with angular frequency approximately $[V^l(r^*)]^{1/2}$. Since $V^l(r^*)$ varies with r^*, the oscillations soon become out of phase from point to point, and the initially smooth perturbation builds up components of ever shortening wavelength. When wavelengths as short as the critical value $2\pi[V^l(r^*)]^{-1/2}$ have developed, the perturbations begin to propagate as free waves out of the region of the potential. For small l (say 2 or 3 or 4), the potential is low and the free propagation is almost immediate (single belch); but for large l the shortening process is gradual, and

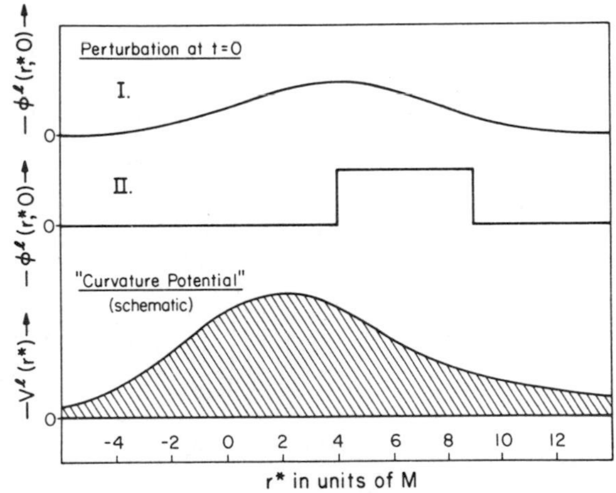

FIG. 1.—Two interesting perturbations of a Schwarzschild black hole. The initial radial "wave forms" of the perturbations are shown in I and II. Their initial time derivatives are assumed zero, and their angular dependence is a spherical harmonic of order l. The perturbations are spread out broadly over the region of strong "curvature potential" $V^l(r^*)$, so spacetime curvature prevents them from propagating until they have developed a wave form containing wavelengths shorter than the characteristic length $2\pi[V^l(r^*)]^{-1/2}$ (see text for discussion).

long wave trains are emitted, of characteristic angular frequency

$$\omega \approx [V^l(r^*)]_{\max}{}^{1/2} \approx \frac{l}{(27)^{1/2}M}\,. \tag{4}$$

Figure 2 shows the profile of the propagating gravitational wave trains at large t for the two inital conditions of Figure 1 and the two multipolarities $l = 20$ and $l = 40$. The estimate of equation (4) is seen to be approximately correct. The length of the wave train depends somewhat on the precise initial conditions chosen, but seems to be rather independent of l. These characteristics are typical of our numerical results in general; but we are able to give no analytic estimate for the train length.

How much of the perturbation radiates down the hole instead of off to infinity? A simple rule of thumb summarizes all our numerical calculations: The quantity

$$\mathfrak{E} = |\varphi^l{}_{,t}|^2 + |\varphi^l{}_{,r^*}|^2 + V^l(r^*)\,|\varphi^l|^2 \tag{5}$$

is a mathematical energy density which is exactly conserved by the evolution of equation (1). Measured in terms of $\mathfrak{E}$, that long-wavelength energy initially located outside the potential maximum is typically radiated outward; that energy initially inside the potential maximum goes down the hole. Thus, the offset of initial conditions I and II from the potential maximum results in most of the energy radiating outward ($\sim$80 or 90 percent).

We emphasize that the phenomenon here exhibited, the "free oscillation of a black

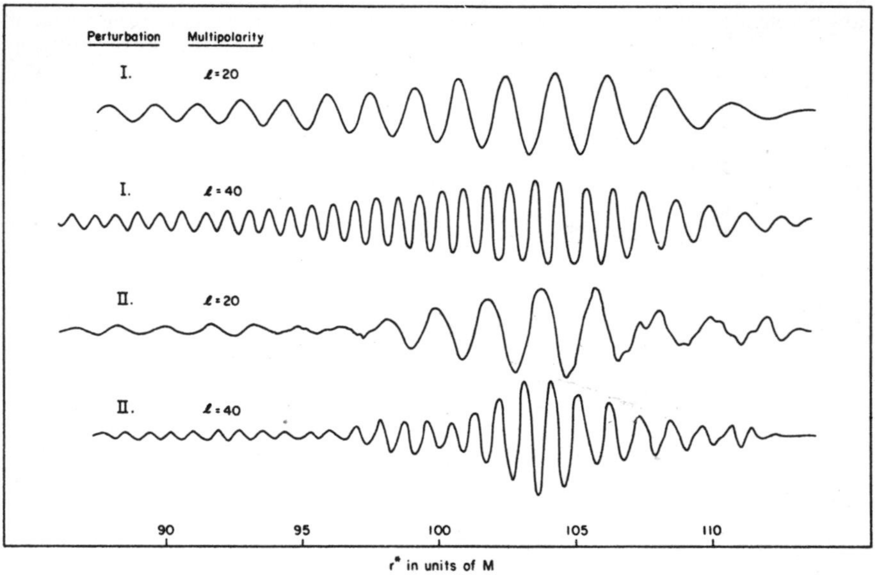

FIG. 2.—Wave forms $\varphi^l(r^*, t)$ of gravitational radiation at large r^* and fixed t, as produced by the initial perturbations I and II of Fig. 1 for $l = 20$ and $l = 40$. The waves are propagating rightward, away from the region of strong "curvature potential," where they originated. Since "wavelength shortening" in the region of the potential proceeds gradually, the waves have the form of a long sinusoidal train of angular frequency $\omega \simeq [V^l(r^*)_{\max}]^{1/2} \simeq (27)^{-1/2}\, l/M$. This frequency can be interpreted as the "vibration" frequency of the black hole (see text).

hole," is distinct from the curvature-potential effect studied by Price (1971) and Fackerell (1971) in which the potential acts as a high-pass filter of gravitational radiation.

The free oscillations of a bell are initiated by a mechanical blow; the Earth's free oscillations are excited by large earthquakes. What processes can induce a black hole to oscillate, i.e., can supply the initial perturbation which we have supplied by fiat in our numerical calculations? Recent calculations by Davis *et al.* (1971) show that vibrational modes are excited—though weakly for high l—by a test particle falling radially into a black hole. (In fact, the entire calculated spectrum can be understood qualitatively as a superposition of such vibrations.) Whether high-l vibrations can be excited *preferentially* by some other pattern of infalling matter is a problem—presently unsolved—of considerable astrophysical relevance. In some cases symmetry considerations can at least inhibit low multipole radiation. For example, the turbulent influx of matter into a black hole might produce perturbations whose dominant multipolarity is determined by the size L of the turbulent cell $l \sim 2\pi M/L \gg 1$.

The essential point of this Letter is that a black hole can be a dynamical entity rather than merely an arena for dynamics. This new point of view suggests new directions of research: How does the rotation of a black hole affect its vibrations? *Are* black-hole vibrations excited significantly by natural astrophysical processes? Might they play a significant role as sources of gravitational radiation?

If Weber's (1969, 1970*a*, *b*) observed gravitational radiation is verified and found to have a highly oscillatory wave form (indicating vibrations of large l as a possible source), black-hole vibrations will become a strong candidate for explaining the observations. Vibration is a mechanism by which "short"-wavelength gravitational radiation can be emitted by a black hole of large mass, so there is no limit in principle on the mass of a black hole which radiates at the frequency of Weber's detection apparatus.

I am pleased to thank Dr. Richard Price and Professor Kip S. Thorne for their invaluable assistance and encouragement. Dr. Remo Ruffini provided helpful suggestions. I thank the Fannie and John Hertz Foundation for their support. This work was supported in part by the National Science Foundation [GP-27304 and GP-28027].

REFERENCES

Davis, M., Ruffini, R., Press, W., and Price, R. 1971, *Phys. Rev. Letters*, **27**, 1466.
Fackerell, E. D. 1971, *Ap. J.*, **166**, 197.
Price, R. H. 1971, paper submitted to *Phys. Rev.*
Regge, T., and Wheeler, J. A. 1957, *Phys. Rev.*, **108**, 1063.
Thorne, K. S. 1971, *Nonspherical Gravitational Collapse—A Short Review* (in press).
Weber, J. 1969, *Phys. Rev. Letters*, **22**, 1320.
———. 1970*a*, *ibid.*, **24**, 276.
———. 1970*b*, *ibid.*, **25**, 180.
Zerilli, F. J. 1970*a*, *Phys. Rev. Letters*, **24**, 737.
———. 1970*b*, *Phys. Rev.*, **D2**, 2141.

PHYSICAL REVIEW D VOLUME 5, NUMBER 12 15 JUNE 1972

Pulses of Gravitational Radiation of a Particle Falling Radially into a Schwarzschild Black Hole*

Marc Davis, Remo Ruffini, and Jayme Tiomno†
Joseph Henry Laboratories, Princeton University, Princeton, New Jersey 08540
(Received 20 December 1971)

Using the Regge-Wheeler-Zerilli formalism of fully relativistic linear perturbations in the Schwarzschild metric, we analyze the radiation of a particle of mass m falling into a Schwarzschild black hole of mass $M \gg m$. The detailed shape of the energy pulse and of the tide-producing components of the Riemann tensor at large distances from the source are given, as well as the angular distribution of the radiation. Finally, analysis of the energy going down the hole indicates the existence of a divergence; implications of this divergence as a testing ground of the approximation used are examined.

In a recent series of investigations Zerilli,[1] Davis and Ruffini,[2] and Davis, Ruffini, Press, and Price[3] have analyzed the problem of a particle falling radially into a Schwarzschild black hole. In this paper this process is analyzed further. We are concerned with the features of the burst of the components of the Riemann tensor significant in the use of a detector and of the angular distribution of gravitational radiation. General and apparently contradictory considerations on the structure of a burst of gravitational radiation in black-hole physics were presented by Gibbons and Hawking[4] and by Press.[5] Some of the major features predicted in these two treatments are found indeed to be present in the detailed analysis of the physical example under consideration. An analysis for the radiation going into the hole is presented and its implications are examined.

We can expand[6] the perturbations $h_{\mu\nu} = g_{\mu\nu} - (g_{\mu\nu})_{\text{Schw}}$ of a Schwarzschild background in spherical harmonics of multipole orders l and m. In our case (a particle falling radially in along the z axis starting at rest from infinity) the "magnetic" and the $m \neq 0$ "electric" terms identically vanish (see Zerilli[1]). We have in this case for large values of the radial distance, in Zerilli's radiation gauge,

$$h_{\mu\nu}(t, r, \theta, \phi) \sim p_{\mu\nu} \sum_l R_l(r, t) \left(\frac{\partial^2}{\partial\theta^2} - \cot\theta \frac{\partial}{\partial\theta} \right) Y_{l0}(\theta, \phi)/2r \qquad (\mu, \nu = 0, 3). \quad (1)$$

Here $p_{\mu\nu}$ is the polarization tensor with the only nonvanishing components $p_{22} = r^2$ and $p_{33} = -r^2 \sin^2\theta$. Thus the outgoing radiation will be totally polarized with the principal axes in the θ and ϕ directions. The function $R_l(r, t)$ satisfies the Zerilli equation which in Fourier-transformed form gives

$$\frac{d^2R_l(r, \omega)}{dr^{*2}} + [\omega^2 - V_l(r)]R_l(r, \omega) = S_l(r, \omega). \qquad (2)$$

Here $r^* = r + 2M \ln(r/2M - 1)$, $V_l(r)$ is the effective curvature potential, and $S_l(r, \omega)$ is the Fourier-transformed electric source term generated by the incoming particle expressed in tensor harmonics.[2,3] Equation (2) has been numerically integrated with the asymptotic boundary conditions

$$R_l(r, \omega) = \begin{cases} A_l^{\text{out}}(\omega) e^{i\omega r^*} & \text{as } r^* \to +\infty \\ A_l^{\text{in}}(\omega) e^{-i\omega r^*} & \text{as } r^* \to -\infty, \end{cases} \qquad (3)$$

where A_l^{out} is given in the Green's function technique by

$$A_l^{\text{out}}(\omega) \propto \int_{-\infty}^{\infty} u_l(r^*, \omega) S_l(r^{*\prime}, \omega)\, dr^{*\prime}, \qquad (4a)$$

$$A_l^{\text{in}}(\omega) \propto \int_{-\infty}^{\infty} v_l(r^{*\prime}, \omega) S_l(r^{*\prime}, \omega)\, dr^{*\prime}. \qquad (4b)$$

Here u_l (v_l) is the solution of the homogeneous equation obtained from (2) specifying a purely ingoing wave at $r^* = -\infty$ (outgoing at $r^* = +\infty$). By a further Fourier transformation we obtain the asymptotic expression

$$R_l^{\text{out}}(r^*, t) = \int_{-\infty}^{\infty} A_l^{\text{out}}(\omega) e^{i\omega(r^* - t)}\, d\omega. \qquad (5)$$

The explicit results of this integration are given in Fig. 1(b) as a function of the retarded time $t - r^*$ for $l = 2$.

The asymptotic expression of the tide-producing components of the Riemann tensor, which is what is measured by gravitational-wave detectors,[7] is easily obtained in the radiation region from

$$R_{\alpha\beta\gamma\delta}=\tfrac{1}{2}(h_{\alpha\delta,\beta\gamma}+h_{\beta\gamma,\alpha\delta}-h_{\alpha\gamma,\beta\delta}-h_{\beta\delta,\alpha\gamma}), \qquad (6)$$

where the comma (,) means ordinary derivative. If we assume the z axis is pointed along the line of propagation of the wave and the principal axes of polarization [see Eq. (1)] are pointed in the x and y directions, the only nonzero Newtonian tide-producing components of the Riemann tensor are, in our problem, $R^{y}{}_{0y0}=-R^{x}{}_{0x0}$, with

$$R^{y}{}_{0y0}(r^*,t)=\sum_{l}\ddot{R}_{l}(r^*,t)W_{l}(\theta,\phi)/2r\,. \qquad (7)$$

Here the dot indicates normal derivative with respect to time. The components of the Riemann tensor for selected l, without their angular dependence factor

$$W_{l}(\theta)=\left(\frac{\partial^2}{\partial\theta^2}-\cot\theta\frac{\partial}{\partial\theta}\right)Y_{l0}(\theta), \qquad (8)$$

and the factors $1/(8\pi)^{1/2}r$ are plotted in Fig. 1(c). Finally, we have also computed the outgoing energy flux from the stress-energy pseudotensor which becomes in the asymptotic region

$$t_{01}\sim\frac{1}{16\pi}\sum_{ll'}\dot{R}_{l}\dot{R}_{l'}W_{l}(\theta,\phi)W_{l'}(\theta,\phi)/4r^2\,. \qquad (9)$$

In Fig. 1(d) we give the outgoing energy flux integrated over all directions for selected values of l. The interference between terms of different l is zero due to the orthogonality of the functions $W_l(\theta,\phi)$. As a check on our entire treatment we have verified that the total energy $\int_{-\infty}^{+\infty}(dE/dt)\,dt$ for every l agrees with the value $\int_{-\infty}^{\infty}(dE/d\omega)\,d\omega$ as given in Ref. 3 within 1%. From (7) we can compute the total flux per steradian; this quantity is plotted in Fig. 2. For pure quadrupole radiation the angular pattern of the radiation has a $\sin^4\theta$ dependence (z axis $\rightarrow\theta=0$). Inclusion of higher multi-

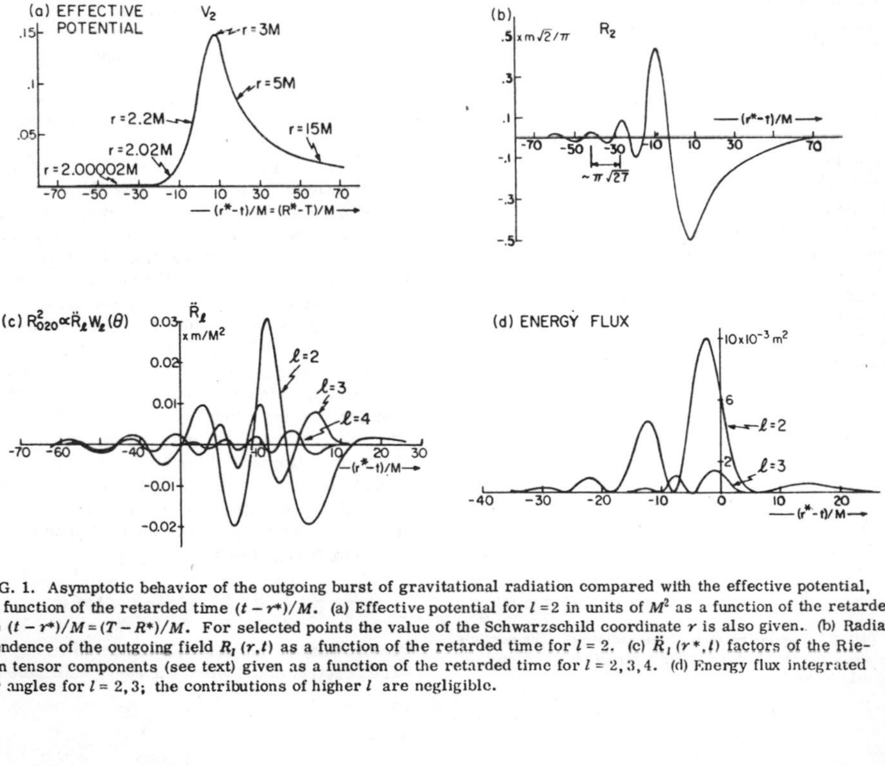

FIG. 1. Asymptotic behavior of the outgoing burst of gravitational radiation compared with the effective potential, as a function of the retarded time $(t-r^*)/M$. (a) Effective potential for $l=2$ in units of M^2 as a function of the retarded time $(t-r^*)/M=(T-R^*)/M$. For selected points the value of the Schwarzschild coordinate r is also given. (b) Radial dependence of the outgoing field $R_l(r,t)$ as a function of the retarded time for $l=2$. (c) $\ddot{R}_l(r^*,t)$ factors of the Riemann tensor components (see text) given as a function of the retarded time for $l=2,3,4$. (d) Energy flux integrated over angles for $l=2,3$; the contributions of higher l are negligible.

poles introduces interference terms which tip the peak of the pattern forward by $7\frac{1}{2}°$. Figure 2 shows that there is no beaming of the radiation.

From the comparison of the different diagrams in Fig. 1 we can distinguish and characterize three different regions in the total energy flux:

(i) $5 \lesssim (r^* - t)/M \lesssim 30$, a precursor,

(ii) $-10 \lesssim (r^* - t)/M \lesssim 5$, a sharp burst,

(iii) $(r^* - t)/M \lesssim -10$, a ringing tail.

The precursor corresponds to the first part of the pulse as produced in the Ruffini-Wheeler approximation.[8] The sharp burst has a width ~$10M$ in agreement with the predictions on qualitative ground by Gibbons and Hawking[4] referring to any emission process taking place during the formation of (or capture by) a collapsed object. However, the present results do not support their suggestion that the "number of zeros" of the Riemann tensor could discriminate between sources of different origin since the ringing tail produces many zero crossings of the Riemann tensor. Finally, the oscillating tail has characteristic frequencies $\omega \sim l/\sqrt{27}$ which correspond to the ringing modes of the black hole found by Press.[5] It is interesting, however, that these ringing modes are energetically significant in this physical example only for low values of l, as is clear from Fig. 1(d).

A deeper insight in the three regions (i), (ii), and (iii) can be gained by the study of the effective potential plotted in Fig. 1 as a function of the retarded time $(t - r^*)/M$ referred to the observer at large distances. This is equal to $(T - R^*)/M$ as "seen" by the ingoing particle as the outgoing wave sweeps past it. R^* and T are the particle's position and Schwarzschild coordinate time computed from the geodesic trajectory (with $T = -\infty$ at $R^* = +\infty$, and $T = +\infty$ at $R^* = -\infty$). The implication is that for radiation "directly" emitted outward from the particle and not reflected, one can specify the radial position of the particle when it supposedly "emits" this radiation. Notice, for example, that the peak of the radiation flux occurs near retarded time $(r^* - t)/M = -2$, when the particle itself is at $2.3M$ (very near the horizon indeed), and just inside the peak of the curvature potential, which peaks at $r = 3M$.

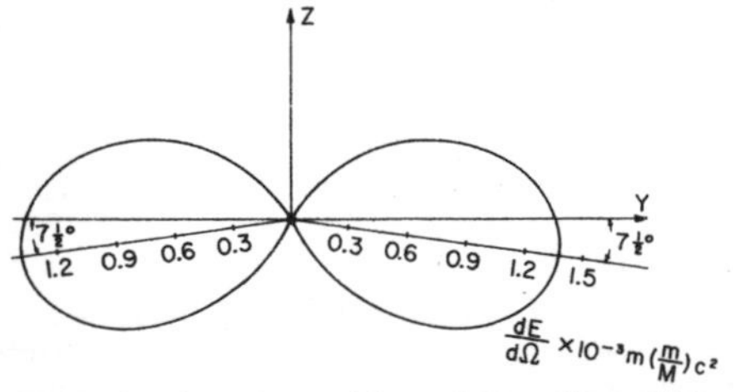

FIG. 2. Angular pattern of the radiation. The particle is supposed to fall down in the z direction which corresponds to $\theta = 0$.

We see that starting from large values of $(r^* - t)/M$ the field $R_l(r, t)$ builds up slowly and thus the energy emission (proportional to $\dot{R}_l^2$) and the Riemann tensor (proportional to $\ddot{R}_l$) gives rise to the very small "precursor" as the particle approaches the effective potential barrier. Concerning the emission of the main burst we have noticed in the evaluation of the integral (4a) for $A_l^{\text{out}}(\omega)$ starting from $r = 2M$ that the main contribution came from the interval $2.1 \lesssim r/M \lesssim 10$. The contributions beyond this point were oscillating, and very slowly damped, since the source decreases only as $r^{-1/2}$ for asymptotic distances. This averaging out of the contributions for large values of r is expected from the linearized theory of gravitation. Note that the ringing comes out after the particle is already inside the barrier. Here we cannot be seeing direct radiation from the particle because the driving source S_l is exponentially decaying, and the contributions to $A_l^{\text{out}}(\omega)$ for $2 \lesssim r/M \lesssim 2.1$ are very small. The wave emitted in this region for a given mode has a characteristic frequency as expected from the frequency spectrum calculated previously.[3] These facts suggest that part of the energy produced in the strong-burst region (ii) was stored in the "resonant cavity" of the geometry and then slowly released in the ringing modes.

We can now briefly summarize the main results of the analysis of the radiation going into the black hole. We have proceeded as follows: (1) Evaluate the amplitude A of the ingoing wave $R_l(r^*, \omega)$ for $r^* \to -\infty$ (purely ingoing waves), (2) solve for the scattering problem of Eq. (2) without source, imposing a purely ingoing wave of amplitude A at $r^* = -\infty$. As a consequence at $r^* = +\infty$ we have an ingoing wave with amplitude B and an outgoing wave with amplitude C. The energy flux going into the black hole is evaluated at $r^* = +\infty$, subtracting from the energy going in (proportional to $|B|^2$) the energy coming out (proportional to $|C|^2$). From the structure of the homogeneous Eq. (2) we also have $|A|^2 = |B|^2 - |C|^2$; therefore we can evaluate the energy going into the black hole simply by the same expression as used to calculate outgoing energy flux[1]:

$$\frac{dE}{d\omega} = \frac{1}{32\pi} \sum_l l(l-1)(l+1)(l+2)\omega^2 |A_l^{\text{in}}|^2 . \tag{10}$$

The results of this analysis are given in Fig. 3. The spectral distribution for every multipole is

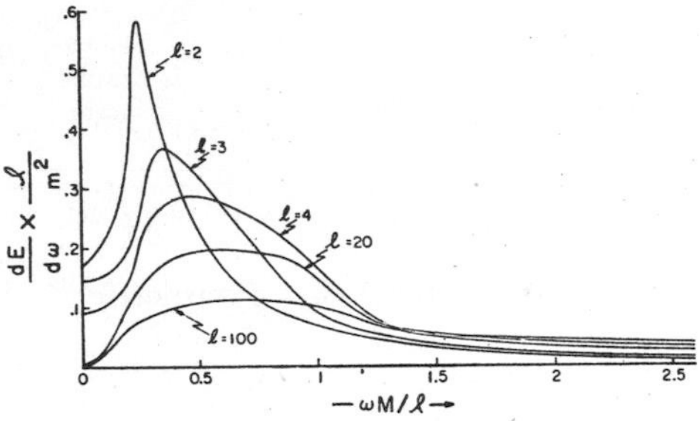

FIG. 3. Energy spectrum of the radiation going into the black hole for selected values of l. The total energy radiated per multipole is roughly constant and $\sim 0.25m^2/M$ for all considered l.

well behaved, and the total energy per multipole is roughly constant $\sim 0.25m^2/M$ (at least up to $l = 100$). This is in contrast to the case of outgoing radiation, where the higher multipoles were exponentially damped.[3] It is plausible to assume that this behavior is indeed valid even for larger l. At first it would seem that the energy summed over all l diverges. This divergence results from the fact that we are treating the incoming object as a point particle. Under this assumption the Regge-Wheeler condition that the perturbation introduced by the particle be small by comparison with the background metric is no longer fulfilled. However, this circumstance is automatically eliminated as soon as a minimum size is assumed for the particle. We take as minimum size $2m$. From the preceding results we have seen that most of the radiation is emitted in the region near the horizon. This and the assumption of the minimum size $2m$ for the particle implies a cutoff in l given by $l_{max} \simeq \frac{1}{2}\pi M/m$. The total energy going in is thus of the order

$$\frac{m^2}{4M}\frac{\pi M}{2m} \sim \frac{\pi m}{8}.$$

It is remarkable that the ingoing radiation does not depend, as does the outgoing radiation, on the ratio m/M, and that the amount of this radiation is a sizable fraction to the total rest mass of the ingoing particle, for an incident particle of minimum size.

Study of the numerical integration of $A_l^{in}(\omega)$ shows that for calculations of ingoing radiation, one can neglect contributions beyond $r = 5M$. Most of the ingoing radiation is generated inside the barrier, where $\omega^2 - V_l(r) < 0$, typically in the interval $2.01 \lesssim r/M \lesssim 3.5$. This large inward burst of energy is therefore generated in a finite Schwarzschild coordinate time interval; it occurs for $-20 < (r^* - t)/M < 15$ and vanishes as $(r^* - t)/M \to -\infty$. It occurs early enough that its reaction on the geodesic path of the incoming particle could affect the nature of the outgoing radiation, because the integral for $A_l^{out}(\omega)$ has significant contributions beginning as $r \gtrsim 2.1M$. Further analysis may give a deeper understanding of this process, and details on exactly how the reaction of ingoing radiation affects the outgoing burst.

*Work partially supported by the National Science Foundation under Grant No. 30799X.

†At the Institute for Advanced Study, Princeton, N. J., when this work was initiated.

[1]F. Zerilli, Phys. Rev. D 2, 2141 (1970).

[2]M. Davis and R. Ruffini, Lett. Nuovo Cimento 2, 1165 (1971).

[3]M. Davis, R. Ruffini, W. Press, and R. Price, Phys. Rev. Letters 27, 1466 (1971).

[4]G. Gibbons and S. Hawking, Phys. Rev. D 4, 2191 (1971).

[5]W. Press, Astrophys. J. Letters 170, L105 (1971).

[6]T. Regge and J. A. Wheeler, Phys. Rev. 108, 1063 (1957).

[7]J. Weber, *General Relativity and Gravitational Waves* (Interscience, New York, 1961).

[8]R. Ruffini and J. A. Wheeler, in *Proceedings of the Cortona Symposium on Weak Interactions*, edited by L. Radicati (Accademia Nazionale die Lincei, Rome, 1971).

A-11

Electromagnetic Field of a Particle Moving in a Spherically Symmetric Black-Hole Background (*).

R. RUFFINI and J. TIOMNO

Joseph Henry Physical Laboratories - Princeton, N. J.

C. V. VISHVESHWARA

New York University - New York, N. Y.

(ricevuto il 22 Novembre 1971)

Very much has been speculated recently on the possible existence of collapsed objects, usually called « black holes », and on their possible observation. CHRISTODOULOU and RUFFINI (¹) have shown that « black holes », which had been considered for a long time to be merely sinks for radiation and other forms of energy, could in fact yield up to 50% of their total energy under favourable circumstances. For these reasons it was justified to think of black holes as the strongest storehouses of energy in the Universe (²). The search for these objects should therefore be directed toward a careful examination of strongly energetic events. From the theoretical point of view a detailed analysis of accretion processes has to be undertaken. Some aspects of this analysis (in the « one-particle approximation »!) have been examined for the emission of gravitational radiation by ZERILLI (³), by DAVIS and RUFFINI (⁴), by DAVIS, RUFFINI, PRESS and PRICE (⁵), and by DAVIS, RUFFINI and TIOMNO (⁶). Some related aspects concerning the scattering of gravitational waves from black holes have been given by VISHVESHWARA (⁷). In this paper the theoretical basis for the analysis of the electromagnetic radiation emitted by a charge moving in the gravitational field of spherically symmetric black holes is presented.

(*) Work partially supported by NSF GP-30799X and NSF GU-3186.

(¹) D. CHRISTODOULOU and R. RUFFINI: *Phys. Rev.*, in press.

(²) D. CHRISTODOULOU and R. RUFFINI: *Back Holes, the largest-energy storehouse in the Universe*, in *Gravity Foundation, 1971*, in press.

(³) F. ZERILLI: *Phys. Rev. D.*, **2**, 2141 (1970).

(⁴) M. DAVIS and R. RUFFINI: *Lett. Nuovo Cimento*, in press.

(⁵) M. DAVIS, R. RUFFINI, W. PRESS and R. PRICE: *Phys. Rev. Lett.*, **21**, 1466 (1971).

(⁶) M. DAVIS, R. RUFFINI and J. TIOMNO: submitted for publication.

(⁷) C. V. VISHVESHWARA: *Nature*, **227**, 936 (1970).

ISRAEL ([8]) has shown that the most general spherically symmetric collapsed objects with closed simply connected horizon is the one given by the Reissner Nordström metric, which in Schwarzschild-like co-ordinate assumes the form

$$ds^2 = g_{\mu\nu}dx^\mu dx^\nu = -\left(1-\frac{2M}{r}+\frac{Q^2}{r^2}\right)dt^2+\left(1-\frac{2M}{r}+\frac{Q^2}{r^2}\right)^{-1}dr^2+r^2(d\theta^2+\sin^2\theta\, d\varphi^2), \tag{1}$$

here $\mu, \nu = 0, 3$ and $G = c = 1$, M is the mass and Q the charge of the given background metric.

We consider here a black hole with $Q \leqslant M$. We analyze the electromagnetic field generated by a particle of mass m and charge q moving in the above metric (1) under the assumptions that the charge $q \ll M$ and the mass $m \ll M$. The electromagnetic vector potential associated with the moving particle can be expanded in terms of four-dimensional vector spherical harmonics obtained from scalar spherical harmonics by means of the following operations:

$$\frac{\boldsymbol{r}}{r}Y^{lm}(\theta,\varphi)\,, \tag{2a}$$

$$\boldsymbol{\nabla} Y^{lm}(\theta,\varphi)\,, \tag{2b}$$

$$\boldsymbol{L} Y^{lm}(\theta,\varphi)\,, \tag{2c}$$

$$e_t Y^{lm}(\theta,\varphi) = (Y^{lm}, 0, 0, 0)\,. \tag{2d}$$

Here as usual $\boldsymbol{L}$ is the angular-momentum operator and $\boldsymbol{\nabla}$ the gradient. The parity of eq. (2.a), (2.b), (2.d) is $(-1)^l$ (electric) and the parity of eq. (2.c) is $(-1)^{l+1}$ (magnetic). We obtain then

$$A_\mu(r,\theta,\varphi,t) = \sum_{lm}\left(\begin{bmatrix} 0 \\ 0 \\ \dfrac{a^{lm}(r,t)}{\sin\theta}\dfrac{\partial Y^{lm}}{\partial\varphi} \\ -a^{lm}(r,t)\sin\theta\dfrac{\partial Y^{lm}}{\partial\theta}\end{bmatrix} + \begin{bmatrix} f^{lm}(r,t)\,Y^{lm} \\ h^{lm}(r,t)\,Y^{lm} \\ k^{lm}(r,t)\dfrac{\partial Y^{lm}}{\partial\theta} \\ k^{lm}(r,t)\dfrac{\partial Y^{lm}}{\partial\varphi}\end{bmatrix}\right) \tag{3}$$

The covariant Maxwell equations to be fulfilled in the given background are given by

$$F^{\mu\nu}{}_{;\nu} = 4\pi J^\mu \qquad \text{or} \qquad (\sqrt{-g}\,F^{\mu\nu})_{,\nu} = \sqrt{-g}\,4\pi J^\mu\,. \tag{4}$$

In the above equation; (,) indicates covariant (ordinary) derivative and $g = \det g_{\alpha\beta}$. Further we have

$$F_{\mu\nu} = A_{\nu,\mu} - A_{\mu,\nu}\,. \tag{5}$$

([8]) W. ISRAEL: *Phys. Rev.*, **164**, 1776 (1967).

The four-current j^μ can itself be expanded in terms of vector harmonics

$$(6)\qquad 4\pi J_\mu = \sum_{l,m}\left(\begin{bmatrix} 0 \\ 0 \\ \dfrac{\alpha^{lm}(r,t)}{\sin\theta}\dfrac{\partial Y^{lm}}{\partial\varphi} \\ -\alpha^{lm}(r,t)\sin\theta\dfrac{\partial Y^{lm}}{\partial\theta}\end{bmatrix} + \begin{bmatrix}\Psi^{lm}(r,t)\,Y^{lm} \\ \eta^{lm}(r,t)\,Y^{lm} \\ \chi^{lm}(r,t)\dfrac{\partial Y^{lm}}{\partial\theta} \\ \chi^{lm}(r,t)\dfrac{\partial Y^{lm}}{\partial\varphi}\end{bmatrix}\right).$$

The nonvanishing components of the electromagnetic-field tensor $F^{\mu\nu}$ of parity $(-1)^{l+1}$ are given by

$$(7)\qquad \begin{cases} F^{0\theta}_{lm} = g^{00}a^{lm}_{,0}Y^{lm}_{,\varphi}/(r^2\sin\theta)\,;\; F^{0\varphi}_{lm} = -g^{00}a^{lm}_{,0}Y^{lm}_{,\theta}/(r^2\sin\theta)\,, \\ F^{r\theta}_{lm} = g^{rr}a^{lm}_{,r}Y^{lm}_{,\varphi}/(r^2\sin\theta)\,;\; F^{r\varphi}_{lm} = -g^{rr}a^{lm}_{,r}Y^{lm}_{,\theta}/(r^2\sin\theta)\,, \\ F^{\theta\varphi}_{lm} = l(l+1)\,a^{lm}/(r^4\sin\theta)\,. \end{cases}$$

Equation (4) gives rise to a single differential equation for the radial function

$$(8)\qquad (g^{rr}a^{lm}_{,r})_{,r} - g_{rr}\frac{\partial^2 a^{lm}}{\partial t^2} - \frac{l(l+1)}{r^2}a^{lm} = \alpha^{lm}\,.$$

The other equations obtained from eqs. (4) are either identically satisfied or equivalent to (8). Similarly, the nonvanishing components of $F^{\mu\nu}$ with parity $(-1)^l$ are

$$(9)\qquad \begin{cases} F^{0r}_{lm} = (f^{lm}_{,r} - h^{lm}_{,0})\,Y^{lm}\,; \qquad F^{0\theta}_{lm} = g^{00}(k^{lm}_{,0} - f^{lm})\,Y^{lm}_{,\theta}/r^2\,, \\ F^{0\varphi}_{lm} = g^{00}(k^{lm}_{,0} - f^{lm})\,Y^{lm}_{,\varphi}/(r^2\sin\theta)\,, \\ F^{r\theta}_{lm} = g^{rr}(k^{lm}_{,r} - h^{lm})\,Y^{lm}_{,\theta}/r^2\,; \qquad F^{r\varphi}_{lm} = g^{rr}(k^{lm}_{,r} - h^{lm})\,Y^{lm}_{,\varphi}/(r^2\sin\theta)\,, \end{cases}$$

and from eqs. (4) we obtain now the following set of differential equations for the radial parts:

$$(10a)\qquad g_{00}[r^2(f^{lm}_{,r} - h^{lm}_{,0})]_{,r} - l(l+1)\,(k^{lm}_{,0} - f^{lm}) = \Psi^{lm}r^2\,,$$

$$(10b)\qquad g_{rr}(h^{lm}_{,0} - f^{lm}_{,r})_{,0} - \frac{l(l+1)}{r^2}(k^{lm}_{,r} - h^{lm}) = \eta^{lm}\,,$$

$$(10c)\qquad [(h^{lm} - k^{lm}_{,r})g^{rr}]_{,r} - (k^{lm}_{,0} - f^{lm})_{,0}g^{00} = \chi^{lm}\,.$$

The remaining equations obtained from (4) are either identically satisfied or equivalent to eqs. (10). The equation of conservation of current to be identically fulfilled is

$$J^\mu_{;\mu} = 0$$

This gives the subsidiary equation

$$(11)\qquad \frac{1}{r^2}(r^2g^{rr}\eta^{lm})_{,r} + \Psi^{lm}_{,0}g^{00} = \frac{l(l+1)}{r^2}\chi^{lm}\,.$$

We introduce a new function $b^{lm}(r, t)$ defined by the following equation:

$$h^{lm}_{,0} - f^{lm}_{,r} = \frac{l(l+1)}{r^2} b^{lm}, \tag{12}$$

and substituting in the expressions (10*a*) and (10*b*) we get

$$g^{rr} b^{lm}_{,0} = k^{lm}_{,0} - f^{lm} + \frac{r^2 \Psi^{lm}}{l(l+1)}, \tag{12a}$$

$$g^{00} b^{lm}_{,0} = h^{lm} - k^{lm}_{,r} - \frac{r^2 \eta^{lm}}{l(l+1)}. \tag{12b}$$

With the help of eqs. (12) we find that the eq. (10*c*) is identically satisfied in view of (11). Further, integrability condition of these equations is summarized in the compact and simple differential equation

$$(g^{rr} b^{lm}_{,r})_{,r} + g^{00} b^{lm}_{,00} - \frac{l(l+1)}{r^2} b^{lm} = \frac{1}{l(l+1)} [(r^2 \Psi^{lm})_{,r} - \eta^{lm}_{,0} r^2]. \tag{13}$$

Equations (13) and (8) take the same form in vacuum, in which case they were first given by WHEELER (9). Equation (13) solves completely the problem of the determination of the electric multipole expansion as $F^{\mu\nu}_{lm}$ given by (9) is expressed in terms of b^{lm}, $b^{lm}_{,0}$, $b^{lm}_{,r}$, Ψ^{lm} and η^{lm}. Both eqs. (13) and (8) have to be solved for any given J^μ by assuming as boundary conditions purely ingoing waves at the surface of the black hole.

The four-current J^μ in the case of a point charge moving in the given background. is given by

$$J^\mu = \frac{q}{\sqrt{-g}} \frac{dz^\mu}{dt} \delta(x - z(t)), \tag{14}$$

$z^\mu \equiv (t, z) \equiv (t, R(t), \Theta(t), \varphi(t))$ describes the trajectory of the particle as given by the equation

$$\frac{d}{ds}\left(\frac{dz^\mu}{dt}\frac{dt}{ds}\right) = F^\mu{}_\alpha \frac{dz^\alpha}{dt}\frac{dt}{ds} + \Gamma^\mu_{\alpha\beta}\frac{dz^\alpha}{dt}\frac{dz^\beta}{dt}\left(\frac{dt}{ds}\right)^2.$$

Both $F^\mu{}_\alpha$ and $\Gamma^\mu_{\alpha\beta}$ are given by the background geometry. From (4) and (6) we obtain

$$\Psi^{lm}(r, t) = \frac{q}{r^2 g^{00}} \delta(r - R) Y^{lm*}(\Theta, \Phi), \tag{15a}$$

$$\eta^{lm}(r, t) = \frac{q}{r^2 g^{rr}} \frac{dR}{dt} \delta(r - R) Y^{lm*}(\Theta, \Phi), \tag{15b}$$

$$\alpha^{lm}(r, t) = \frac{q}{l(l+1)} \left[-\frac{d\Phi}{dt} \sin\Theta\, Y^{lm*}{}_{,\Theta} + \frac{1}{\sin\Theta}\frac{d\Theta}{dt} Y^{lm*}{}_{,\Phi}\right] \delta(r - R). \tag{15c}$$

These are the relevant source terms to be used when solving eqs. (8) and (13).

(9) J. A. WHEELER: *Geometrodynamics* (New York, 1962), p. 203. Similar results for the sourceless case were also obtained by L. FAVELLA - *Tesi di Laurea*, 1957, Torino University. We thank Prof. T. REGGE to have pointed out this interesting work.

Finally we have to give the expression for the energy radiated. The tensor momentum energy for the electromagnetic field is

$$T^{\mu}{}_{\nu} = (F^{\mu\varrho} F_{\varrho\nu} + \tfrac{1}{4}\delta^{\mu}{}_{\nu} F^{\varrho\sigma} F_{\varrho\sigma})/4\pi \,.$$

The flux of energy through a surface S is given by the expression

$$\left(\frac{\mathrm{d}E}{\mathrm{d}t}\right)_S = \int_S T^r_0 r^2 \sin\theta \, \mathrm{d}\theta \, \mathrm{d}\varphi \,. \tag{16a}$$

Here

$$T^r_0 r^2 \sin\theta = -g^{rr} \sum_{lm} \sum_{l'm'} [a^{lm}{}_{,r} a^{l'm'}{}_{,0} + (k^{lm}_{,r} - h^{lm})(k^{l'm'}{}_{,0} - f^{l'm'})] \, (Y^{lm}_{,\theta} Y^{l'm'}_{,\theta} \sin\theta + \tag{16b}$$
$$+ Y^{lm}_{,\varphi} Y^{l'm'}_{,\varphi}/\sin\theta) + g^{rr} \sum_{lm} \sum_{l'm'} [a^{lm}_{,r}(f^{l'm'} - k^{l'm'}_{,0}) + (k^{lm}_{,r} - h^{lm}) \, a_{l'm',0}](Y^{lm}_{,\varphi} Y^{l'm'}{}_{,\theta} - Y^{lm}{}_{,\theta} Y^{l'm'}{}_{,\varphi}) \,.$$

The spectral distribution of the radiation is immediately obtained by substituting the Fourier transform of eqs. (3) into (16) and taking the usual time average. Detailed computations of the energy spectrum and amount of radiation emitted by a charged particle falling into a black hole have been done by two of us (R.R. and J.T.) and will be published elsewhere.

* * *

It is a pleasure to acknowledge discussions with F. ZERILLI.

SOVIET ASTRONOMY – AJ VOL. 15, NO. 3 NOVEMBER-DECEMBER, 1971

A-12

HALOS AROUND "BLACK HOLES"

V. F. Shvartsman

Shternberg Astronomical Institute, Moscow
Institute of Applied Mathematics, Academy of Sciences of the USSR
Translated from Astronomicheskii Zhurnal, Vol. 48, No. 3, pp. 479-488, May-June, 1971
Original article submitted July 20, 1970

"Black holes" – bodies confined within their gravitational radius – will necessarily attract interstellar gas. If $M_{hole} > 0.03\ M_{\odot}$, at least $0.01\ mc^2$ of the infalling material should be converted into radiation. The corresponding luminosity would be of order $10^{32}\,(M/10\,M_{\odot})^{3/2}\,(\rho_{gas}/10^{-24}\,g\cdot cm^{-3})^{1/2}$ erg/sec, and would result from synchrotron radiation by magnetized plasma that would be heated to $T \approx 10^{12}$°K during the infall process. The spectrum would have a very mild slope extending from optical to radio wavelengths. Black holes might be observable as faint optical stars with no lines; they could be distinguished by intensity fluctuations on a time scale $\Delta t \approx 10^{-5}$-10^{-2} sec, with no periodic component whatever. In many cases accretion by massive holes ($10 \leqslant M \leqslant 10^4\ M_{\odot}$) should engender hard radiation (x and/or γ rays) exhibiting a flare behavior on a time scale ranging from a few months to tens of years; the peak intensity should be 10-100 times the synchrotron intensity. This phenomenon is associated with the turbulence of the interstellar medium: the angular momentum of the gas would halt the infall near the gravitational radius of the hole, and the momentum would be "annihilated" when the object moves into the adjacent turbulence cell. Only if $M > 10^6\ M_{\odot}$ would the angular momentum of the gas be capable of diminishing the mass falling into an isolated hole (that is, its luminosity). During infall toward a hole of $M \approx 10^5\ M_{\odot}$, the gas will initially remain cool ($T \approx 5000$°K) and will display an emission spectrum similar to the optical spectra of quasars. Because of accretion, a hole in a binary-star system might be observable as a visible secondary component. Accretion by a hole can be distinguished observationally from accretion by a neutron star. Possible candidates for black holes that may actually have been detected include certain type-Dc white dwarfs, the γ-ray star Sgr γ-1, the x-ray flare stars Cen X-2 and Cen X-4, and such objects as Sco X-1 and Cyg X-2.

1. INTRODUCTION

Perhaps the most interesting implication of general relativity theory is the prediction that the universe may contain masses that are confined within their gravitational radius $r_g = 2GM/c^2$. To an external observer such objects should appear rather like "black holes," drawing matter and radiation into themselves. We will recall that although a collapsed body will of itself radiate nothing at all – neither light, neutrinos, nor gravitational waves – it will nevertheless possess a static gravitational field which will influence its surroundings. Matter that is drawn in will reach r_g only asymptotically, after an infinite time; the region $r < r_g$ could not be observed at all by an external observer, and would thereby "drop out" of our space [1].

How might "black holes" be formed? There are at least five ways:

1) holes may have been present in the universe "from the beginning," that is, left over from the epoch of singularity;

2) holes may have developed from density fluctuations during the prestellar stage;

3) holes may have formed from supermassive first-generation stars;

4) holes may have resulted from the relativistic evolution of close clusters, galaxies, and the like;

5) finally, holes may have developed from ordinary but sufficiently massive stars.

It is highly probable that this last possibility has been realized. The discovery of pulsars, as is now well recognized, has confirmed the old theoretical prediction that massive stars are unstable and should suffer a catastrophic collapse at the end of their evolution. But theory predicts two final states after collapse: a neutron star in the event that the mass of the object is less than 1.5 $M_\odot$; and an uninterrupted infall, or a "collapsed" star, if M > 1.5 $M_\odot$.

Neutron stars have been discovered because of their activity: the generation of relativistic particles and radio waves. Collapsed stars, however, are passive by their very nature. Consequently, it is usually considered that "black holes" might most likely be discovered through their influence on radiating matter: on the motion of the normal component in a binary star [2-4], on stars in globular clusters [5] or galaxies [6], on the motion of the galaxies themselves [7] and so on.

However, the method of the "excluded third" is risky in astronomy. And furthermore, when configurations with an invisible component are selected the objects sought might be left out of the list altogether. As we shall demonstrate in this paper, holes whose mass exceeds the solar mass ought to be surrounded by highly luminous halos.

The question of the energy released through accretion by collapsed bodies was first posed by Zel'dovich [8] and Salpeter [9]. The accretion of gas by "superholes" ($M > 10^7 M_\odot$) located at the centers of galaxies has been discussed by Lynden-Bell [10]. We shall be interested primarily in "stellar" masses, with $10^{-1} M_\odot \lesssim M \lesssim 10^6 M_\odot$. Their luminosity will turn out to be given by the single equation $L \approx 0.1\, c^2\, dM/dt$, while their spectra can be highly diversified.

2. INTENSITY OF THE ACCRETION AND FLOW REGIMES

Let us imagine a massive object moving at a velocity u through a gas possessing no angular momentum. On the "back" side of such an object a conically shaped shock wave will be formed in which the gas will lose the component of its velocity perpendicular to the shock front. After compression in the shock wave, particles having sufficiently small impact parameters will fall into the star [1]. One can show that the infall of gas will begin at the characteristic distance

$$r_c = \alpha GM/v_c^2 = \alpha 10^{14} M v_{c(10)}^{-2}\,\text{cm}, \tag{1}$$

the distance at which the potential energy of the particles becomes comparable to the kinetic energy; and the flux of mass at the star will be

$$\frac{dM}{dt} = \beta 10^{11} M^2 v_{c(10)}^{-3} n_c \,\text{g/sec}. \tag{2}$$

In Eqs. (1) and (2) the mass M is expressed in solar masses, the velocity $v_c = (u^2 + a_c{}^2)^{1/2}$ is expressed in units of 10 km/sec, and the sound velocity a_c in the gas and the density n_c [atoms/cm^3] of the material drawn in refer to distances $r > r_c$. The dimensionless coefficients α and β are functions of the adiabatic index of the gas and the ratio u/a. We may take the rough approximations $\alpha \approx 1$, $\beta \approx 1$. Numerical values for various γ and u/a have been given by Salpeter [9]. Historically, Eq. (2) was first obtained by Hoyle, Lyttleton, and Bondi [11-13].

Note that for any reasonable value of the adiabatic index γ the diameter of the tail will be $d \approx r_c \gg r_g$, and the pressure will be finite. It therefore seems to us beyond doubt that a symmetrization of the flow should take place during infall, so that for $r \ll r_c$ the motion of the plasma may be considered radial (see Fig. 1).

In the spherically symmetric case, with back pressure absent, a supersonic flow of the gas will be established at $r \ll r_c$, that is, practically free fall [1]. To find the temperature variation, we shall write the second law of thermodynamics, dE = −pdV + dQ, in the form

$$\frac{3}{2} R^* \frac{dT}{dt} = R^* \frac{T}{\rho}\frac{d\rho}{dt} - 5 \cdot 10^{20} T^{1/2} \rho \varkappa + \frac{dQ'}{dt}. \tag{3}$$

Here R* is the gas constant; the second term on the right-hand side describes the losses of 1 g of plasma to radiation ($\varkappa = 1$ corresponds to bremsstrahlung from a fully ionized plasma), and the third term represents the energy variation due to other nonadiabatic processes. Since $\rho \propto r^{-3/2}$, we have

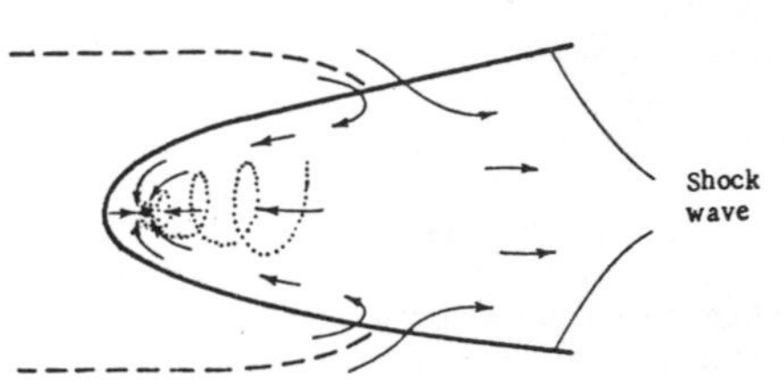

Fig. 1. The pattern of hydrodynamic accretion in the case where the star's own velocity u is much greater than the sound velocity a in the gas. The dashed curves correspond to the critical impact parameter. The dotted curve denotes the trajectory (helical) that the particles describe if an angular momentum relative to the star is present. In the figure the momentum vector is oriented parallel to the direction of motion of the object.

$$\frac{dT}{dr} = -\frac{T}{r} + [2 \cdot 10^{-4} M v_{c(10)}^{-3/2} n_c] \frac{\sqrt{T}}{r} \varkappa + \frac{dQ'/dt}{v \cdot \frac{3}{2} R^*}. \quad (4)$$

As it falls in toward a massive object the gas will be efficiently cooled through radiation ($T = [A \ln (r/r_0) + T_0^{1/2}]^2$); when $T \approx 5000°K$ is reached, recombination will begin and a temperature $T \approx$ const will be established. In the case of infall to an object of low mass, radiation will play a minor role; if the last term is neglected, the temperature variation will approach a $T \propto r^{-1}$ law, and the adiabatic index of the gas $\gamma \to 5/3$. The physical explanation of this difference is clear: the smaller the object, the smaller its gravitational radius, that is, its characteristic scale and the time for infall, and the radiation processes will become slow compared to the contraction process. Taking $dQ'/dt = 0$ (see the Appendix), we obtain for the critical mass, by Eq. (4),

$$M_{cr} \sim 10^4 M_{\odot} [T'_{(4)}]^{1/2} v_{c(10)}^3 n_c^{-1} \varkappa_{(2)}^{-1}. \quad (5)$$

Here $T'_{(4)}$ denotes the temperature at the distance of interest to us in units of 10^4°K, and $\varkappa_{(2)} \equiv \varkappa/100$.

3. THE GAS-DYNAMICS APPROXIMATION

The high temperatures developed during infall toward a mass $M < M_{cr}$ cast some doubt on the applicability of the gas-dynamical approach. At $r < r_c$, the mean free path with respect to Coulomb collisions,

$$l = kT^2 / ne^4 \cdot L_{\text{Coul}} \approx 10^{12} T_{(4)} n^{-1} \text{ cm}, \quad (6)$$

will in this case far exceed the characteristic size r of the region. For $M \approx M_{\odot}$, the free path $l \approx r$ even at the critical radius. We recall that the mass flux at the star in the approximation of noninteracting particles is smaller than the gas-dynamical flux (2) by a factor $(c/a_{\text{sound}})^2 \approx 10^9$ [1].

However, the material drawn in will contain magnetic fields. The Larmor radius of protons moving at thermal velocities will be smaller than the dimensions of the region of motion even at field strengths $H > 1.3\, T^{1/2}\, r^{-1}$ gauss. Inasmuch as $T \le T_c (r_c/r)$, we have the condition for capture

$$H(r) \gg 10^{-12} T_{c(4)}^{3/2} M^{-1} (r_c/r)^{3/2} \text{ gauss}. \quad (7)$$

The field at infinity is of order 10^{-6} gauss, so that in order for the condition (7) to be satisfied out to $r = r_g$ it is necessary that H increase with depth at least as $r^{-0.8}$. We shall see in Sec. 5 that there are grounds for expecting a far steeper rise in H(r), so that the gas-dynamics approach would be applicable.

The rise in the magnetic field would prevent any heat exchange between the different layers (see the Appendix). It would also affect the radiation and motion of the plasma. But first let us digress to consider one other topic.

4. LUMINOSITY AND EFFICIENCY IN THE IDEALIZED PROBLEM

For $M > M_{cr}$ the gas will be practically isothermal, so that the luminosity

$$L = \frac{dM}{dt} R^* \int_{r_1}^{r_2} T \frac{dn}{n} \approx 10^{34} M_{(5)}^2 v_{c(10)}^{-3} n_c \text{ erg/sec}. \quad (8)$$

Here $M_{(5)} \equiv M/10^5 M_{\odot}$, $T \approx 5000°K$, and r_1 is the radius below which radiation will rapidly be quenched by general-relativity effects (gravitational plus the Doppler red shift) [1]. Roughly speaking, $r_1 \approx 2r_g$. The luminosity (8) will fall mainly in the optical range and will result from line and recombination emission. If $M < M_{cr}$, the plasma temperature will increase inward along an adiabatic curve, and the luminosity of the spherical layer bounded by r and $r/2$ will be

$$L(r) \approx r^3 \varkappa 10^{-27} \cdot n^2 \cdot T^{1/2}(r)$$

$$\approx 10^{21} M^3 v_{c(10)}^{-6} n_c^2 [T'_{12}(r)]^{1/2} \text{ erg/sec}. \quad (9)$$

The value $T(r) \approx 10^{12}$°K may be regarded as an upper limit. The optical depth

$$\tau(r) = \int_r^{r_c} \sigma n_c (r_c/r)^{3/2} dr \approx 10^{-6} (\sigma/\sigma_k) M (r_g/r)^{1/2} n_c T_{c(4)}^{-3/2} \quad (10)$$

(where σ_k is the Compton cross section) will always be much less than unity if $M < M_{cr}$, but if $M > M_{cr}$ the interior regions may become opaque, despite the Doppler shift.

In the approximation considered here, the efficiency of black holes would be extremely low. If accretion by an object with $M > M_{cr}$ takes place, only $\approx 10^{-8}$ of the mass of the infalling material will be converted into radiation; if $M < M_{cr}$ the efficiency will be even lower. The actual situation, however, is more favorable than this idealized one.

5. THE INFLUENCE OF MAGNETIC FIELDS[1]

Any plasma that is drawn in will necessarily contain magnetic fields. The observable effects will depend in an essential way on the law for the growth of the fields during the infall process. In our opinion there is only one real possibility: the

[1] See the note added in proof.

magnetic energy ε_m and the gravitational energy ε_{gr} per unit volume should be of the same order not only at the critical radius (where all four energies – gravitational, kinetic, thermal, and magnetic – would be of the same order) but also in the zone $r < r_c$. We can prove this statement by assuming the contrary. For suppose that in some region $\varepsilon_m \ll \varepsilon_{gr}$; then the field would not affect the infall, and because of strict freezing-in (it is readily seen that during infall the magnetic viscosity will always be far smaller than the kinematic viscosity) its radial component H_r would increase according to the law $H_r \propto r^{-2}$, so that $\varepsilon_m = H^2/8\pi \propto r^{-4}$. However, $\varepsilon_{gr} \propto r^{-5/2}$, and an equality $\varepsilon_m \approx \varepsilon_{gr}$ would therefore be established very rapidly. On the other hand, the inequality $\varepsilon_m \gg \varepsilon_{gr}$ also would not be possible, because the field energy (due to freezing-in) would be derived from the energy of contraction, that is, from the kinetic energy, which would be smaller than ε_{gr}. We are therefore left with the condition $\varepsilon_m \approx \varepsilon_{gr}$. The behavior of the infall will be highly complicated in this situation: the field will have a predominantly radial character, and will be annihilated at the boundaries of sectors in which it is oppositely directed; the plasma will be inhomogeneous; the rate of infall will be nonuniform; very strong gutter instabilities will develop periodically in regions where $H \perp r$; and so on.

We shall nevertheless assume that despite the magnetic field, on the average the infall of gas does not depart seriously from free fall; that is, we shall suppose that $\bar{v} \approx (2GM/r)^{1/2}$ and $\bar{n} \propto r^{-3/2}$. Admittedly, both our idealizations may seem far-reaching, but it is natural to make them in our first steps toward a solution of the problem. We shall, then, let $H^2/8\pi = anGMm_p/r$, with $a \approx 1$.

In the case of infall toward a mass $M < M_{cr} \approx 10^4 M_\odot$, the plasma will be strongly heated, and the synchrotron losses will rapidly become decisive. In the range $T > 10^{10}$ °K, Eq. (4) will take the form ($dQ'/dt = 0$)

$$\frac{dT}{dr} = -\frac{T}{r} + [3 \cdot 10^{-8} M^2 v_{c(10)}^{-3/2} n_c] \frac{T^2}{r^2} a. \qquad (11)$$

Its solution is $T = (Cr + A/2r)^{-1}$; after the value

$$T_{max} = 3 \cdot 10^{12} M^{-1/2} v_{c(10)}^{3/4} n_c^{-1/2} a^{-1/2} \,{}^\circ\mathrm{K} \qquad (12)$$

is reached the "temperature" will begin to fall. The efficiency of radiant energy release will be

$$\eta = 0.2\, mc^2 T_{max(12.5)}, \quad \text{if} \quad T_{max} < 3 \cdot 10^{12}, \qquad (13a)$$

$$\eta = 0.2 mc^2 T_{max(12.5)}^{-8}, \quad \text{if} \quad T_{max} > 3 \cdot 10^{12}. \qquad (13b)$$

Of course, since equilibrium will not have been established during the accretion process [see Eq. (6)], the "temperature" T should be interpreted not as the Maxwellian parameter but as the mean energy of motion of the electrons across the magnetic field. Equations (11-13) as well as (17-19) below are, to some extent, merely illustrative in character. (In a separate paper [17] we have given a more detailed discussion of the circumstances of magnetic accretion, a theorem regarding the "equipartition of energy," and a demonstration that in the case of accretion by "black holes" the synchrotron mechanism should convert about 0.1 mc^2 of the infalling material into radiation.) We shall defer to Sec. 8 a description of the corresponding spectrum; we first wish to point out one possible factor that might raise the efficiency.

6. THE INFLUENCE OF ANGULAR MOMENTUM

Interstellar gas, generally speaking, would possess a certain angular momentum K relative to the line of motion of a hole. It is well recognized [1] that, in general relativity, particle capture is possible if $K < K_g = 2mcr_g$. After it approaches a gravitating center an isolated particle with $K > K_g$ will again recede to infinity. According to Sec. 3, however, plasma at any distance from a collapsed object may be regarded as a continuous medium. Its compression at the tail of the flow, in the shock wave, will be accompanied by radiant energy release; if this energy becomes negative, then matter can never leave the neighborhood of the hole. The corresponding criterion is evident: it is necessary that the momentum K be much smaller than the quantity $K_c = mr_c v_c$. Thus if the momentum of the material drawn in satisfied the inequality

$$2mr_g c < K \ll mr_c v_c, \qquad (14)$$

then before they fall into the hole the particles will go into a stationary orbit. The minimum radius of a stable stationary orbit would be $r_{min} = 3r_c$ [18]. In order for a particle in Keplerian motion to reach r_{min}, it should emit radiation of 0.07 mc^2; if the role of viscosity is appreciable and nearly solid-body rotation prevails, the efficiency would be twice as great.

Might we expect the condition (14) to be satisfied under actual conditions? Denoting the tangential velocity component by v_t, let us write the inequality in the form

$$1.2 \cdot 10^{-4} v^{2}_{c(10)} < v_{t(10)}(r = r_c) \ll v_{c(10)}. \qquad (14a)$$

First let us consider the left-hand inequality. Estimates indicate that the internal scale of interstellar turbulence, $l_0 = \mathrm{Re}_0\, \nu'/v_t$, will always be less than or of the order of the radius r_c (Re_0 is the critical Reynolds number, and ν' is the magnetic viscosity). Hence $v_t(r_c) = v_t(L_t) \cdot (r_c/L_t)^x$, where $L_t \approx 100$ pc and $v_t(L_t) \approx 10$ km/sec [19]. In the interstellar medium, because of suppression of turbulence by the magnetic field and the formation of shock waves, the spectrum $v(l)$ should be steeper than a Kolmogorov spectrum ($x = 1/3$); evidently $1/2 \lessdot x < 1$ [19, 20]. Let $x \equiv 2/(3 + y)$. It is then easily seen than the first of the inequalities (14) will be satisfied for masses

$$M > M_t^{I} \approx 3.5 M_{\odot} v^{3}_{c(10)} [1.2 \cdot 10^{-2} v_{c(10)}]^{y}. \qquad (15)$$

Thus for objects with stellar masses, spherically symmetric accretion will evidently be realized in many cases, while in some cases the plasma will be halted not far from r_g.

In the case of accretion of gas by very massive objects, rotation will undoubtedly play a role, and the radiant efficiency $\eta \approx 0.1\ mc^2$. It is now appropriate to inquire as to the influence of the momentum on dM/dt, that is, the right-hand inequality (14). For $L_t = 100$ pc and $v_t(L_t) = 10$ km/sec this inequality will become

$$M \ll M_t^{II} \simeq 3 \cdot 10^{6} M_{\odot} v^{2}_{c(10)}. \qquad (16a)$$

The criterion (16a) refers to the case where $v_c \approx v_t \approx 10$ km/sec;[2] if $v_c \gg 10$ km/sec, only differential galactic rotation would be capable of preventing accretion. For a body 10 kpc away from the galactic center we would have in place of the criterion (16a):

$$M \ll M_t^{II} \simeq 3 \cdot 10^{9} M_{\odot} v^{3}_{c(50)}. \qquad (16b)$$

The role of the momentum of the gas in accretion by objects with $M > 10^6 M_{\odot}$ has been pointed out qualitatively by Salpeter [9]. He has also indicated a fundamental mechanism for momentum loss: the gas in adjacent turbulence cells would have opposite directions of rotation. We shall return to this topic presently, but we first with to call the reader's attention to an important case in which the sign of the momentum might not change.

7. ACCRETION IN BINARY SYSTEMS

Our motive in discussing this situation is clear: in addition to a high efficiency, binary systems would ensure an enormous loss rate dM/dt. Thus, variables of the β Lyrae type lose as much as 10^{-5} $M_{\odot}$/yr; under favorable conditions about one-half of this mass could fall into the second component. Hence the luminosity of holes in binary systems would in principle be capable of reaching 10^{38}-10^{39} erg/sec.

Around the dense component gas streams should form a disk with an approximately Keplerian velocity distribution. The mechanism for momentum loss would be viscosity, which would tend to produce solid-body rotation. The inner regions of the disk would be retarded, and the outer regions accelerated. Some of the gas would leave the binary system, while the rest would sink in toward the dense component. Very high temperatures would evidently be attained during this settling process (see the model calculation by Prendergast and Burbidge [22]).

Accretion by holes, neutron stars, and white dwarfs in close binary systems has been proposed on several occations [8, 22-24] as a source of energy. An important question arises here: would it be possible to distinguish accretion by a hole from accretion by a star? The answer will be clear from the considerations above. In the first case the radiation would all come from a hot, thin disk; in the second case half the radiation would be due to the disk and half to emission from the surface of the star, in which an appreciable contribution should be present from equilibrium radiation [25] and/or plasma oscillations [26]. Furthermore, in the case of accretion by a neutron star the radiation ought to contain a strictly periodic term (with $p \approx 1$ sec) arising from the rotation of the star. In the case of accretion by a hole, only random luminosity variations would be expected, with $\Delta t \approx 10^{-5}$-10^{-2} sec (see Sec. 9 below). Also, the emission spectrum of the disk itself probably would, in general, be quite flat. Thus a search for a hole in a binary system, based on the absence of an optically visible component [2-4], might actually exclude some of the objects sought.

8. GAMMA-RAY, X-RAY, AND OPTICAL STARS?

We shall now subdivide accretion into spherical, helical, and disk types, depending on the character of particle motion. The first two types would be realized around isolated objects; the last type, in binary systems. In the case of disk accretion, most of the energy should be released in the form of x-ray photons. Helical accretion would lead

[2] Note that under the conditions prevailing in the Galaxy, supermassive objects necessarily would rapidly diminish their velocity u [but not $v = (u^2 + a^2_{\mathrm{sound}})^{1/2}$] to about 10 km/sec [21].

to the appearance of γ-ray stars as well as x-ray stars. Indeed, as we have been in Sec. 6, the spectrum of interstellar turbulence is such that in many instances the plasma would be halted (transferred from a helical to a circular orbit) in the vicinity of r_g. When a hole moves into the adjacent turbulence cell, where the gas is rotating in the opposite direction, particle collisions will begin to take place near r_g, and unlike the case of settling onto a disk, the process of momentum loss would here be very rapid.

It is worth noting that in the case of accretion by an isolated neutron star the role of the momentum of the gas would probably be small because of the small value of r_c (see Sec. 6). On the other hand, even if a disk were to develop, it would be located near the surface of the neutron star and would rapidly sink into the star. The integrated luminosity of the objects in this situation would be very low, of order 10^{30} ergs/sec [see Eq. (2)]; and most of the radiation would fall in the ultraviolet [27]. If the neutron star has a magnetic field, the infall of gas would be stopped, in general, far beyond r_g, and would be accompanied by Langmuir oscillations at radio frequencies [26].[3] Evidently, then, only "black holes" could be "pure" γ-stars.

In this connection it is interesting to note the recent discovery [28] of a discrete source of γ rays ($E_\gamma > 50$ MeV) which has not yet been identified with an x-ray object. Yet, the sensitivity of the x-ray equipment was one or two orders higher than the sensitivity of the γ-ray counters.

The intensity of disk-type accretion should remain unchanged over tens of thousands of years, apart from fluctuations associated with the inhomogeneity of the gas flow. (Possible eclipses of the disk by the normal component would be of special interest.) The radiation of isolated objects due to spiral-type accretion would undoubtedly exhibit a flare behavior. The minimum characteristic time for a flare would be of order r_c/v_c, that is, a few months. Curiously enough, among the known sources of hard x rays there are definitely two classes of objects: those whole luminosity has remained constant, on the average, throughout the entire period of observation (for example, Sco X-1 and Cyg X-2); and those whose luminosity has varied by tens of times within a year, either appearing or disappearing from the field of view (such as Cen X-2 and Cen X-4 [29, 30]). Observers have often suggested that Sco X-1 and Cyg X-2 might belong to binary systems [31-34].

The accretion of gas by an isolated hole of mass $M \approx 1$-$100\ M_\odot$ should, as mentioned in previous sections, be primarily spherical. Magnetic fields should play a definite role in the radiation. According to Eqs. (2) and (13a), the corresponding luminosity should be

$$L \sim 0.1 T_{\max(12.3)} c^2 dM/dt$$
$$\approx 2 \cdot 10^{33} M_{100}^{3/2} v_{10}^{-9/4} n^{1/2} a^{-1/2}\ \text{erg/sec}; \qquad (17)$$

here $M_{100} = M/100\ M_\odot$. The spectrum of the radiation should have a distinctive form: the intensity should remain nearly constant over a wide frequency range near

$$\nu_{\max} \sim 10^{14} \cdot M_{100}^{-3/8} v_{10}^{-9/16} n^{-9/8} a^{-11/8}\ \text{Hz}, \qquad (18)$$

an exponential decline should set in at $\nu \gg \nu_{\max}$, and at $\nu \ll \nu_{\max}$ there should be an extremely slow decline to frequencies at which absorption of the radiation becomes appreciable.

$$L(\nu) = \int_{\nu/2}^{\nu} (dL/d\nu)\, d\nu \propto \nu^{4/13}, \qquad dL/d\nu \propto \nu^{-9/13}, \qquad (19)$$

Coherent mechanisms and negative self-absorption might be operative.

In cases where γ rays are generated, they would be in addition to synchrotron radiation. However, because of the "accumulation" of material in circular orbits, the "peak" γ-ray luminosity should be one or two orders higher than given by Eq. (17).

The radiation of a "black hole" would of course heat the gas at $r > r_c$, thereby influencing the intensity of the accretion [see Eq. (2)]. One can show that for most cases of interest this effect would be insignificant, but in certain cases it could be decisive. A more thorough discussion of this topic, together with a solution of the self-consistent problem as exemplified by a neutron star, has been given elsewhere [27].

Conceivably, then, individual "black holes" might already be observable today, as faint optical stars with no lines but with a nonthermal spectrum extending far into the low-frequency range (as far as radio frequencies). Perhaps some of the objects heretofore regarded as type DC white dwarfs are actually "black holes."

9. A CRITICAL EXPERIMENT

What properties of the radiation would allow "black holes" of stellar mass to be distinguished *reliably* from other objects? In our view, such properties would be the exceptionally small size

---

[3] From the observational standpoint such objects might appear as "second generation" pulsars. Further details have been given elsewhere [26].

of a hole together with the absence of any rotation. In other words, because of the development of instability in a magnetized plasma the luminosity of a hole would fluctuate on a time scale $\Delta t \approx 10^{-5}$–10^{-2} sec, but a periodic term should be entirely absent.

10. QUASAR-LIKE STARS?

Equations (5) and (16) imply that in the case of accretion by a hole with $M \approx 10^5 M_\odot$ the infalling gas will be cooled up to the onset of the "spiral" mode (somewhere in the zone $r \approx 10^{-4}$–$10^{-3}\ r_c$); $T \approx 5000°K$. The corresponding luminosities are given by Eq. (8), and are small compared to the integrated luminosity, but nevertheless they fall wholly in the optical range. The spectrum of the radiation – broadened emission lines, well-developed recombination bands, the presence of a nonthermal continuum and absorption lines – should to a large extent resemble the optical spectrum of quasars. On the other hand, holes with masses of order $10^5 M_\odot$ are interesting in that they might have come from "first generation" stars that developed from entropy perturbations in the pregalactic medium [35] (the remainder of these perturbations would presumably have served as the origin of the globular clusters [36]).

In the case of accretion by "superholes" with $M \gg 10^5 M_\odot$, the momentum of the gas would be expected to play a role from the very outset; but here too it would seem that in many cases a regime of "cold accretion" would develop, accompanied by a spectrum similar to that observed for quasars. We intend to examine this possibility in a separate paper.

APPENDIX

THE ABSENCE OF HEAT CONDUCTIVITY BETWEEN DIFFERENT LAYERS

Why have we consistently neglected the term dQ'/dt, representing the heat conductivity? For radiative conductivity, we have done so because the optical depth of the plasma is negligible [see Eq. (10)]; for ion conductivity, because the infall of the gas is supersonic; and finally, for electron conductivity, because the magnetic fields grow rapidly during the infall process. Let us consider this last point more carefully. According to Eq. (6), the time required for exchange of energy between different electrons will far exceed the time required for infall toward the hole. Thus heat conductivity could only arise from the migration of electrons. However, such migration will be severely limited by the small Larmor radius of the electrons ($l_L/r < 10^{-6}$), by the strictness with which the freezing-in conduction is satisfied, and by the tangling of the lines of force during infall (see Sec. 5). In the interior, where the temperatures and fields are large, there will in addition be a small "mean free path" relative to synchrotron losses, as compared to the characteristic scale of the motion.

The author is indebted to Ya. B. Zel'dovich for suggesting the topic and for much valuable counsel. He is also grateful to G. S. Bisnovatyi-Kogan and L. M. Ozernoi for their comments.

NOTE ADDED IN PROOF

Kardashev [37] was the first to point out the circumstance that in accretion by a "black hole" a substantial fraction of the rest mass of the infalling material might be liberated because of the presence of a magnetic field; in this connection see also the remark on page 360 of the Russian edition of Zel'dovich and Novikov's book [1]. The synchrotron radiation emitted upon accretion by the magnetosphere of a neutron star has been considered in recent papers [14,15]. Bisnovatyi-Kogan and Syunyaev, in discussing the problem of infrared sources [16], consider that in accretion by collapsed stars the annihilation of the magnetic field may lead to the formation near r_g of a shock wave at whose front a substantial part of the energy will be transformed into plasma oscillations with the emission of radiation. Our views concerning magnetic accretion have been set forth more fully in a separate paper [17].

LITERATURE CITED

1. Ya. B. Zel'dovich and I. D. Novikov, Relativistic Astrophysics [in Russian], Nauka, Moscow (1967); English translation, Vol. 1, Univ. Chicago Press (1971).
2. O. Kh. Guseinov and Ya. B. Zel'dovich, Astron. Zh., 43, 313 (1966) [Sov. Astron.–AJ, 10, 251 (1966)]; Ya. B. Zel'dovich and O. Kh. Guseinov, Astrophys. J., 144, 840 (1965).
3. V. L. Trimble and K. S. Thorne, Astrophys. J., 156, 1013 (1969).
4. O. Kh. Guseinov and Kh. I. Novruzova, Astron. Tsirk., No. 560 (1970).
5. A. A. Wyller, Astrophys. J., 160, 443 (1970).
6. A. M. Wolfe and G. R. Burbidge, Astrophys. J., 161, 419 (1970).
7. S. van den Bergh, Nature, 224, 891 (1969).
8. Ya. B. Zel'dovich, Dokl. Akad. Nauk SSSR, 155,

67 (1964) [Sov. Phys.-Dokl., 9, 195 (1964)].
9. E. E. Salpeter, Astrophys. J., 140, 796 (1964)].
10. D. Lynden-Bell, Nature, 223, 690 (1969).
11. F. Hoyle and R. A. Lyttleton, Proc. Cambridge Phil. Soc., 35, 405 (1939).
12. H. Bondi and F. Hoyle, Mon. Not. Roy. Astron. Soc., 104, 273 (1944).
13. H. Bondi, Mon. Not. Roy. Astron. Soc., 112, 195 (1952).
14. P. R. Amnuél' and O. Kh. Guseinov, Izv. Akad. Nauk Azerbaidzhan SSR, Ser. Fiz. Tekh. Mat., No. 3, 70 (1968).
15. G. S. Bisnovatyi-Koyan and A. M. Fridman, Astron. Zh., 46, 721 (1969) [Sov. Astron. -AJ, 13, 566 (1970)].
16. G. S. Bisnovatyi-Kogan and R. A. Syunyaev, Preprinty Inst. Priklad. Matem. Akad. Nauk SSSR, No. 31 (1970); Astron. Zh. [Sov. Astron. -AJ (in press)].
17. V. F. Shvartsman, Preprinty Inst. Priklad. Matem. Akad. Nauk SSSR, No. 42 (1970).
18. S. A. Kaplan, Zh. Éksp. Teor. Fiz., 19, 951 (1949).
19. S. A. Kaplan and S. B. Pikel'ner, The Interstellar Medium, Harvard Univ. Press (1970).
20. S. A. Kaplan, Dokl. Akad. Nauk SSSR, 94, 33 (1954).
21. W. H. McCrea, Mon. Not. Roy. Astron. Soc., 113, 162 (1953).
22. K. H. Prendergast and G. R. Burbidge, Astrophys. J., 151, L83 (1968).
23. I. S. Shklovskii, Astron. Zh., 44, 930 (1967) [Sov. Astron.-AJ, 11, 749 (1968)]; Astrophys. J., 148, L1 (1967).
24. A. G. W. Cameron and M. Mock, Nature, 215, 464 (1967).
25. Ya. B. Zel'dovich and N. I. Shakura, Astron. Zh., 46, 225 (1969) [Sov. Astron.-AJ, 13, 175 (1969)].
26. V. F. Shvartsman, Radiofizika, 13, 1852 (1970).
27. V. F. Shvartsman, Preprinty Inst. Priklad. Matem. Akad. Nauk SSSR, No. 57 (1969); Astron. Zh., 47, 824 (1970) [Sov. Astron.-AJ, 14, 662 (1971)].
28. G. M. Frye, J. A. Staib, A. D. Zych, V. D. Hopper, W. R. Rawlinson, and J. A. Thomas, Nature, 223, 1320 (1969).
29. W. H. G. Lewin, J. E. McClintock, and W. B. Smith, Astrophys. J., 159, L193 (1970).
30. W. D. Evans, R. D. Belian, and J. P. Conner, Astrophys. J., 159, L57 (1970).
31. Yu. N. Efremov, Astron. Tsirk., No. 401 (1967).
32. E. M. Burbidge, C. R. Lynds, and A. N. Stockton, Astrophys. J., 150, L95 (1967).
33. R. P. Kraft and M.-H. Demoulin, Astrophys. J., 150, L183 (1967).
34. J. Kristian, A. R. Sandage, and J. A. Westphal, Astrophys. J., 150, L99 (1967).
35. A. G. Doroshkevich, Ya. B. Zel'dovich, and I. D. Novikov, Astron. Zh., 44, 295 (1967) [Sov. Astron.-AJ, 11, 233 (1967)].
36. P. J. E. Peebles and R. H. Dicke, Astrophys. J., 154, 891 (1968).
37. N. S. Kardashev, Astron. Zh., 41, 807 (1964) [Sov. Astron.-AJ, 8, 643 (1965)].

Commun. math. Phys. 25, 152—166 (1972)

A-13

Black Holes in General Relativity

S. W. Hawking
Institute of Theoretical Astronomy, University of Cambridge, Cambridge, England

Received October 15, 1971

Abstract. It is assumed that the singularities which occur in gravitational collapse are not visible from outside but are hidden behind an event horizon. This means that one can still predict the future outside the event horizon. A black hole on a spacelike surface is defined to be a connected component of the region of the surface bounded by the event horizon. As time increase, black holes may merge together but can never bifurcate. A black hole would be expected to settle down to a stationary state. It is shown that a stationary black hole must have topologically spherical boundary and must be axisymmetric if it is rotating. These results together with those of Israel and Carter go most of the way towards establishing the conjecture that any stationary black hole is a Kerr solution. Using this conjecture and the result that the surface area of black holes can never decrease, one can place certain limits on the amount of energy that can be extracted from black holes.

1. Introduction

It has been known for some time that a non-rotating star of more than about two solar masses has no low temperature equilibrium configuration. This means that such a star must undergo catastrophic collapse when it has exhausted its nuclear fuel unless it has managed to eject sufficient matter to reduce its mass to less than twice that of the sun. If the collapse is exactly spherically symmetric, the metric is that of the Schwarzschild solution outside the star and has the following properties (see Fig. 1):

1. The surface of the star will pass inside the Schwarzschild radius $r = 2Gc^{-2}M$. After this has happened there will be closed trapped surfaces [1, 2] around the star. A closed trapped surface is a spacelike 2-surface such that both the future directed families of null geodesics orthogonal to it are converging. In other words, it is in such a strong gravitational field that even the outgoing light from it is dragged inwards.

2. There is a space-time singularity.

3. The singularity is not visible to observers who remain outside the Schwarzschild radius. This means that the breakdown of our present physical theory which one expects to occur at a singularity cannot affect

what happens outside the Schwarzschild radius and one can still predict the future in the exterior region from Cauchy data on a spacelike surface.

One can ask whether these three properties of spherical collapse are stable, i.e. whether they would still hold if the initial data for the collapse were perturbed slightly. This is vital because no real collapse situation will ever be exactly spherical. From the stability of the Cauchy problem in general relativity [3] one can show that a sufficiently small perturbation of the initial data on a spacelike surface will produce a perturbation of the solution which will remain small on a compact region in the Cauchy development of the surface. This shows that property (1) is stable, since there is a compact region in the Cauchy development of the initial surface which contains closed trapped surfaces. It then follows that property (2) is stable provided one makes certain reasonable assumptions such as that the energy density of matter is always positive. This is because the existence of a closed trapped surface implies the occurrence of a singularity under these conditions [4]. There remains the problem of the stability of property (3). Since the question of whether singularities are visible from outside depends on the solution at arbitrarily large times, one cannot appeal to the result on the stability of the Cauchy problem referred to above. Nevertheless it seems a reasonable conjecture that property (3) is indeed stable. If this is the case, we can still predict what happens outside collapsed objects, and we need not worry that something unexpected might occur every time a star in the galaxy collapsed. My belief in this conjecture is strengthened by the fact that Penrose [5] has tried and failed to obtain a contradiction to it, which would show that naked singularities must occur. Penrose's method has been generalised by Gibbons [6] who has shown that it cannot lead to a contradiction at least in some cases. This paper will be written therefore on the assumption that property (3) holds.

In Section 2 a black hole is defined in terms of a event horizon and it is shown that the surface area of a black hole cannot decrease with time. In Section 3 it is shown that a rotating stationary black hole must be axisymmetric, and in Section 4 it is shown that any stationary black hole must have a topologically spherical boundary. Together with the results of Israel and Carter, this strongly supports the conjecture that a black hole settles down to a Kerr solution. This conjecture is used in Section 5 to relate the surface area of a black hole to its mass, angular momentum and electric charge. Using the result that the surface area cannot decrease with time one can then place upper bounds on the amounts of energy that can be extracted from black holes. These limits suggest that there may be a spin dependent force between two black holes analogous to that between magnetic dipoles.

2. The Event Horizon

In order to discuss the region outside a collapsed object one needs a precise notion of infinity in an asymptotically flat space-time. This is provided by Penrose's concept of a weakly asymptotically simple space [2]; the spacetime manifold $\mathcal{M}$ of such a space can be imbedded in a larger manifold $\tilde{\mathcal{M}}$ on which there is a Lorentz metric $\tilde{g}_{ab}$ which is conformal to the spacetime metric g_{ab}, i.e. $\tilde{g}_{ab} = \Omega^2 g_{ab}$ where Ω is a smooth function which is zero and has non-vanishing gradient on the boundary of $\mathcal{M}$ in $\tilde{\mathcal{M}}$. This boundary consists of two null hypersurfaces $\mathscr{I}^+$ and $\mathscr{I}^-$ which each have topology $S^2 \times R^1$ and which represent future and past null infinity respectively. One can then interpret property (3) as saying that it should be possible to predict events near $\mathscr{I}^+$. I shall therefore say that a weakly asymptotically simple space is (future) asymptotically predictable if there is a partial Cauchy surface $\mathscr{S}$ such that $\mathscr{I}^+$ lies in the closure in $\tilde{\mathcal{M}}$ of $D^+(\mathscr{S})$, the future Cauchy development of $\mathscr{S}$. (A partial Cauchy surface is a spacelike surface without edge which does not intersect any non-spacelike curve more than once. $D^+(\mathscr{S})$ is the set of all points p such that every past directed non-spacelike curve from p intersects $\mathscr{S}$ if extended far enough.)

Roughly speaking one would expect a space to be asymptotically predictable if there are no singularities in $J^+(\mathscr{S})$, the future of $\mathscr{S}$, which are naked, i.e. which lie in $J^-(\mathscr{I}^+)$, the past of future null infinity. One can make this more precise. Consider an asymptotically predictable space in which there are no singularities to the past of $\mathscr{S}$. Suppose there is a closed trapped surface $\mathscr{T}$ in $D^+(\mathscr{S})$. Then there will be a singularity to the future of $\mathscr{T}$, i.e. there will be a nonspacelike geodesic in $J^+(\mathscr{T})$ which is future incomplete. Can this geodesic be seen from $\mathscr{I}^+$? The answer is no. For suppose $\mathscr{T}$ intersected $J^-(\mathscr{I}^+)$. Then there would be a point $p \in \mathscr{I}^+$ in $J^+(\mathscr{T})$. The past directed null geodesic generator of $\mathscr{I}^+$ through p would eventually leave $J^+(\mathscr{T})$ and so would contain a point q of the boundary $\dot{J}^+(\mathscr{T})$. Now the boundary of the future of any closed set $\mathscr{W}$ is generated by null geodesic segments which either have no past end-points or have past end-points on $\mathscr{W}$ [2,4]. Since the generator λ of $\dot{J}^+(\mathscr{T})$ through q would enter $D^+(\mathscr{S})$ it would have to have an end point on $\mathscr{T}$ since otherwise it would intersect $\mathscr{S}$ and pass into the past of $\mathscr{S}$ which would be impossible, as $\mathscr{T}$ is to the future of $\mathscr{S}$. The generator λ would intersect $\mathscr{T}$ orthogonally. However, as $\mathscr{T}$ is a closed trapped surface, the null geodesics orthogonal to $\mathscr{T}$ are converging. Together with the weak energy condition: $T_{ab}K^aK^b \geqq 0$ for any timelike vector K^a, this implies that there will be a point conjugate to $\mathscr{T}$ within a finite affine length on any null geodesic orthogonal to $\mathscr{T}$ [4]. Points on such a geodesic beyond the conjugate point will lie in

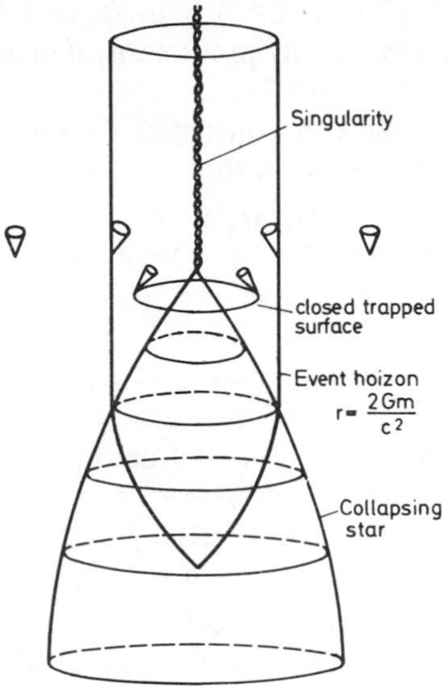

Fig. 1. Spherical collapse

the interior of $J^+(\mathscr{T})$ and not on its boundary [2, 4]. However the generator λ of $\dot{J}^+(\mathscr{T})$ would have infinite affine length from $\mathscr{T}$ to $\mathscr{I}^+$ since $\mathscr{I}^+$ is at infinity. This establishes a contradiction which shows that $\mathscr{T}$ does not intersect $J^-(\mathscr{I}^+)$. Thus the future incomplete geodesic in $J^+(\mathscr{T})$ is not visible from $\mathscr{I}^+$.

Since $J^-(\mathscr{I}^+)$ does not contain $\mathscr{T}$, its boundary $\dot{J}^-(\mathscr{I}^+)$ must be non-empty. This is the *event horizon* for $\mathscr{I}^+$ and is the boundary of the region from which particles or photons can escape to infinity. It is generated by null geodesic segments which have no future end-points. The convergence ϱ of these generators cannot be positive. For suppose it were positive on some open set $\mathscr{U}$ of $\dot{J}^-(\mathscr{I}^+)$. Let $\mathscr{F}$ be a spacelike 2-surface in $\mathscr{U}$. Then the outgoing null geodesics orthogonal to $\mathscr{F}$ would be converging. One could deform a small part of $\mathscr{F}$ so that it intersected $J^-(\mathscr{I}^+)$ but so that the outgoing null geodesics orthogonal to $\mathscr{F}$ were still converging. This again would lead to a contradiction since the null geodesics orthogonal to $\mathscr{F}$ could not remain in $\dot{J}^+(\mathscr{F})$ all the way out to $\mathscr{I}^+$.

If there were a point on the event horizon which was not in $D^+(\mathscr{S})$, the future Cauchy development of $\mathscr{S}$, a small perturbation could result in there being points near $\mathscr{I}^+$ which were not in $D^+(\mathscr{S})$. Since I am assuming that asymptotic predictability is stable, I shall slightly extend the definition to exclude this kind of situation. In an asymptotically predictable space $J^+(\mathscr{S})\cap J^-(\mathscr{I}^+)$ is in $D^+(\mathscr{S})$. I shall say that such a space is *strongly* asymptotically predictable if in addition $J^+(\mathscr{S})\cap \dot{J}(\mathscr{I}^+)$ is in $D^+(\mathscr{S})$.

In such a space one can construct a family $\mathscr{S}(t)$ $(t>0)$ of partial Cauchy surfaces in $D^+(\mathscr{S})$ such that

(a) for $t_2>t_1$, $\mathscr{S}(t_2)\subset J^+(\mathscr{S}(t_1))$;

(b) each $\mathscr{S}(t)$ intersects $\mathscr{I}^+$ in a 2-sphere $\mathscr{A}(t)$;

(c) for each $t>0$, $\mathscr{S}(t)\cup[\mathscr{I}^+\cap J^-(\mathscr{A}(t))]$ is a Cauchy surface for $D^+(\mathscr{S})$.

The construction is as follows. Choose a suitable family $\mathscr{A}(t)$ of 2-spheres on $\mathscr{I}^+$. Put a volume measure on $\mathscr{M}$ so that the total volume of $\mathscr{M}$ in this measure is finite [7]. Define the functions $f(p)$ and $h(p,t)$, $p \in D^+(\mathscr{S})$ as the volumes of $J^+(p)\cap D^+(\mathscr{S})$ and $[J^-(p)-J^-(\mathscr{A}(t))]\cap D^+(\mathscr{S})$ respectively. They will be continuous in p and t. The surface $\mathscr{S}(t)$ is then defined to be the set of points p such that $h(p,t)=t\,f(p)$.

For sufficiently large t, the surfaces $\mathscr{S}(t)$ will intersect the event horizon and so $\mathscr{B}(t)$ defined as $\mathscr{S}(t)-J(\mathscr{I}^+)$ will be nonempty. I shall define a *black hole* on the surface $\mathscr{S}(t)$ to be a connected component of $\mathscr{B}(t)$. In other words, it is a region of $\mathscr{S}(t)$ from which there is no escape to $\mathscr{I}^+$. As time increases, black holes may merge together and new black holes may be created by further bodies collapsing but a black hole can never bifurcate. For suppose the black hole $\mathscr{B}_1(t_1)$ on the surface $\mathscr{S}(t_1)$ divided into two black holes $\mathscr{B}_2(t_2)$ and $\mathscr{B}_3(t_2)$ by a later surface $\mathscr{S}(t_2)$. Then $\mathscr{B}_2(t_2)$ and $\mathscr{B}_3(t_2)$ would each have to contain points of $J^+(\mathscr{B}_1(t_1))$. However every nonspacelike curve which intersected $\mathscr{B}_1(t_1)$ would intersect $\mathscr{S}(t_2)$. Therefore $J^+(\mathscr{B}_1(t_1))\cap\mathscr{S}(t_2)$ would be connected and would be contained in $\mathscr{B}_2(t_2)\cup\mathscr{B}_3(t_2)$.

Suppose that $\mathscr{M}$ is initially nonsingular in the sense that $J^-(\mathscr{S})$, the region to the past of $\mathscr{S}$, is isometric to a region to the past of a Cauchy surface in an asymptotically simple space [2]. Then for small values of t, the surfaces $S(t)$ will be compact. However the surfaces $\mathscr{S}(t)$ are homeomorphic to each other for all $t>0$ and so they will all be compact. This implies that the boundary $\partial\mathscr{B}_1(t)$ of a black hole $\mathscr{B}_1(t)$ will be compact.

Since the generators of $\dot{J}^-(\mathscr{I}^+)$ have no future end points and have convergence $\varrho\leqq 0$, the surface area of $\partial\mathscr{B}_1(t)$ cannot decrease with t. If two black holes $\mathscr{B}_1(t_1)$ and $\mathscr{B}_2(t_1)$ on the surface $\mathscr{S}(t_1)$ merge to form a single black hole $\mathscr{B}_3(t_2)$ on a later surface $\mathscr{S}(t_2)$, then the area of $\partial\mathscr{B}_3(t_1)$ must be at least the sum of the areas of $\partial\mathscr{B}_1(t_1)$ and $\partial\mathscr{B}_2(t_1)$. In fact it must be strictly greater than this sum because $\partial\mathscr{B}_3(t_2)$ contains two disjoint closed sets which correspond to the generators of $\dot{J}(\mathscr{I}^+)$ which intersect $\partial\mathscr{B}_1(t_1)$ and $\partial\mathscr{B}_2(t_1)$. Since $\partial\mathscr{B}_3(t_2)$ is connected, it must also contain an open set of points which correspond to generators which have past end points between t_2 and t_1. These results will be used in the next section to place certain limits on the possible behaviour of black holes.

3. Stationary Black Holes

In a collapse that was strongly asymptotically predictable one would expect the solution outside the event horizon to become stationary eventually. This suggests that one should study exactly stationary solutions containing black holes in the hope that one of these will represent the final state of a collapsed system. In this section I shall therefore consider spaces which satisfy the following conditions:

(i) They are weakly asymptotically simple.

(ii) There exists a one parameter isometry group $\phi_t : \mathcal{M} \to \mathcal{M}$ whose Killing vector K^a is timelike near $\mathscr{I}^+$ and $\mathscr{I}^-$

(iii) There exist both a *past* event horizon $\dot{J}^+(\mathscr{I}^-)$ and a *future* event horizon $\dot{J}^-(\mathscr{I}^+)$.

(iv) There exists a partial Cauchy surface $\mathscr{S}$ from which the *exterior region* $\bar{J}^+(\mathscr{I}^-) \cap \bar{J}^-(\mathscr{I}^+)$ can be determined, i.e. the exterior region is contained in $D^+(\mathscr{S}) \cup D^-(\mathscr{S})$, the Cauchy development of $\mathscr{S}$.

(v) The two event horizons $\dot{J}^+(\mathscr{I}^-)$ and $\dot{J}^-(\mathscr{I}^+)$ intersect in a compact surface $\mathscr{F}$.

In a real situation where an initially nonsingular body collapses there will not of course be any past event horizon, since that part of the space will be inside the body and so will be different and will be nonstationary. The statement that a past horizon exists is therefore to be understood as a condition on the analytic continuation of the stationary solution to which the real solution tends. Except in certain limiting cases, one can prove that provided there are no singularities in the exterior region, there will indeed be a past horizon which will intersect the future horizon. Penrose however has questioned whether it is reasonable to assume the nonexistence of naked singularities in the unphysical past region of the analytic extension. In fact the results that will be given in this paper can be obtained without assuming anything about the behaviour of the solution in the past though the proofs are then considerably more complicated. They will be given in [3].

Israel [8] has shown that a space which satisfies conditions (i)–(v) must be the Schwarzschild solution if it also satisfies the following conditions:

(vi) It is empty, i.e. $T_{ab}=0$.

(vii) It is *static* and not merely stationary, i.e. the Killing vector K^a is hypersurface orthogonal so $\eta^{abcd} K_b K_{c;d}=0$.

(viii) The gradient of $K^a K_a$ is not zero anywhere outside the horizon. This means that there is no neutral point at which a particle can remain at rest outside the black hole.

Carter [9] has considered spaces which satisfy conditions (i)–(vi) and

(ix) are axisymmetric.

(x) The surface $\mathscr{F}$ which is the intersection of the two horizons has the topology S^2.

He shows that such spaces fall into disjoint families, each depending on at most two parameters m and a. The parameter m represents the mass of the black hole as measured from infinity and the parameter a is hc/Gm where h is the angular momentum as measured from infinity. One such family is known, namely the Kerr solutions with $a^2 < m^2$ [10]. It seems unlikely that there are any others. It has therefore been conjectured that, at least in a collapse that does not deviate too much from spherical symmetry, the solution outside the horizon tends to one of the Kerr solutions with $a^2 < m^2$. I shall here give further support to this conjecture by justifying condition (ix), that is, I shall show that a rotating black hole must be axisymmetric if it satisfies conditions (i)–(v). Condition (x) will be justified in the next section.

I shall assume that the matter in the space satisfies the weak energy condition and obeys well behaved hyperbolic equations such as Maxwell's equations or those for a scalar field. It then follows that the solution must be analytic near infinity where the Killing vector is timelike [11]. I shall take the solution elsewhere to be the analytic continuation of this region near infinity. The idea is now to consider the solution immediately to the future of the surface F of intersection of the two horizons (Fig. 2). This is determined by Cauchy data on the two horizons. By analyticity the solution in this region determines the local nature of the solution everywhere else. I shall show that the Cauchy data on the intersecting horizons is invariant under a continuous group which leaves the intersection surface $\mathscr{F}$ invariant but moves points on one horizon along the null generators of the horizon towards $\mathscr{F}$ and moves points on the other horizon along the generators away from $\mathscr{F}$ (Fig. 3). From the uniqueness of the Cauchy problem it then follows that the solution must admit a Killing vector field which on the horizon is directed along the null generators. However, in general, the Killing vector field K^a which is timelike near infinity will be spacelike on the horizon. There must thus be two independent Killing vector fields and a two parameter isometry group. Near infinity the extra symmetry will have the character of a Poincaré transformation. Since a black hole in asymptotically flat space is not invariant under a space translation or a Lorentz boost, the extra symmetry must correspond to a spatial rotation and so the solution will

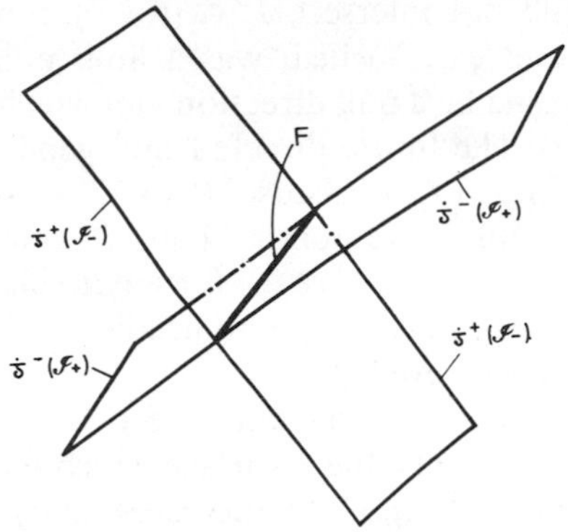

Fig. 2. The intersecting horizons

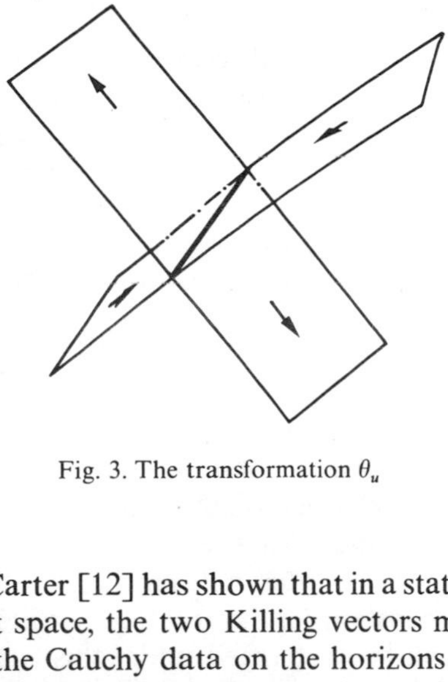

Fig. 3. The transformation θ_u

be axisymmetric. Carter [12] has shown that in a stationary axisymmetric asymptotically flat space, the two Killing vectors must commute.

To show that the Cauchy data on the horizons is invariant under a group, I shall first show that the null generators of the horizons have zero convergence ϱ and shear σ (see [13] for definitions). For this purpose consider a spacelike 2-sphere $\mathscr{C}$ on $\mathscr{I}^-$. The family of surfaces $\mathscr{C}(t) \equiv \phi_t(\mathscr{C})$ obtained by moving $\mathscr{C}$ along the Killing vector K^a will cover $\mathscr{I}^-$. Let λ be a null geodesic generator of the future horizon $\dot{J}^-(\mathscr{I}^+)$ which intersects $\mathscr{F}$. The strong causality condition [14] will hold on the compact set $\mathscr{F}$ since it is in the Cauchy development of $\mathscr{S}$. This means that λ must leave $\mathscr{F}$ and so must leave $\dot{J}^+(\mathscr{I}^-)$ and enter $J^+(\mathscr{I}^-)$. Suppose there were some t' such that λ did not intersect $J^+(\mathscr{C}(t'))$. Then for any t, the generator $\phi_t(\lambda)$ would not intersect $\dot{J}^+(\mathscr{C}(t'+t))$. Since $\mathscr{F}$ is compact there would be a point $q \in \mathscr{F}$ which was a limit point of $\mathscr{F} \cap \phi_t(\lambda)$ as $t \to -\infty$ and there would be a null direction at q which was a limit of the direction of the $\phi_t(\lambda)$. The future directed null geodesic from q in this direction would lie in $\dot{J}^-(\mathscr{I}^+)$ because $\dot{J}^-(\mathscr{I}^+)$ is a closed set and its generators have no future end-points. Thus it would enter $J^+(\mathscr{I}^-)$. However it could not intersect $J^+(\mathscr{C}(t))$ for any t since it was a limit of the $\phi_t(\lambda)$ as $t \to -\infty$. This establishes a contradiction which shows that λ must intersect $J^+(\mathscr{C}(t))$ for every t.

The generators of $\dot{J}^-(\mathscr{I}^+)$ which intersect $\mathscr{F}$ form a compact set invariant under ϕ_t. Let $\mathscr{L}(t)$ be the 2-surface which is the intersection of these generators with $\dot{J}^+(\mathscr{C}(t))$. As t increases, $\mathscr{L}(t)$ will move to the future along the generators of $\dot{J}^-(\mathscr{I}^+)$. The convergence ϱ of these generators is less than or equal to zero. If it were less than zero anywhere the area of $\mathscr{L}(t)$ would increase. However the area of $\mathscr{L}(t)$ must remain the same since it is moving under the isometry ϕ_t. Thus $\varrho = l_{a;b} m^a \bar{m}^b = 0$ where l^a is the null tangent vector to the generators of $\dot{J}^-(\mathscr{I}^+)$ and m^a and $\bar{m}^a$ are complex conjugate null vectors orthogonal to l^a with $m^a \bar{m}_a = -1$. It then follows from the weak energy condition that the shear $\sigma = l_{a;b} m^a m^b$, Ricci tensor component $\Phi_{00} = -\frac{1}{2} R_{ab} l^a l^b$ and Weyl tensor *component* $\psi_0 = -\mathscr{C}_{abcd} l^a m^b l^c m^d$ are zero on $\dot{J}^-(\mathscr{I}^+)$ [13]. Similarly the corresponding quantities $\mu = -n_{a;b} \bar{m}^a m^b$, $\lambda = -n_{a;b} \bar{m}^a \bar{m}^b$, $\Phi_{22} = -\frac{1}{2} R_{ab} n^a n^b$ and $\psi_4 = -C_{abcd} n^a \bar{m}^b n^c \bar{m}^d$ must be zero on $\dot{J}^+(\mathscr{I}^-)$ where n^a is the null vector tangent to the generators of $\dot{J}^+(\mathscr{I}^-)$ and m^a and $\bar{m}^a$ are orthogonal to n_a. For the empty space Einstein equations the Cauchy data is determined by ψ_0 on $\dot{J}^-(\mathscr{I}^+)$, ψ_4 on $\dot{J}^+(\mathscr{I}^-)$ and ϱ, μ and $\psi_2 = -\frac{1}{2} C_{abcd}(l^a n^b l^c n^d + l^c n^b m^c \bar{m}^d)$ on the intersection surface $\mathscr{F}$ [19, 16]. Apart from ψ_2, these quantities are all zero in the present case. At $\mathscr{F}$ one can normalise l^a and n^a by $l^a n_a = 1$ and one can then parallelly propagate them and m^a and $\bar{m}^a$ up the horizons. There remains the freedom $l^a \to e^u l^a$, $n^a \to e^{-u} n^a$ where u is a function on $\mathscr{F}$. If u is not constant on $\mathscr{F}$ the rotation coefficient $\alpha + \bar{\beta} = l_{a;b} n^a \bar{m}^b$ will be changed by this transformation. However if u is constant all the rotation coefficients and Riemann tensor components will be unchanged under this transformation if one moves from a point on $\dot{J}^-(\mathscr{I}^+)$ an affine distance v from $\mathscr{F}$ to a point on the same generator an affine distance $e^{-u} v$ from $\mathscr{F}$. Similarly everything is unchanged on $\dot{J}^+(\mathscr{I}^-)$ if one moves from an affine distance of w to one of $e^u w$. This means that the Cauchy data on the horizon is invariant the one parameter transformation θ_u which leaves $\mathscr{F}$ invariant and is generated by the vector fields $-v l^a$ and $w n^a$ on $\dot{J}^-(\mathscr{I}^+)$ and $\dot{J}^+(\mathscr{I}^-)$ respectively. The solution determined by the Cauchy data will therefore admit an isometry group θ_u whose Killing vector $\hat{K}^a$ will coincide with $-v l^a$ or $w n^a$ on the horizons.

In the case of the Einstein equations with a scalar field, the additional Cauchy data needed is the value of the field on the two horizons. As Φ_{00} and Φ_{22} are zero, the field must be constant along each generator. The Cauchy data is thus invariant under the transformation θ_u in this case also. For the Einstein-Maxwell equations the additional data is $F_{ab}l^a m^b$ on $\dot{J}^-(\mathscr{I}^+)$, $F_{ab}n^a\bar{m}^b$ on $\dot{J}^+(\mathscr{I}^-)$ and $F_{ab}(l^a n^b + m^a\bar{m}^b)$ on $\mathscr{F}$. However the first two are zero as Φ_{00} and Φ_{22} are zero and the third is invariant under θ_u. Similar results will hold for other well behaved fields.

The Killing vector K^a will be tangent to the horizons since they are invariants under the isometry ϕ_t. Thus on the horizons it will either be directed along the null generators or it will be spacelike. The former will be the situation if the solution is not only stationary but *static* [17]. In this case the Killing vectors K^a and $\hat{K}^a$ will coincide. However one can appeal to Israel's results [8, 18] to show that the solution must be spherically symmetric if it is either empty or contains only a Maxwell field.

If the solution is not static but only stationary and is empty, one can generalise a result of Lichnerowitz [19] to show that there will be a region where K^a is spacelike. (Details will be given in [3].) Part, at least, of this region will be outside the horizon. If the horizon is contained in this region, the Killing vectors K^a and $\hat{K}^a$ will be distinct and so the solution will be axisymmetric. A particle travelling along a null geodesic generator of the horizon would be moving with respect to the stationary frame, i.e. the integral curves of K^a. Thus, in a sense, the horizon would be rotating with respect to infinity. One can therefore say that any *rotating* black hole which satisfies conditions (i)–(v) must be axisymmetric.

There remains the possibility that there could be stationary, nonstatic solutions in which the horizon was not rotating but in which K^a was spacelike in a region outside and disjoint from the horizon. In such a situation there would have to be neutral points outside the horizon where K^a was timelike and the gradient of $K^a K_a$ was zero. I have not been able to rule out this possibility but it seems to be unstable in that one could extract an indefinitely large amount of energy by a method proposed by Penrose [20]: consider a small particle with momentum $p_1^a = m_1 v_1^a$ where m_1 is the mass and v_1^a is the future directed unit tangent vector to the world-line. If the particle moves on a geodesic its energy $E_1 = p_1^a K_a$ will be constant. Suppose that the particle were to fall from infinity into the region where K^a was spacelike and there divided into two particles momenta p_2^a and p_3^a. By local conservation, $p_1^a = p_2^a + p_3^a$. Since K^a was spacelike, one could arrange that $E_2 = p_2^a K_a$ was negative. Then $E_3 > E_1$ and particle 3 could escape to infinity, where its total energy (rest-mass plus kinetic) would be greater than that of the original particle. Particle 2 on the other hand would have to remain in the region where K^a was spacelike and so could neither escape to infinity nor fall through the horizon. By repeating this process many times one could gradually extract energy from the solution. As one did this the solution would presumably change gradually. However the region where K^a is spacelike could not disappear since there would have to be somewhere for the negative energy particles to go. If it remained disjoint from the horizon, one would apparently extract an indefinite amount of energy. If, on the other hand, the region moved so that it included the horizon, K^a and $\hat{K}^a$ would be distinct and so, suddenly, the solution would be axisymmetric. I therefore feel that such a situation will not occur and that any stationary, nonstatic, black hole will be axisymmetric.

It is worth noting that the field equations did not play a very important part in the above argument. Thus one can apply it also to the case of a rotating black hole surrounded by a ring of matter. The solution will not be analytic at the ring but it will be elsewhere, and so the above result will still hold. At first sight this seems to lead to a paradox. Consider a rotating star surrounded by a non-rotating square frame of rods. Suppose that the star collapsed to produce a rotating black hole. From the above it appears that the solution ought to become axisymmetric. However the presence of the square frame will prevent this.

The answer seems to be that the field of the frame will distort the rotating black hole slightly. The reaction of the rotating black hole back on the frame will cause it to start rotating and so to radiate gravitational waves. Eventually the rotation of both the black hole and the frame will be damped out and the solution will approach a *static* state in which K^a and $\hat{K}^a$ coincide and which therefore does not have to be axisymmetric. I am grateful to James B. Hartle and Brandon Carter for suggesting this solution to the paradox.

4. The Topology of the Event Horizon

In this section I shall prove that each connected component of the 2-surface $\mathscr{L}(t)$ in the event horizon has the topology S^2. This is done by showing that, if it had any other topology, one could deform it outwards into the exterior region in such a way that the future directed outgoing null geodesics orthogonal to it would be converging. As in Section 2, this would lead to a contradiction.

Let l^a be the future directed null vector tangent to the future event horizon and let n^a be the other future directed null vector orthogonal to $\mathscr{L}(t)$ normalized so that $l^a n_a = 1$ but with the freedom $l^a \to e^a l^a$, $n^a \to e^{-a} n^a$ not restricted for the moment. Now move each point of $\mathscr{L}(t)$ an affine distance $-w$ along the null geodesics with tangent vector n^a. This moves $\mathscr{L}(t)$ to the past into $J^-(\mathscr{I}^+)$. To keep l^a orthogonal to the 2-surface, it has to be propagated so that

$$\tau = \bar{\alpha} + \beta$$

where $\tau = l_{a;b} m^a n^b$. The change of the convergence ϱ produced by this movement is given by one of the Newman-Penrose equations [13]:

$$\frac{d}{dw} \varrho = \bar{\delta}\tau + (\bar{\beta} - \alpha)\tau - \bar{\tau}\tau - \psi_2 - 2\Lambda \tag{1}$$

where $\Lambda = R/24$, $\alpha - \bar{\beta} = \bar{m}_{a;b} m^a \bar{m}^b$ and $\bar{\delta}\tau = \bar{m}^a \tau_{;a}$. The combination $\bar{\delta} - (\alpha - \bar{\beta})$ acting on the spin weight $+1$ quantity τ is the covariant Newman-Penrose operator $\bar{\eth}$ [21].

Under a rescaling transformation $l'^a = e^a l^a$, $n'^a = e^{-a} n^a$, the quantity τ changes to $\tau' = \tau + \delta a$ and so $\frac{d}{dw} \varrho$ changes to

$$\frac{d}{dw'} \varrho' = -\tau'\bar{\tau}' + \bar{\eth}\tau + \bar{\eth}\,\delta a - \psi_2 - 2\Lambda. \tag{2}$$

The term $\bar{\eth}\,\delta a$ is the Laplacian of a in the 2-surface. One can choose a so that the sum of the last four terms on the right of Eq. (2) is constant on $\mathscr{L}(t)$. The sign of this constant value will be determined by that of the integral of $-(\psi_2 + 2\Lambda)$ over $\mathscr{L}(t)$ ($\bar{\eth}\tau$, being a gradient, has zero integral). This integral can be evaluated from another Newman-Penrose equation which can be written as

$$\eth(\alpha + \bar{\beta}) - \bar{\eth}(\bar{\alpha} + \beta) + \eth(\alpha - \bar{\beta}) + \bar{\eth}(\bar{\alpha} - \beta) = -2\psi_2 + 2\Lambda + 2\Phi_{11} \tag{3}$$

where $\Phi_{11} = -\frac{1}{4} R_{ab}(l^a n^b + m^a \bar{m}^b)$. When integrated over the 2-surface the terms in $\bar{\alpha} + \beta$ disappear but there is in general a contribution from the $\bar{\alpha} - \beta$ terms because the vector field m^a will have singularities on the 2-surface. The contribution from these singularities is determined by the Euler number χ of the 2-surface. Thus

$$\int (-\psi_2 + \Phi_{11} + \Lambda)\, dS = \pi\chi. \tag{4}$$

(The real part of this equation is, in fact, the Gauss-Bonnet theorem.)

Therefore

$$-\int (\psi_2 + 2\Lambda)\, dS = \pi\chi - \int (\Phi_{11} + 3\Lambda)\, dS. \tag{5}$$

For the electromagnetic and other reasonable matter fields, obeying the Dominant Energy condition [25] $\Phi_{11} + 3\Lambda \geqq 0$. The Euler number χ is $+2$ for the sphere, 0 for the torus and negative for other compact orientable 2-surfaces. ($\mathscr{L}(t)$ has to be orientable as it is a boundary.)

Suppose that the right hand side of Eq. (5) was negative. Then one could choose a so that $\frac{d}{dw'} \varrho'$ was negative everywhere on $\mathscr{L}(t)$. For a small negative value of w' one would obtain a 2-surface in $J^-(\mathscr{I}^+)$ such that the outgoing null geodesics orthogonal to the surface were converging. This would lead to a contradiction.

Suppose now that χ was zero and that $\Phi_{11}+3\Lambda$ was zero on the horizon. Then one could choose a so that the sum of the last four terms in Eq. (2) was zero everywhere on $\mathscr{L}(t)$. If ψ_2 was nonzero somewhere on $\mathscr{L}(t)$, the term $-\tau'\bar{\tau}'$ in Eq. (2) would be nonzero and one could change a slightly so as to make $\frac{d}{dw'}\varrho'$ negative everywhere on $\mathscr{L}(t)$. This would again lead to a contradiction. If, on the other hand, ψ_2 was zero everywhere on $\mathscr{L}(t)$, the Cauchy data would be zero and so the solution would be flat. One could not have a flat toroidal horizon in a flat space which is Euclidean at infinity.

This shows that $\mathscr{L}(t)$ and hence $\mathscr{F}$ has spherical topology. It has not been excluded that $\mathscr{L}(t)$ might have several disconnected components, each topologically spheres, which represent separate black holes at constant distance from each other along the axis of symmetry. For reasons explained in the next section I think this could occur only in the limiting case where the black holes have a charge $q=\pm G^{\frac{1}{2}}m$.

5. Energy Limits

In view of the previous two sections, it seems reasonable to assume that in a gravitational collapse the solution outside the horizon settles down into a Kerr solution with $a^2<m^2$. If the collapsing body had a net electrical charge of q e.s.u., one would expect the solution to settle down to a charged Kerr solution [10] with $a^2+e^2<m^2$ where $e=G^{-\frac{1}{2}}q$. One would therefore expect the area of the 2-surface $\partial\mathscr{B}(t)$ in the horizon to approach the area of a 2-surface in the horizon of the Kerr solution. This area is

$$4\pi G^2c^{-4}(2m^2-e^2+2m(m^2-a^2-e^2)^{\frac{1}{2}})\,. \tag{6}$$

Consider a black hole which by a surface $\mathscr{S}(t)$ has settled down to a Kerr solution with parameters m_1, a_1 and e_1. Suppose the black hole now interacts with various particles or fields and then settles down again by a surface $\mathscr{S}(t')$ to a Kerr solution with parameters m_2, a_2 and e_2. From Section 2 it follows that the area of $\partial\mathscr{B}(t')$ is greater than or equal to the area of $\partial\mathscr{B}(t)$. In fact it must be strictly greater if there is any disturbance at the event horizon. Thus

$$2m_2^2-e_2^2+2m_2(m_2^2-a_2^2-e_2^2)^{\frac{1}{2}}>2m_1^2-e_1^2+2m_1(m_1^2-a_1^2-e_1^2)^{\frac{1}{2}}\,. \tag{7}$$

Note that m_2 can be less than m_1 if a_1 or e_1 are nonzero One can interpret this as meaning that one can extract rotational or electrostatic energy from a black hole. One way of extracting rotational energy would be to construct an asymmetric frame of rods around the black hole. According to the argument in Section 3, the frame would start rotating and would emit gravitational waves until the rotation of the black hole was damped out. Another way of extracting energy would be to use the fact that the Killing vector K^a is spacelike on and in a region just outside the event horizon. One could then employ the procedure described in Section 3 of sending in a particle and getting one back with greater energy. In this case the negative energy particles could fall through the horizon. Christodoulou [22] has shown that using this process one can get arbitrarily near the limit set by inequality (7). Using charged particles one can extract electrostatic energy from a black hole. By lowering a particle of the opposite charge down the axis on a rope one can get arbitrarily near the limit set by (7).

Consider now a situation in which two stars a long way apart collapse to form black holes $\mathscr{B}_1(t)$ and $\mathscr{B}_2(t)$. One can neglect the interaction between them and take the areas of $\partial\mathscr{B}_1(t)$ and $\partial\mathscr{B}_2(t)$ to be given by formula (6) with the values of the parameters m_1, a_1, e_1 and m_2, a_2, e_2 respectively. Suppose the two black holes now collide and merge to form a single black hole $\mathscr{B}_3(t')$. This process will give rise to a certain amount of gravitational radiation and, if the black holes are charged, of electromagnetic radiation as well. By the conservation law for weakly asymptotically simple space-times [23], the energy of this radiation is $(m_1+m_2-m_3)\,c^2$. This is limited by the requirement that the area of $\partial\mathscr{B}_3(t')$ must be greater than the sum of the areas of $\partial\mathscr{B}_1(t)$ and $\partial\mathscr{B}_2(t)$. This gives the inequality

$$\begin{aligned} &2m_3^2-e_3^2+2m_3(m_3^2-a_3^2-e_3^2)^{\frac{1}{2}} \\ &> 2m_2^2-e_2^2+2m_2(m_2^2-a_2^2-e_2^2)^{\frac{1}{2}}+2m_1^2-e_1^2+2m_1(m_1^2-a_1^2-e_1^2)^{\frac{1}{2}}\,. \end{aligned} \tag{8}$$

The fraction $\varepsilon=(m_1+m_2)^{-1}(m_1+m_2-m_3)$ of energy radiated is always less than $1-2^{-\frac{3}{2}}$. If e_1 and e_2 are zero or have the same sign, then $\varepsilon<\frac{1}{2}$. If, in addition, $a_1=a_2=0$, then $\varepsilon<1-2^{-\frac{1}{2}}$.

By the conservation of charge, $e_3=e_1+e_2$. Angular momentum, on the other hand, can be carried off by the radiation. This cannot happen however if the situation is axisymmetric, i.e. if the black holes have their rotation axes aligned along their direction of approach to each other. Then $m_3a_3=m_1a_1+m_2a_2$. It can be seen that if the angular momenta have the same sign there is less energy available to be radiated than if they have opposite signs. This suggests that there may be a spin dependent force between two black holes as indeed one might expect from the analogy between angular momentum and magnetic dipole moment. Unlike the electrodynamic case, the force would be attractive if the angular momenta had opposite directions and repulsive if they had the same direction. However, even in the limiting case $m_1 = m_2 = a_1 = a_2$, it is still possible to radiate some energy. This suggests that the repulsive force is never strong enough to balance the attractive force between the masses. It seems that the only way to obtain a strong enough repulsive force is to go to the other limiting case in which $m_1 = e_1$, $m_2 = e_2$ and $a_1 = a_2 = 0$. Hartle and I [24] have found static solutions containing two or more such black holes.

Acknowledgment. I have been greatly helped by discussions with Brandon Carter, George Ellis, James Hartle and Roger Penrose.

References

1. Penrose, R.: Phys. Rev. Letters **14**, 57 (1965).
2. — In: de Witt, C. M., Wheeler, J. A. (Eds.): Battelle Rencontres (1967). New York: Benjamin 1968.
3. Hawking, S. W., Ellis, G. F. R.: The large scale structure of space time. Cambridge: Cambridge University Press (to be published).
4. — Penrose, R.: Proc. Roy. Soc. A **314**, 529 (1970).
5. Penrose, R.: Seminar at Cambridge University, January 1971 (unpublished).
6. Gibbons, G. W., Penrose, R.: To be published.
7. Geroch, R. P.: J. Math. Phys. **11**, 437 (1970).
8. Israel, W.: Phys. Rev. **164**, 1776 (1967).
9. Carter, B.: Phys. Rev. Letters **26**, 331 (1971).
10. — Phys. Rev. **174**, 1559 (1968).
11. Müller zum Hagen, H.: Proc. Cambridge Phil. Soc. **68**, 199 (1970).
12. Carter, B.: Commun. math. Phys. **17**, 233 (1970).
13. Newman, E. T., Penrose, R.: J. Math. Phys. **3**, 566 (1962).
14. Hawking, S. W.: Proc. Roy. Soc. A **300**, 187 (1967).
15. Sachs, R. K.: J. Math. Phys. **3**, 908 (1962).
16. Penrose, R.: Characteristic initial data for zero rest mass including gravitation, preprint (1961).
17. Carter, B.: J. Math. Phys. **10**, 70 (1969).
18. Israel, W.: Commun. math. Phys. **8**, 245 (1968).
19. Lichnerowicz, A.: Théories relativistes de la gravitation et de l'électromagnétisme. Paris: Masson 1955.
20. Penrose, R.: Nuovo Cimento Serie 1, **1**, 252 (1969).
21. Newman, E. T., Penrose, R.: J. Math. Phys. **7**, 863 (1966).
22. Christodoulou, D.: Phys. Rev. Letters **25**, 1596 (1970).
23. Penrose, R.: Phys. Rev. Letters **10**, 66 (1963).
24. Hartle, J. B., Hawking, S. W.: Commun. math. Phys. To be published.
25. Hawking, S. W.: Commun. math. Phys. **18**, 301 (1970).

S. W. Hawking
Institute of Theoretical Astronomy
University of Cambridge
Madingley Road
Cambridge, England